Advances in Intelligent Systems and Computing

Volume 751

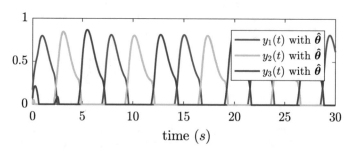

$$\boldsymbol{x}_t = \left(1 - a\right)\boldsymbol{x}_{t-1} + a\left(\lambda \boldsymbol{W}_{rec}\boldsymbol{r}_{t-1} + \boldsymbol{W}_{in}\boldsymbol{u}_t + \boldsymbol{W}_{fb}z_{t-1}\right), \tag{1}$$

where $a = 0.1$ is a constant value called leaking rate that determines the time scale of reservoir dynamics. $\boldsymbol{W}_{rec} \in \mathbb{R}^{N_x \times N_x}$ is the recurrent connection weight matrix of the reservoir, and each component is set to a value that is randomly generated from a Gaussian distribution with zero mean and variance $1/pN_x$. $\lambda = 1.2$ is a scale of recurrent weights of the reservoir. Larger λ makes the dynamics of the reservoir neurons more chaotic. $\boldsymbol{r}_t$ is the output vector of reservoir neurons. $\boldsymbol{W}_{in} \in \mathbb{R}^{N_x \times N_i}$ is the weight matrix from the input to the reservoir neurons. $\boldsymbol{W}_{fb} \in \mathbb{R}^{N_x \times N_f}$ is the weight matrix from the MLR or readout units to the reservoir neurons, and z_t is the feedback vector. Each component of $\boldsymbol{W}_{in}$ and $\boldsymbol{W}_{fb}$ is set to a uniformly random number between -1 and 1. The activation function of every neuron in the reservoir is the *tanh* function.

The MLR is a four-layer NN with static neurons in the order of 100, 40, 10 and 3 from the bottom layer; the activation function of each neuron is the *tanh* function. Each neuron in the bottom layer of MLR receives outputs from all the reservoir neurons. The outputs of $N_f = 10$ neurons in the hidden layer, one lower than the output layer, are fed back to every neuron of the reservoir as feedback vector $z_t \in \mathbb{R}^{N_f}$. $N_i = 7$ inputs are derived from the environment; all these inputs are given to each neuron in the reservoir and the bottom layer of MLR. Each initial weight of MLR is set to a randomly generated value from a Gaussian distribution with zero mean and variance $0.01/n$, where n is the number of inputs in each layer. The MLP is trained by BP and the stochastic gradient descent algorithm with a learning rate of 0.01.

The network outputs are critic V_t and actor vector $\boldsymbol{A}_t$. The sum of $\boldsymbol{A}_t$ and the exploration component vector $\boldsymbol{rnd}_t$ is used as the motion signal of the agent at time t. Each component of $\boldsymbol{rnd}_t$ is set to a uniformly random value between -1 and 1.

2.2 Learning

In this network, only the weights of the paths indicated by the red arrows in Fig. 1 are trained using the BP based actor-critic algorithm. The training signal for critic at time $t - 1$ is given as

$$V_{t-1}^{train} = V(\boldsymbol{u}_{t-1}) + \hat{r}_{t-1} = r_t + \gamma V(\boldsymbol{u}_t), \tag{2}$$

where $\hat{r}_{t-1}$ is TD-error at time $t - 1$, which is given as

$$\hat{r}_{t-1} = r_t + \gamma V(\boldsymbol{u}_t) - V(\boldsymbol{u}_{t-1}), \tag{3}$$

where r_t is a reward received by the agent at time t and $\gamma = 0.99$ is the discount rate. The training signal for actor at time $t - 1$ is given as

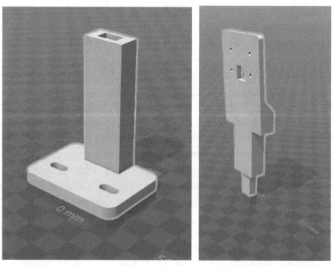

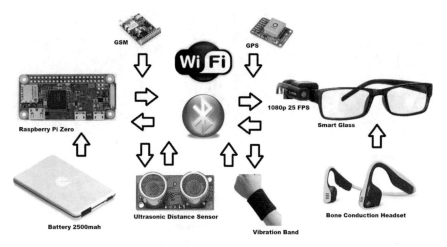

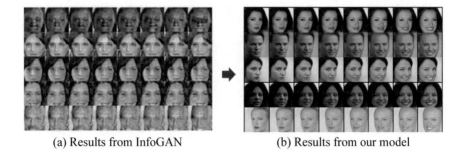

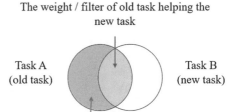

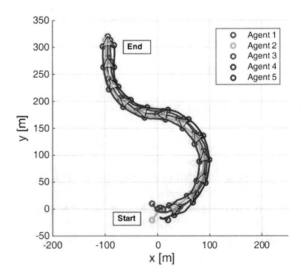

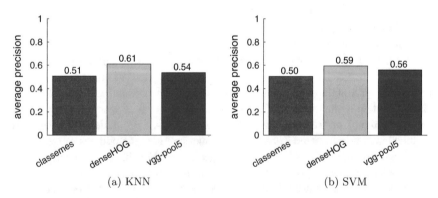

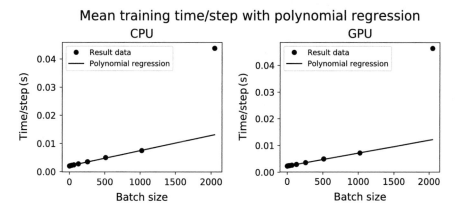

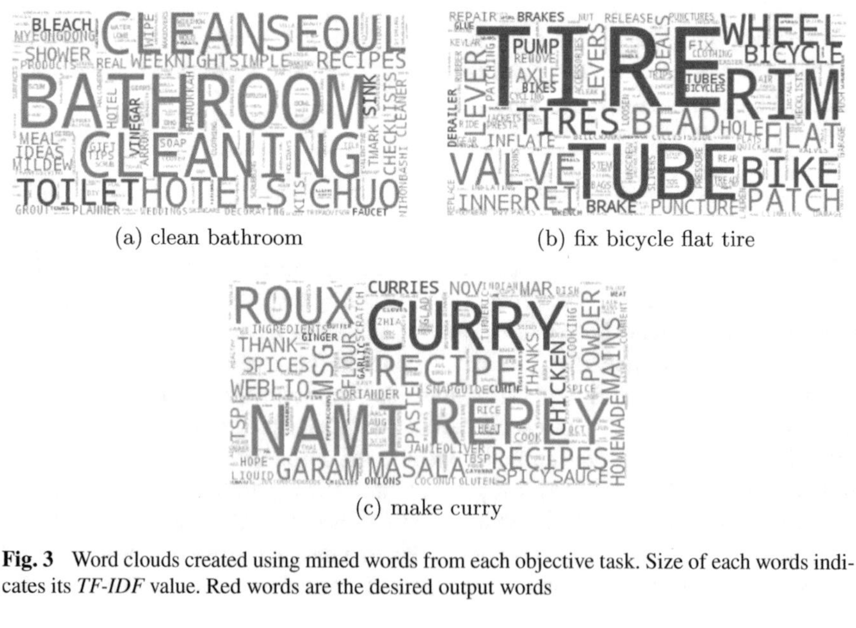

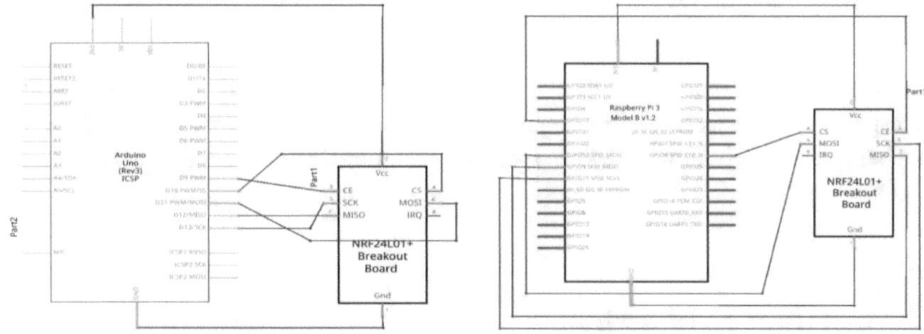

Advances in Intelligent Systems and Computing

Volume 751

Series editor

Janusz Kacprzyk, Polish Academy of Sciences, Warsaw, Poland
e-mail: kacprzyk@ibspan.waw.pl

The series "Advances in Intelligent Systems and Computing" contains publications on theory, applications, and design methods of Intelligent Systems and Intelligent Computing. Virtually all disciplines such as engineering, natural sciences, computer and information science, ICT, economics, business, e-commerce, environment, healthcare, life science are covered. The list of topics spans all the areas of modern intelligent systems and computing such as: computational intelligence, soft computing including neural networks, fuzzy systems, evolutionary computing and the fusion of these paradigms, social intelligence, ambient intelligence, computational neuroscience, artificial life, virtual worlds and society, cognitive science and systems, Perception and Vision, DNA and immune based systems, self-organizing and adaptive systems, e-Learning and teaching, human-centered and human-centric computing, recommender systems, intelligent control, robotics and mechatronics including human-machine teaming, knowledge-based paradigms, learning paradigms, machine ethics, intelligent data analysis, knowledge management, intelligent agents, intelligent decision making and support, intelligent network security, trust management, interactive entertainment, Web intelligence and multimedia.

The publications within "Advances in Intelligent Systems and Computing" are primarily proceedings of important conferences, symposia and congresses. They cover significant recent developments in the field, both of a foundational and applicable character. An important characteristic feature of the series is the short publication time and world-wide distribution. This permits a rapid and broad dissemination of research results.

More information about this series at http://www.springer.com/series/11156

Jong-Hwan Kim · Hyun Myung
Junmo Kim · Weiliang Xu
Eric T Matson · Jin-Woo Jung
Han-Lim Choi
Editors

Robot Intelligence Technology and Applications 5

Results from the 5th International Conference on Robot Intelligence Technology and Applications

Springer

Editors
Jong-Hwan Kim
School of Electrical Engineering
Korea Advanced Institute of Science and Technology (KAIST)
Daejeon
Korea (Republic of)

Hyun Myung
Department of Civil and Environmental Engineering
Korea Advanced Institute of Science and Technology (KAIST)
Daejeon
Korea (Republic of)

Junmo Kim
School of Electrical Engineering
Korea Advanced Institute of Science and Technology (KAIST)
Daejeon
Korea (Republic of)

Weiliang Xu
Department of Mechanical Engineering
The University of Auckland
Auckland
New Zealand

Eric T Matson
Department of Computer and Information Technology
Purdue University
West Lafayette, IN
USA

Jin-Woo Jung
Department of Computer Science and Engineering
Dongguk University
Seoul
Korea (Republic of)

Han-Lim Choi
Department of Aerospace Engineering
Korea Advanced Institute of Science and Technology (KAIST)
Daejeon
Korea (Republic of)

ISSN 2194-5357 ISSN 2194-5365 (electronic)
Advances in Intelligent Systems and Computing
ISBN 978-3-319-78451-9 ISBN 978-3-319-78452-6 (eBook)
https://doi.org/10.1007/978-3-319-78452-6

Library of Congress Control Number: 2018936632

Printed on acid-free paper

This Springer imprint is published by the registered company Springer International Publishing AG part of Springer Nature
The registered company address is: Gewerbestrasse 11, 6330 Cham, Switzerland

Preface

This is the fifth edition that aims at serving the researchers and practitioners in related fields with a timely dissemination of the recent progress on robot intelligence technology and its applications, based on a collection of papers presented at the 5th International Conference on Robot Intelligence Technology and Applications (RiTA), held in Daejeon, Korea, December 13–15, 2017. For better readability, this edition has the total of 47 articles grouped into 5 parts: Part I: Artificial Intelligence, Part II: Autonomous Robot Navigation, Part III: Intelligent Robot System Design, Part IV: Intelligent Sensing and Control, and Part V: Machine Vision.

The theme of this year's conference is: "Robots revolutionizing the paradigm of human life by embracing artificial intelligence." The gap between human and robot intelligence is shrinking fast, and this is changing the way we live and work. It also affects how we conduct research in AI, recognizing the need for robotic systems that can interact with their environments in human-like ways, and embracing intelligent robots and the new paradigms they create in business and leisure. This book will bring their new ideas in these areas and leave with fresh perspectives on their own research.

We do hope that readers find the Fifth Edition of Robot Intelligence Technology and Applications, RiTA 5, stimulating, enjoyable, and helpful for their research endeavors.

Daejeon, Korea (Republic of) Jong-Hwan Kim
Honorary General Chair

Daejeon, Korea (Republic of) Hyun Myung
General Chair

Daejeon, Korea (Republic of) Junmo Kim
Program Chair

Auckland, New Zealand Weiliang Xu
Organizing Chair

West Lafayette, USA Eric T Matson
Organizing Chair

Seoul, Korea (Republic of) Jin-Woo Jung
Publications Chair

Daejeon, Korea (Republic of) Han-Lim Choi
Publications Chair

Contents

Part I
Artificial Intelligence

Artificial Intelligence Approach to the Trajectory Generation and Dynamics of a Soft Robotic Swallowing Simulator

Dipankar Bhattacharya, Leo K. Cheng, Steven Dirven and Weiliang Xu

Abstract Soft robotics is an area where the robots are designed by using soft and compliant modules which provide them with infinite degrees of freedom. The intrinsic movements and deformation of such robots are complex, continuous and highly compliant because of which the current modelling techniques are unable to predict and capture their dynamics. This paper describes a machine learning based actuation and system identification technique to discover the governing dynamics of a soft bodied swallowing robot. A neural based generator designed by using Matsuoka's oscillator has been implemented to actuate the robot so that it can deliver its maximum potential. The parameters of the oscillator were found by defining and optimising a quadratic objective function. By using optical motion tracking, time-series data was captured and stored. Further, the data were processed and utilised to model the dynamics of the robot by assuming that few significant non-linearities are governing it. It has also been shown that the method can generalise the surface deformation of the time-varying actuation of the robot.

Keywords Soft robotics · Swallowing robot · Peristalsis · Matsuoka's oscillator Machine learning · Optimisation

D. Bhattacharya (✉) · W. Xu
Department of Mechanical Engineering, University of Auckland, Auckland, New Zealand
e-mail: dbha483@aucklanduni.ac.nz
URL: https://unidirectory.auckland.ac.nz/profile/dbha483

W. Xu
e-mail: p.xu@auckland.ac.nz

L. K. Cheng
Auckland Bioengineering Institute, University of Auckland, Auckland, New Zealand
e-mail: l.cheng@auckland.ac.nz

S. Dirven
School of Advanced Technology and Engineering, Massey University, Auckland, New Zealand
e-mail: S.Dirven@massey.ac.nz

J.-H. Kim et al. (eds.), *Robot Intelligence Technology and Applications 5*,
Advances in Intelligent Systems and Computing 751,
https://doi.org/10.1007/978-3-319-78452-6_1

1 Introduction

In the field of robotics, phenomena involving natural and biological process always continue to be an active source of inspiration. Although the capabilities of the traditional robots have seen a lot of significant development, the demand for studying various biological processes and developing systems which are capable of soft and continuous interaction with the environment has led many researchers to open the gateway of a new field of research known as the soft robotics. In previous couple of years, the area of soft robotics has progressed toward becoming a very much characterised discipline with working practices.

Soft robotics is an area where the robots are designed by using soft and compliant modules. The advantage of soft robots over rigid body counterparts are involving infinite degrees of freedom and different movements [1]. Robots which can completely mimic different biological processes can be developed by combining various functional units strategically to act as a single unit, which generates smooth actions and adaptability to different environmental conditions. Engineered mechanisms like robots and machines have been created by utilising stiff or fully rigid segments associated with each other using joints to change input energy to mechanical energy. Some of the advantages of soft robots are: safe human–robot cooperation, minimal effort production, and basic manufacture with negligible coordination [2]. Researchers have developed many soft actuators, sensors and robots that have possible applications in biomedical surgery [3], rehabilitation [4], biomimicking biological processes [5] and research in this area have shown flexibility, agility as well as sensitivity. One of the excellent examples of a soft rehabilitation robot is the swallowing robot (SR) developed by the University of Auckland to evaluate human swallow process under different bolus swallow conditions [6, 7]. The robot could help the researchers in developing texture modified foods for patients suffering from swallow related disorders like dysphagia. But to get to a certain point where soft robots like the SR can deliver its maximum potential, improvements in the following areas must be realised: [8]

- different actuation schemes inspired by various biological phenomena like the peristalsis in the human swallowing process.
- novel dynamic modelling approaches by using machine learning techniques as the current ones are unable to capture their dynamics.
- new data-driven based control algorithms that are adaptive to their material properties.
- developing a user-friendly environment for the researchers from other areas to use the soft robots efficiently.

In this paper, a novel approach for generating the peristalsis in the conduit of the SR is developed. In addition to it, a data-driven technique has been implemented to determine the governing differential equations (DE) of the robot. The novelty of this work in relation to the previous papers is the aid of powerful machine learning tools to model the soft-robot and a thorough search of the relevant literature yielded no related article. The peristalsis scheme is inspired by the biological swallowing process in human beings by keeping the anatomy of the oesophagus in mind. The conduit of the robot is manufactured by using a soft elastomer powered by air and hence, it shows a continuum behaviour [9, 10]. Due to the robot's complex, and compliant nature, traditional modelling and control approaches are difficult to apply. Hence, this paper emphasises more on machine learning based regression techniques over other conventional methods. The technique can be further extended to implement control algorithms which utilises powerful regression techniques like Support Vector Machines (SVM), and Relevance Vector Machines (RVM) [11, 12].

2 Generating Different Peristalsis Pattern in the Swallowing Robot

2.1 *Matsuoka Oscillator*

A group of neurons or oscillators coupled with each other to produce the desired set of a rhythmic pattern is known as a Central pattern generator (CPG) neural circuit. One of the examples of a CPG neural circuit is Matsuoka Oscillator (MO). Matsuoka introduced the neural oscillator as shown in Fig. 1 which includes two different types of neurons, flexor and extensor neuron where an individual neuron can be represented by two simultaneous first order nonlinear differential equations [13]. The fundamental idea behind the working of MO is mutual inhibition of neurons where each neuron is generating a periodic signal to inhibit the other neuron. In Fig. 1, extensor and flexor neurons are represented by the blue and green ellipse respectively. The variable x_i is known as the membrane potential or firing rate of the ith neuron, variable v_i represents the degree of adaptation or fatigue, c is the external tonic input which should always be greater than zero, τ and T represents the rising and adaptation

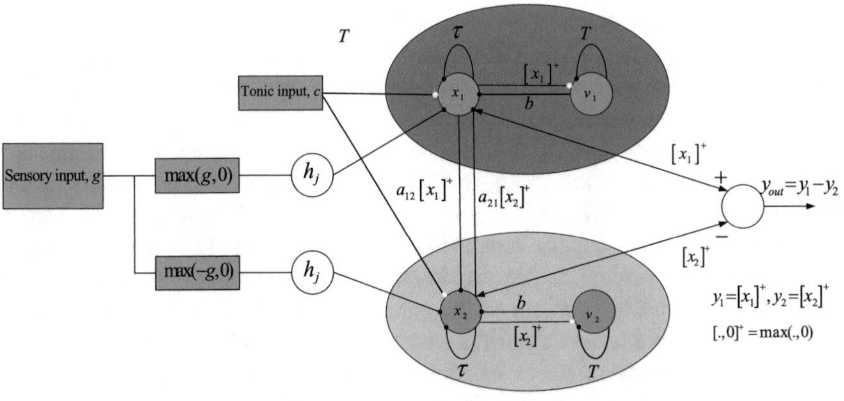

Fig. 1 Schematic diagram of a two-neuron Matsuoka oscillator

time constant for x_i and v_i respectively, a_{ij} is the synaptic weight which represents the coupling strength or strength of inhibition between the two neurons, b specifies the steady-state firing rate, and g_j is the sensory feedback weighted by a gain which is generally the output of a system to which the oscillator is trying to entrain such that the positive part of the sensory feedback is applied to one neuron and negative part to the other. The final output of the oscillator is represented by y_{out} which is the difference between the output of the two neurons. Compared to other oscillators, MO is efficient regarding the order of computation, and the tuning of parameters. Also, there is no requirement of implementing any post-processing algorithms, and MO model has a large region of stability as compared to other oscillator. The mathematical model of a three-neuron MO is expressed in Eq. 1. The structure of the connections among the three neuron in the oscillator remains similar to that of the two neuron oscillator as shown in Fig. 1. The only difference in case of the former one as compared to the latter one is the mutual inhibition which is now occurring between the three neurons instead of two. Due to the complexity in the diagram of the three neuron MO, only the two neuron counterpart has been illustrated.

$$
\begin{aligned}
&\left[\begin{array}{c}\dot{\mathbf{x}}_1(t)\\ \hline \dot{\mathbf{x}}_2(t)\end{array}\right] = -\left[\begin{array}{cc}\frac{1}{\tau}\mathbf{I}_{3\times3} & \frac{b}{\tau}\mathbf{I}_{3\times3}\\ \hline \mathbf{0}_{3\times3} & \frac{1}{T}\mathbf{I}_{3\times3}\end{array}\right] + \left[\begin{array}{c}\mathbf{x}_1(t)\\ \hline \mathbf{x}_2(t)\end{array}\right] + \left[\begin{array}{c}-\frac{1}{\tau}\mathbf{C}\\ \hline \frac{1}{T}\mathbf{I}_{3\times3}\end{array}\right]\mathbf{y}(t) + \left[\begin{array}{c}\frac{c}{\tau}\,[[1\ 1\ 1]]^T\\ \hline \mathbf{0}_{3\times1}\end{array}\right]\\
&\text{where,}\mathbf{C} = \begin{bmatrix}0 & a_{12} & a_{13}\\ a_{12} & 0 & a_{23}\\ a_{13} & a_{23} & 0\end{bmatrix}, \dot{\mathbf{x}}_1 = [\dot{x}_1, \dot{x}_2, \dot{x}_3]^T, \dot{\mathbf{x}}_2 = [\dot{v}_1, \dot{v}_2, \dot{v}_3]^T, \text{and } \dot{x}_1, \dot{x}_2 \in \mathbb{R}^{1\times3},\\
&\mathbf{y} = \max\left(\mathbf{x}_1, \mathbf{0}\right), \mathbf{y} \in \mathbb{R}^3 \ .
\end{aligned}
\tag{1}
$$

2.2 *Problem Formulaton for the Parameter Estimation*

The problem of designing a MO network is to predict its parameters for generating a peristaltic wave of desired frequency and amplitude. How the parameters of the oscillator control the wave is not well-defined all the time [14, 15]? A three neuron MO defined by Eq. 1, has been implemented on MATLAB and the parameter values were chosen randomly in such a way that they satisfy the oscillator stability constraints. The estimation of parameters of the three neuron MO falls under the category grey box modelling as the system of differential equations representing the oscillator network are known. The parameters govern the shape of the oscillation (amplitude and frequency), but as the differential equations are non-linear in nature, so it is hard to find out the particular set of parameter values which represents the desired amplitude and frequency of oscillation. One way to obtain the parameters is by formulating a quadratic programming problem on the parameters of the MO and then, solving it by applying an optimisation algorithm. By using Lagrange's multiplier in the cost function, the limit on the size of the parameters can be constrained. The fundamental concept behind applying such algorithm is to optimise the parameters of the oscillator to return the user-defined shape (amplitude and frequency) of the peristaltic waveform (reference trajectory). Let $\boldsymbol{\theta}$ and $J(\boldsymbol{\theta})$ denote the parameter vector and the formulated quadratic objective function respectively given in Eq. (2). The method of optimising the quadratic cost function is illustrated in the block diagram in Fig. 2. A modified Newton's method known as the trust region reflective algorithm (TRRA) is used to determine the parameter values required for generating the peristalsis trajectory characterised by the user at the input of Fig. 2.

$$\begin{aligned}
\hat{\theta} = \arg\min_{\boldsymbol{\theta}} J(\boldsymbol{\theta}) = \frac{1}{2}\sum_{i=1}^{\mathrm{m}} ||(\mathbf{y}^{(i)}{}_{ref} - \mathbf{y}^{(i)}(\boldsymbol{\theta}))||_2{}^2, \mathbf{y}^{(i)}{}_{ref} \text{ and } \mathbf{y}^{(i)}(\boldsymbol{\theta}) \in \Re^3 \\
\text{subject to, } 0.2 \le T \le 5, \\
0 \le K_n \le 1, \\
0 < b, \\
0 < c \le 10, \\
0 < a_{ij} \le 1 + b \text{ (MO stability condition)} \\
\text{where, } \boldsymbol{\theta} = [T, K_n, a_{12}, a_{13}, a_{23}, b, c]^T, \\
\mathbf{y}^{(i)}{}_{ref} = \left[y^{(i)}{}_{ref1}, y^{(i)}{}_{ref2}, y^{(i)}{}_{ref3}\right]^T .
\end{aligned} \tag{2}$$

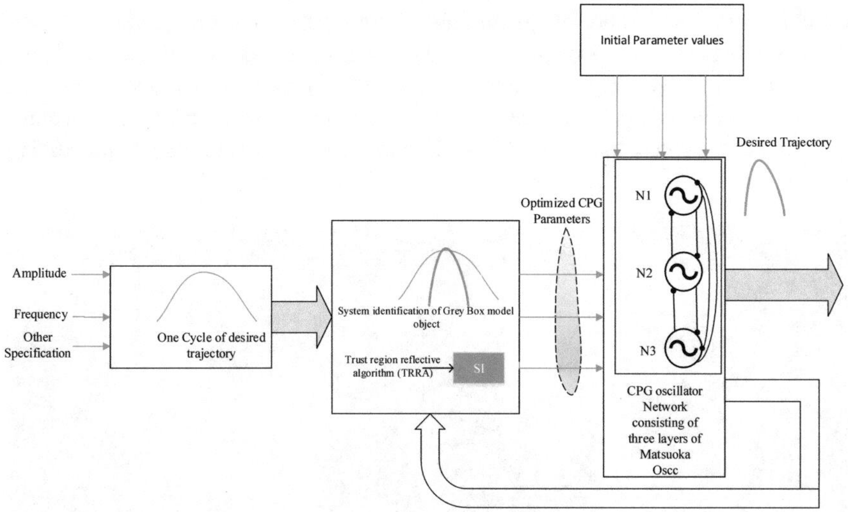

Fig. 2 Block diagram representing the process adopted for estimating the parameters a three neuron Matsuoka oscillator network to generate a well-defined peristalsis trajectory in a soft-bodied swallowing robot

3 Optical Motion Tracking and System Identification of the Swallowing Robot

3.1 Swallowing Robot's Motion Capture by Vicon

To capture the deformation of the conduit of the SR, a flat version of the robot was developed to approximate the original robot. We used optical motion capture system by installing the Vicon cameras at the several locations of the motion capture lab. In total, twelve cameras were utilised out of which four were set next to the robot. An array of 20 hemispherical markers of size 3 mm was glued on the actuator. The marker array was covering almost one-third of the entire deformation surface. A spacing of 2.5 cm was maintained between them so that during the actuation the markers do not interfere with each other's path. In addition to the small markers, three big reference markers of diameter 7 mm were placed at three corners of the plastic frame, so that the location, orientation, and alignment of the robot can be observed on the TV screen. A tape was used to mask the plastic frame of the robot as illustrated in Fig. 3a so that the cameras do not capture any insignificant objects surrounding the actuator of the robot. The motion capture experiment starts after the calibration of the cameras and setting the origin of the coordinate system. The centre of one of the large markers was chosen for the before mentioned purpose. Once the setup and the calibration have been done, data capturing was started by inflating the chambers of the robot by generating different peristaltic trajectories discussed in the previous sections. Figure 3 shows the images from the experiment conducted on the flat swallowing robot. The Vicon tracker software provided with the system can plot the marker's position data in real time as well as it can be used to generate a spreadsheet file where the entire time series data can be stored for further processing.

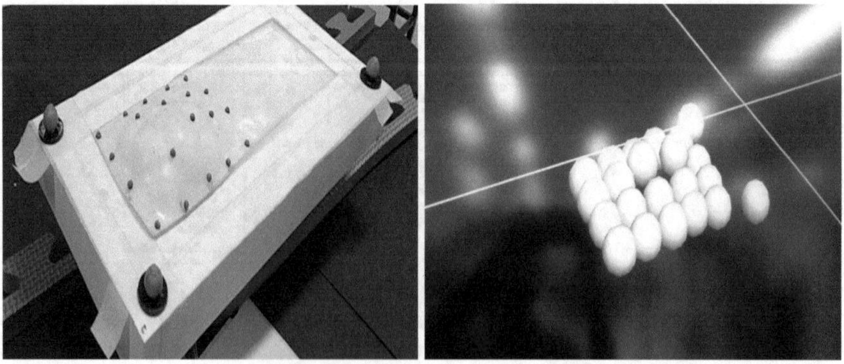

(a) Flat Swallowing Robot, (b) LCD Screen

Fig. 3 Image of the flat swallowing robot and an array of optical motion capture markers which are placed on top of it to visualise and record the displacement of the chambers of the robot on a TV screen

3.2 Processing the Captured Time-Series Data

The time-series data recorded from the Vicon tracking experiment needs to be processed before using it for the system identification of the SR. The first step is to interpolate the missing frames (cells) in the data which the cameras were not able to capture. Further processing on the data can only be done, once the missing data were generated. The next step is to remove any particular trend or pattern from the data which are not any interest to us. Such kinds of aspects are the major cause of some distortion which can eventually lead to incorrect identification of the dynamics of the robot. By using system identification toolbox in the MATLAB, both the processing tasks can be performed satisfactorily. The final step is to design a filter that can minimise the effect of noise which has been added during the data capture experimentation. The nature of the noise was taken out to be additive, and distributed normally. A sixth order Butterworth filter with suitable cut-off frequency was designed to remove the ripples from the captured time-series data.

3.3 State Initialisation

The Vicon experiment is an example of data capturing in the Lagrangian frame of reference. The preliminary tasks before conducting the Vicon experiments are setting the origin and the coordinate axis. For an initial modelling purpose, the second column of markers as shown in Fig. 4 was chosen from the array. As the movement of the markers in the x and y direction of the chosen coordinate system was negligible hence, only z-direction displacement was selected. Let $\{x_i\}_{i=1}^{3}$ be the chosen states defining the displacement of the markers in z-direction and $\mathbf{f}$ is the unidentified governing dynamics of the SR respectively expressed by Eqs. (3) and (4). The movement of the three markers along the z-axis was considered for a preliminary testing purpose. A training dataset of size 401×3 was taken out from the original processed dataset of size 2923×3 (see Sect. 3.1) for modelling the dynamics of the actuator where the row index denotes the time stamp and column index denotes the

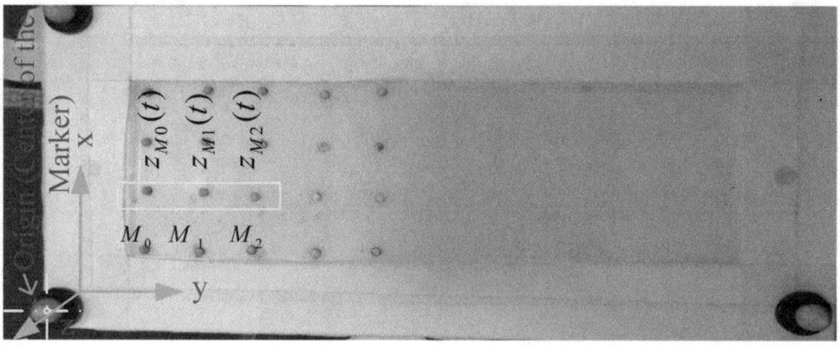

Fig. 4 State initialisation scheme

number of states. The first step is to evaluate the derivative of the states by using numerical differentiation. The rest of the dataset is kept for validating the model. Let $\dot{\mathbf{x}}(t) = [\dot{x}_1, \dot{x}_2, \dot{x}_3]^T$ be the first order time derivative of the initialised states, computed by applying central difference method.

$$\begin{aligned} x_1 &= z_{M_0}(t) \\ x_2 &= z_{M_1}(t) \\ x_3 &= z_{M_2}(t) \\ \mathbf{x}(t) &= [x_1, x_2, x_3]^T \mid \mathbf{x} \in \Re^3 \end{aligned} \tag{3}$$

$$\text{and, } \dot{x}_i(t) = f_i(\mathbf{x}(t)), i \in \{1, 2, 3\} \tag{4}$$

3.4 Defining Regression Problem

After defining the state vector $\mathbf{x}(t)$ as per the Eqs. (3) and (4) and its time derivative $\dot{\mathbf{x}}(t)$ (computed numerically in Sect. 3.3) for the training dataset $\{t_i\}_{i=1}^{m}$ ($m = 401$), arrange $\mathbf{x}(t)$ and $\dot{\mathbf{x}}(t)$ in two large matrices $\mathbf{X}$ and $\dot{\mathbf{X}}$ such that the order of each matrix is $m \times n$ (401×3) expressed by the Eq. (5).

$$\mathbf{X} = \begin{bmatrix} x_1^{(1)} & x_2^{(1)} & x_3^{(1)} \\ x_1^{(2)} & x_2^{(2)} & x_3^{(2)} \\ \vdots & \vdots & \vdots \\ x_1^{(m)} & x_2^{(m)} & x_3^{(m)} \end{bmatrix}, \dot{\mathbf{X}} \mid \mathbf{X}, \dot{\mathbf{X}} \in \Re^{m\times 3} \tag{5}$$

The next step is to construct a library of functions consisting of constant, different orders of polynomial, and trigonometric terms. Eq. (6) gives the expression for the library function [15, 16]. If the number of states are $n = 3$ and order of the polynomial is $k = 5$ then the dimension of $\mathbf{X}^{P_k}$ is $m \times {}^{k+n-1}C_k$ (401×21) and dimension of $\boldsymbol{\Theta}(\mathbf{X})$ is $m \times p$, where $p = \sum_{i=0}^{k} {}^{i+n-1}C_i$ for zero trigonometric terms. Each column of $\boldsymbol{\Theta}(\mathbf{X})$ represents the potential candidate for the final right-hand side expression for Eq. (4) and p represents the number of features or the considered candidate functions. There is an enormous opportunity for decision in building the library function of non-linearities. Since only a few of the candidates will appear in the final expression of $\mathbf{f}(.)$ ($[f_1, f_2, f_3]^T$), a regression problem can be defined to determine the coefficients of sparse vector given by (7).

$$\boldsymbol{\Theta}(\mathbf{X}) = \begin{bmatrix} | & | & | & & | & & | & | \\ 1, & \mathbf{X}^{\mathbf{P}_1}, & \mathbf{X}^{\mathbf{P}_2}, & \ldots, & \sin(\mathbf{X}), & \ldots & & \\ | & | & | & & | & & | & | \end{bmatrix} \mid \boldsymbol{\Theta}(\mathbf{X}) \in \Re^{m\times p} \tag{6}$$

$$\dot{\mathbf{X}} = \boldsymbol{\Theta}(\mathbf{X})\mathbf{V}$$
$$\text{where, } \mathbf{V} = [\mathbf{v}_1, \mathbf{v}_2, \mathbf{v}_3] | \mathbf{V} \in \Re^{p\times 3} \quad (7)$$

Each column $\mathbf{v}_k$ of $\mathbf{V}$ in Eq. (7) denotes a sparse vector coefficient responsible for the terms that are present on the right-hand side for one of the row equations in Eq. (4). Once the regression problem is defined, the next step is to determine $\mathbf{V} = \mathbf{V}^*$ by implementing sequentially threshold least square algorithm (STLSA) to promote sparsity due to the use of $l1$ regularisation. The sparsity of the coefficients determined by STLSA can be controlled by a threshold parameter λ. After optimising the regression problem, the model was validated by using the entire dataset.

4 Results

4.1 Parameter Estimation Results

A sinusoidal waveform with an amplitude of 1 V ($y_{ref}(t)$) and a period of 4.5 s was taken as a reference (Fig. 5a). As the oscillator network will actuate a single layer of air pressure chambers at a time, the neurons in the oscillator must mutually inhibit which means when one of the neuron is active the other two neurons must be inactive. The trajectories ($y_{ref1}(t)$, $y_{ref2}(t)$, $y_{ref3}(t)$) as shown in Fig. 5b was considered as references for determining the desired parameter values of each neuron i.e. y_{ref1} for neuron N_1 and so on. Trust region reflective algorithm (TRRA) was used to solve the nonlinear least square regression problem defined in the Eq. (2), and hence 67.57%, 67.00% and 60.22% of fitness values were achieved for the reference specified for neuron N_1, N_2, and N_3 respectively. A mean square error (MSE) of 0.0394 and final prediction error (FPE) of 2.72×10^{-6} were found. Fig. 6 shows the comparison of the oscillator output with initial random parameter values ($\boldsymbol{\theta}_{initial}$) and estimated values ($\hat{\boldsymbol{\theta}}$). The final estimated parameter values are given in Table 1 and the trajectory generated by the oscillator network corresponding to the $\hat{\boldsymbol{\theta}}$ is shown in Fig. 7.

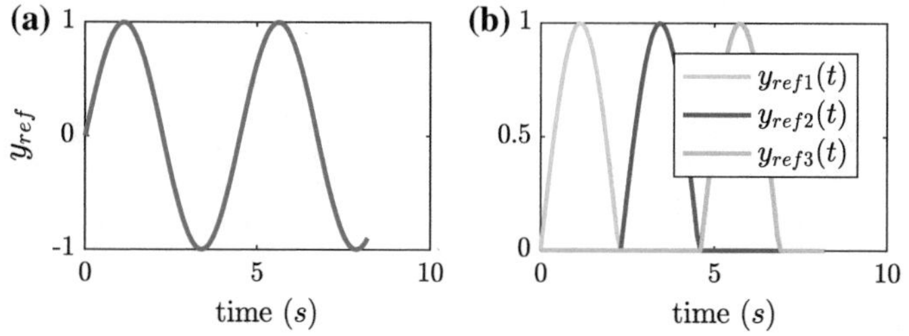

Fig. 5 **a** A sinusoidal reference of amplitude 1 V and a period of 4.5 s. **b** Reference trajectory extracted from the sinusoidal wave used for the estimation of the parameters of the three neurons in the Matsuoka oscillator network

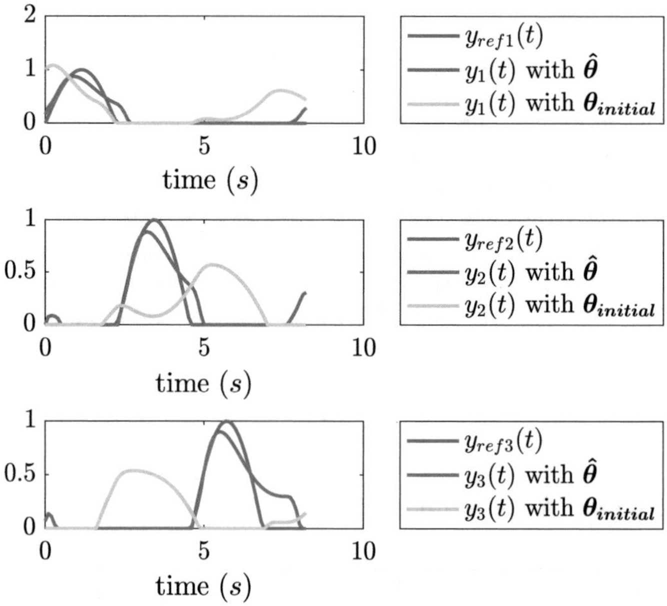

Fig. 6 Plots showing one cycle of the reference trajectory, and one cycle of the trajectory generated by the Matsuoka Oscillator network for initial set of parameters ($\boldsymbol{\theta}_{initial}$) and estimated set of parameters ($\hat{\boldsymbol{\theta}}$) for the three neurons

Table 1 Initial and final estimates of $\boldsymbol{\theta}$

Parameters	$\boldsymbol{\theta}$	Range	$\hat{\theta}_i$
K_n	0.7	0–1	0.4372
a_{13}	2.697	0–(1 + b)	2.8660
a_{23}	2.697	0–(1 + b)	2.8660
b	2.5	0–∞	3.8877
c	1.689	0.1–10	1.5975
$\mathbf{x}_0$	$[1, -.1, 0, 0, 0, 0]$		0.168, $2.6\times^{-5}$, 0.069, $3.3\times^{-5}$, 0.0230, $1.93\times^{-6}$

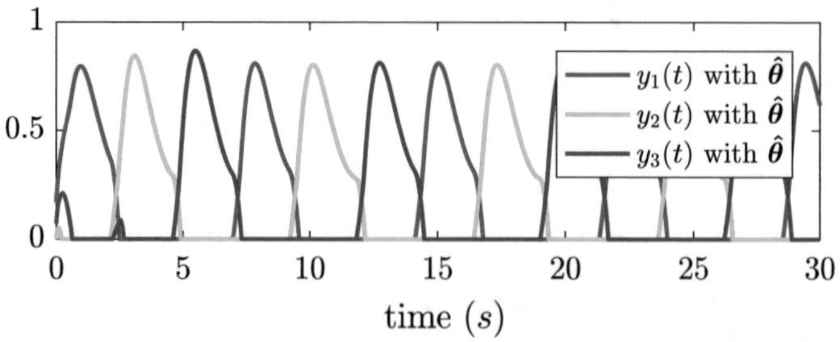

Fig. 7 Peristalsis trajectory generated by the Matsuoka oscillator network for actuating the swallowing robot

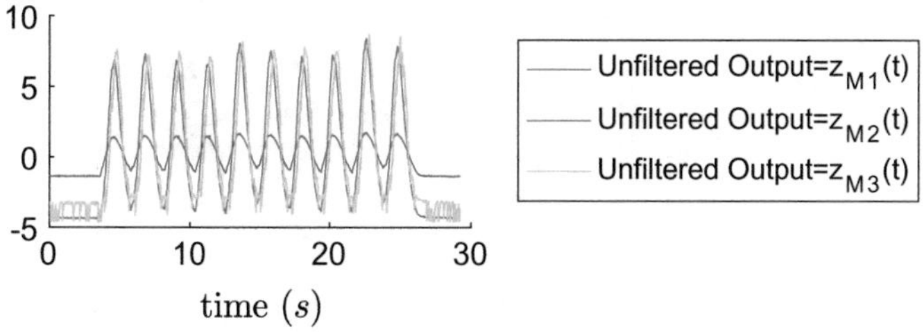

Fig. 8 Plots of the unfiltered recorded time series data of the displacement of three markers indicated by the boxed region in Fig. 4 moving along the z-axis

Fig. 9 Magnitude plot of the designed Butterworth filtering dB

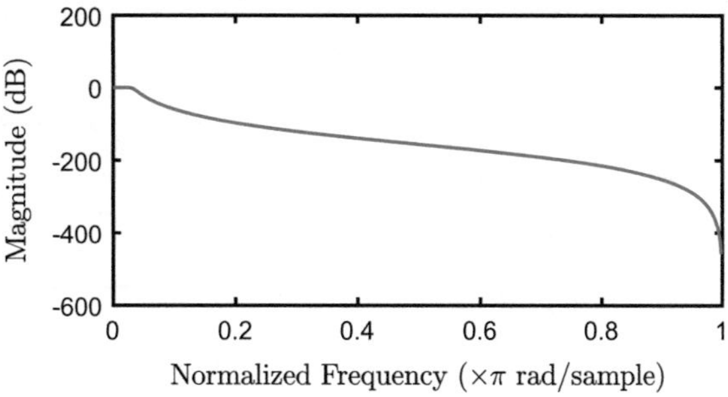

4.2 Preprocessing Results

A sixth order Butterworth low-pass filter with a cut-off frequency of $\Omega = 10$ rad/s (normalised frequency, $\omega = 0.032\pi$) was chosen based on NPSD response to filter out the high-frequency noise present in the collected time-series dataset which is saved as a spreadsheet in Sect. 3.1 (Fig. 8). In Fig. 8, it can be clearly seen that the movement of the marker 2 and 3 are similar and much higher than the marker 1 because the displacement of the top layer of the robot was restricted by the plastic mold. Figure 9 shows the magnitude plot of the filter transfer function w.r.t the normalised frequency which was used to filter the movement of the markers along the z-axis represented by a time-series dataset. Once the filter was designed according to the specified filter parameters, the time-series data was applied to it to get the filtered output ($\{z_{M_i}\}_{i=1}^{3}$). The response of the filter when the detrended data was applied to it as an input is shown in Fig. 10.

4.3 Regression Results

The x_1^{actual}, x_2^{actual}, and x_3^{actual} is the training dataset extracted from the time-series displacement of the three markers moving along the z- axis as shown in Fig. 4. The dataset is then used to find out the solution for the defined regression problem and

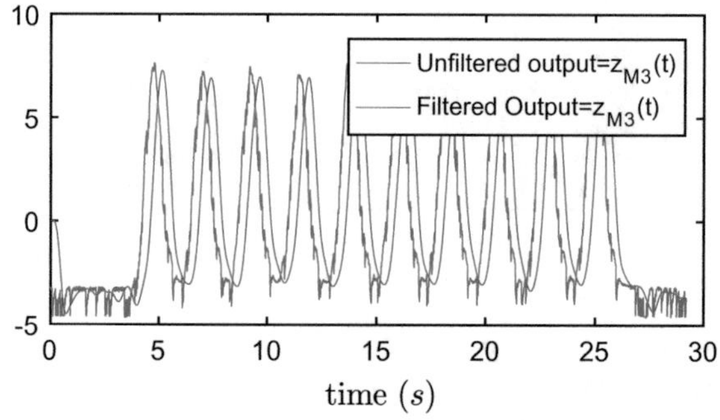

Fig. 10 Plot showing the comparison between the unfiltered and filtered time-series data for the third marker

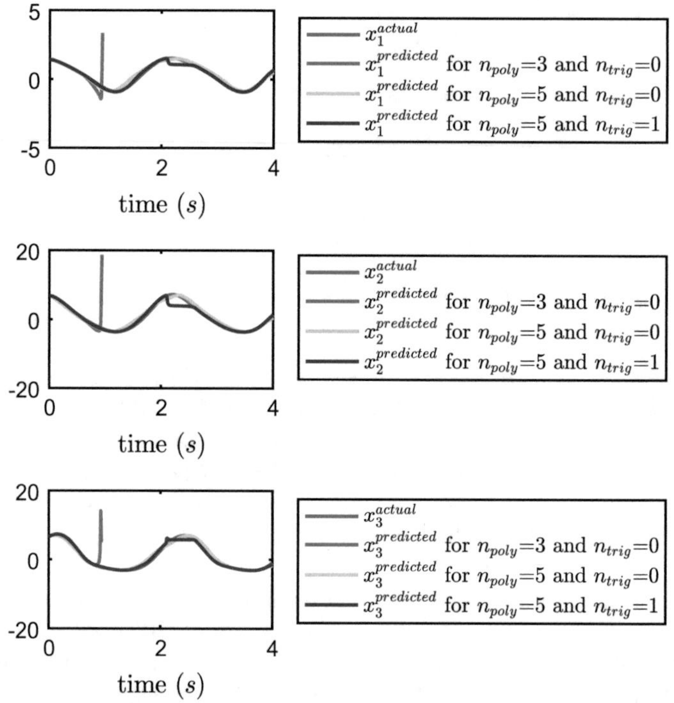

Fig. 11 Plot showing the comparison between the actual and the predicted values of the state trajectories validated over the training dataset

hence, to discover the dynamics of the flat robot. The first step is to select the library of the candidate function ($\boldsymbol{\Theta}(\mathbf{X})$) having nonlinear terms of different types. The library must contain terms of different polynomial order as well as trigonometric terms such that it can explore the corners of the possible nonlinear space. Figure 11 shows the result of the system identification for zero trigonometric terms and λ equal to 0.07. Once the model has been generalised, the initial validation was done on the training dataset, and it was found that the error between the actual and the predicted value ($x_1^{predicted}$, $x_2^{predicted}$, and $x_3^{predicted}$) was very high, even though the order of the polynomial (n_{poly}) raised from three to five as shown in the Fig. 11. As soon as the trigonometric terms (n_{trig}) were admitted, the predicted values began to match the actual values as shown in the Fig. 11 by the purple coloured plot.

5 Conclusion

The soft bodied swallowing robot's actuator was made fully from soft elastomer, powered by air. Due to the compliant nature of the material, the conventional modelling approaches were hard to apply. In summary, this paper describes a powerful data-driven technique which is implemented successfully to model the dynamics of the robot. The dynamics were derived from the processed data captured from the Vicon experiments. To perform the experiments, artificial peristalsis was generated and applied to actuate the robot, by implementing a neural oscillator scheme. The parameters of the oscillator governing the actuation were determined by applying a modified Newton's algorithm.

Acknowledgements The work presented in this paper was funded by Riddet Institute Centre of Research Excellence, New Zealand.

References

1. Yap, H.K., Ng, H.Y., Yeow, C.-H.: High-force soft printable pneumatics for soft robotic applications. Soft Robot. **3**(3), 144–158 (2016)
2. Katzschmann, R.K., Marchese, A.D., Rus, D.: Hydraulic autonomous soft robotic fish for 3d swimming. In: Proceedings Conference on Experimental Robotics, pp. 405–420. Springer (2016)
3. Ranzani, T., Cianchetti, M., Gerboni, G., Falco, D., Menciassi, A.: A soft modular manipulator for minimally invasive surgery: Design and characterization of a single module. **32**(1), 187–200 (2016)
4. Song, Y.S., Sun, Y., Van Den Brand, R., Von Zitzewitz, J., Micera, S., Courtine, G., Paik, J.: Soft robot for gait rehabilitation of spinalized rodents. In: IEEE/RSJ International Conference on Intelligent Robots and Systems (IROS), 2013 Conference Proceedings. pp. 971–976. IEEE (2013)
5. Wei, Y., Chen, Y., Ren, T., Chen, Q., Yan, C., Yang, Y., Li, Y.: A novel, variable stiffness robotic gripper based on integrated soft actuating and particle jamming. Soft Robot. **3**(3), 134–143 (2016)
6. Chen, F.J., Dirven, S., Xu, W.L., Li, X.N.: Soft actuator mimicking human esophageal peristalsis for a swallowing robot. IEEE/ASME Trans. Mechatron. **19**(4), 1300–1308 (2014)
7. Zhu, M., Xie, M., Xu, W., Cheng, L.K.: A nanocomposite-based stretchable deformation sensor matrix for a soft-bodied swallowing robot. IEEE Sens. J. **16**(10), 3848–3855 (2016)
8. Rus, D., Tolley, M.T.: Design, fabrication and control of soft robots. Nature **521**(7553), 467–475 (2015). https://doi.org/10.1038/nature14543

9. Dirven, S., Chen, F., Xu, W., Bronlund, J.E., Allen, J., Cheng, L.K.: Design and characterization of a peristaltic actuator inspired by esophageal swallowing. IEEE/ASME Trans. Mechatron. **19**(4), 1234–1242 (2014)
10. Dirven, S., Xu, W., Cheng, L.K.: Sinusoidal peristaltic waves in soft actuator for mimicry of esophageal swallowing. IEEE/ASME Trans. Mechatron. **20**(3), 1331–1337 (2015)
11. Bhattacharya, D., Nisha, M.G., Pillai, G.: Relevance vector-machine-based solar cell model. Int. J. Sustain. Energy **34**(10), 685–692 (2015). https://doi.org/10.1080/14786451.2014.885030
12. Iplikci, S.: Support vector machines–based generalized predictive control. Int. J. Robust Nonlinear Control **16**(17), 843–862 (2006). https://doi.org/10.1002/rnc.1094
13. Matsuoka, K.: Sustained oscillations generated by mutually inhibiting neurons with adaptation. Biol. Cybern. **52**(6), 367–376 (1985). https://doi.org/10.1007/BF00449593
14. Fang, Y., Hu, J., Liu, W., Chen, B., Qi, J., Ye, X.: A cpg-based online trajectory planning method for industrial manipulators. In: Conference Proceedings 2016 Asia-Pacific Conference on Intelligent Robot Systems (ACIRS), pp. 41–46 (2016)
15. Bhattacharya, D., Cheng, L.K., Dirven, S., Xu, W: Actuation planning and modeling of a soft swallowing robot. In: Mechatronics and Machine Vision in Practice (M2VIP), 2017 24th International Conference on. pp. 1–6. IEEE (2017)
16. Brunton, S.L., Proctor, J.L., Kutz, J.N.: Discovering governing equations from data by sparse identification of nonlinear dynamical systems. Proc. Natl Acad. Sci. **113**(15), 3932–3937 (2016)

Reinforcement Learning of a Memory Task Using an Echo State Network with Multi-layer Readout

Toshitaka Matsuki and Katsunari Shibata

Abstract Training a neural network (NN) through reinforcement learning (RL) has been focused on recently, and a recurrent NN (RNN) is used in learning tasks that require memory. Meanwhile, to cover the shortcomings in learning an RNN, the reservoir network (RN) has been often employed mainly in supervised learning. The RN is a special RNN and has attracted much attention owing to its rich dynamic representations. An approach involving the use of a multi-layer readout (MLR), which comprises a multi-layer NN, was studied for acquiring complex representations using the RN. This study demonstrates that an RN with MLR can learn a "memory task" through RL with back propagation. In addition, non-linear representations required to clear the task are not observed in the RN but are constructed by learning in the MLR. The results suggest that the MLR can make up for the limited computational ability in an RN.

Keywords Echo state network · Reservoir network · Reservoir computing
Reinforcement learning

1 Introduction

Deep learning (DL) has surpassed existing approaches in various fields. It suggests that a successfully trained large scale neural network (NN) that is used as a massive parallel processing system is more flexible and powerful than the systems that are carefully designed by engineers. In recent years, end-to-end reinforcement learning (RL), wherein the entire process from sensors to motors is composed of one NN without modularization and trained through RL, has been a subject of focus [1]. For quite some time, our group has suggested that the end-to-end RL approach is critical

T. Matsuki (✉) · K. Shibata
Oita University, 700 Dannoharu, Oita, Japan
e-mail: matsuki@oita-u.ac.jp

K. Shibata
e-mail: shibata@oita-u.ac.jp

J.-H. Kim et al. (eds.), *Robot Intelligence Technology and Applications 5*,
Advances in Intelligent Systems and Computing 751,
https://doi.org/10.1007/978-3-319-78452-6_2

for developing a system with higher functions and has demonstrated the emergence of various functions [2]. Recently, deep mind succeeded in training an NN to play Atari video games using the end-to-end RL approach [3]. A feature of this approach is that the system autonomously acquires purposive and general internal representations or functions only from rewards and punishments without any prior knowledge about tasks.

When the system learns proper behaviors with time, it has to handle time series data and acquire necessary internal dynamics. A recurrent structure is required for the system to learn such functions using an NN. We demonstrated that a recurrent NN (RNN) trained by back propagation through time (BPTT) using autonomously produced training signals based on RL can acquire a function of "memory" or "prediction" [4, 5]. However, it is difficult for a regular RNN to acquire complex dynamics such as multiple transitions among states through learning.

BPTT is generally used to train an RNN, but factors such as slow convergence, instability, and computational complexity can cause problems. A reservoir network (RN) such as a liquid state machine, proposed by Jaeger [6], or an echo state network (ESN), proposed by Maass [7], is often used to overcome such issues. The RN uses an RNN called "reservoir," which comprises many neurons that are sparsely connected with each other in a randomly chosen fixed weight. The reservoir captures the history of inputs, receives its outputs as feedback, and forms dynamics including rich information. The outputs of RN are generated as the linear combinations of the activations of reservoir neurons by readout units, and the network is trained by updating only the readout weights from the reservoir neurons to generate the desired values. Therefore, it is easy for the RN to learn to process the time series data and generate complex time series patterns. We believe that the RN can be a key to solving the problems associated with acquiring complex dynamics through learning. In this study, an ESN, which is a kind of RN having rate model neurons, is used. The ESN has been used in various studies, including motor control [8] and dynamic pattern generation [9, 10].

To generate outputs that cannot be expressed as linear combinations of dynamic signals in reservoir and inputs from the environment, an approach using more expressive multi-layer readout (MLR), which uses a multi-layer NN (MLNN) trained by back propagation (BP) instead of regular readout units for output generation was studied [11]. Bush and Anderson showed that ESN with MLR can approximate the Q-function in a partially observable environment through Q-learning [12]; Babinec and Pospíchal showed that the accuracy of time series forecasting was improved with this approach [13]. These studies were conducted more than a decade ago. However, we believe that such an architecture will be vital in the future as more complex internal dynamics and computation are required to follow the trend of the increasing importance of end-to-end RL.

In this study, we focus on learning to memorize necessary information from the past and utilize this information to generate appropriate behaviors using an ESN with MLR. We also demonstrate that such functions can be learned by simple BP that does not involve trace back to the past as in BPTT.

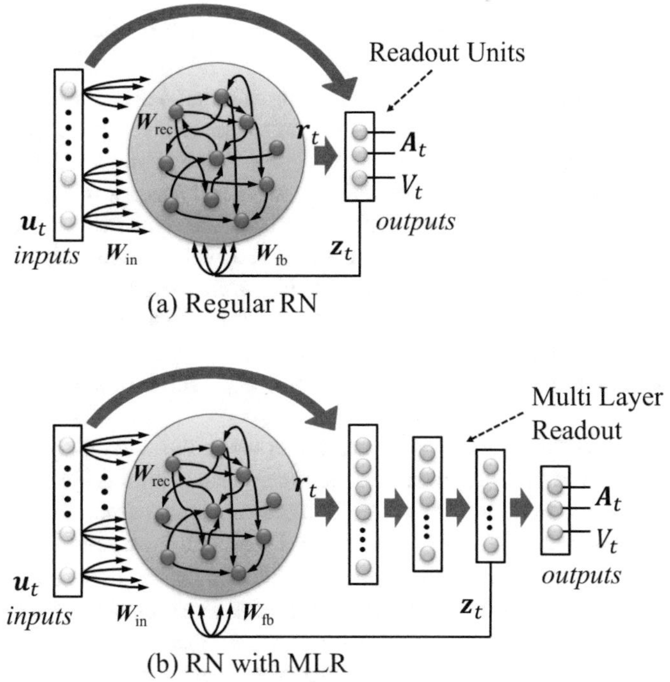

Fig. 1 Network architectures of **a** a regular reservoir network (RN) and **b** an RN with multi-layer readout (MLR)

2 Method

2.1 Network

The network architectures used in this study are shown in Fig. 1. Instead of single layer readout units, as shown in Fig. 1a, an RN comprises a multi layer neural network (MLNN) such as multi-layer readout (MLR), as shown in Fig. 1b, to generate outputs. The inputs from the environment are provided to the reservoir and readout units or MLNN, and the outputs of the reservoir neurons are provided to the MLNN. The reservoir's capacity for storing large volumes of information and the MLNN's ability of flexibly extracting the necessary information from large volumes of information and generating appropriate outputs are combined. Therefore, it is expected that tracing back to the past with BPTT is no longer required, and memory functions can be acquired only with BP.

The number of reservoir neurons is $N_x = 1000$. The reservoir neurons are dynamical model neurons and are recurrently connected with the connection probability $p = 0.1$. The internal state vector of reservoir neurons at time t, $\boldsymbol{x}(t) \in \mathbb{R}^{N_x}$ is given as

$$\boldsymbol{x}_t = \left(1 - a\right)\boldsymbol{x}_{t-1} + a\left(\lambda \boldsymbol{W}_{rec}\boldsymbol{r}_{t-1} + \boldsymbol{W}_{in}\boldsymbol{u}_t + \boldsymbol{W}_{fb}z_{t-1}\right), \tag{1}$$

where $a = 0.1$ is a constant value called leaking rate that determines the time scale of reservoir dynamics. $\boldsymbol{W}_{rec} \in \mathbb{R}^{N_x \times N_x}$ is the recurrent connection weight matrix of the reservoir, and each component is set to a value that is randomly generated from a Gaussian distribution with zero mean and variance $1/pN_x$. $\lambda = 1.2$ is a scale of recurrent weights of the reservoir. Larger λ makes the dynamics of the reservoir neurons more chaotic. $\boldsymbol{r}_t$ is the output vector of reservoir neurons. $\boldsymbol{W}_{in} \in \mathbb{R}^{N_x \times N_i}$ is the weight matrix from the input to the reservoir neurons. $\boldsymbol{W}_{fb} \in \mathbb{R}^{N_x \times N_f}$ is the weight matrix from the MLR or readout units to the reservoir neurons, and z_t is the feedback vector. Each component of $\boldsymbol{W}_{in}$ and $\boldsymbol{W}_{fb}$ is set to a uniformly random number between -1 and 1. The activation function of every neuron in the reservoir is the *tanh* function.

The MLR is a four-layer NN with static neurons in the order of 100, 40, 10 and 3 from the bottom layer; the activation function of each neuron is the *tanh* function. Each neuron in the bottom layer of MLR receives outputs from all the reservoir neurons. The outputs of $N_f = 10$ neurons in the hidden layer, one lower than the output layer, are fed back to every neuron of the reservoir as feedback vector $z_t \in \mathbb{R}^{N_f}$. $N_i = 7$ inputs are derived from the environment; all these inputs are given to each neuron in the reservoir and the bottom layer of MLR. Each initial weight of MLR is set to a randomly generated value from a Gaussian distribution with zero mean and variance $0.01/n$, where n is the number of inputs in each layer. The MLP is trained by BP and the stochastic gradient descent algorithm with a learning rate of 0.01.

The network outputs are critic V_t and actor vector $\boldsymbol{A}_t$. The sum of $\boldsymbol{A}_t$ and the exploration component vector $\boldsymbol{rnd}_t$ is used as the motion signal of the agent at time t. Each component of $\boldsymbol{rnd}_t$ is set to a uniformly random value between -1 and 1.

2.2 Learning

In this network, only the weights of the paths indicated by the red arrows in Fig. 1 are trained using the BP based actor-critic algorithm. The training signal for critic at time $t-1$ is given as

$$V_{t-1}^{train} = V(\boldsymbol{u}_{t-1}) + \hat{r}_{t-1} = r_t + \gamma V(\boldsymbol{u}_t), \tag{2}$$

where $\hat{r}_{t-1}$ is TD-error at time $t-1$, which is given as

$$\hat{r}_{t-1} = r_t + \gamma V(\boldsymbol{u}_t) - V(\boldsymbol{u}_{t-1}), \tag{3}$$

where r_t is a reward received by the agent at time t and $\gamma = 0.99$ is the discount rate. The training signal for actor at time $t-1$ is given as

$$A_{t-1}^{train} = A(u_{t-1}) + \hat{r}_{t-1} rnd_{t-1}. \tag{4}$$

The weights in reservoir W_{rec}, W_{in} and W_{fb} are not trained.

3 Experiment

A memory task was employed to examine the capability of an RN with MLR. Comparing a regular RN and an RN with MLR, we examined the practicality of the parallel and flexible processing capability of MLNN in the memory task.

The outline of the task is shown in Fig. 2. An agent is placed on a 15.0×15.0 plane space. At every step, the agent moves according to the two actor outputs each of which determines the moving distance in either x or y directions respectively. The purpose of this agent is to learn the actions needed to first enter the switch area, and then go to the goal. The radius of the goal or switch area is 1.5.

From the environment, an agent receives $N_i = 7$ signals as an input vector

$$u_t = [d'_g,\ sin\theta_g,\ cos\theta_g,\ d'_s,\ sin\theta_s,\ cos\theta_s,\ signal], \tag{5}$$

where d'_g, d'_s are distances to the goal and switch areas, respectively, and are normalized into the interval $[-1, 1]$. θ_g and θ_s are the angles between the x-axis and the goal or switch direction from the agent. Only when the agent is in the switch area, a *signal* is given as

$$signal = \begin{cases} 0 & d_s > R_s \\ 10 & d_s \leq R_s, \end{cases} \tag{6}$$

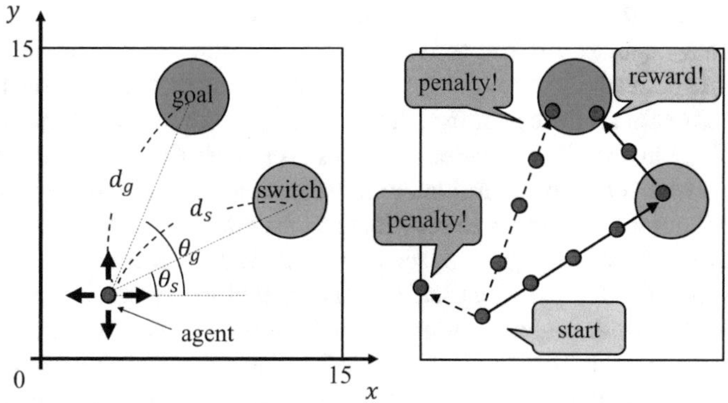

Fig. 2 Outline of the memory task. An agent must first enter the switch area, and then go to the goal area

where R_s is the radius of the switch area. In each trial, the agent, goal and switch are randomly located in the field, and their areas do not overlap with each other. A punishment $r_t = -0.1$ is set for an agent when it contacts the wall, and a punishment $r_t = -0.5$ is set when the agent enters the goal area before the switch area. A reward $r_t = 0.8$ is set when it enters the goal area after entering the switch area. One trial is terminated after the agent acts 200 steps or enters the goal area. After 50,000 trials, the system stops learning.

4 Result

After 50,000 learning trials, the agent behaviors were observed in two cases wherein the sign of each actor output should be changed before and after entering the switch area. The trajectories of the agent in the test trial are shown in Fig. 3.

For comparison, the results in the case of RN with MLR are shown in Fig. 3a, b and those in the case of regular RN are shown in Fig. 3c, d. As shown in Fig. 3a, b, the agent with MLR first entered the switch area after which it entered the goal area. The result shows that this network succeeded in learning a memory task through RL without tracing back to the past as in BPTT. In addition, the network acquired functions to memorize necessary information and generate the desired action signal only with BP. In contrast, as shown in Fig. 3c, d, the agent with a regular RN failed to learn the desired behavior. Without entering the switch area, the agent entered the goal area either after wandering in the field (case 1) or remained continually struck to the wall (case 2).

To observe the activation of the reservoir neurons, the switch was fixed at the center of the field, and the goal or the agent was located at one of the four points: $(3, 3)$, $(3, 12)$, $(12, 3)$ and $(12, 12)$; the test trial was then implemented. Various activations were found among the reservoir neurons, but in most cases, it was difficult to find a clear regularity in the activation. In Fig. 4, the network outputs and two characteristic activations of reservoir neurons, in certain cases, are shown with the agent trajectory. In all the cases, the activation of the neuron (1) in Fig. 4 decreases after switching in a similar way. Such neurons remember that the agent has already entered the switch area and contribute to reflect the memory to the outputs of actor. Some other neurons that seems to contribute to the memory function were found.

A non-linear function of present sensor signals and the memorized information that represents whether the agent has already entered the switch area are required to generate appropriate outputs. In the case of Fig. 4b, by entering the switch area, the y-motion should be changed from "go up" to "go down," whereas in the case of Fig. 4c, it should be changed from "go down" to "go up". Then, to determine whether such outputs are generated in the reservoir, we attempted to find the reservoir neurons with the same sign as the output as that of the actor output for y-axis motion before and after the agent was inside the switch area. The neuron (2) in Fig. 4 is the only one found among the total of 1,000 reservoir neurons. However, the activation pattern of neuron (2), shown in Fig. 4, seems to lag behind the actor output

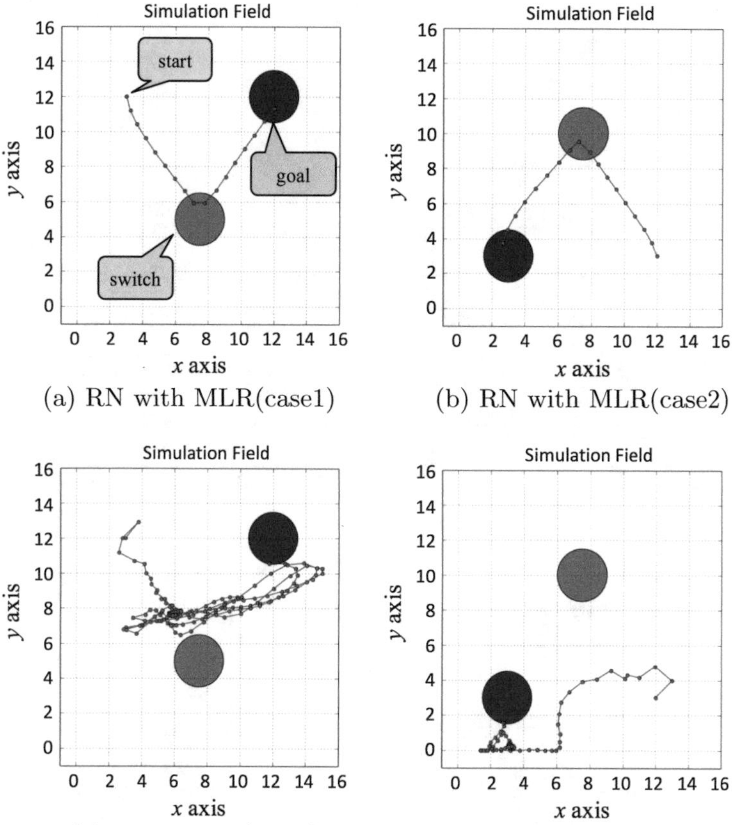

(a) RN with MLR(case1) (b) RN with MLR(case2)

(c) regular RN(case1) (d) regular RN(case2)

Fig. 3 Comparison of agent trajectory for two cases between RN with MLR and regular RN

for the *y*-axis. Then, the feedback connection weight matrix $\boldsymbol{W}_{fb}$ is set to zero and the same test was performed to eliminate the influences from the MLR to the reservoir. In that case, the activation which has a similar feature to the neuron (2) was not observed in the reservoir but the agent could clear the task. This suggests that the activation in Fig. 4(2) appeared under the influences of the MLR through the feedback connections. Considering that the regular RN could not learn the memory task, the non-linear function of memorized information in the reservoir and present sensor signals required to clear the task are not generated in the reservoir but are constructed through learning in the MLR. In other words, the learning of MLNN in MLR is necessary to non-linearly integrate the outputs of reservoir and sensor signals, which enables the switching of actor outputs based on memory.

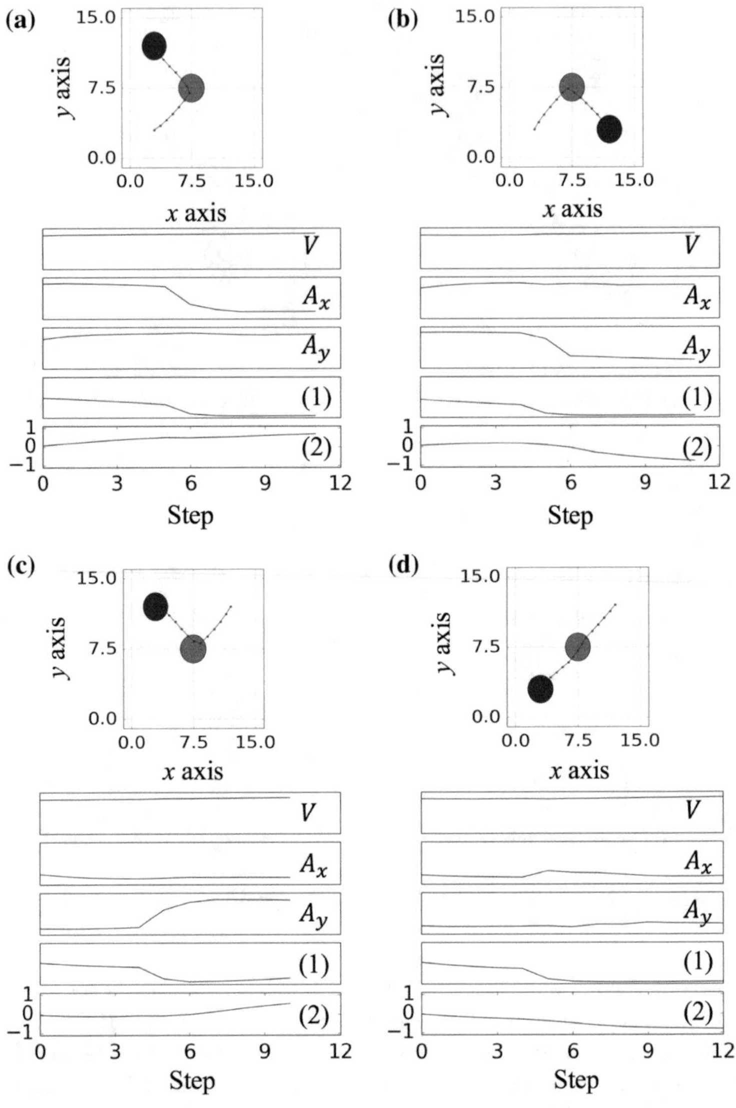

Fig. 4 Outputs of the network and two characteristic reservoir neurons during one trial for 4 cases of goal and initial agent locations. V is critic, and A_x and A_y are the actor outputs for x- and y-axis, respectively

5 Conclusion

This study demonstrated that an RN with MLR, which generates outputs with an MLNN instead of readout units, can learn a memory task using RL with simple BP without using BPTT. Various activations were observed among the neurons in the reservoir. There were neurons whose activation decreases after switching in a similar way in all cases, and it is considered that these neurons greatly contribute to the function of memory. It was confirmed that non-linear representations, such as changes in actor outputs before and after reaching the switch area, were not observed in the reservoir dynamics; however, owing to the expressive MLNN, the MLR learns to construct the non-linear representations and successfully generate appropriate actor outputs. Our future study includes applying this frame to more complex tasks; analyzing the length of time the network can continue to store necessary information; analyzing how the feedback from a hidden layer of MLR to the reservoir influences reservoir dynamics; developing a method to train input weights of the reservoir to extract necessary information from inputs; and verifying whether our already proposed RL method [14, 15], in which chaotic internal dynamics in a NN are used as exploration components, is applicable.

Acknowledgements This work was supported by JSPS KAKENHI Grant Number 15K00360.

References

1. LeCun, Y., Bengio, Y., Hinton, G.: Deep learning. Nature **521**, 436–444 (2015)
2. Shibata, K.: Functions that emerge through end-to-end reinforcement learning (2017). arXiv:1703.02239
3. Mnih, V., Kavukcuoglu, K., Silver, D., Graves, A., Antonoglou, I., Wierstra, D., Riedmiller, M.: Playing Atari with deep reinforcement learning (2013). arXiv:1312.5602
4. Shibata, K., Utsunomiya, H.: Discovery of pattern meaning from delayed rewards by reinforcement learning with a recurrent neural network. In: Proceedings of IJCNN, pp. 1445–1452 (2011)
5. Shibata, K., Goto, K.: Emergence of flexible prediction-based discrete decision making and continuous motion generation through Actor-Q-Learning. In: Proceedings of ICDL-Epirob, ID 15 (2013)
6. Jaeger, H.: The "echo state approach to analysing and training recurrent neural networks-with an erratum note. Bonn, Germany: German National Research Center for Information Technology GMD Technical Report 148.34, p. 13 (2001)
7. Maass, W., Natschlger, T., Markram, H.: Real-time computing without stable states: a new framework for neural computation based on perturbations. Neural Comput. **14**(11), 2531–2560 (2002)
8. Salmen, M., Ploger, P.G.: Echo state networks used for motor control. In: Proceedings of the 2005 IEEE International Conference on Robotics and Automation, ICRA 2005. IEEE (2005)
9. Sussillo, D., Abbott, L.F.: Generating coherent patterns of activity from chaotic neural networks. Neuron Article **63**(4), 544–557(2009)
10. Hoerzer, G.M., Legenstein, R., Maass, W.: Emergence of complex computational structures from chaotic neural networks through reward-modulated Hebbian learning. Cerebral Cortex **24**(3), 677–690 (2014)

11. Lukusevičius, M., Jaeger, H.: Reservoir computing approaches to recurrent neural network training. Comput. Sci. Rev. **3**(3), 127–149 (2009)
12. Bush, K., Anderson, C.: Modeling reward functions for incomplete state representations via echo state networks. In: Proceedings. 2005 IEEE International Joint Conference on Neural Networks, 2005. IJCNN '05, vol. 5. IEEE (2005)
13. Babinec, Š., Pospchal, J.: Merging echo state and feedforward neural networks for time series forecasting. In: Artificial Neural Networks ICANN 2006, pp. 367–375 (2006)
14. Goto, Y., Shibata, K.: Emergence of higher exploration in reinforcement learning using a chaotic neural network. In: Proceedings of International Conference on Neural Information Processing (ICONIP) 2016, LNCS 9947, pp. 40–48 (2016)
15. Matsuki, T., Shibata, K.: Reward-based learning of a memory-required task based on the internal dynamics of a chaotic neural network. In: Proceedings of International Conference on Neural Information Processing (ICONIP) 2016, LNCS 9947, pp. 376–383 (2016)

Selective Filter Transfer

Minju Jung and Junmo Kim

Abstract Today deep learning has become a supreme tool for machine learning regardless of application field. However, due to a large number of parameters, tremendous amount of data is required to avoid over-fitting. Data acquisition and labeling are done by human one by one, and therefore expensive. In a given situation, the dataset with adequate amount is difficult to acquire. To resolve the problem, transfer learning is adopted (Yosinski et al, Adv Neural Inf Process Syst, 2014, [1]). The transfer learning is delivering the knowledge learned from abundant dataset, e.g. ImageNet, to the dataset of interest. The fundamental way to transfer knowledge is to reuse the weights leaned from huge dataset. The brought weights can be either frozen or fine-tuned with respect to a new small dataset. The transfer learning definitely showed a improvement on target performance. However, one drawback is that the performance depends on the similarity between the source and the target dataset (Azizpour et al, Proceedings of the IEEE conference on computer vision and pattern recognition workshops, 2015, [2]). In other words, the two datasets should be alike to be effective. Then finding the similar source dataset becomes another difficulty. To alleviate the problems, we propose a method that maximizes the effectiveness of the transferred weights regardless of what source data is used. Among the weights pre-trained with source data, only the ones relevant to the target data is transferred. The relevance is measured statistically. In this way, we improved the classification accuracy of downsized 50 sub-class ImageNet 2012 by 2%.

1 Introduction

Today is the deep-learning era. The machine learning technique is used not only in field of image processing, but also the field with other intricate data, i.e. natural language, industrial or economic data. Often performance of the machine exceeds that

M. Jung · J. Kim (✉)
School of Electrical Engineering, KAIST, Daejeon, South Korea
e-mail: junmo.kim@kaist.ac.kr

M. Jung
e-mail: alswn0925@kaist.ac.kr

J.-H. Kim et al. (eds.), *Robot Intelligence Technology and Applications 5*,
Advances in Intelligent Systems and Computing 751,
https://doi.org/10.1007/978-3-319-78452-6_3

of the human's, for example, it has been years that machine's classification accuracy of ImageNet 2012 dataset outperformed human level. With a huge machine capability with an enormous amount of parameters, the machine identifies fine details of image, as well as coarse categories.

However, this characteristic also can be an obstacle to learning. Due to the large number of parameters, learning gets over-fitted. To meet the learning capacity, an overwhelmed amount of data is required. At the lowest, the common datasets used in the field have approximately 50000–60000 training images, i.e. cifar 10, 100 or MNIST. And the standard database in image processing, called ImageNet, has 10 million images for training. In reality, however, constructing such massive database is impracticable. Gathering proper images is laborious, not to mention labeling each. Only human resource is capable, namely, the process becomes very costly.

To mitigate the problem, several approaches are attempted. One way is transfer learning. Transfer learning is delivering knowledge learned from huge database to another small database. In this case, the dataset trained earlier to a network is called 'source' dataset and the small one trained next is called 'target' dataset. Also the process of the training with source dataset is called 'pre-training'. Generally, the method is regarded as essential and fundamental. However, there is a drawback to the method, i.e. the source data needs to be similar to the target. Searching for a similar huge dataset is yet another problem.

In this paper, we propose a way to enhance efficiency of transfer learning given any source data. As the similarity level of source to target data differs, the pre-trained weights vary in terms of usefulness to the target task. Here, each weight's importance to the target task is measured statistically. Then according to the measure, only the weights which are highly target-relevant are preserved and fine-tuned with respect to target data. The left filters are re-initialized and the whole network is fine-tuned to the new task. With the proposed method, we improved the classification accuracy of downsized 50 sub-class ImageNet 2012 by 2%.

2 Method

The proposed method consists of 4 steps. First, pretrain a network with source data. Second, target data is given to the network and evaluated without updating. One epoch of target data is given sequentially and absolute value of each filters activation is summed respectively. Then the summed value is regarded as the importance measure of each filter. Third, a threshold is set with respect to the measure and only the filters whose absolute sums of outputs are above the value are fine-tuned with the target data. Other filters with smaller activation are re-initialized. Lastly, the whole network is fine-tuned to the new task (Fig. 1).

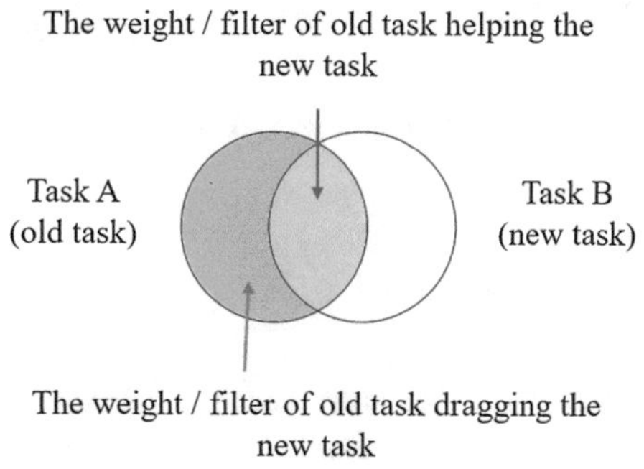

Fig. 1 One of the concept of this paper. Among the pre-trained filters, there are ones which are helpful to the target task and the others who are dragging the performance

2.1 Pretraining

To boost the performance regarding a small target data, a network is pre-trained with a source data. The pre-training helps the network grasp deeper understanding of real world. Among the knowledge learned from the source data, a part of it is target-friendly, while others are not.

2.2 Evaluating the Pre-trained Filters

One epoch of the target data is given and evaluated with the pre-trained network. No weight update is performed in this step. Each filters absolute value of response with respect to the target data is summed sequentially. The value is regarded as an importance measure of a filter since the more the filter is relevant to target task, the bigger activation comes out. For example, if a filter is for classifying natural-images, it will show higher activation with a problem, e.g. whether a given image is cat or dog, than a filter adapted with man-made images.

2.3 Isolated Fine-Tuning and Re-initialization

Only the filters whose activation values are above the threshold are fine-tuned with respect to the target data. This process is to recover connectivity among them. An ablation study is given in Sect. 3. And the other filters with smaller activation are re-initialized.

2.4 *Whole Network Fine-Tuning*

Lastly, the whole network is fine-tuned with the target data.

3 Experiment

The proposed transfer learning technique was implemented in Caffe [3]. All of the experiments were done with Nvidia TitanX.

3.1 *Database*

The two main datasets used in this experiment are from ImageNet 2012. The original ImageNet 2012 dataset has 1000 categories and 1.2 million images for training. The class categories vary from natural to man-made object/scene. The class categories are not only various but also very specific, e.g. English setter or cocker spaniel regarding only the dog classes.

In this work, the two subsets of ImageNet 2012 were used. A subset is composed of 50 man-made categories and another consists of 50 natural object/scene classes. The classes in each subset were randomly chosen among 1000 categories of original dataset. Generally, ImageNet 2012 dataset has about 1000 images per class and so does the source data, however, the target data has only 100 images per class, i.e. 1/10 of the source data size. In this way, the problem is more advanced and realistic.

3.2 *Modified AlexNet*

For experiment, an architecture modified from AlexNet [4] was used. As the AlexNet was designed for 1000 class ImageNet, the structure needed to be downsized. Generally, the architecture is determined by experience. The Fig. 2 is the one chosen. Also, we implemented it as a fully convolutional network. Designing a network without fully-connected layers became a trend ever since the appearance of ResNet [5] and it makes the implementation easier.

3.3 *Transfer Learning*

The natural image set, one of the subset of ImageNet 2012 explained above, was pre-trained to the network. A base learning rate was chosen empirically, i.e. 0.01

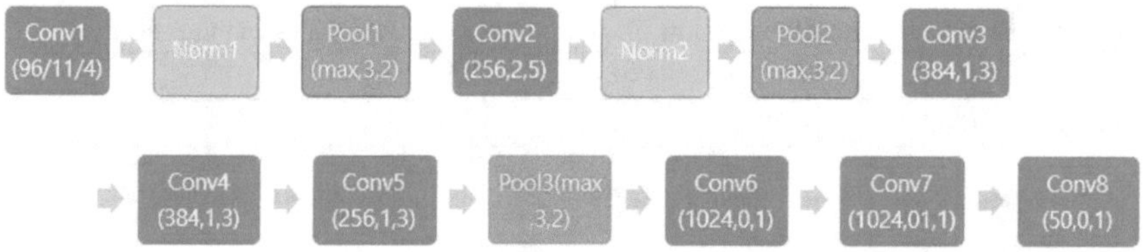

Fig. 2 A diagram of the architecture used in the experiment. 'Conv' denotes convolutional layer and the following numbers in parenthesis mean number of filters, zero-padding dimension, and filter size, respectively. 'Norm' is a local response normalization. 'Pool' is a pooling layer with the numbers in the parenthesis meaning type, padding, and size, respectively

Table 1 The ratio of re-initialized filters and final accuracy

Ratio (%)	0	5	10	15	20	...
Accuracy	0.6043	0.6019	0.6247	0.5998	0.6108	...

and then trained for 18000 iteration until loss and accuracy saturate. And then 1/10 of the base learning rate was applied, 0.001 for 5500 more iterations. The final top-5 accuracy got from this pre-training was 0.86.

Then the pre-trained weights are fixed and a target dataset is given. During one epoch of feed-forwarding the target dataset, $l1$ norm of each filter's activation is summed. Then regarding the statistic, a threshold is set. The filters with activation bigger than the value are fine-tuned for 1500 iterations. It was empirically enough to recover the connections among the preserved filters. On the other hand, the other left filters were re-initialized. The initialize scheme used in this experiment was Gaussian. Finally, the whole network is fine-tuned with the target data. Table 1 is the result when the threshold varies from 5 to 20.

As a baseline, we experimented a basic fine-tuning setting. After the network was pre-trained under the same policy, the network was fine-tuned with the target data for 24450 iterations with a learning rate 0.001. After the network's performance got saturated, the learning rate was degraded to 0.0001 and the machine was tuned for other 150 iterations. The final accuracy of the target data was 0.604. By contrast, when the target data is trained from scratch, it only got 0.551 with the similar learning rate policy and 0.58 when trained for 340000 iterations—which was more than 13 times of the iteration number.

4 Discussion

The Table 1 in Sect. 3 shows effectiveness of the proposed method. When the 10% of the filters are re-initialized, the top-5 accuracy increased by 2%.

Table 2 The result of the ablation study. The ratio of re-initialized filters and final accuracy

Ratio (%)	0	5	10	15	20	...
Accuracy	0.6043	0.5873	0.602	0.6007	0.6062	...

4.1 An Ablation Study

We did an ablation study to prove the usefulness of the isolated fine-tuning stage, i.e. to fine-tune only the filters with bigger activation to the target data before fine-tuning the whole network. The exactly same experiments were done without the stage and the Table 2 shows the result.

Generally, the accuracy of 1 was higher than 2. As the filters whose activation is smaller than threshold get re-initialized, the connectivity among the pre-trained filters is damaged. Therefore the effect of the transferred filters gets weakened. When compared to the accuracy of basic fine-tuning experiment, no result in the Table 2 is higher than 0.604. The need to perform the isolated fine-tuning stage is shown.

5 Conclusion

Though the transfer learning boosted the performance, there is still a weakness that the source and target datasets should be alike to be effective. To resolve the problem, we proposed a method that improves the transfer learning performance regardless of what source data is used. Among the weights pre-trained with source data, only the ones relevant to the target data is transferred. The relevance is measured statistically. In this way, we improved the classification accuracy of 50 sub-class ImageNet 2012 by 2%.

Acknowledgements This work was supported by Institute for Information communications Technology Promotion (IITP) grant funded by the Korea government (MSIT) (2017-0-01780, The technology development for event recognition/relational reasoning and learning knowledge based system for video understanding).

References

1. Yosinski, J. et al.: How transferable are features in deep neural networks? Adv. Neural Inf. Process. Syst. (2014)
2. Azizpour, H. et al.: From generic to specific deep representations for visual recognition. In: Proceedings of the IEEE Conference on Computer Vision and Pattern Recognition Workshops (2015)
3. Jia, Y. et al.: Caffe: convolutional architecture for fast feature embedding (2014). arXiv:1408.5093

4. Krizhevsky, A., Sutskever, I., Hinton, G.E.: Imagenet classification with deep convolutional neural networks. Adv. Neural Inf. Process. Syst. (2012)
5. He, K. et al.: Deep residual learning for image recognition. In: Proceedings of the IEEE Conference on Computer Vision and Pattern Recognition (2016)

Improving Deep Neural Networks by Adding Auxiliary Information

Sihyeon Seong, Chanho Lee and Junmo Kim

Abstract As the recent success of deep neural networks solved many single domain tasks, next generation problems should be on multi-domain tasks. To its previous stage, we investigated how auxiliary information can affect the deep learning model. By setting the primary class and auxiliary classes, characteristics of deep learning models can be studied when the additional task is added to original tasks. In this paper, we provide a theoretical consideration on additional information and concluded that at least random information should not affect deep learning models. Then, we propose an architecture which is capable of ignoring redundant information and show this architecture practically copes well with auxiliary information. Finally, we propose some examples of auxiliary information which can improve the performance of our architecture.

1 Introduction

Deep neural networks (DNNs) can be seen as universal function approximators [5]. In this scheme, many domains of deep learning has been trained to fit the distribution between input and output for each application. Until now, output (i.e. target distribution) is simply a target value for a task; For example, target output of the classification task is a class label. Even though some multi-task learning models has been proposed [10], there are not much studies for how to cope with huge auxiliary information in DNNs.

Even though DNNs are capable of fitting any training data [11], recent success of deep learning is based on understanding of superb generalization techniques [3, 4].

S. Seong · C. Lee · J. Kim (✉)
Statistical Inference and Information Theory Laboratory, School of Electrical Engineering, KAIST, Daejeon, South Korea
e-mail: junmo.kim@kaist.ac.kr

S. Seong
e-mail: sihyun0826@kaist.ac.kr

C. Lee
e-mail: yiwan99@kaist.ac.kr

J.-H. Kim et al. (eds.), *Robot Intelligence Technology and Applications 5*, Advances in Intelligent Systems and Computing 751,
https://doi.org/10.1007/978-3-319-78452-6_4

By rethinking the auxiliary information problem, additional information may act as constrains on optimization problems and thus may act as regularization. For example, adding human domain knowledge to deep learning (i.e. Artificial intelligence and human collaborating) may improve the performance of solving problems. Therefore, we study possibilities of auxiliary information to improve the performance of deep learning classifiers.

The scope of this paper are (1) to theoretically investigate the effect of additional information, (2) to propose a deep learning models for heavy auxiliary information and (3) to improve the performance of DNNs using large auxiliary information.

2 The Effect of Additional Information

Intuitively, it can be taken for granted that if the additional information is not malicious, it does not compromise the original information because additional information can be ignored. This common sense can be represented in information theory [1] as follows:

$$H(y) \geq H(y|y_{add}) \tag{1}$$

where y is a target output, y_{add} is an auxiliary information and H is entropy of the information.

The cross-entropy loss is the most common objective functions for deep learning and we present how additional information can affect deep learning models.

$$H(p(y), q(y)) = H(p(y)) + D_{KL}(p(y)||q(y)) \tag{2}$$

$$= -\sum_{y} p(y) \log q(y) \tag{3}$$

$$H(p(y|y_{add}), q(y|y_{add})) \tag{4}$$

$$= H(p(y|y_{add})) + D_{KL}(p(y|y_{add})||q(y|y_{add})) \tag{5}$$

$$= -\sum_{y_{add}} p(y_{add}) \sum_{y} p(y|y_{add}) \log q(y|y_{add}) \tag{6}$$

where $p(y)$ is a target distribution and $q(y)$ is a output distribution of a deep learning model. Here, we show even random information does not harm the original information. If y_{add} is random, i.e. $p(y_{add}) = \frac{1}{|y_{add}|}$ and $p(y|y_{add}) = p(y)$,

$$H(p(y|y_{add}), q(y|y_{add})) = H(p(y), q(y)) \tag{7}$$

Therefore theoretically, adding auxiliary information at least should not harm the deep learning models once the additional information is not malicious. This motivates us to try to find a deep neural network architecture which is not harmed by auxiliary information.

3 Model Architecture

We propose an deep learning classifier architecture which can ignore additional information and thus possibly not be harmed by it. Our module can be added to the top layer of common models so that models can deal with class-specific (or task-specific) information as in Fig. 1. Here, additional information is dealt with auxiliary modules which is independent of primary class-specific layer. However, since these auxiliary layers can affect the lower non-class-specific layers, the common feature layers can possibly be improved by additional information. Similar architecture is introduced in [8]. Our loss function can be written as follows:

$$\mathcal{L}_{primary} = H(p_{primary}, q_{primary}) \tag{8}$$

$$\mathcal{L}^i_{auxiliary} = H(p^i_{auxiliary}, q^i_{auxiliary}) \tag{9}$$

$$\mathcal{L}_{total} = \lambda \mathcal{L}_{primary} + (1 - \lambda) \sum_N \mathcal{L}^i_{auxiliary} \tag{10}$$

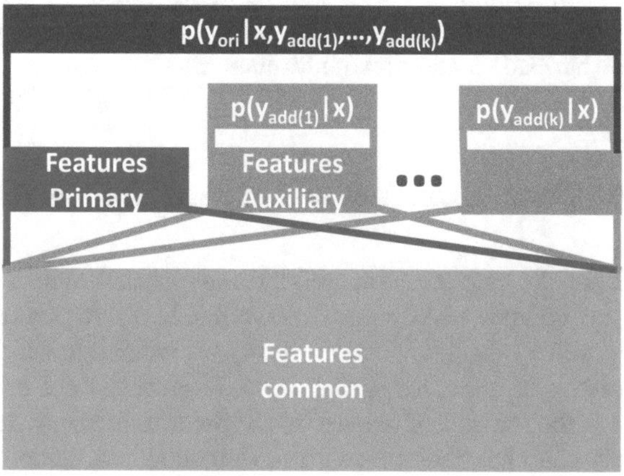

Fig. 1 Proposed model architecture

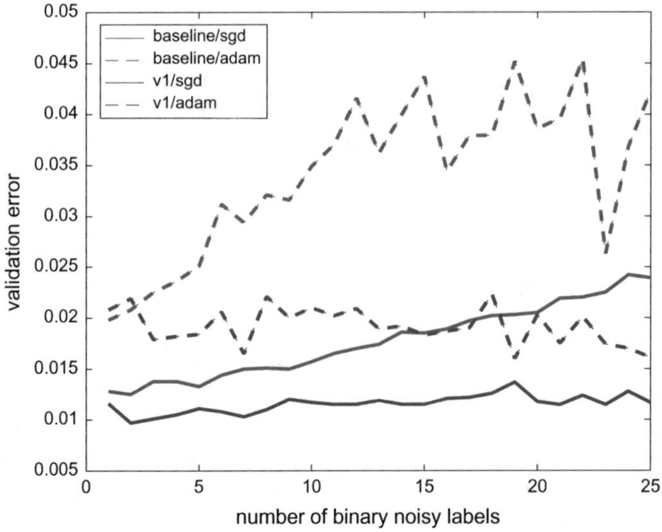

Fig. 2 An experiment on redundant additional information. As number of binary noisy labels increases, general multi-task learning techniques are compromised by redundant information. However, ours shows good resistance to redundant information. We adopted both standard stochastic gradient descend and ADAM optimizer [7] for comparison. Baseline is DNNs on MNIST dataset and v1 is the same version added with our modules

where $\mathscr{L}$ is the loss, $\mathscr{L}^i_{auxiliary}$ is the loss for i-th auxiliary module and N is the number of auxiliary module. We set $\lambda = 0.5$ but set learning rates of primary features $\times 10$ of auxiliary features to maintain the balance of learning.

We also experimentally show that our model does not be harmed by additional information in Fig. 2. We build a toy DNN architecture with MNIST dataset and compare conventional multi-task classifier settings as in [6] to ours. In this experiment, additional information is composed of random labels therefore additional information should be effectively removed. Figure 2 shows that our model does not be harmed by additional information.

4 Some Examples of Auxiliary Information

Here, we propose some examples of auxiliary information which may improve the performance of deep neural networks. First, we propose a direct method which tries to solve mis-classification results on validation set. We calculated the confusion matrix of inference probabilities. This makes a n by n dimensional matrix where n is the number of classes. Each row of this matrix represents that which class the classifier has determined given validation data belonging to a particular class. If the two rows of this matrix show similar values, then this means the classifier is confused

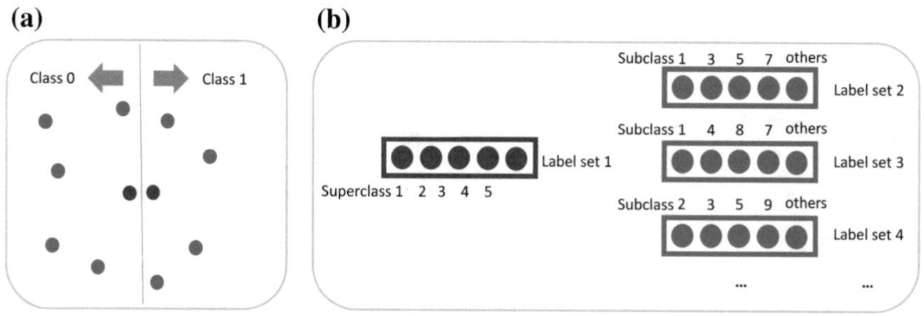

Fig. 3 **a** Labeling techniques for dividing whole classes into binary which maximizes margin between two confusing classes. **b** Super/sub-class division techniques for our auxiliary information schemes

with two classes. To fix this confusion, if inference probabilities of two classes as similar, draw a hyperplane which is perpendicular to a line connecting two points. Then the whole class is divided into binary which can maximizes margin between two confusing classes as in Fig. 3a. We run this process for several times to obtain multiple binary class information.

In addition, we adopted a super/sub-classes labeling scheme for auxiliary information. Based on the confusion matrix, we run the K-means clustering techniques [2, 9]. Our purpose is to provide auxiliary information so randomness of K-means clustering method is not problematic. We run the clustering algorithm several times and consider each of them as additional information as in Fig. 3b.

5 Experimental Results

Performances of our model is evaluated on CIFAR-100 and Places365 dataset. CIFAR-100 is a dataset for object classification which consists of 60000 32×32 color images in 100 classes. Places365 is a dataset for scene classification which consists of 8 million color images and we use the Places365-Challenge 2016 version and 256×256 small images version of them. For the CIFAR-100, we set up a baseline 10-layer convolutional neural networks and compared results with and without our modules. For the Places365, we adopted baseline deep residual networks (ResNet) [3] results in [12] and compared with our implementations of ResNet. The classification accuracies are given in Table 1. All accuracies are on validation dataset.

Table 1 Validation accuracies for CIFAR-100 and Places365 datasets

Method	Accuracies (%)
CIFAR-100	
Baseline CNN	63.99
Ours	65.89
Places365	
Baseline ResNet	54.74
Ours	56.58

6 Conclusion and Future Works

In this paper, we theoretically showed that additional random information does not compromise the original information. Based on this idea, we proposed a deep learning classifier which does not easily compromised by the additional information. Furthermore, we proposed some examples of novel methods to obtain auxiliary information which can improve the performance of deep learning classifiers under our architectures. Our method showed superb performances on some image classification datasets.

However, this paper did not investigate the possibility of performance degradation by malicious information both theoretically and experimentally. Also, proposed architecture is able to ignore auxiliary information and seems to practically work well, but not guaranteed to exclude malicious information in lower feature layers. Also, our future work would be on optimal auxiliary information to improve the performance of deep learning methods.

Acknowledgements This work was supported by the ICCTDP (No. 10063172) funded by MOTIE, Korea.

References

1. Cover, T.M., Thomas, J.A.: Elements of information theory. John Wiley & Sons (2012)
2. Forgy, E.W.: Cluster analysis of multivariate data: efficiency versus interpretability of classifications. Biometrics **21**, 768–769 (1965)
3. He, K., Zhang, X., Ren, S., Sun, J.: Deep residual learning for image recognition. In: Proceedings of the IEEE Conference on Computer Vision and Pattern Recognition, pp. 770–778 (2016)
4. Hinton, G.E., Salakhutdinov, R.R.: Reducing the dimensionality of data with neural networks. Science **313**(5786), 504–507 (2006)
5. Hornik, K.: Approximation capabilities of multilayer feedforward networks. Neural Netw. **4**(2), 251–257 (1991)
6. Jung, H., Lee, S., Yim, J., Park, S., Kim, J.: Joint fine-tuning in deep neural networks for facial expression recognition. In: Proceedings of the IEEE International Conference on Computer Vision, pp. 2983–2991 (2015)

7. Kingma, D., Ba, J.: Adam: a method for stochastic optimization (2014). arXiv:1412.6980
8. Li, Z., Hoiem, D.: Learning without forgetting. In: European Conference on Computer Vision, pp. 614–629. Springer (2016)
9. MacQueen, J., et al.: Some methods for classification and analysis of multivariate observations. In: Proceedings of the Fifth Berkeley Symposium on Mathematical Statistics and Probability, vol. 1, pp. 281–297. Oakland, CA, USA (1967)
10. Ruder, S.: An overview of multi-task learning in deep neural networks (2017). arXiv:1706.05098
11. Zhang, C., Bengio, S., Hardt, M., Recht, B., Vinyals, O.: Understanding deep learning requires rethinking generalization (2016). arXiv:1611.03530
12. Zhou, B., Lapedriza, A., Khosla, A., Oliva, A., Torralba, A.: Places: a 10 million image database for scene recognition. IEEE Trans. Pattern Anal. Mach. Intell. (2017)

Improved InfoGAN: Generating High Quality Images with Learning Disentangled Representation

Doyeon Kim, Haechang Jung, Jaeyoung Lee and Junmo Kim

Abstract Deep learning has been widely used ever since convolutional neural networks (CNN) have shown great improvements in the field of computer vision. Developments in deep learning technology have mainly focused on the discriminative model; however recently, there has been growing interest in the generative model. This study proposes a new model that can learn disentangled representations and generate high quality images. The model concatenates the latent code to the noise in the training process and maximizes mutual information between the latent code and the generated image, as shown in InfoGAN, so that the latent code is related to the image. Here, the concept of balancing between discriminator and generator, which was introduced in BEGAN, is adapted to create better quality images under high-resolution conditions.

1 Introduction

Recently, there has been a resurgence of interest in generative models. A generative model can learn the data distribution that is given to it and generate new data. One of the generative models, Variational Autoencoders (VAEs) [1], can formulate images in a framework of probabilistic graphical models that maximize the lower bound on the log likelihood of data. The base model of the present paper,

D. Kim · H. Jung · J. Lee · J. Kim (✉)
School of Electrical Engineering, KAIST 291 Daehak-ro, Yuseong-gu, Daejeon, Republic of Korea
e-mail: junmo.kim@kaist.ac.kr
URL: https://sites.google.com/site/siitkaist

D. Kim
e-mail: doyeon_kim@kaist.ac.kr

H. Jung
e-mail: jhcsmile@kaist.ac.kr

J. Lee
e-mail: hellojy@kaist.ac.kr

J.-H. Kim et al. (eds.), *Robot Intelligence Technology and Applications 5*, Advances in Intelligent Systems and Computing 751,
https://doi.org/10.1007/978-3-319-78452-6_5

Generative Adversarial Networks (GANs) [2], can learn real data distribution in an unsupervised manner using a discriminator and a generator. The generator produces the data, which takes the noise variable z, and the discriminator determines what is real between generated data and real input data. Radford et al. [3] introduced deep convolutional generative adversarial networks (DCGANs) that used GANs with convolutional architectures. DCGANs have shown that the model can produce unseen images.

However, GANs are widely known to be unstable during training, often generating noisy and incomprehensible images despite the application of some tricks [4]. While GANs can learn data distribution, the noise variable used to generate data does not contain any meaningful data attributes. The noise variable is "entangled representation," and the opposite concept is "disentangled representation," which is introduced in InfoGAN [5]. If we could learn disentangled representation, we would be able to output the data that we want by manipulating them. However, the results of the high-resolution images in InfoGAN would not be realistic enough. Therefore, we have designed a model that learns interpretable and disentangled representations on datasets while improving the quality of the image.

2 Motivation

While GANs learn the data distribution of a real image, a noise variable used as the input of a generator does not have any information about the attributes of the image. To find and control this attribute variance in the generated image, a latent code is adapted in InfoGAN. InfoGAN learns to have mutual information between the latent code and the generated image. The desired output is obtained with various attributes by changing this latent code. InfoGAN works well on small and simple datasets like MNIST or SVHN, however, on a high quality dataset like CelebA, the quality of the generated image is not preserved. For this reason, a model is proposed with a small branch network to improve the quality of the generated image that is disentangled by using latent code.

3 Model Description

We propose a GAN structure based on BEGAN [6], which uses an encoder and a decoder in a discriminator and achieves high quality results. We put a latent code vector c and a typical input noise vector z as an input. This means that a deterministic information vector is concatenated with the input of the generator to learn semantic representations, the method applied from InfoGAN where the latent code vector determines different attributes, such as rotation and orientation of the lighting and thickness of the character. However, InfoGAN does not generate high quality images because it focuses only on maximizing the mutual information between the

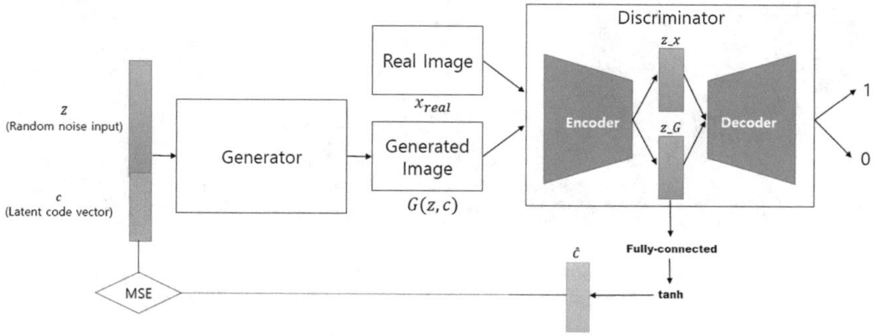

Fig. 1 The structure of the proposed GAN with a small branch network model. This architecture is based on BEGAN using an autoencoder in the discriminator. In the proposed model, a fully connected layer with a tanh function is added to the encoded vector from the generated image in order to infer what the input latent code means

latent code vector and the generated image. To achieve both inferring latent code vector and high resolution images simultaneously, we designed a small branch network composed of a fully connected layer with a tanh function as a classifier and calculated an L2 loss, L_c, between the predicted $\widehat{c}$ from an encoded vector in the discriminator and the input latent code vector. During the training steps, L_c is added to update both the generator and discriminator parameters, θ_D and θ_G. The loss functions consist of three parts, like the loss functions in BEGAN. Equation 1 is our modified loss functions. Our architecture is illustrated in Fig. 1.

$$\begin{cases} L_D = L(x) - k_t L(G(z_D, c)) + L_c(c, \widehat{c}) & \mathit{for}\,\theta_D \\ L_G = L(G(z_G, c)) + L_c(c, \widehat{c}) & \mathit{for}\,\theta_G \\ k_{t+1} = k_t + \lambda_k(\gamma L(x) - L(G(z_G, c))) & \mathit{for\,each\,training\,step}\,t \end{cases} \quad (1)$$

4 Experiments

We experimented with the model described above and the Large-scale CelebFaces Attributes (CelebA) dataset [7], which is a large-scale face attribute dataset with more than 200 K celebrity images. This dataset has 10,177 identities, 202,599 images, and 40 binary attribute annotations per image. We trained the model with 100 k iterations; the training time was about 10 h on a single GTX1080 GPU. The proposed code was implemented using Python and Tensorflow. In the training and test steps, the sizes of all input images were scaled to 64 × 64. The noise z had 64 dimensions and code c had 2 dimensions. The learning rate was the same for both the discriminator and the generator—0.00008. The Adam optimizer was used to train the model.

Fig. 2 The images generated from the proposed architecture. The size of each image is 64 × 64. These images indicate that the proposed architecture generates high quality images

Figure 2 shows some output images of the generator with random vector z at a resolution of 64 × 64 pixels. All faces are realistic and natural, as with the results from BEGAN. This means that applying a latent code to BEGAN does not have a negative effect on the convergence of the original model.

4.1 Varying Code c and Fixed Noise z

Figure 3 shows the results for facial images with varying code c and fixed noise z. Two types of codes were used for training: one for the vertical axis and the other for the horizontal axis. It can be assumed that the first code learned the smile attribute and the second code learned the hair attribute by observing the result images. With the lowest value for latent code c, the person's teeth are not showing in the image, but in the images on the right, the person in the image has a bigger smile. In addition, as you look down row by row, you can see that the generated images have more hair. Therefore, it is apparent that the second code has a hair attribute.

4.2 Comparison of the Results with the Proposed Model to InfoGAN Results

Figure 4 shows that the proposed model resulted in images that were superior to those obtained with InfoGAN whose main contribution was to learn disentangled representation from the data, without the guarantee of high quality results. On the left side of Fig. 4, the results indicate many blurs and noises, and the image resolution is quite small (32 × 32 pixels). However, the proposed model improved the resolution to a 64 × 64 pixel size with more accurate and disentangled representations.

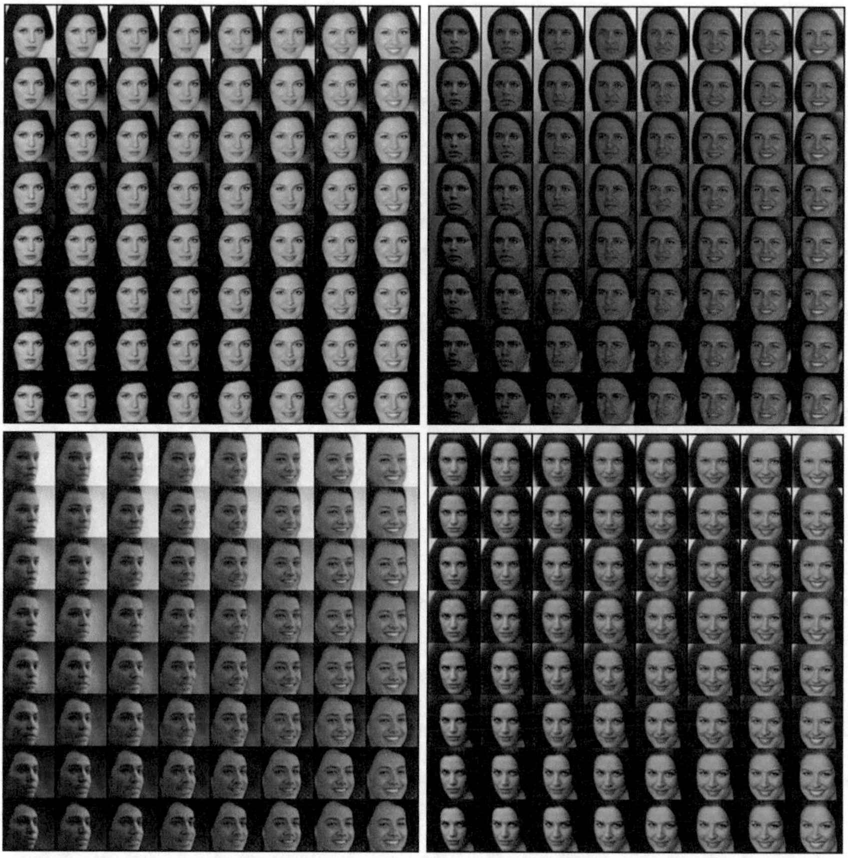

Fig. 3 Results obtained by changing the latent code on the horizontal and vertical axes. There are changes in facial expressions along the horizontal axis and in the amount of hair along the vertical axis

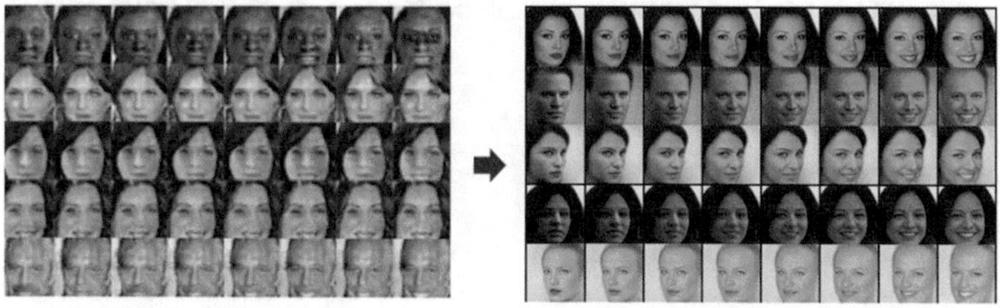

(a) Results from InfoGAN (b) Results from our model

Fig. 4 Qualitative comparison of the InfoGAN results and the high quality facial image results from the proposed model. In **a** there are some blurs and noises. In **b** high quality images are generated with variations in facial expression

5 Conclusion

The purpose of this paper was not only to learn interpretable and disentangled representations on some datasets but also to improve the quality of the generated images. We proposed a GAN with a small branch network to infer representations of attributes for high quality images. Both semantic disentangled representations and high quality images were obtained by utilizing the encoded vector from the generated image in the discriminator. Therefore, this work can be used to enhance data augmentation skills in deep learning and to construct general representation in image embedding. However, the present work has some limitations. For example, it was not known how many attributes would be found, and there is no guarantee that each attribute will match each latent code. In future work, our model will be extended to deal with these problems.

References

1. Kingma, D.P., Welling, M.: Auto-Encoding Variational Bayes. arXiv preprint arXiv:1312.6114 (2013)
2. Goodfellow, I., Pouget-Abadie, J., Mirza, M., Xu, B., Warde-Farley, D., Ozair, S., Bengio, Y.: Generative adversarial nets. In: Advances in Neural Information Processing Systems, pp. 2672–2680 (2014)
3. Radford, A., Metz, L., Chintala, S.: Unsupervised Representation Learning With Deep Convolutional Generative Adversarial Networks. arXiv preprint arXiv:1511.06434 (2015)
4. Salimans, T., Goodfellow, I., Zaremba, W., Cheung, V., Radford, A., Chen, X.: Improved techniques for training gans. In: Advances in Neural Information Processing Systems, pp. 2234–2242 (2016)
5. Chen, X., Duan, Y., Houthooft, R., Schulman, J., Sutskever, I., Abbeel, P.: Infogan: interpretable representation learning by information maximizing generative adversarial nets. In: Advances in Neural Information Processing Systems, pp. 2172–2180 (2016)
6. Berthelot, D., Schumm, T., Metz, L.: Began: Boundary Equilibrium Generative Adversarial Networks. arXiv preprint arXiv:1703.10717 (2017)
7. Liu, Z., Luo, P., Wang, X., Tang, X.: Deep learning face attributes in the wild. In: Proceedings of the IEEE International Conference on Computer Vision, pp. 3730–3738 (2015)

Task-related Item-Name Discovery Using Text and Image Data from the Internet

Putti Thaipumi and Osamu Hasegawa

Abstract There is a huge number of data on the Internet that can be used for the development of machine learning in a robot or an AI agent. Utilizing this unorganized data, however, usually requires pre-collected database, which is time-consuming and expensive to make. This paper proposes a framework for collecting names of items required for performing a task, using text and image data available on the Internet without relying on any dictionary or pre-made database. We demonstrate a method to use text data acquired from Google Search to estimate term frequency-inverse document frequency (TF-IDF) value for task-word-relation verification, then identify words that are likely to be an item-name using image classification. We show the comparison results of measuring words' item-name likelihood using various image classification settings. Finally, we have demonstrated that our framework can discover more than 45% of the desired item-names on three example tasks.

Keywords Artificial intelligence · Computer vision · Web mining
Image classification

1 Introduction

In this information age, people share much knowledge and experiences globally through the Internet. With new innovations to generate and perceive these experiences, e.g. smartphones and virtual reality headsets, amount of information available is growing with increasing rate every second. This information can be used by others to enjoy or learn from conveniently. If a robot or an AI can also utilize this information to learn to perform new tasks like humans do, it would help reduce the cost

P. Thaipumi (✉) · O. Hasegawa
Department of Computational Intelligence and Systems Science,
Tokyo Institute of Technology, Tokyo, Japan
e-mail: thaipumi.p.aa@m.titech.ac.jp

O. Hasegawa
e-mail: oh@haselab.info

J.-H. Kim et al. (eds.), *Robot Intelligence Technology and Applications 5*,
Advances in Intelligent Systems and Computing 751,
https://doi.org/10.1007/978-3-319-78452-6_6

of the data collecting process enormously. However, harvesting useful information from web data is challenging. Many studies extracted the data relying on knowledge base in someway, e.g. a natural language processing database or a dictionary [4, 6, 13].

In this work, we propose a framework that enables a robot or an AI system to, without pre-created database, identify words that describe required items for a task of interest. It is difficult to verify a relationship between words and the task without using any statistics of text database or dictionary. We overcome this difficulty by using *term frequency-inverse document frequency* (*TF-IDF*) value calculated on documents collected from the Internet via Google Search. Likewise, we remove dictionary dependency for identifying words part-of-speed tag by using images from Google Image to verify their item-name likelihood instead.

By removing database dependency, apart from developmental cost reduction, our framework has several other benefits. For example, the performance of the system is not limited or biased by the knowledge of the developer. Our framework is also robust to data out-of-date problem because there will always be new data on the Internet. We believe that our study is a step towards realizing a learning system that enables a robot to learn real world knowledge using purely data on the Internet. The contribution of the study is as follow.

- We propose a method for mining words and measuring their relationship with the task using data from Google Search result. We discuss in detail in Sect. 3.1.
- By using results from the word mining step, we propose a method for measuring likelihood of being item-name for each word using image data from Google Image. We discuss in detail in Sect. 3.3.

By utilizing these two proposed methods together, our framework can discover names of items in a task without relying on any human-prepared database.

2 Related Works

Gathering images from the Internet has become increasingly easier because of improvement in quality of Image Search websites, e.g. Google Image Search and Flickr. A number of studies take advantage of these websites, and propose systems which discover visual knowledge from the Internet. Fergus et al. [8] suggested using Google Image to acquire images, and trained their image recognition model on those images. Chen et al. [5] proposed a two-step training method for a convolutional neural network (CNN). They initialized the CNN weight by training on "easy" sample images from Google Image, then generalized the CNN by fine tuning its weight on "hard" sample images from Flickr.

Divvala et al. [6] used both image data and text data for visual aspect discovery. To ensure that they can discover every visual aspect of the topic of interest, they relied on a huge amount of text collected in Google Books Ngram 2012 English corpora [11]. Then, by using Google Image Search result of each term as filter, the non-visual related aspect terms were removed. The discovered aspect terms were first pruned out using term occurrence count and part of speech tag provided by the corpora. While their method can discover a large number of related aspects, they cannot find new unseen aspects which are not included in the corpora.

Similarly, Chen et al. [4] proposed a semantic concept discovery method, for a target event using Flickr images and their associated tags. Using tags promotes the flexibility of the system to work with unseen concept. But, at the same time, their method relies on natural language toolkit (NLTK) [1] and WordNet [12] to identify part-of-speech tag of event description words and discover tags, which suffers from the same problem as [6]. Chen et al. also used collected images to confirm visually significance of each discovered concept.

3 Proposed Method

The scenario of our study begins with a robot or an AI agent receiving a task input from its owner, e.g. "clean bathroom" and "fix bicycle flat tire". We call this task *objective task*. Its goal is to automatically collect words that are most likely to be names of items required to perform the *objective task* from the Internet, without relying on any pre-made database or dictionary. The proposed framework achieve this goal by, first perform text data mining using Google Search. After that, using collected words to gather images from Google Image. Finally, using image classification process to identify item-name words in the original word list.

3.1 Text Data Mining

In this section, we describe how our system automatically collects words from the Internet and measure how much they relate to *objective task*.

Firstly, we need to collect a number of *objective task* related documents. For this, we search on Google using *objective task* as search query. Then, from the first search result page, we collect visible text in a web page of each result and form a document. Visible text is the remaining text in a web page after we remove all graphic and HTML tags. We also discard the documents collected from www.youtube.com, which usually appear at the top of the result page, where relevant data is mostly in the video, not in the visible text. Let D^T be a set of these collected documents for an *objective task* T. In our experiments, the cardinality $|D^T|$, i.e. a number of collected documents, is between 8 and 12.

We begin by removing all numbers, symbols and words with less than three characters to reduce number of candidate words. We then create set W^T, a set of all remaining words in the task documents D^T. The number of collected words $|W^T|$ is usually large (around 2500–3500 words) and mostly contains common words ("the", "is" or "should") and unrelated words (words from advertisement or web page user

(a) *TF* weighted (b) *IDF* weighted

Fig. 1 Word cloud of words mining from "fix bicycle flat tire" task, Size of the word correspond to its *TF* or *IDF* value. Blue colored words are example of not related common words. Green colored words are example of not related rare words

interface). Learning visual concept about every word in W^T is time-consuming and does not necessary provide benefits to the system. Thus, we filter out less relevant words using their *TF-IDF* values.

TF-IDF is a statistic value that shows how important each word is to a document or a topic. The calculation involves two frequency values. The first value is the *term frequency* (*TF*). As the name implies, *TF* value reflects how often word appears in the topic of interest. For simplicity, we use match count of word $w_i \in W^T$ in all task documents $d \in D^T$, instead of using its actual frequency. The calculation can be formulated as follow.

$$TF_i^T = \sum_{d \in D^T} CountWord(w_i, d), \tag{1}$$

where $CountWord(w, d)$ is the number of times word w appears in document d.

Because common words have high *TF* regardless of document topic, *TF* value alone is not a good indicator of word-topic relation. Figure 1a shows word cloud of W^T from "fix bicycle flat tire" task, where word size corresponds to its *TF* value. Words in blue are examples of not-related but frequently-used words with high *TF* value.

To accurately use *TF* value, we need to combine it with *inverse document frequency* (*IDF*) value. *IDF* value indicates how uncommon the word is in all topics, thus it can be used to filter out common words in W^T. To calculate actual *IDF* value for each word, we need to have a large number of documents which preferably cover all possible topics. For the best of our knowledge, collecting documents from all topics requires using pre-collected list, which we want to avoid. Instead, we estimate *IDF* value using documents from random topics we can acquire using existing data. We use words $w_i \in W^T$ as Google Search query and collect visible text from the result web pages. Let D^{w_i} be a set of documents retrieved using word w_i and D^W be a set of all retrieved documents ($D^W = \bigcup_{w_i \in W} D^{w_i}$). Then, for word $w_i \in W^T$, its IDF_i^T value is estimated using Eq. 2.

$$IDF_i^T = \ln\left(\frac{|W^T|}{\sum_{w_j \in W^T} FindWord(w_i, D^{w_j})}\right), \tag{2}$$

where $FindWord(w, D)$ returns 1 if word w appears in any documents in D and returns 0 otherwise. Even though, the appearance of word w_i in D^{w_i} does not reflect its frequency, we include $FindWord(w_i, D^{w_i})$, which always be 1, in the equation to avoid divided-by-zero problem.

In this IDF_i^T estimation, instead of using document count, we use the number of query words whose search result includes the word w_i. In other words, we regard all documents in each set D^{w_j} as pages of one big document. This help reduce *IDF* values of words that appear in a lot of topics but only mentioned in some websites. We compare our proposed estimation with document counting estimation in Sect. 4.1.

Figure 1a shows word cloud whose word size corresponds to its *IDF* value. Green words are examples of rare but unrelated words. Their *IDF* values reflect their rarity but do not reflect their relation with the *objective task*. Using both *TF* and *IDF* value, we calculate $TF\text{-}IDF_i^T$ which can effectively identify *objective task* related words in W^T. The $TF\text{-}IDF_i^T$ value of word w_i can be calculated as follows,

$$TF\text{-}IDF_i^T = TF_i^T \times IDF_i^T. \tag{3}$$

After we calculated *TF-IDF* values, we create two sets W_{top}^T and W_{bot}^T which include N_{top} words with highest *TF-IDF* value and N_{bot} words with lowest *TF-IDF* value respectively.

3.2 Images Collection

In the next phase of our framework, we collect images of words in W_{top}^T and W_{bot}^T from the Internet. We first use words in W_{top}^T and W_{bot}^T as Google Image search query to retrieve 100 image URLs from search result page. Then we download full-size images from those URLs.

In our observation, images retrieved from using an item-name as search query will visually represent one item class, possibly with a small number of subclasses. Furthermore, Google Image search results tend to show only small sets of viewpoint of an item. In contrast, non-item-name words represent action or abstract concept, which causes their collected images to have almost no visual relation between each other. Thus, testing images of an item-name word with an image classification should yield a good result, while the classification results should be considerably worse for images of non-item-name words. Figure 2 shows example images of some item-name words and non-item-name words.

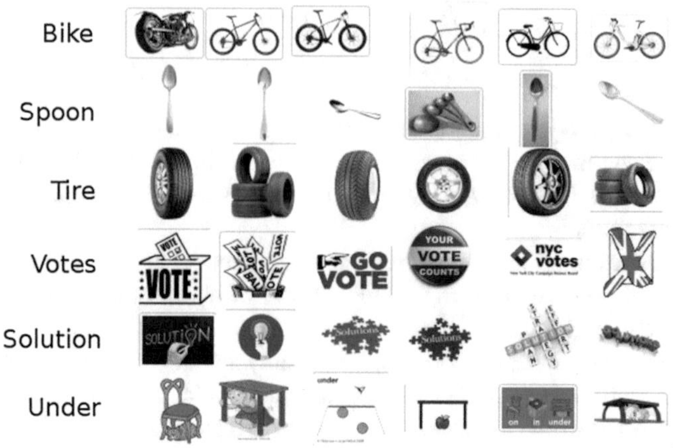

Fig. 2 Example images of item-name and non-item-name words. We consider "bike", "spoon" and "tire" as item-name words because their images consistently show one item class. On the other hand, "votes", "solution" and "under" are considered non-item-name words

3.3 Item-Name Likelihood Measurement

We use image classification score as a measurement item-name likelihood of word, or the possibility of the word representing an item class. These scores are calculated using cross-validation scheme with images of $w_i \in W^T_{top}$ as positive samples and randomly selected images from words in W^T_{bot} as negative samples. Because the randomness of negative sample, the classification score is almost entirely dependent on images of each word in W^T_{top}.

Before performing image classification process for measuring item-name likelihood of each word, we calculate a dense HOG representation vector for each image, using the feature extraction algorithm in [6]. We follow the same process and resize all images to 64×64 pixels before extracting feature. The resulting HOG representation vector is very effective for object detection task [7], which can boosts the classification score for item-name images.

The classifier used for measurement should have robustness to intra-class variation and could be trained and evaluated in a short time. Based on this, we decide to use K-nearest neighbors (KNN) as the classifier. We set its parameter K to 5. The effect of K value is insignificant in our framework, but if we set the value too high, it will slow the algorithm down. Since we only use classification accuracy for comparison between images of collected words, we find that higher performance of the classifier does not contribute directly to the final result. Therefore, we do not use more powerful classifiers, such as `Exemplar-SVMs` [10], because they will increase the complexity without improve overall performance of our framework. Item-likelihood score *item-score*^T_i of word $w_i \in W^T_{top}$ can be calculated using Eq. 4.

$$item\text{-}score_i^T = \frac{\sum_{j=1}^{N_{rand}} crosval(I^{w_i}, \bar{I}^j)}{N_{rand}}, \tag{4}$$

where $crosval(I^{w_i}, \bar{I}^j)$ is a cross-validation score using images of words w_i and randomly selected negative sample images set $\bar{I}^j$. N_{rand} is a number of times we calculate classification scores for each words. In our experiments, we use $N_{rand} = 20$.

For each *objective task*, the final item-name list output is N_{item} words with highest *item-score* in W_{top}^T.

4 Experimental Results

4.1 Text Data Mining

In this section, we investigate three samples *objective task* for text mining process: "clean bathroom", "fix bicycle flat tire" and "make curry". We follow the mining process explained in Sect. 3.1. We perform two experiments, the first experiment is to compare the *IDF* estimation methods, while the second experiment is to identify desired output words for the final result.

4.1.1 Comparing IDF Estimations

We compare the *IDF* values estimated using two difference equations. The first, for the clarity of explanation, is called IDF_{query} which is estimated using Eq. 2 in Sect. 3.1 The other is IDF_{page} estimated as follow,

$$IDF_i^T = \ln\left(\frac{|D^W \setminus D^{w_i}|}{CountDoc(w_i, D^W \setminus D^{w_i}) + 1}\right), \tag{5}$$

where $CountDoc(w, D)$ is the number of documents in D which include word w. In other words, Eq. 5 estimates IDF_{page} by regarding text from each website as one document. We evaluate both estimations by calculating *mean absolute error*(*MAE*) using IDF_{book} as baseline. IDF_{book} is calculated using book count from Google Books Ngram 2012 English corpora [11] by replacing the term $CountDoc(w_i, D^W \setminus D^{w_i})$ and $|D^W \setminus D^{w_i}|$ in Eq. 5 with $CountDoc(w_i, D^{book})$ and $|D^{book}|$ respectively, where D^{book} is a set of books in the corpora. Because the corpora contains a large number of books data ($|D^{book}| = 4,541,627$), we assume IDF_{book} to be more accurate than both web data estimated values and suitable to be used as evaluation baseline. We do not use Google Books Ngram corpora in our framework because we want to avoid pre-collected database, therefore, we only use it for evaluation process. Table 1 shows

Table 1 *Mean absolute error* of two web data estimated *IDF* values, using Google Books Ngram data as baseline. Numbers in parentheses are percentages over average baseline values

Evaluated value	Clean bathroom	Fix bicycle flat tire	Make curry
IDF_{query}	1.47(85.39%)	1.54(78.50%)	1.35(73.62%)
IDF_{page}	2.56(148.84%)	2.58(131.79%)	2.38(129.50%)

the *MAE* of both estimations for the three *object tasks*. We can see that IDF_{query} is more accurate than IDF_{page} in every case.

4.1.2 Desired Output Words Identifying

To evaluate our proposed framework, we manually select the desired output words(or desired words) from W^T. We set two criteria for the collecting process. First, the word must describes an item used or related to the objective task in the sentence. If the word appears in the website but in a non-related context (e.g. word "knife" in advertisement in make curry website), it will be discarded. The second criteria is, the word must clearly describe the item by itself, i.e. it must be a one-word item-name. This is because the method we use for text mining can discover only one-word term. For example, the desired words for "fix bicycle flat tire" task includes "bicycle", "screwdriver" and "tire", but does not include "disk break" and "inner tire".

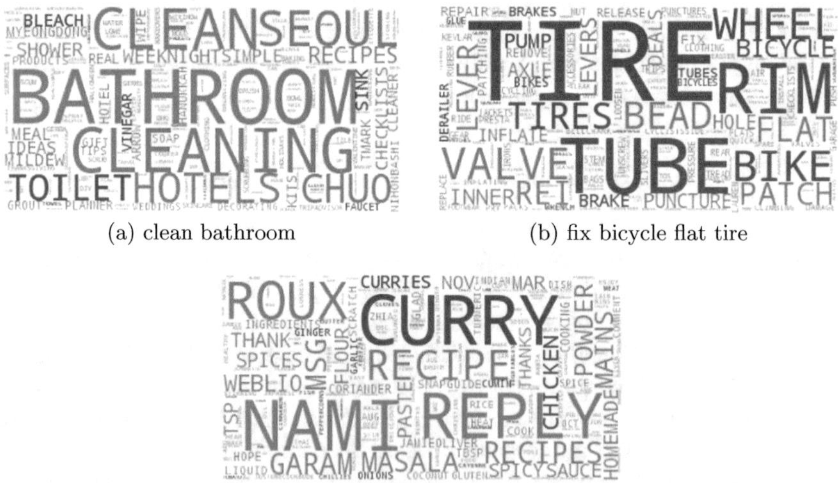

(a) clean bathroom (b) fix bicycle flat tire

(c) make curry

Fig. 3 Word clouds created using mined words from each objective task. Size of each words indicates its *TF-IDF* value. Red words are the desired output words

Figure 3 shows word clouds of W^T for each *objective task*. The size of words relates to their *TF-IDF* values. Red colored words are the desired words of respective *objective task*. As we can see, the biggest words in all three word clouds are included in the *objective task* string. This is because search query words usually have high *TF* value. For "fix bicycle flat tire" task, most of the big words are desired words. In the case of the "clean bathroom" task, although the four largest words ("bathroom","cleaning","clean" and "hotel") are undesired words, because they are not item-name, they are highly related to the task. The cause of retrieving only small number of the desired words is that, "clean bathroom" task requires a small number of tools. As for the "make curry" task, because its search result mostly consists of forum-style web pages, which usually discuss about cooking other different dishes, the acquired visible text consists of a lot of unrelated contents. These contents lead to lower *TF* value of the desired words.

4.2 Item-Name Likelihood Measurement Experiment

The goal of this section is to demonstrate the effectiveness of our proposed likelihood measurement method on text mining results from experiment in Sect. 4.1. We conduct two experiments. The first experiment is for comparing results of different image classification settings, while the second experiment is to show the results of overall proposed framework.

4.2.1 Classifier Settings Experiment

In this experiment, for all words in W^T_{top} and W^T_{bot} of the three tasks, we use NLTK [2] with WordNet to identify their part-of-speech tags and select the words that can only be noun. Finally we manually check if the downloaded images of each word visually represent an item class or not. In this experiment, the word labels are considered based on its images only, we ignore the word's meaning and its context. We remove any words with ambiguous images from the test. The created *test word list* consists of 93 item-name words and 380 non-item-name words. The action-name words, e.g. "swimming" or "running", and place-name words, e.g. "garden" "kitchen", are included in non-item-name list to test if the measurement method has robustness to consistent visual features of non-item images or not. For this experiment, we randomly select 100 words from W^T_{top} and W^T_{bot} for negative image samples.

For comparison, we use three image representation methods. The first method is the dense HOG feature explained in Sect. 3.3. The second method is `classemes` [14] used in [4]. The third method is a modified `VGG_M_1024` [3, 9]. We remove the last few network layers until the last layer is ROI pooling layer called `roi_pool5`. The resulting network has 15 layers. We use pre-trained network weights from ImageNet Model trained by Girshick et al. [9]. For the purpose of our framework, using fixed pre-trained weight without fine-tuning has shown to be sufficiently effective.

For the consistency of the features, we resize all images so that the smallest dimension is 256 pixels long, then we select region of interest to be 16 pixels inside each image's edges.

For the classifier, we compare between KNN and SVM with RBF kernel. Both of these classifiers are selected because of their simplicity and robustness to intra-class variation. Then, using *test word list* we previously mentioned, we evaluate item-likelihood measurement using all combinations of image representations and classifiers. Figures 4 and 5 show precision-recall curves and average precision values respectively. As we can see, the HOG feature and KNN classification setting yields the most promising results for our framework. Not only it has highest AP score, but it is also the fastest to process among the 6 settings.

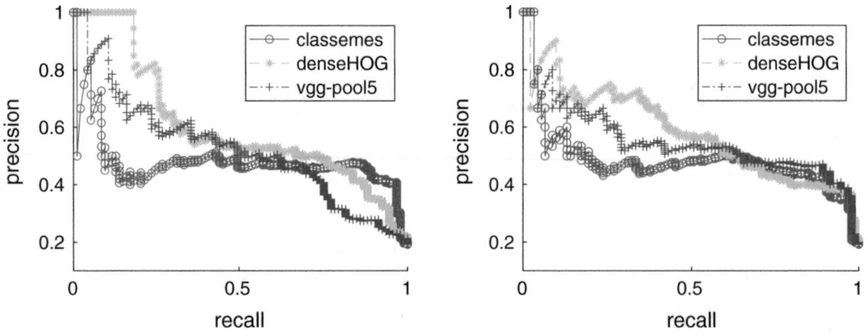

(a) KNN (b) SVM

Fig. 4 Precision-recall curve of item-likelihood measurement using difference classification settings

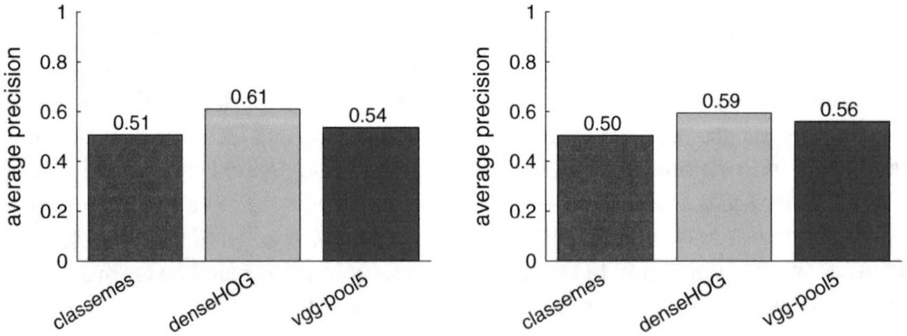

(a) KNN (b) SVM

Fig. 5 Average precision of item-likelihood measurement using difference classification settings

4.2.2 Overall Framework Result

For each objective task, with $N_{top} = |W^T_{top}| = 1000$ and $N_{bot} = |W^T_{bot}| = 50$, we follow the process described in Sects. 3.2 and 3.3. Table 2 shows the experimental results for $N_{it} = 100$ and $N_{it} = 200$. The right-most column shows the precision rate of using W^T_{top} as final output. It can be seen that, choosing 200 highest Item-likelihood scored words as output, provides significantly better precision rate while contains more than 45% of the desired words. Table 3 shows example words after each word selection step. Words in the right-most column is words that are not included in W^T_{top}

Table 2 Item-name count, recall rate and precision rate result after likelihood image scoring for each objective task

Objective task	Item-name count			Recall rate		Precision rate		
	$N_{it} = 100$	$N_{it} = 200$	In W^T	$N_{it} = 100$	$N_{it} = 200$	$N_{it} = 100$	$N_{it} = 200$	W^T_{top}
Clean bathroom	17	26	43	0.40	0.60	0.17	0.13	0.036
Fix bicycle flat tire	22	33	56	0.39	0.59	0.22	0.17	0.050
Make curry	14	35	75	0.19	0.47	0.14	0.18	0.063

Table 3 Examples words result from each word selection step. The word "chain" is an example of words that appear on two tasks but only highly relate to one

Task name	Label	High-*TF-IDF* high-img-score (final output)	High-*TF-IDF* low-img-score	Low-*TF-IDF*
Clean bathroom	Desired words	Squeegee, broom, gloves, bathtub, trashcan, mirror	Towel, cloth, disinfectant, sponge, showerheads	Hamper, rug, drawers, ladder, cups
	Undesired words	Dustpan, bed, clogs, toaster	Bathroom, cleaning, clean, hotels	Urking, quieter, tablespoon, color
Fix bicycle flat tire	Desired words	Pumps, wrench, gloves, bicycle, handle, pedals	Tube, rim, tires, pump, tubes, **chain**, sandpaper	Chainring, rasp, chains, skewers, rag, nails
	Undesired words	Strollers, helmet, pliers, helmets	Valve, flat, rei, patch, lever	Training, only, things, poor
Make curry	Desired words	Bowl, lid, cauliflower, pork, spoon, tomatoes	Chicken, ginger, garlic, chillies, butter, fish	Chilies, blender, saucepan, apple, cream, cilantro
	Undesired words	Fishcakes, hbgafm, pleaser	Nami, reply, roux, recipe, garam	Myriad, optimised, **chain**

because low *TF-IDF*. Words in the next column to the left is words in W_{top}^{T} that have low Item-likelihood scored. Finally, words in the middle column are included in the final output. The word "chain" appears in task documents of two *objective tasks*. It passed the *TF-IDF* filter for the task "fix bicycle flat tire" because it was mentioned often. But for the task "make curry", it appears only once in a term "restaurant chain", thus it was not included in the W_{top}^{T}. This word is a good example of the text mining process's result.

5 Conclusion

In this study, we propose a framework for collecting task related item-names from the Internet without relying on pre-collected database. By using text data from Google Search, our framework can collect task-related words using *TF-IDF* values. The result in Sect. 4.1 shows that, *IDF* value estimated by combining resulting web pages of each search query into one document is more accurate than counting each web pages separately. We also show how to identifies which word is more likely to be item-name using image classification. Section 4.2 shows that using HOG feature with KNN classifier for measuring word's item-name likelihood gives about 60% average precision on our 483-word test data. Finally, we demonstrate that, by collecting 200 words from the Internet, the proposed framework can discover more than 45% of the desired words for the three example *objective tasks*.

It is possible to improve the result by adding more extensions to the framework. For example, using web wrapping technique to remove unrelated content on the web pages before the word extraction. Or, we can collect more documents by searching with difference *objective task* phase. The additional documents will increase the number of discovered desired words as well as improve *TF-IDF* estimation accuracy. Also, adding more specific information (words) into Google Image search query can improve result images' consistency of the related words, e.g. words in "high-*TF-IDF* low-img-score" column in Table 3. In our future works, we will aim toward realizing these improvements and develop a system that utilize its output. For example, an object-base task recognition system that uses collected item-name and images for its object detection training process.

References

1. Bird, S.: Nltk: the natural language toolkit. In: Proceedings of the COLING/ACL on Interactive Presentation Sessions, pp. 69–72. Association for Computational Linguistics (2006)
2. Bird, S., Klein, E., Loper, E.: Natural language processing with Python: analyzing text with the natural language toolkit, O'Reilly Media, Inc. (2009)
3. Chatfield, K., Simonyan, K., Vedaldi, A., Zisserman, A.: Return of the devil in the details: delving deep into convolutional nets. In: BMVC (2014)

4. Chen, J., Cui, Y., Ye, G., Liu, D., Chang, S.F.: Event-driven semantic concept discovery by exploiting weakly tagged internet images. In: Proceedings of International Conference on Multimedia Retrieval, ICMR '14, pp. 1:1–1:8. ACM, New York, NY, USA (2014). http://doi.acm.org/10.1145/2578726.2578729
5. Chen, X., Gupta, A.: Webly supervised learning of convolutional networks. In: Proceedings of the IEEE International Conference on Computer Vision, pp. 1431–1439 (2015)
6. Divvala, S.K., Farhadi, A., Guestrin, C.: Learning everything about anything: Webly-supervised visual concept learning. In: The IEEE Conference on Computer Vision and Pattern Recognition (CVPR) (June 2014)
7. Felzenszwalb, P.F., Girshick, R.B., McAllester, D., Ramanan, D.: Object detection with discriminatively trained part-based models. IEEE Trans. Pattern Anal. Mach. Intell. **32**(9), 1627–1645 (2010)
8. Fergus, R., Fei-Fei, L., Perona, P., Zisserman, A.: Learning object categories from internet image searches. Proc. IEEE **98**(8), 1453–1466 (2010)
9. Girshick, R.: Fast R-CNN. In: The IEEE International Conference on Computer Vision (ICCV) (Dec 2015)
10. Malisiewicz, T., Gupta, A., Efros, A.A.: Ensemble of exemplar-SVMS for object detection and beyond. In: 2011 IEEE International Conference on Computer Vision (ICCV), pp. 89–96. IEEE (2011)
11. Michel, J.B., Shen, Y.K., Aiden, A.P., Veres, A., Gray, M.K., Pickett, J.P., Hoiberg, D., Clancy, D., Norvig, P., Orwant, J., et al.: Quantitative analysis of culture using millions of digitized books. Science **331**(6014), 176–182 (2011)
12. Miller, G.A.: Wordnet: a lexical database for english. Commun. ACM **38**(11), 39–41 (1995)
13. Riboni, D., Murtas, M.: Web mining and computer vision: new partners for object-based activity recognition. In: 2017 IEEE 26th International Conference on Enabling Technologies: Infrastructure for Collaborative Enterprises (WETICE), pp. 158–163. IEEE (2017)
14. Torresani, L., Szummer, M., Fitzgibbon, A.: Efficient object category recognition using classemes. In: Computer Vision-ECCV 2010, pp. 776–789 (2010)

An Exploration of the Power of Max Switch Locations in CNNs

Jeongwoo Ju, Junho Yim, Sihaeng Lee and Junmo Kim

Abstract In this paper, we present the power of max switch locations in convolutional neural networks (CNNs) with two experiments: image reconstruction and classification. First, we realize image reconstruction via a convolutional auto-encoder (CAE) that includes max pooling/unpooling operations in an encoder and decoder, respectively. During decoder operation, we alternate max switch locations extracted from another image, which was chosen from among the real images and noise images. Meanwhile, we set up a classification experiment in a teacher-student manner, allowing the transmission of max switch locations from the teacher network to the student network. During both the training and test phases, we let the student network receive max switch locations from the teacher network, and we observe prediction similarity for both networks while the input to the student network is either randomly shuffled test data or a noise image. Based on the results of both experiments, we conjecture that max switch locations could be another form of distilled knowledge. In a teacher-student scheme, therefore, we present a new max pool method whereby the distilled knowledge improves the performance of the student network in terms of training speed. We plan to implement this method in future work.

Keywords Max pool · Max transfer pool · Max encourage pool · Max switch locations · Convolutional neural networks

J. Ju · S. Lee
Division of Future Vehicle, KAIST, 291 Daehak-ro, Yuseong-gu,
Daejeon, Republic of Korea
e-mail: veryju@kaist.ac.kr
URL: https://sites.google.com/site/siitkaist

S. Lee
e-mail: haeng@kaist.ac.kr

J. Yim · J. Kim (✉)
School of Electrical Engineering, KAIST, 291 Daehak-ro, Yuseong-gu,
Daejeon, Republic of Korea
e-mail: junmo.kim@kaist.ac.kr

J. Yim
e-mail: creationi@kaist.ac.kr

J.-H. Kim et al. (eds.), *Robot Intelligence Technology and Applications 5*,
Advances in Intelligent Systems and Computing 751,
https://doi.org/10.1007/978-3-319-78452-6_7

1 Introduction

In recent years, convolutional neural networks (CNNs) have attracted much attention due to their astonishing performance in various applications. One of the essential elements in CNNs, a pooling layer, summarizes a preceding feature map, average/max pooling a receptive field. Max pooling is the most common; and then, its theoretical analysis has been conducted extensively [1–3]. However, prior to work [4], few studies had focused on the locations of maxima whose major role had been to backpropagate the gradient. Since Zeiler et al. [4] introduced a max switch that records the locations of maxima in deconvolutional networks, several works have exploited max switch to address their own problems (e.g., semantic segmentation [5], visualizing CNNs [6], object contour detection [7], and constructing encoder and decoder [8–10]). All of them employ max unpooling layer in their CNNs, which consequently require a max switch to perform unpooling operation. Moreover, Andre et al. [8], the most recent study, has proven that image reconstruction is possible using only max switch locations.

Inspired by previous work [8] as well as our own observation (peculiar phenomena that are described in Sect. 3), we further investigated the power of max switch locations in two experiments: image reconstruction and classification. Based on our observations of the experiments, we conjectured that max switch locations could enable knowledge transfer; thus, we invented a new way whereby CNNs provide their knowledge to another solely through max switch locations. We hope to fully realize max switch locations-based transfer learning in future work.

2 Related Work

Since Zeiler et al. [4] introduced the max switch to address a issue of learning image representation in a hierarchical manner, successive works [5–7, 9, 10] have made use of max switch locations for visualizing CNNs, semantic segmentation, object contour detection, building auto-encoder, etc. However, the usage of max switch locations in above works is limited to enable training deconvolutional networks or decoder. In contrast, Andre et al. [8] enjoy the fruitful benefits of max switch locations, and they introduced another ability of max switch locations to researchers. They demonstrated that image reconstruction is possible using solely max switch locations—what they refer to as the tunnel effect—which had not been discovered until their report. They constructed a convolutional auto-encoder (CAE) by inserting a particular type of layer between the encoder and decoder; it was designed to block feature transmission in order to train a decoder with only max switch locations coming from an encoder. Astonishingly, the quality of the reconstructed image is as good as the original image. The difference between their CAE and our CAE is that we insert a bottleneck layer so that information can pass back and forth through it.

3 The Power of Max Switch Locations

3.1 Image Reconstruction

We constructed the encoder by stacking layers in the order of convolution (conv), batch normalization (bn), ReLU, max pool, and fully-connected (fc), and we repeated it multiple times. The decoder has a similar structure to the encoder; however, convolution is replaced with deconvolution (deconv), and the max pool is replaced with max unpool. We inserted the bottleneck layer between them (Fig. 1). In fact, this form of encode-decoder architecture is almost the same as [5, 7, 8], in which the max unpool layer required the max switch locations to be transmitted from the max pool layer in an encoder. After auto-encoder training was completed with L2 reconstruction loss, we established experiment protocol as follows: First, as depicted in Fig. 2, we presented an image as an input to the encoder, and during the process of decoding, we gave decoder different max switch locations extracted from another image (hereafter, we will call this image the max switch supplier). Second, we replaced the input image with a noise image to access the effect of the input image on reconstruction. We conducted the experiment on two dataset: MNIST and CIFAR-10. MNIST has grayscale images with a size of 28×28, 60K training examples, 10K test examples, and 10 classes (0 to 9) in total. CIFAR-10 consists of 32×32 natural images, 50K training images, and 10K test images that belong to one of 10 classes. As for the result of MNIST, regardless of what input is given, the reconstructed image resemble the max switch supplier rather than the input image. However, there seemed to be a tendency in CIFAR-10's results for the max switch supplier and the input image play a role in reforming its shape and drawing its color pattern on a reconstructed

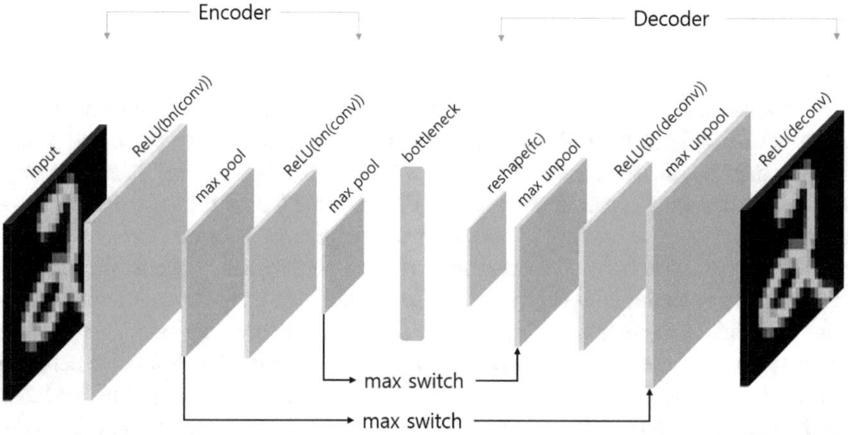

Fig. 1 The architecture of our CAE. It consists of an encoder, decoder, and bottleneck. The transmission of the max switch between the encoder and decoder is necessary to enable the max unpooling operation

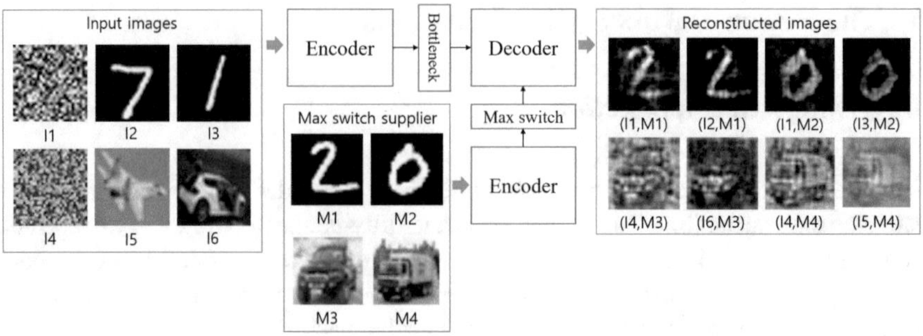

Fig. 2 Image reconstruction results. When a particular input image is given, we insert max switch locations from "max switch supplier" into a decoder. Two encoders in this figure are identical. (Ii, Mj) denotes the reconstructed image when the i-th input image and j-th max switch supplier are provided

image, respectively. Judging by the work [8], one may consider those to be obvious consequences. However, what they [8] have done is represent the pixel of an image on a space of max switch locations, which we have not done. To the best of our knowledge, no literature has already carried out our trial. Thus, we are the first to reconfirm the power of max switch locations along with [8]; even though we allow feature information to pass, the effect of the max switch on reconstruction is still dominant in terms of reconstruction.

3.2 Image Classification

To further investigate the power of max switch locations on another task, image classification, we built a new experiment protocol. First, we introduced the teacher network and the student network; the former is regular CNNs whose training is done in a standalone manner, while the latter receives max switch locations from teacher network during both the training and test phases. To this end, we presented a new pooling method, the max transfer pool by which the teacher network passes its max switch location onto the student network. Figure 4 show how the max transfer pool works; it chooses—according to max switch locations transferred from the teacher network—corresponding activation value within a pool region, ignoring the location of maximum value on its receptive field. Concerning the usage of max switch locations to assign one activation value onto the following feature map, max transfer pool and max unpool (used in Sect. 3.1) are the same in principle. The difference between the two is that the former reduces the resolution of the feature map, but the latter enlarges it. Figure 5 describes the overall structure used in this experiment. Both the teacher and student networks have an identical structure. The student network starts training with softmax loss while each max transfer pool layer obtains max switch locations from the corresponding max pool layer in the teacher network. This also

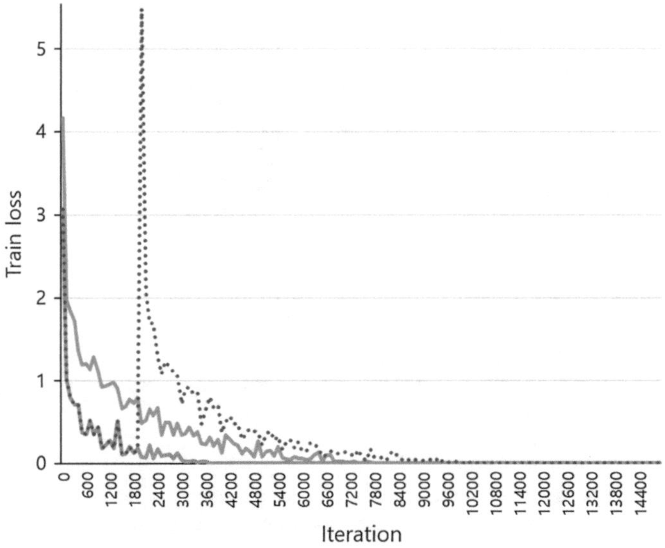

Fig. 3 Train losses for the CIFAR-10 classification. The blue line is the loss from the standalone version of the student network. The orange line represents the loss when the student network learns via the max transfer pool for whole training phase. The green dotted line depicts the loss when we release the connection between the teacher network and the student network in the middle of training—at the 2,000-th iteration—and replace the max transfer pool layer of the student network with a regular max pool layer

occurs during the test phase. In Fig. 3, one can see a quick decrease in the train loss of the student network, which is trained via the max transfer pool (orange line), compared to the standalone version (blue line). In the test phase, the student network with a max transfer pool shows 75.1%, which is substantially below its standalone performance, 81.38% (this is also the test accuracy of the teach network since they have and identical structure). We assume that a strong dependency on max switch locations results in such great degradation.

As in Sect. 3.1, similar experiments were undertaken to access the power of max switch locations in image classification. First, we presented a set of test data to the teacher network, and we showed randomly shuffled version of the test data to the student network while transmitting the max switch location between both networks (Fig. 6 illustrates how it proceeds).

Second, we simply fed the noise image—not a noisy image—to the student network. For both cases, we then measured prediction similarity for both networks by metric: $similarity = \frac{1}{N}\sum_{i=1}^{N} I(Tpred\_i == Spred\_i)$. $I(\cdot)$ denotes an indicator function that yields 1 as an input argument is true, and 0 otherwise. N is the total number of test data. *Tpred_i* and *Spred_i* are predictions of the teacher and student networks yield given i-th input image (e.g., in Fig. 6, Img9 and Img5 correspond to *Spred_*1 and *Spred_*2, respectively). As we can see in Table 1, no matter what kind of input image is provided, the classification results of the student network heavily depend on

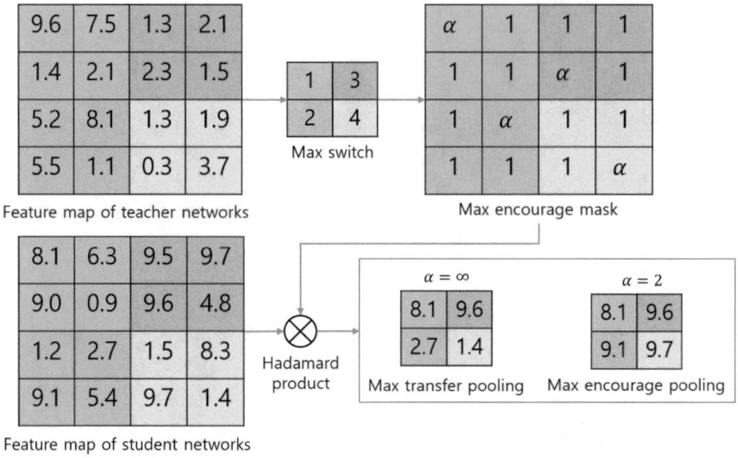

Fig. 4 An operation of the max transfer pool and the max encourage pool. α is the encourage rate. The max transfer pool is a special case of the max encourage pool where α is set to infinity

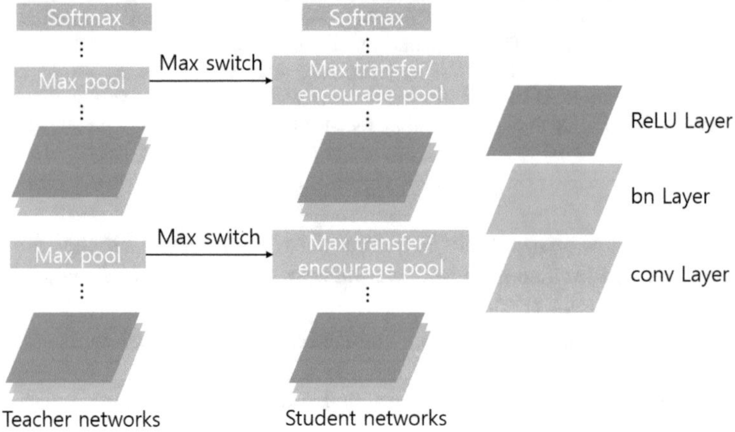

Fig. 5 A schematic of the max switch transmission between the teacher and student networks

Table 1 Prediction similarity for the teacher and student networks given two types of input

	Noise (%)	Shuffled (%)
Similarity	73.04	71.04

the class of the max switch supplier predicted by the teacher network—in line with image reconstruction (Sect. 3.1).

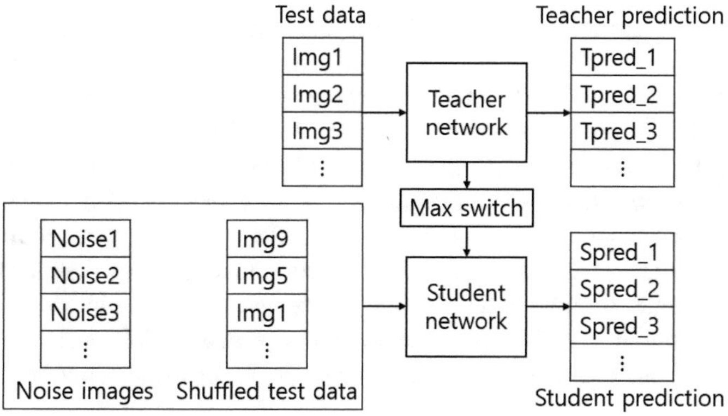

Fig. 6 A schematic of the experimental protocol used to evaluate the prediction similarity for both the teacher and student network. First, we present a set of test data to the teacher network, and we provide a randomly shuffled version of the test data to the student network. Second, we simply input a set of noise images into the student network

4 Future Work and Discussion: Knowledge Transfer

On account of the results in Sect. 3, it could be concluded that the max switch location has—to some extent—the capability to carry information. After searching for another field in which max switch locations can be used, we finally have arrived at knowledge transfer. In order to enable knowledge transfer, one has to consider two key questions: (1) what type of knowledge (in our case, max switch locations) should be transferred? and (2) how do we transfer the knowledge? To answer these questions, we introduced a new transfer approach via two types of the pool method, the max transfer pool and max encourage pool. As is common with existing approaches [11], the entire training consists of two steps: stage 1 followed by stage 2. During stage 1, the priori learned teacher network transmits max switch locations to a student network via a max transfer pool or max encourage pool. In stage 2, we let the student network resume training alone with initial weights that resulted from stage 1. However, due to the limitations of the max transfer pool, we introduced, alternatively, the max encourage pool that is a general version of the max transfer pool. By accomplishing this knowledge transfer, we expect to obtain the benefit of fast optimization.

4.1 The Max Transfer Pool

As we depict in Fig. 4 and explained in Sect. 3.2, the max transfer pool chooses a feature map value within a pooling area in accordance with the max switch transferred from a teacher network, regardless of the location of maximum value on its input feature map. We started stage 1 via the max transfer pool, and before we proceeded to stage 2, we replaced the max transfer pool layer with a regular max pool layer at a certain iteration due to the absent of a teacher network. As plotted in Fig. 3, there was a rapid rise in the train loss as soon as the replacement occurred at 2,000-th iteration, which could cause severe problems in stage 2 (green dotted line). We believe that such a sudden increase is caused by guidance of the teacher networks via max transfer pool. During stage 1, the student network is trained to comply well with the order of teacher; however, when an abrupt disconnection takes place, the student network seems to struggle to find a solution itself, as though it is restarting learning at the beginning of training, which results in a low performance. This observation led us to devise a softened version of the max transfer pool: the max encourage pool.

4.2 The Max Encourage Pool

We proposed a max encourage pool, which, as its name implies, encourages the student network to select feature values placed in the transferred max switch locations—not force it to do what was done in the max transfer pool—as much as a specified value, we call this value the encourage rate. It is important to note that the max encourage pool is a general version of the max transfer pool, whose encourage rate is set to infinity (in the case of positive feature map values). To avoid abrupt disconnection with the teacher network, we let the encourage rate gradually decrease until the

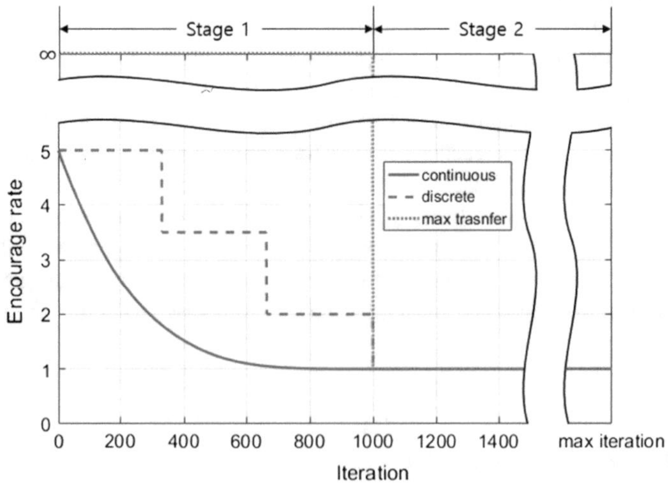

Fig. 7 A graph of the encourage rate decay. The blue line and orange dotted line represent the continuous and discrete change of the encourage rate in stage 1, respectively. They are designed to fall to 1 at the end of stage 1. Afterwards, both values remain at 1 in stage 2. If the encourage rate value equals infinity during stage 1, that is the max transfer pool (green dotted line)

number of training iteration reached a specified value. We attempted two strategies on how to schedule the encourage rate as iterations increased: continuous, discrete. Figure 7 demonstrates how the two schedule strategies are different; the requirement they have to satisfy is arriving at 1—no further encouragement—at the end of stage 1. In the case of a continuous encourage rate, we calculated a corresponding encourage rate at each iteration using the following formula: $base_{er}(1 - \frac{iter}{stage1\_iter})^{power} + 1$. $base_{er}$ is the base encourage rate (e.g., $base_{er} + 1$ is the initial value of the encourage rate), *iter* is the number of the current iteration, *stage*1_*iter* is the total number of iterations for stage 1, *power* is the exponent that decides the form of the schedule function (i.e., the speed of the drop of the encourage rate). If a encourage rate follows discrete change, we let it drop at fixed particular intervals within stage 1. We eagerly hope to fully realize the max encourage pool method in future work.

5 Conclusion

In this paper, we have demonstrated the power of max switch locations through two experiments: image reconstruction and classification. The experimental results have shown that a network trained by a receiving max switch seems to ignore the input image but depends heavily on the max switch locations. This phenomenon has driven us to regard max switch locations as distilled knowledge. Thus, we developed a new pooling method, the max encourage rate, which will be explored in more detail remain in future work.

References

1. Boureau, Y.L., Ponce, J., LeCun, Y.: A theoretical analysis of feature pooling in visual recognition. In: Proceedings of the 27th International Conference on Machine Learning (ICML-10), pp. 111–118 (2010)
2. Boureau, Y.L., Bach, F., LeCun, Y., Ponce, J.: Learning mid-level features for recognition. In: 2010 IEEE Conference on Computer Vision and Pattern Recognition (CVPR), pp. 2559–2566. IEEE (2010)
3. Boureau, Y.L., Le Roux, N., Bach, F., Ponce, J., LeCun, Y.: Ask the locals: multi-way local pooling for image recognition. In: 2011 IEEE International Conference on Computer Vision (ICCV), pp. 2651–2658. IEEE (2011)
4. Zeiler, M.D., Taylor, G.W., Fergus, R.: Adaptive deconvolutional networks for mid and high level feature learning. In: 2011 IEEE International Conference on Computer Vision (ICCV), pp. 2018–2025. IEEE (2011)
5. Noh, H., Hong, S., Han, B.: Learning deconvolution network for semantic segmentation. In: Proceedings of the IEEE International Conference on Computer Vision, pp. 1520–1528 (2015)
6. Zeiler, M.D., Fergus, R.: Visualizing and understanding convolutional networks. In: European Conference on Computer Vision, pp. 818–833. Springer (2014)

7. Yang, J., Price, B., Cohen, S., Lee, H., Yang, M.H.: Object contour detection with a fully convolutional encoder-decoder network. In: Proceedings of the IEEE Conference on Computer Vision and Pattern Recognition, pp. 193–202 (2016)
8. de La Roche Saint Andre, M., Rieger, L., Hannemose, M., Kim, J.: Tunnel effect in CNNS: image reconstruction from max switch locations. IEEE Signal Process. Lett. **24**(3), 254–258 (2017)
9. Zhang, Y., Lee, K., Lee, H., EDU, U.: Augmenting supervised neural networks with unsupervised objectives for large-scale image classification. In: International Conference on Machine Learning (ICML) (2016)
10. Zhao, J., Mathieu, M., Goroshin, R., LeCun, Y.: Stacked what-where auto-encoders (2015). arXiv:1506.02351
11. Romero, A., Ballas, N., Kahou, S.E., Chassang, A., Gatta, C., Bengio, Y.: Fitnets: Hints for thin deep nets (2014). arXiv:1412.6550

An Empirical Study on the Optimal Batch Size for the Deep Q-Network

Minsuk Choi

Abstract We empirically find the optimal batch size for training the Deep Q-network on the cart-pole system. The efficiency of the training is evaluated by the performance of the network on task after training, and total time and steps required to train. The neural network is trained for 10 different sizes of batch with other hyper parameter values fixed. The network is able to carry out the cart-pole task with the probability of 0.99 or more with the batch sizes from 8 to 2048. The training time per step for training tends to increase linearly, and the total steps for training decreases more than exponentially as the batch size increases. Due to these tendencies, we could empirically observe the quadratic relationship between the total time for training and the logarithm of batch size, which is convex, and the optimal batch size that minimizes training time could also be found. The total steps and time for training are minimum at the batch size 64. This result can be expanded to other learning algorithm or tasks, and further, theoretical analysis on the relationship between the size of batch or other hyper-parameters and the efficiency of training from the optimization point of view.

1 Introduction

Deep learning has been successfully applied to various fields with various algorithms, ranging from computer vision problems such as image classification [1], image detection [2, 3], to control problems such as autonomous driving [4], classic control problems and video games [5]. Deep Q-Network(DQN) is one of the popular network architectures based on reinforcement learning, and applied to solve various of tasks with human level performance [6, 7].

It is crucial to design proper network architecture and set appropriate hyper-parameters for training to achieve such a high-level performance [8]. Especially,

M. Choi (✉)
School of Electrical Engineering, Korea Advanced Institute of Science and Technology, Daejeon, South Korea
e-mail: minsukchoi@kaist.ac.kr

J.-H. Kim et al. (eds.), *Robot Intelligence Technology and Applications 5*,
Advances in Intelligent Systems and Computing 751,
https://doi.org/10.1007/978-3-319-78452-6_8

the hyper-parameter plays significant role in training the network. The network performance diminishes or the network might not converge with improper hyper-parameters, even for the networks that have identical architectures.

However, it is still a mystery how to find the optimal values of hyper-parameters. There are some recommendations on hyper-parameter settings [9], which are obtained from numerous repetitive trainings, but yet there are no well organized laws. Furthermore, it becomes more complicated for reinforcement learning algorithms, because the network is trained with the reward from the environment which is more arbitrary than the predefined labeled datasets in supervised learning algorithms, and also the state and action can be continuous. Under these circumstances, the hyper-parameters are set based on the general standard values and usually be tuned through trials and errors.

Batch size is one of the hyper-parameters that is important in increasing the efficacy of neural network training [10]. In this paper, we will train a DQN with various sizes of batch and analyze the tendency of the efficiency of training as the batch size changes, and empirically find the optimal batch size for the DQN in the cart-pole environment.

2 Optimal Batch Size

The optimality of the batch size can be evaluated by the efficiency of training. Efficiency of training can be represented with two factors: a performance on task after the training and the time or number of steps required for the training to reach a sufficient performance to solve the task. But an explicit way to find the optimal size of a batch for efficient training before actual training is not known yet. If the batch size is too small, the number of steps for training until the agent achieving good performance will tend to be large. With the small size of batch, the amount of updating network parameters for each step is small. Therefore, the more steps are needed to train the network sufficiently. If the batch size is too large, the time for single step will be long because the larger size of batch takes more computation time. Hence, there is a trade-off between the number of steps for training and the training time for single step, which makes it complicated to choose the proper size of batch for training.

Moreover, the number of steps or episodes for training is unpredictable yet. In other words, it is complicated to presume the total steps and time to train the agent sufficiently, which implies that the optimality of the batch size is also very complicated to find before actual training.

In this paper, we empirically find the optimal size of batch for training an agent for a well defined task and environment. The agent was trained repeatedly for various sizes of batch, and the optimality of batch size is evaluated with the total time and the number of steps taken until the agent completes the task to a prescribed degree.

3 Experiment

3.1 Environment

Cart-Pole The cart-pole problem provided by OpenAI Gym, "CartPole-v0" is selected as the task for the experiment. The cart-pole problem is one of the classic control problems, which is well-defined and frequently used as a standard task for both control and RL studies [11]. OpenAI Gym is a library that provides interfaces to develop and test RL algorithms, and allows simple implementation of RL algorithms on various predefined environments. In the cart-pole environment, a cart moves on horizontal frictionless track with a pole attached to the cart. The pole would tend to fall due to the gravity. The goal of the task is to maintain the pole upward by moving the cart left or right along the track and bring the cart to the origin on the track.

Based on this description, OpenAI Gym cart-pole environment defines the state of the environment by 4 values, which are the cart position, the cart velocity, the pole angle, and the pole angular velocity. The action of the environment is defined by pushing the cart to the left or right. The reward is 1 for every step until the termination of the episode, which represents how long the agent has maintained the pole upwards and the cart in a proper range of displacement. The episode is terminated when the pole angle exceed $\pm 24.9°$, the cart position exceeds ± 2.4, or the total number of steps in the episode exceeds 200. The task is considered as solved when the average reward over 100 consecutive episodes is greater or equal to 195.0 (Fig. 1).

Deep Q-Network A DQN is used as an agent. Algorithm 1 presents the specific algorithm that we use. The neural network is a simple feed-forward neural network that consists of 2 fully connected layers with 16 and 2 dimension each. We choose the simple architecture with small dimensions in order to avoid excessive calculations and to reduce the running time for a single experiment for repetition.

Settings Each experiment consists of 10 trainings, which vary in batch size: 1, 8, 16, 32, 64, 128, 256, 512, 1024, and 2048. Each training terminates when the condition for solving the task defined in cart-pole environment is satisfied, or the total number of episodes for training exceeds 5000, which is considered as failure to solve the task. For each experiment, a random seed for the network parameters and the cart-pole environment initialization is fixed. The experiments are executed on both CPU and GPU.

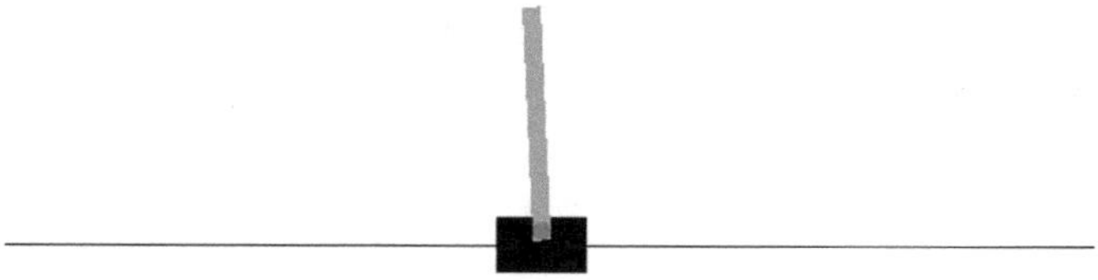

Fig. 1 Cart-pole environment provided by OpenAI Gym

Algorithm 1 Deep Q-learning

Randomly initialize the action-value function Q
Initialize the replay memory with size N
for $episode = 1 \rightarrow max\ episode$ **do**
  Randomly initialize the environment
  Observe the state s_t
  for $step = 1 \rightarrow max\ step$ **do**
    Generate random p such that $0 \leq p \leq 1$
    if $p < e$ **then**
      Select a random action a
    else
      Select a action a_t from $argmax_a Q(s_t, a)$
    end if
    Do action a_t, observe the state s_{t+1}, and get reward r_t
    Store the memory (s_t, a_t, s_{t+1}, r_t) to replay memory
    if *the number of stored memory* $>$ *batch size B* **then**
      Sample random B memories (s_i, a_i, s_{i+1}, r_i) from the replay memory
      for *memory* in *sampled memories* **do**
        $x_i = max_{a_i} Q(s_i, a_i)$
        $$y_i = \begin{cases} r_i & \text{for terminal state } s_{i+1} \\ r_i + \gamma max_a Q(s_{i+1}, a) & \text{for non-terminal state } s_{i+1} \end{cases}$$
      end for
      Update Q with x and y
    end if
    if task is solved **then**
      break
    end if
  end for
end for

3.2 Result

We ran 400 experiments in total: 200 on CPU and 200 on GPU. The average completion rate is 98.5%. The completion rate means the rate of the trainings that the agent solves the cart-pole task out of the total number of trainings. The completion rate for each batch size is on Table 1. The mean training time per step is calculated by dividing the mean training time by the mean training steps. The training time per step by batch size and its linear regression are presented in Fig. 2. The mean values of the total training steps are presented in Fig. 3. The mean training time and its quadratic regression are presented in Fig. 4. The batch sizes that produced the minimum value of the mean total time and the mean total number of steps for trainings and those minimum values are presented in Table 2.

Table 1 Completion rate

Batch size	Completion rate
1	0.885
8	0.995
16	0.9975
32	0.99
64	0.9975
128	1.0
256	0.995
512	0.9975
1024	0.9925
2048	1.0

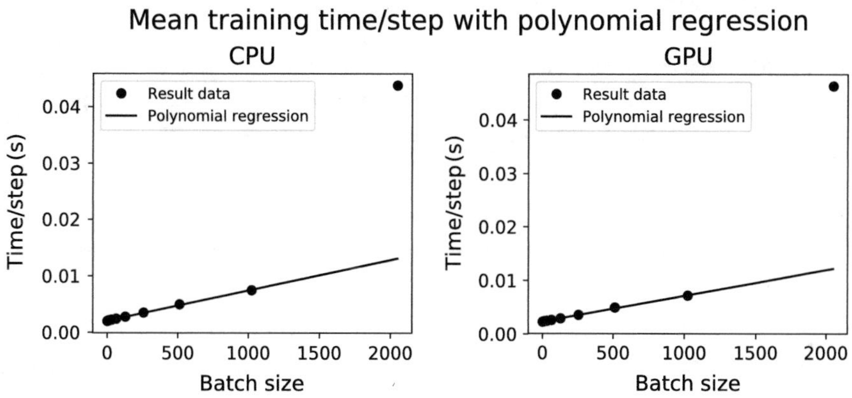

Fig. 2 Mean time/step from the experiments on CPU(left) and GPU(right) with polynomial regression for the batch size from 1 to 1024

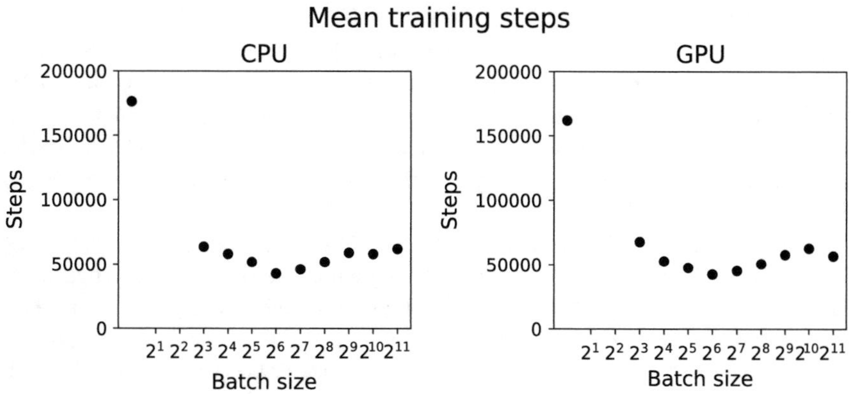

Fig. 3 Mean training steps from the experiments on CPU(left) and GPU(right)

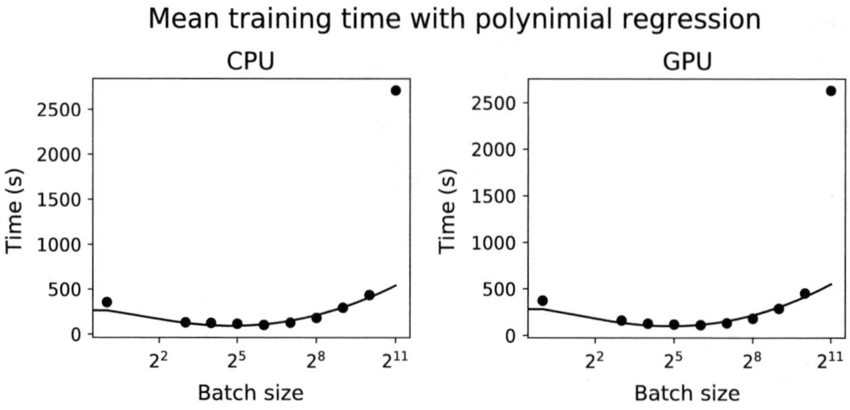

Fig. 4 Mean training time from the experiments on CPU(left) and GPU(right) with polynomial regression for the batch size from 1 to 1024

Table 2 Minimum mean training time, steps

		Mean time (s)	Mean # of steps
CPU	Minimum value	103	42,733
	Batch size	64	64
GPU	Minimum value	111	42,511
	Batch size	64	64

4 Discussion

4.1 Completion Rate

The completion rate is higher than 0.99 for the batch sizes from 8 to 2048, which means less than 1% of the training failed to solve the cart-pole problem. The differences in completion rate between the batch sizes from 8 to 2048 is less than 0.01, which is negligible. For the batch size 1, the completion rate is 0.885, which is 0.1 or much lower than other results. With the batch size of 1, the agent failed to solve the problem in the rate of 11.5% of training, which is 10 times higher than other batch sizes.

There can be 2 reasons to explain this. First, the small size of batch obviously means fewer optimizations are operated in single update. The amount of optimizations in single update is proportional to the size of the batch. The input batch of batch size 8 operates 8 times more optimizations for single update than the input batch of batch size 1. The less the network is optimized, the lower the performance. Second, while training with the small batch, there are fewer chances to sample a good experience with a high reward. A large batch has more chances to sample a good experience than a small batch for the identical replay memory. Also, when the network is trained with the smaller batch, good experiences with high rewards will be more delayed relative to the training with the larger batch, because of the first reason. This implies that at the same iteration of the training, the average rewards of experiences in the replay memory of the smaller batch training will be lower than the replay memory of the larger batch training. It is more difficult to train the agent to achieve high rewards by updating the agent with the samples containing low rewards.

From these results, it seems that there is a threshold for the batch size to obtain a decent completion rate, or a high performance on the task. In our experiment it is 8, and the batch size larger than 8 is sufficient to achieve a completion rate no less than 0.99.

4.2 Mean Training Time, Steps, and Time/Step

As presented in Fig. 2, the mean training time per step and the batch size have a linear relationship except for the case of 2048 batches. The computations to update network parameters are directly proportional to the size of batch. Hence the computation time will also be proportional to the batch size, under the constraints such as memory or computing capacity. If the batch size exceeds a certain threshold, those constraints cause bottlenecks. The following equation presents the relationship between the mean training time per step and the batch size.

$$\frac{\text{(Mean Training Time)}}{\text{(Mean Training Steps)}} = a\text{(Batch Size)} + b \tag{1}$$

if (Batch Size) < threshold. Table 3 contains the values of a and b in (1).

The mean value of training steps becomes maximum at batch size 1, and the minimum at batch size 64, in both CPU and GPU experiments. The mean steps for training tend to decrease more than exponentially for batch sizes from 1 to 64, logarithmically increase for batch sizes from 64 to 512, and maintain a certain value for batch sizes larger than 512. As the batch size increases, the amount of network updates increases, but the start time of the update is delayed because the update begins when the number of memory is greater than or equal to the batch size. Also, the network tends to be overfitted to the memories at the early stage of training for larger sizes of batch. There are more chances to sample memories that are sampled in previous updates as the batch size increases.

Table 3 Coefficients for (1) from a linear regression of the mean training time/step for batch sizes from 1 to 1024 in CPU and GPU experiments

	CPU	GPU
a	0.000005	0.000005
b	0.0021	0.0023

Table 4 Coefficients for (2) from a quadratic regression of the mean training time for batch sizes from 1 to 1024 in CPU and GPU experiments

	CPU	GPU
a	11.74	12.04
b	−113.04	−117.76
c	365.19	388.29

With these tendencies of the mean training time per step and the mean training steps, the mean training time and the logarithm of batch size show a quadratic relationship except for batch size 2048, where the mean value of training time per step is exceptionally large and considered as the outlier. For batch sizes smaller than 64, the decreasing tendency of the mean training steps is dominant because the mean values of training steps decrease faster than exponentially while the mean values of training time per step increase linearly. For batch sizes larger than 64, the increasing tendency of the mean training time per step is dominant because the mean values of training time per step increase linearly but the mean values of training steps increase logarithmically or even maintain a certain level. The quadratic relationship between the mean training time and the logarithm of batch size from 1 to 1024 is as follows:

$$(\text{Mean Training Time}) = a(\log(\text{Batch Size}))^2 + b\log(\text{Batch Size}) + c \quad (2)$$

if (Batch Size) < threshold, where the coefficients a, b and c are in Table 4.

The completion rate of the agent is no less than 0.99 for batch sizes greater than or equal to 8, and the mean training time and steps are minimized at batch size 64. The function of the mean training time for the logarithm of batch size is convex, thus the optimal batch size minimizing the mean training time exists.

5 Conclusion

In this paper, we have shown that the optimal batch size for the DQN agent in the OpenAI cart-pole exists and we have empirically found it. The performance of the agent improves as the batch size increases, but saturates early. The training time per step increases linearly as the batch size increases until a certain threshold. The required number of steps for training steeply decreases as the batch size increases until the optimal batch size is reached, and logarithmically increases or maintains a certain level for batch sizes larger than the optimal batch size. The training time has a quadratic relationship to the logarithm of batch sizes less than the threshold value that is obtained from the relationship of the training time per step and the batch size. We can plan this study on more complex agents or tasks with larger state and action dimensions.

Acknowledgements This work was supported by the ICT R&D program of MSIP/IITP [2016-0-00563, Research on Adaptive Machine Learning Technology Development for Intelligent Autonomous Digital Companion].

References

1. Krizhevsky, A., Sutskever, I., Hinton, G.E.: Imagenet classification with deep convolutional neural networks. In: Advances in Neural Information Processing Systems (2012)
2. Redmon, J., et al.: You only look once: unified, real-time object detection. In: Proceedings of the IEEE Conference on Computer Vision and Pattern Recognition (2016)
3. Girshick, R., et al.: Rich feature hierarchies for accurate object detection and semantic segmentation. In: Proceedings of the IEEE Conference on Computer Vision and Pattern Recognition (2014)
4. Bojarski, M., et al.: End to end learning for self-driving cars (2016). arXiv:1604.07316
5. Lillicrap, T.P., et al.: Continuous control with deep reinforcement learning (2015). arXiv:1509.02971
6. Mnih, V., et al.: Playing Atari with deep reinforcement learning (2013). arXiv:1312.5602
7. Mnih, V., et al.: Human-level control through deep reinforcement learning. Nature **518**(7540), 529–533 (2015)
8. Sprague, N.: Parameter selection for the deep q-learning algorithm. In: Proceedings of the Multidisciplinary Conference on Reinforcement Learning and Decision Making (RLDM) (2015)
9. Bengio, Y.: Practical recommendations for gradient-based training of deep architectures. In: Neural Networks: Tricks of the Trade, pp. 437–478. Springer, Berlin, Heidelberg (2012)
10. Loshchilov, I., Hutter, F.: Online batch selection for faster training of neural networks (2015). arXiv:1511.06343
11. Barto, A.G., Sutton, R.S., Anderson, C.W.: Neuronlike adaptive elements that can solve difficult learning control problems. IEEE Trans. Syst. Man Cybern. **5**, 834–846 (1983)

Design and Development of Intelligent Self-driving Car Using ROS and Machine Vision Algorithm

Aswath Suresh, C. P. Sridhar, Dhruv Gaba, Debrup Laha and Siddhant Bhambri

Abstract The technology of self-driving cars is quite acclaimed these days. In this paper we are describing the design and development of an Intelligent Self Driving Car system. The system is capable of autonomously driving in places where there is a color difference between the road and the footpath/roadside, especially in gardens/parks. On the basis of the digital image-processing algorithm, which resulted into optimal operation of the self-driving car is based on a unique filtering and noise removal techniques implemented on the video feedback via the processing unit. We have made use of two control units, one master and other is the slave control unit in the control system. The master control unit does the video processing and filtering processes, whereas the slave control unit controls the locomotion of the car. The slave control unit is commanded by the master control unit based on the processing done on consecutive frames via Serial Peripheral Communication (SPI). Thus, via distributing operations we can achieve higher performance in comparison to having a single operational unit. The software framework management of the whole system is controlled using Robot Operating System (ROS). It is developed using ROS catkin workspace with necessary packages and nodes. The ROS was loaded on to Raspberry Pi 3 with Ubuntu Mate. The self-driving car could distinguish between the grass and the road and could maneuver on the road with high accuracy. It was

A. Suresh (✉) · C. P. Sridhar · D. Gaba · D. Laha
Department of Mechanical and Aerospace Engineering,
New York University, New York, USA
e-mail: as10616@nyu.edu; aswathashh10@gmail.com

C. P. Sridhar
e-mail: cp.sridhar.91@gmail.com

D. Gaba
e-mail: dg3035@nyu.edu

D. Laha
e-mail: dl3515@nyu.edu

S. Bhambri
Department of Electronics and Communication Engineering,
Bharati Vidyapeeth's College of Engineering, Pune, India
e-mail: siddhantbhambri@gmail.com

J.-H. Kim et al. (eds.), *Robot Intelligence Technology and Applications 5*,
Advances in Intelligent Systems and Computing 751,
https://doi.org/10.1007/978-3-319-78452-6_9

able to detect frames having false sectors like shadows, and could still traverse the roads easily. Thus, self- driving cars have numerous advantages like autonomous surveillance, car- parking, accidents avoidance, less traffic congestion, efficient fuel consumption, and many more. For this purpose, our paper describes a cost-effective way for implementing self- driving cars.

Keywords Autonomous · Drive · Image processing · ROS
SPI

1 Introduction

An autonomous car is a vehicle that is capable of sensing its environment and navigating without human input. Many such vehicles are being developed, but as of May 2017 automated cars permitted on public roads are not yet fully autonomous.

All self-driving cars still require a human driver at the wheel who is ready to take control of the vehicle at any moment. Autonomous cars use a variety of techniques to detect their surroundings, for path planning and efficient decision-making they uses radar, laser light, GPS, odometer, and computer vision techniques and many more for correct behavior of the car. Advanced control systems interpret sensory information to identify appropriate navigation paths, as well as obstacles and relevant road signs. Path planning and obstacle avoidance utilizing the information obtained by the above-mentioned sensors and techniques to drive to the goal without intervention of a driver [1].

Most of the driver assistance systems for the self-driving cars are developed to assist the drivers and to improve the driver's' safety such as in lane keeping [2], autonomous parking [3], and collision avoidance [4]. Such systems are used heavily to reduce traffic and accident rates. And currently rigorous Research is going on in this field to improve the accuracy of the self-driving cars.

The major impetus which leads to the development of the self-driving car technology was the Grand Challenge and Urban Challenge organized by DARPA (Defense Advanced and Research Projects Agency) [5, 6]. These were first organized in 2004, then in 2005 (Grand Challenge), and then followed in 2007 (Urban Challenge). This has led to numerous studies regarding accurate path planning, which are categorized into three major areas namely, Potential-field approach [7], Roadmap based approach [8–11], and Cell decomposition based approach [12].

In paper [13], Voronoi cell based approach is used for path representation, collision detection and the path modification rather than waypoint navigation. But the inaccuracy in predicting the size of these cells in a dynamic environment makes the approach ineffective. On the other hand, our color segregation based approach (explained in Sect. 4) that does not require cells and path planning done in real time environment, has a higher performance.

The complete system is subdivided into the drive mechanism and the control system of the car. The drive mechanism is responsible for the locomotion of the car as it has the motors, tires, mount for RaspiCam, chassis etc. for the operation of the car. But, the control system is the most important part of the system because the entire intelligence of the system is residing in this part. The control system is subdivided into a master control unit and a slave control unit, which are explained in detail in the Sect. 3. The reason for choosing distributed control systems for our self-driving car is to improve the overall performance of the system.

In this paper we are explaining the algorithm that was created for the autonomous car and its implementation on the self-driving car prototype to verify its capabilities. The key aspect of our algorithm behind the autonomous behavior of the car is the detection of color difference between the road and the green grass to determine the direction in which the car must turn based on the run-time camera input and removal of noise for error free decision making is also done and explained in the Sect. 4. The programming of the prototype was done in "OpenCV (Python)".

The upcoming sections explains the following information:

Section 2: Mechanical Structure and Design of the system
Section 3: Electrical connections and architecture of data flow
Section 4: Algorithm used for decision making.

2 Mechanical Design

The mechanical design of the car is shown in the Fig. 1. The car in the image is a prototype for representing our algorithm for effective determination of the direction of the car's movement.

For the prototype of the RC car, at first the circuit of the RC car is removed, and the control of the system is given to the Arduino development board using the Adafruit shield. In this mechanical structure, Ping Sensors (Ultrasonic Sensor) are attached for the autonomous braking to avoid collision.

The RC car used for the prototype is based on differential drive mechanism. Differential drive robots are based on differential drive mechanism in which the motion of the system is divided into two sections which are right and left parts. The wheels of both segments are attached to common axis and have independent motion and rotation capabilities. If the system consists of two wheels, both the wheels can independently be driven either forward or backwards. In differential drive, we can vary the velocity of each wheel independently and change the direction of motion of the whole unit.

Further, the camera mount is designed for mounting the RaspiCam on the front end of the car to get the required field of vision for the efficient image processing. The Camera angle from the car must be 20° at the height of 20 cm. For this purpose, we have designed and 3D printed the camera mount, as shown in the Figs. 2 and 3. Also, prototype of car driving autonomously in park is as shown in Fig. 4.

The camera mount is fixed on the front end of the car as shown in the Fig. 3.

Fig. 1 Assembled autonomous system

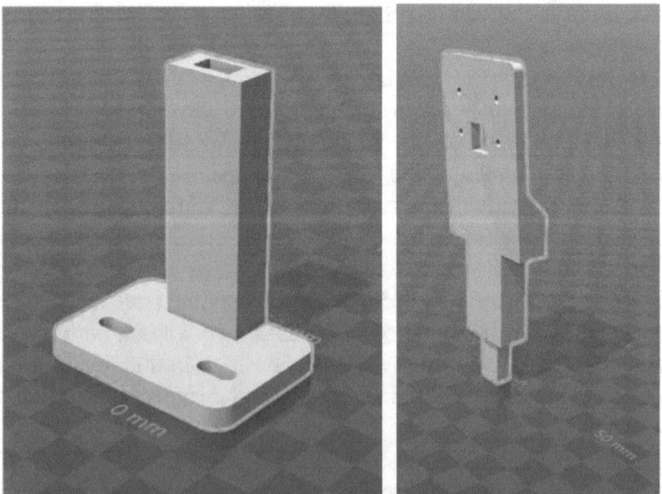

(a) Isometric View of Lower part

(b) Isometric View of Upper Part

Fig. 2 Camera mount design

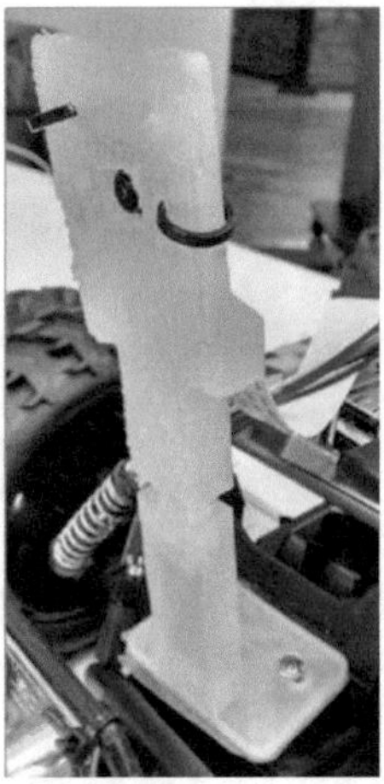

Fig. 3 3D printed camera mount

Fig. 4 Car driving autonomously in park

3 Electrical Design

The control system of the autonomous car is subdivided into master and slave control unit. The master control unit is the Raspberry Pi B3 microprocessor and the slave control unit is based on Arduino development platform. We separated the two control units and distributed the task allocated to them because by this manner we can achieve higher responsiveness and better performance of the system.

Architecture of Control system: Control and Data flow

The architecture and flow of control as well as data in our system is explained in the following subsection as shown in Fig. 5.

3.1 RaspiCam Video Camera

The video is captured using the RaspiCam, which is a 5MP camera attached with the Raspberry Pi B3 such that the video (i.e. image frames) are recorded and then sent to the Raspberry Pi microprocessor for processing of data and decision making.

3.2 Master Control Unit: Raspberry Pi B3 Microprocessor

The master control unit is used for video processing in the system. The video recorded by RaspiCam is converted into image frames and processed by Raspberry Pi B3 microprocessor because of its higher processing power and additional capabilities which makes it the most suitable choice for this application. The algorithm used for the processing of the image frames is explained in detail in Sect. 4. At the end of the algorithm a message array of 8 bits is created which is transmitted wirelessly to the Arduino platform for further executions.

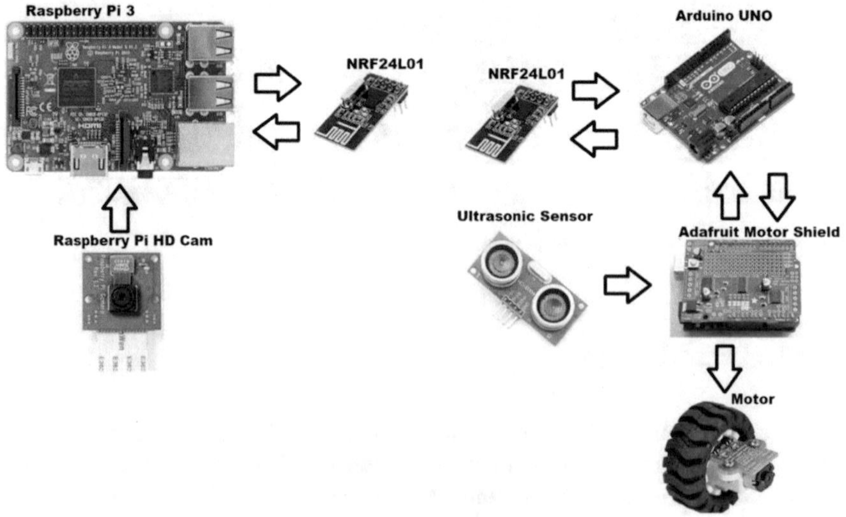

Fig. 5 Architecture of data flow in the system

3.3 Slave Control Unit: Arduino Development Platform

The slave micro-controller is used for controlling the drive system of the car. The Arduino is connected to the Adafruit motor shield which drives power from the Lithium Polymer battery. Arduino commands the motor shied and then it controls the operation of the motors for the drive of the car.

3.4 Wireless Serial Peripheral Interface Communication: nRF24L01 Module

The results generated after the processing of data by implementing the algorithm in the master control unit, are to be transmitted to the slave control unit (Arduino Development Platform) for maneuvering the car. The communication between the master and the slave control unit is achieved using the Serial Peripheral Interface (SPI) communication. SPI wireless communication is achieved using the nRF24L01 (2.4 GHz) transceiver modules and the data is transmitted in the form of 8-bit signal. The signal bits are interpreted in the following manner at the slave control unit (Fig. 6):

1. Bits 1–3: Driving motor speed
2. Bits 4–6: Steering motor speed
3. Bit 7: Driving motor direction
4. Bit 8: Steering motor direction.

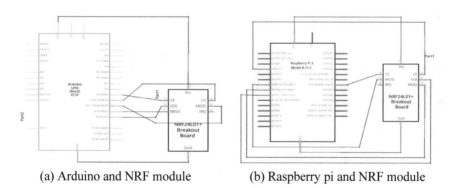

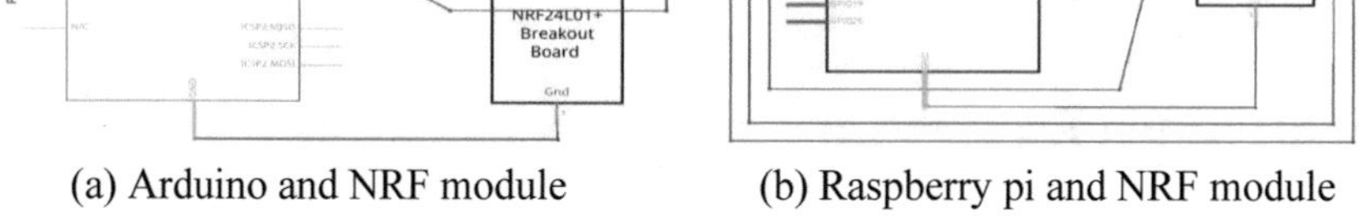

(a) Arduino and NRF module (b) Raspberry pi and NRF module

Fig. 6 Circuit diagram of wireless SPI connection

3.5 *Ping Sensors (Ultrasonic Sensor)*

For emergency breaking system of the autonomous self-driving car, we have used Ping sensors. The slave control unit monitors the readings from the Ping Sensors, such that if by any means something or someone comes in the way of the car the breaks are applied immediately to avoid collision.

3.6 *Adafruit Motor Drive Shield*

The Adafruit Motor Shied is attached to the Arduino development platform. Arduino commands the motor shied which further controls the motors of the car. We have used motor shield to segregate the high current battery supply for the motor from the low current supply for the electronic systems.

3.7 *Power Source: Lithium Polymer Battery*

We are making use of Lithium Polymer Battery (11.1 V–3 s) for providing power supply to the entire system. Direct supply from the battery is given to the motors as they have high current rating and regulated voltage (5 V from 7805IC) for the electronics controllers as they operate on 5 V. This is the common source of power for the system.

3.8 *ROS: Robot Operating System*

All the software frame work of the system is well developed using ROS catkin workspace with necessary packages and nodes. The ROS was loaded on to Raspberry Pi 3 with Ubuntu Mate. Python scripts had been used for programming in ROS.

4 Algorithm Explanation

We extract the video feedback from the RaspiCam, which is recorded by the Raspberry Pi 3, and dynamically store in it for processing. Raspberry Pi processes the input frame by frame, on which we sequentially implement the following filters for decision making.

4.1 *Resizing of Image Frame*

The RaspiCam provides us with highly detailed images. It has 5MP camera for HD recording. For rapid response of our system, we have chosen our image to have 600 Pixel of width because processing on large matrices (derived from the images) will take larger amount of time to process it. So, to avoid lag in our system we are reducing the size of our image by using "imutils.resize()" command of "OpenCV" library as shown in Fig. 7. This way all our frames are resized to 600 pixels width, which makes our filtering procedure more convenient and efficient.

4.2 *Gaussian Blur Algorithm*

Every image, which we get from the RaspiCam is having some or the other form of high frequency noise. So, to remove this noise we are making use of low pass filters. One of the most common forms of algorithm which will not only remove the sharp noises, but also blur the image is the Gaussian Blur filters.

For operation of this filter we need to specify the width and height of the kernel to the function, which should be positive and odd. Here we have used a kernel of 11 sizes. Further, we will have to specify the standard deviation in X and Y direction, i.e. σ_X and σ_Y; here we are using both as zeros, thus they are calculated from the kernel size. Therefore, via Gaussian blur we are able to remove Gaussian noise very effectively as shown in Fig. 8.

The decision making algorithm of the self-driving car is as shown in flow chart in Fig. 9.

```
image = imutils.resize(frame, width=600)
```

Fig. 7 Function for resizing the frame width

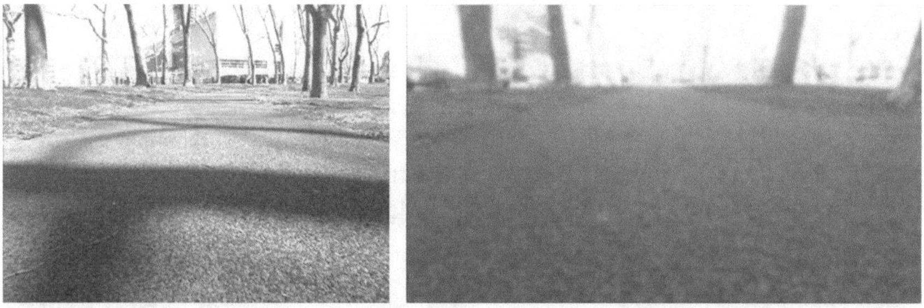

```
image = cv2.GaussianBlur(image,(11,11),0)
```

Fig. 8 Illustrating implementation of Gaussian blur and its function code

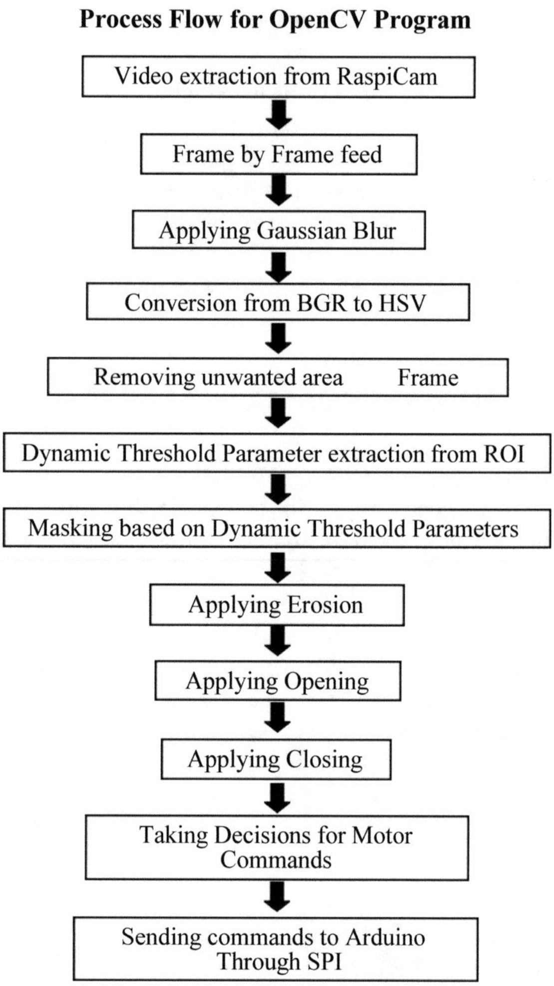

Fig. 9 Flowchart for the decision making algorithm

4.3 *Conversation from BGR (Blue, Green, Red) to HSV (Hue, Saturation, Value)*

We are changing the color space from BGR to HSV such that we can effectively segregate the road from the green grass. In this step we are simply converting the entire BGR image/frame (width = 600) to HSV color space so that just by simply making a range mask, we are able to segregate the desired color range which is grey road from the green grass in the case of a park as shown in Fig. 10.

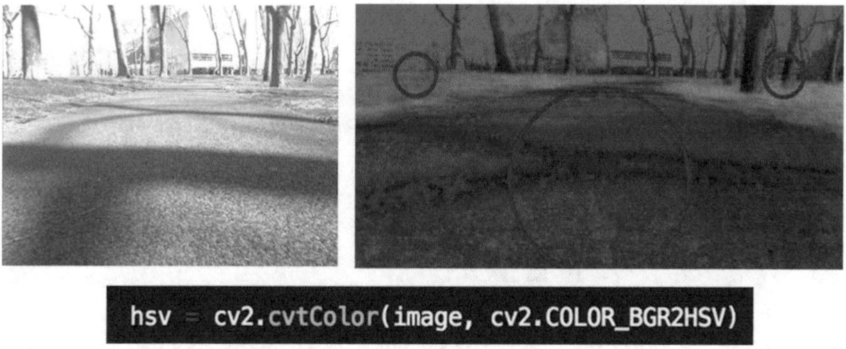
hsv = cv2.cvtColor(image, cv2.COLOR_BGR2HSV)

Fig. 10 Illustrating the conversion of BGR to HSV color space

4.4 Removing Unwanted Area from Image Frame

After the image is converted from BGR color space to HSV color space, we find the area having the most information present and remove the unwanted area from the frame. For our system to perform with utmost accuracy in any unstructured environment, we need to remove all unnecessary information from the frame. For example, we will have to remove those regions which do not contain the road and grass information. These unwanted areas like the sky, the trees, twigs, sticks, etc. which do not have road and grass, are to be removed. This parameter of new height of the frame to be used is dependent on the height of the camera and the vertical inclination of the camera. Therefore, after this step we get the updated frame for higher performance.

Here we are making use of np.size() function to get the height and width dimension of the HSV color space. Using this, we adjusted the new frame for our processing as shown in Fig. 11.

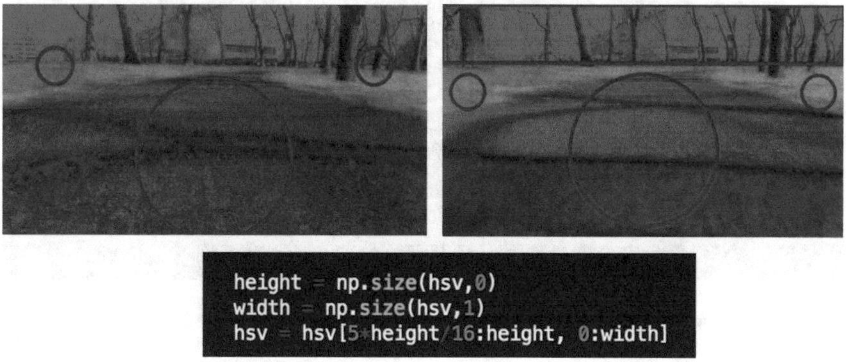
height = np.size(hsv,0)
width = np.size(hsv,1)
hsv = hsv[5*height/16:height, 0:width]

Fig. 11 Illustrating removal of unwanted areas

4.5 Dynamic Threshold Parameter

In this step we are adjusting our frame size such that we could determine the required Region of Interest (ROI). As shown in the Fig. 12, we have used the following command to get it.

4.6 Dynamic Threshold Setting

In this code we could take values from the path [] matrix (created from hsv [] matrix) and assign its corresponding values to h [], s [], and v [] arrays. Then we simply find the values which are the maximum and minimum of the h [], s [], and v [] arrays and save them in hmax, hmin, smax, smin, vmax, and vmin. And then correspondingly using them to create a two "numpi" arrays ("rangemin[] and rangemax[]"), one having the maximum ranges and the other having the minimum ranges as shown in Fig. 13.

4.7 Masking Based on Dynamic Threshold Parameters

The following code creates the mask for transforming the image frame in accordance to the min and max range provided from the previous steps as shown in Fig. 14. We have made us of cv2.inRange () function to create this mask.

```
path = hsv[1*height/8:height,width/2-(2*width/8):width/2+(2*width/8)]
```

Fig. 12 Dynamic thresholding of the parameter ROI

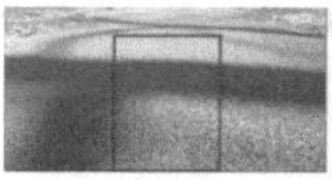

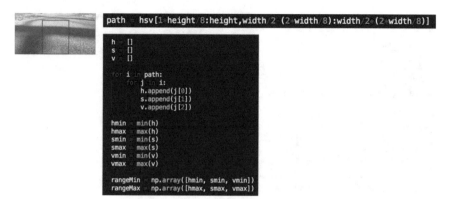

```
path = hsv[1*height/8:height,width/2-(2*width/8):width/2+(2*width/8)]
```

```
h = []
s = []
v = []

for i in path:
    for j in i:
        h.append(j[0])
        s.append(j[1])
        v.append(j[2])

hmin = min(h)
hmax = max(h)
smin = min(s)
smax = max(s)
vmin = min(v)
vmax = max(v)

rangeMin = np.array([hmin, smin, vmin])
rangeMax = np.array([hmax, smax, vmax])
```

Fig. 13 Code for finding the minimum and maximum ranges of hue, saturation, and value in HSV color space

```
mask = cv2.inRange(hsv, rangeMin, rangeMax)
```

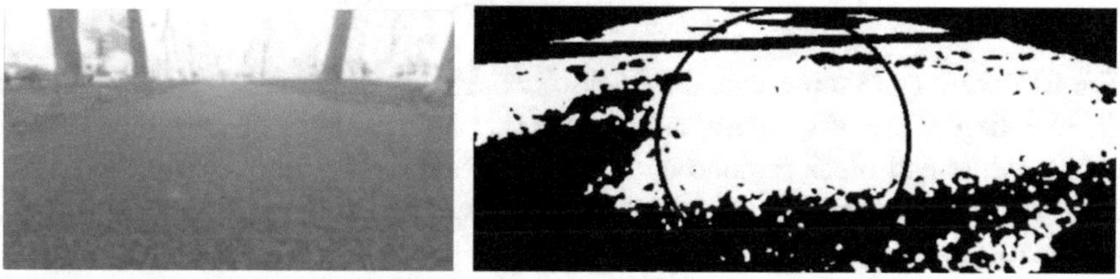

Fig. 14 Illustrating mask creation from the max and min ranges of HSV

4.8 Applying Erosion, Opening, and Closing

In this step we are performing Erosion, Opening and Closing functions one by one to remove various kinds of noises from our image frames, which is the basis of our decision making.

- Erosion will remove the boundaries of the foreground object (displayed in white) as the kernel (25 × 25) slides over the image (convolution), and the pixel in the original image is considered 1 only if all the pixels in the kernel are 1, otherwise it is eroded 0.
- Opening is another function to remove noise and it functions in a manner of erosion followed by dilation, thus removing small particle noise from the image frame.
- Closing is reverse of Opening, i.e. Dilation followed by Erosion; it is useful for closing small holes, or small black points on the object.

Therefore, implementation of these filters one by one helps the robot to render the clean and noise free image frame as shown in the Fig. 15.

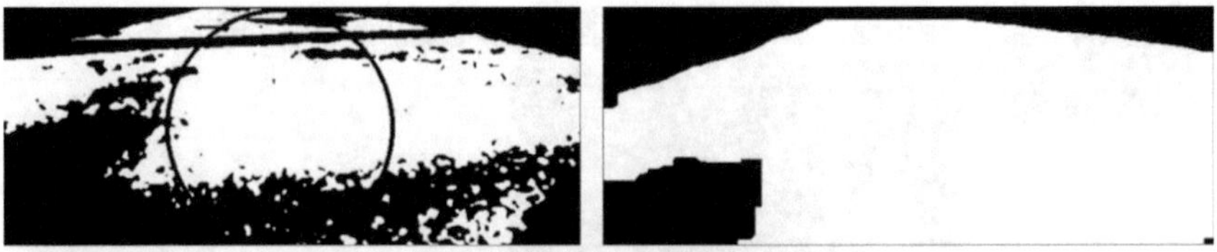

```
kernel = np.ones((25,25),np.uint8)
erosion = cv2.erode(mask,kernel,iterations = 1)
cv2.imshow("Erosion",erosion)
opening = cv2.morphologyEx(erosion, cv2.MORPH_OPEN, kernel)
cv2.imshow("opening",opening)
kernel = np.ones((15,15),np.uint8)
closing = cv2.morphologyEx(opening, cv2.MORPH_CLOSE, kernel)
cv2.imshow("closing",closing)
```

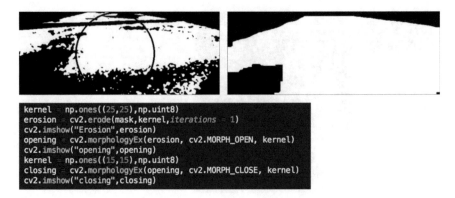

Fig. 15 Implementation of erosion, opening, and closing filters in code and frames

4.9 New Regions of Interest for Turning Decisions/Taking Decisions for Motor Commands

The following code represents the 'if-else' conditions which are used to determine the direction of the drive motor and the steering motor. These 'if-else' loops check for the white and black regions on the top left and right side of the image frame. This way we can decide in which direction the car has to move, and then transmit the data to the slave controller (Arduino).

Also, the second image Fig. 16 shows the formatting of the 8-bit data encoding which is transmitted to the slave controller. In Fig. 16 the first three bits represent the Drive Wheel Velocity, the next three consecutive bits represent the Steer Wheel Velocity, the second last bit represents the direction of the Drive Wheel, and the last bit represents the direction of the Steer Wheel.

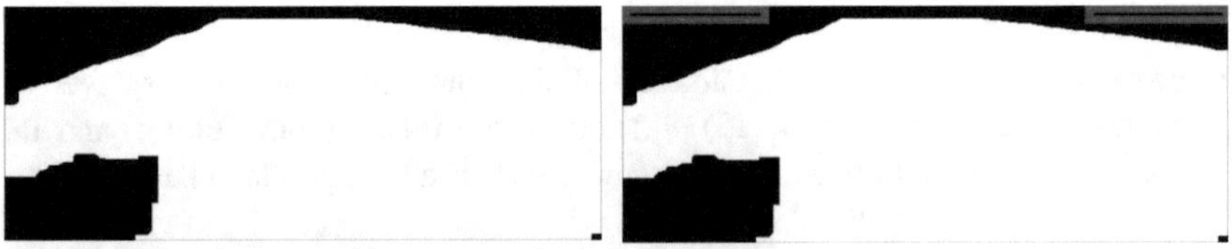

```
leftTop = closing[0][2*wMask/16]
rightTop = closing[0][14*wMask/16-1]
if leftTop == 0 and rightTop == 0:
    message = "10000011"
elif leftTop == 255 and rightTop == 255:
    message = "10000011"
elif leftTop == 255 and rightTop == 0:
    message = "10012011"
elif leftTop == 0 and rightTop == 255:
    message = "10012010"
print message
```

```
8 Bit data encoding

1 2 3 - Drive Wheel Velocity

4 5 6 - Steer Wheel Velocity

7 - Drive Direction

8 - Steer Direction
```

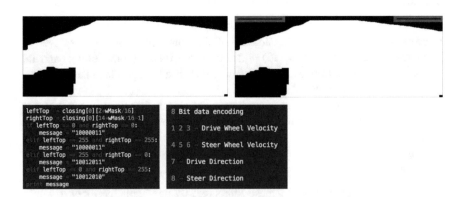

Fig. 16 Determination of the direction of turning

```
dataTX = list(message)
while len(dataTX) < 8:
    dataTX.append(0)

start = time.time()
radio.write(dataTX)
print(format(dataTX))
```

```
drive_motor->setSpeed(drive);
if (driveDirection == 1){
drive_motor->run(FORWARD);
}
else if(driveDirection == 0){
  drive_motor->run(BACKWARD);
}

steering_motor->setSpeed(steer);
 if (steerDirection == 1){
 steering_motor->run(FORWARD);
}
else if(steerDirection == 0){
   steering_motor->run(BACKWARD);
}
```

Fig. 17 Illustration of message signal transmission from Raspberry Pi and decoding of the message signal at Arduino

4.10 Sending Commands to Arduino Through SPI

Once the algorithm has determined the direction in which the car has to move, the master controller (Raspberry Pi) will transmit the data via wireless serial communication to the slave controller (Arduino).

The dataTX takes the message generated from the previous step and it will save it in the dataTX which is transmitted serially to Arduino. Also, in this step we check if the length of the dataTX is less than 8. If it is less than 8, we will have to append a 0 to make it 8-bit to avoid any errors which could happen during wireless communication.

Now, serial communication begins and we record time using time.time() function into the start variable. After that we write this data to the radio transmitter which will transmit it wirelessly to Arduino for movement and simultaneously print it onto the console as well.

The transmitted code is then received by the Arduino, which is further decoded by it for implementation of the locomotion commands. For example, if the drive or steering direction bit is 1 then move the respective motor in forward direction, whereas if the drive or steering direction bit is 0 then move the respective motor is moved in the backward direction as shown in Fig. 17.

5 Conclusion

Thus, for reducing the cost of self-driving cars we use normal cameras and sensors instead of complicated sensors, which tend to be quite expensive. So, our system resulted into a more cost-effective system as compared to the current ones. Additionally, the digital image processing techniques used in the algorithm of the self-driving car had satisfactory results. We implemented and tested the self-driving car in Cadman Plaza Park.

Future Scope

As for future development, we could use machine learning and deep learning techniques for even higher accuracy of the system and dynamic upgradation of the systems parameters according to its surroundings could also be implemented onto the system.

Also, for those roads which do not have significant color difference between road and the side roads, we could use the road dividers as markers to stay on the road as the object recognition algorithms will track these road dividers for driving the car on the road.

Acknowledgements The authors would like to thank Makerspace, New York University to provide support and resources to carry out our research and experiments.

References

1. Wit, J., Crane, C.D., Armstrong, D.: Autonomous ground vehicle path tracking. J. Robot. Syst. **21**(8), 439–449 (2004)
2. Wang, J., Schroedl, S., Mezger, K., Ortloff, R., Joos, A., Passegger, T.: Lane keeping based on location technology. IEEE Trans. Intell. Transp. Syst. **6**(3), 351–356 (2005)
3. Li, T.S., Yeh, Y.-C., Wu, J.-D., Hsiao, M.-Y., Chen, C.-Y.: Multifunctional intelligent autonomous parking controllers for carlike mobile robots. IEEE Trans. Ind. Electron. **57**(5), 1687–1700 (2010)
4. Cho, Kwanghyun, Choi, Seibum B.: Design of an airbag deployment algorithm based on pre-crash information. IEEE Trans. Veh. Technol. **60**(4), 1438–1452 (2011)
5. Özgüner, Ü., Stiller, C., Redmill, K.: Systems for safety and autonomous behavior in cars: the DARPA Grand Challenge experience. Proc. IEEE **95**(2), 397–412 (2007)
6. Thrun, S., et al.: Stanley: the robot that won the DARPA Grand Challenge. J. Field Robot. **23** (9), 661–692 (2006)
7. Khatib, O.: Real-time obstacle avoidance for manipulators and mobile robots. Int. J. Robot. Res. **5**(1), 90–98 (1986)
8. Koenig, S., Likhachev, M.: Fast replanning for navigation in unknown terrain. IEEE Trans. Robot. **21**(3), 354–363 (2005)
9. Rao, N.S.V., Stoltzfus, N., Iyengar, S.S.: A retraction method for learned navigation in unknown terrains for a circular robot. IEEE Trans. Robot. Autom. **7**(5), 699–707 (1991)
10. Kavraki, L.E., Švestka, P., Latombe, J.-C., Overmars, M.H.: Probabilistic roadmaps for path planning in high-dimensional configuration spaces. IEEE Trans. Robot. Autom. **12**(4), 566–580 (1996)
11. Choset, H.M.: Principles of Robot Motion: Theory, Algorithms, and Implementation. MIT Press, Cambridge, MA (2005)
12. Zhang, L., Kim, Y.J., Manocha, D.: A hybrid approach for complete motion planning. In: IROS, Oct 2007, pp. 7–14
13. Lee, U., Yoon, S., Shim, H.C., Vasseur, P., Demonceaux, C.: Local path planning in a complex environment for self-driving car. In: The 4th Annual IEEE International Conference on Cyber Technology in Automation, Control and Intelligent Systems, 4–7 June 2014, Hong Kong, China

Intelligent Smart Glass for Visually Impaired Using Deep Learning Machine Vision Techniques and Robot Operating System (ROS)

Aswath Suresh, Chetan Arora, Debrup Laha, Dhruv Gaba and Siddhant Bhambri

Abstract The Smart Glass represents potential aid for people who are visually impaired that might lead to improvements in the quality of life. The smart glass is for the people who need to navigate independently and feel socially convenient and secure while they do so. It is based on the simple idea that blind people do not want to stand out while using tools for help. This paper focuses on the significant work done in the field of wearable electronics and the features which comes as add-ons. The Smart glass consists of ultrasonic sensors to detect the object ahead in real-time and feeds the Raspberry for analysis of the object whether it is an obstacle or a person. It can also assist the person on whether the object is closing in very fast and if so, provides a warning through vibrations in the recognized direction. It has an added feature of GSM, which can assist the person to make a call during an emergency situation. The software framework management of the whole system is controlled using Robot Operating System (ROS). It is developed using ROS catkin workspace with necessary packages and nodes. The ROS was loaded on to Raspberry Pi with Ubuntu Mate.

A. Suresh (✉) · C. Arora · D. Laha · D. Gaba
Department of Mechanical and Aerospace Engineering, New York University, New York, USA
e-mail: as10616@nyu.edu; aswathashh10@gmail.com

C. Arora
e-mail: ca1941@nyu.edu

D. Laha
e-mail: dl3515@nyu.edu

D. Gaba
e-mail: dg3035@nyu.edu

S. Bhambri
Department of Electronics and Communication Engineering, Bharati Vidyapeeth's College of Engineering, Pune, India
e-mail: siddhantbhambri@gmail.com

J.-H. Kim et al. (eds.), *Robot Intelligence Technology and Applications 5*,
Advances in Intelligent Systems and Computing 751,
https://doi.org/10.1007/978-3-319-78452-6_10

1 Introduction

28,950,000 visually impaired people in the US only shows the importance of this project regarding both engineering and commercializing purpose. 24.8% of these people have access to smartphone and can afford a wearable device which can improve their life. Therefore, by calculations, a total of 7,200,000 patients could be addressed. But this issue is on a global level, so if we also consider developed areas of Japan and Europe, this project could address up to 2,400,000 and 8,800,000 people respectively. A device which can help them in doing daily tasks like walking on the road without any bodies help, identifying objects kept in front of them, recognizing people surrounded by them, detecting some danger causing object or situation. This device would work as a sixth sense for them and eliminate the need of visual senses anymore. It could be very beneficial for a blind person as he would be able to know what is going on his surrounding and can interact with it with more enthusiasm and would not feel left out anymore.

The Mechatronic kit can solve the challenge of blindness to some extent by giving more knowledge about the surrounding of the blind person to him. This task could be achieved by some ultrasonic sensors, vibrating band and computer vision glass. Firstly, the glasses would detect the objects nearby like a person standing, a dog, or even a chair and table and inform about this to the user. Then, multiple ultrasonic sensors are used to give two types of feedback. In the first place, they provide an approximate distance of the object detected by the glasses by making a sound, and second by identifying some nearby object in all directions, and the user would be able to know that by the vibration band. Moreover, this device consists of a GPS and GSM module which could be very valuable when the user is in an emergency and want to contact to his relatives or nearby emergency center.

1.1 Related Work

In the field of wearable assistive technology, many works are being carried out with several features embedded into it. The [1] author discusses minimizing the use of hands for the assists and allows the users to wear the device on the body such as head-mounted devices, wristbands, vests, belts, shoes, etc.

The portability of the device makes it compact, lightweight, and can be easily carried. In another product [2] BuzzClip, acts as a mobility tool for the blind people rather than a completely autonomous device. It is majorly used to detect the upper body and head level obstacles. It uses the sound waves for detection of obstacles and provides a vibration feedback to notify the users of the presence of obstacles. Not many features are added into this. Another product [3] OrCam, pioneered a compact device with insightful functionality. It can hear any text, appearing on any surface, it can recognize the faces of people, and it can even identify supermarket products and money notes. It can be attached to the side of the glass. But it lacks the significant features of other designs like detection of obstacles.

The Orcam, Exsight and Pivothead Aira smart glasses are well known in the market, but as from the comparison as shown in Fig. 1 it could be interpreted that better the functionality, higher the cost. And vice versa, lower the cost, lesser the functionality included in the product. But the Smart Glass, presented in this paper, meets both expectations of the user, i.e. low cost and better functionality and this makes the Smart Glass earmark in the market from other products available in the market.

1.2 Total Servable Market

Based on the calculations we did as per Table 1, we found out that there are nearly 16.3 million people who are visually impaired across North America (7.2 Million), Europe (6.8 Million) and Japan (2.4 Million) as shown in Fig. 2. Based on the trend from Table 1 the visually impaired population will grow from 16.3 million to 17.1 million with a market growth of about 5% in 2 years' time. So, according to the

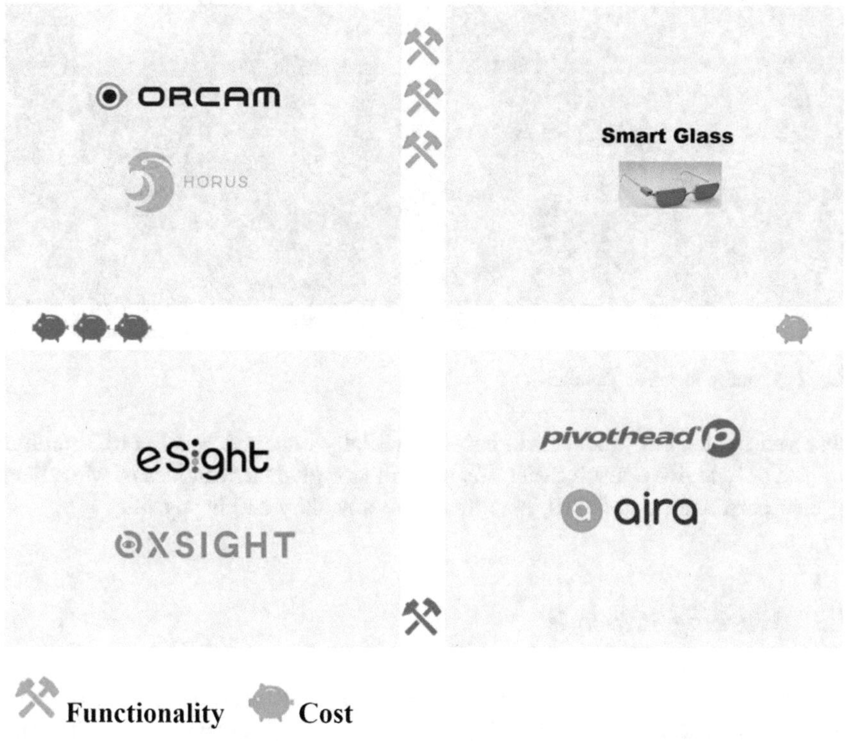

Fig. 1 Cost and functionality comparison of various product with smart glass

Table 1 Visually impaired population across North America, Europe and Japan

	Northern America	Europe	Japan	Total
Population (million)	579	743.1	127.3	
% of smartphone users 2017	42%	35%	55%	
% of smartphone users 2019	44%	37%	58%	
% living in urban areas	82%	73%	93%	
% of visually impaired	5%	5%	5%	
Visually impaired above poverty line	72%	72%	72%	
Total 2017	7.2	6.8	2.4	16.3
Total 2019	7.5	7.1	2.5	17.1

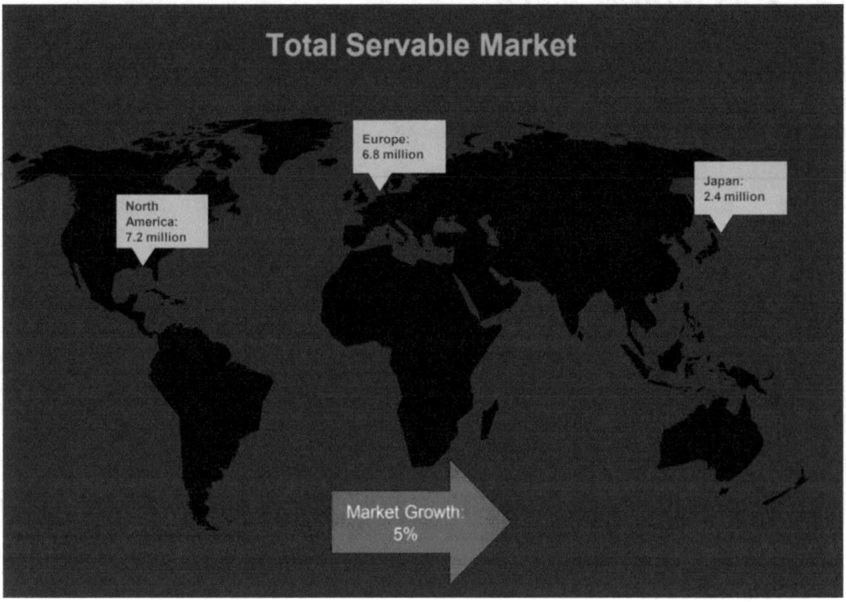

Fig. 2 Visually impaired population

observed trends, our product will help the visually impaired people (16.3 million) not just only to live a much easier life, but will also give them an idea of who all are in their surrounding and thus, help them live a healthy and happy life.

1.3 Working Principle

The smart glass prototype consists of components like raspberry pi zero, bone conduction headset, 1080p 25 FPS HD camera, 2500 mAh battery, ultrasonic distance sensor, vibration band made using vibration disk, GSM module, GPS

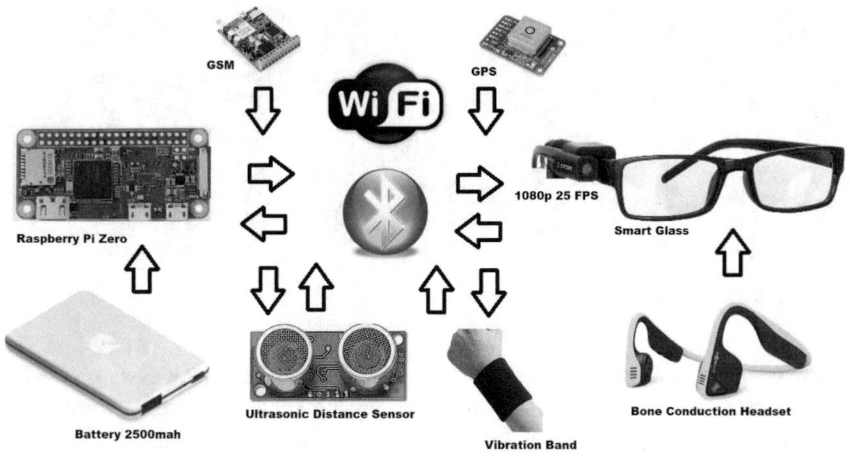

Fig. 3 Working principle of the smart glass

module and Bluetooth–Wi-Fi modules. It's essential that all components are connected to each other using Bluetooth/Wi-Fi interface and give a response without any significant delay. The brain of this project is Raspberry Pi zero which is a credit card sized ARM processor, and all the components are connected to it. First of all, the smart glass with which HD camera is attached is communicating to Raspberry Pi through Bluetooth/Wi-Fi and gives live feed to it so that it can process every frame. Ultrasonic detects the distance of the objects and send it to Raspberry which eventually sends signals to the Vibration Band accordingly. Though the bone conduction headset, the micro-controller gives the response and warning to the user so that user can act accordingly (Fig. 3).

2 Design and Fabrication of Smart Glass

Since visually impaired people do not want products that seem to be designed particularly for them, we have taken great care to make sure that the glasses were designed to look like the ones which ordinary people use. Figure 4 shows the CAD

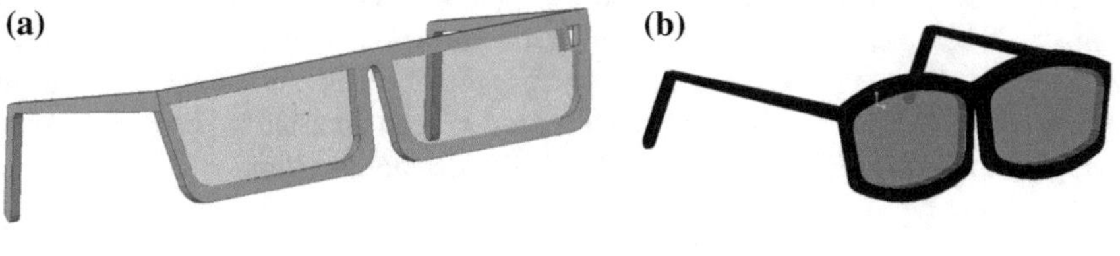

Fig. 4 Model of smart glass version1 (**a**) and version2 (**b**) using PTC Creo

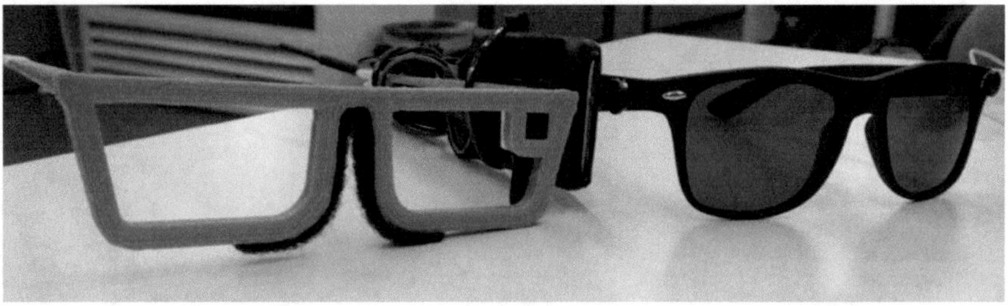

Fig. 5 Prototype of the smart glass

model of the glasses designed using PTC Creo. The CAD model was then saved as a STL file after which it was sliced using Ultimaker Cura Software. For setting up the printer, the infill density was set as 17% with the print speed and travel speed at 65 mm/s and 120 mm/s respectively and with the layer height of 0.15 mm. Once the G-code was generated, it was saved in an SD Card and the SD Card was inserted in Ultimaker 2+ 3D printer to 3D print the glasses using ABS material. Figure 5 shows two final versions of the prototype. This emphasizes the fact that the product is closer to the typical design and does not differentiate the visually impaired people.

3 Proposed Framework

Mechatronics kit consist mainly three components, and each of them is briefly described in the below section:

3.1 Ultrasonic Hand Device

For visually impaired people, it's challenging to detect the object distance near him. His stick can only reach up to some feet, but after that, he has no idea. But, the mechatronics kit resolves this by using multiple ultrasonic sensors, which could give real-time feedback to him and warn him if a fast-moving object is approaching to him from any direction.

The ultrasonic sensors work on the principle of reflection of sound waves and distance is being calculated by multiplying the time taken to receive the sound wave and speed of sound, and then dividing it by 2. And all this could be done in microseconds. Therefore, if multiple ultrasonic is been attached to the person, they can give instant feedback about the object distance around him.

There are vibration band on both the hands of the blind person, which have a connection with the Raspberry Pi wirelessly and would vibrate according to the ultrasonic sensor input. Each band consist of two vibration strips, one in front and one in back. If a close by object is in front of the person, the strips in the front of both the hand vibrates and if the object is on the left or right-hand side, then accordingly the vibration will be there in left or right hand, accordingly. Moreover, if the object is moving closer to the user, it will give warning by making an appropriate sound in the user's ear.

3.2 *Computer Vision Glasses (Real-Time Object Detection Using Deep Learning)*

The Computer Vision Glasses consist of 2 major hardware components, and those are Raspberry Pi and an HD camera which is compatible with that. The camera works as an input sensor which takes the live feed of the surrounding and then gives the feed to the brain, i.e. Raspberry Pi, which computes all the algorithm, detect and recognize all the objects in the surrounding. After recognizing the objects, this gives feedback to the user about the object description by making a sound.

Now elaborating the most crucial part, detecting and recognizing the objects. The project uses deep learning with MobileNets and Single Shot Detector. When these both modules are combined, they give very fast, real-time detection of the devices which are resource constrained for example Raspberry Pi or Smartphone.

3.2.1 Single Shot Detection

To detect the object using deep learning, usually three primary methods are encountered:

- Faster R-CNNs [4]
- You Only Look Once (YOLO) [5]
- Single Shot Detectors (SSDs) [6]

The first one, Faster R-CNN is the most famous method to detect objects, but this technique is hard to implement and further it can be slow as the order of 7 FSP.

To accelerate the process, one can use YOLO as it can process up to 40–90 FSP, however in this case accuracy has to be compromised.

Therefore, this project uses SSD, originally developed by Google, is a balance between both CNN and YOLO. It is simply because it completely removes proposal generation and subsequent pixel or feature resampling stages but includes all computational in a single network. This makes it easy to train and integrate with a system requiring object detection. [6] also, shows that SSD has competitive accuracy as compared to Faster R-CNN and other related frameworks.

3.2.2 MobileNets

The traditional way to develop an object detection network is to use already existing architecture such as ResNet but this is not the best path for resource-constrained devices because of their large size (200–500 MB).

To eliminate the above problem, MobileNets [7], can be used because they are designed for those type of resources. The difference between the MobileNets and traditional CNN is the incorporation of depth wise separable convolution in MobileNets. This splits the convolution into two parts, namely: 3 * 3 depth wise convolution and 1 * 1 pointwise convolution, which reduces the number of parameters in the network keeping it more resource productive.

The MobileNet SSD was being trained first time by COCO dataset (Common Objects in Context) and then improved by PASCAL VOC reaching 72.7% mAP (mean average precision). This dataset is based on a basic object generally present in our environment, and this includes 20 + 1(background) objects, which can be recognized through this technique [8].

The Fig. 6, represents an algorithm to detect and recognize the objects through the computer vision glasses. To start, the live video captured by the camera positioned on the glasses gives the input to the raspberry-pi, where it takes one frame after another and convert it into Blob with the aid of Deep Neural Network (DNN) module. This gives out all the detected objects(n) in one image. Then confidence of each object is being calculated that whether it is an object from the database or not, and if yes, then it is being labelled with the same. When all the objects in one frame are covered, then it takes another frame and processes it.

3.3 Emergency Situation

When a visually blind person gets to know some idea about his environment using ultrasonic band and CV glasses, then he will try to explore more of his surrounding, and therefore he could confront to danger situations while doing so. Moreover, a blind person is exposed to dangerous situations whether he is using a mechatronic kit or not. In these circumstances, he should be able to immediately call for help to nearest emergency center and his relatives.

Mechatronics kit solves this problem by encapsulating a GSM, GPS and voice recognition module connected to the Arduino. Whenever the person is in these type of situations, the device can quickly alert predefined contacts and the emergency center nearest to him by tracing his current GPs location. All this could be done by just a giving a command through voice. The microcontroller responds immediately by sending a message to current GPS location of the person.

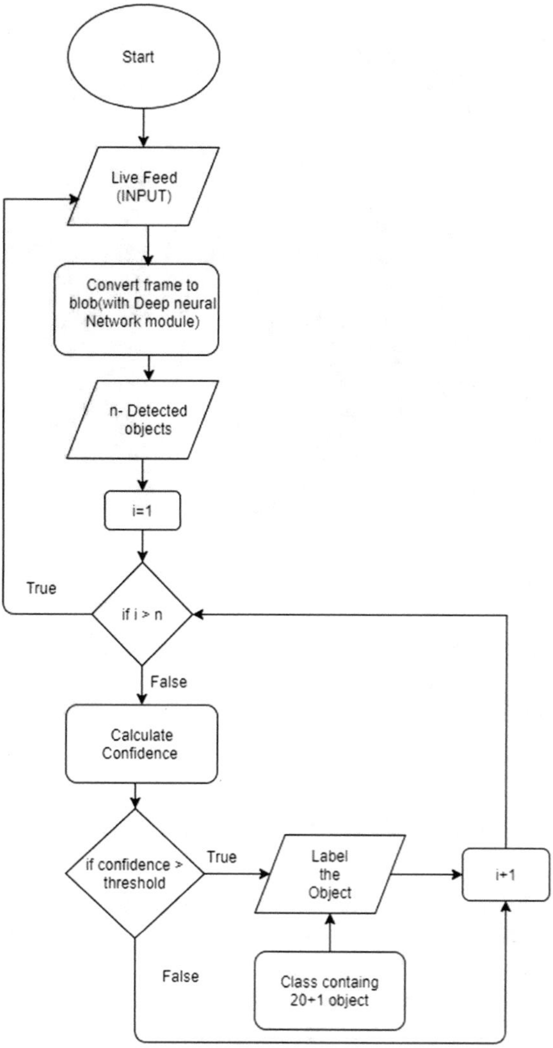

Fig. 6 Graph showing the object detection and recognition algorithm

4 Result and Discussion

As shown in the Fig. 7a, b, the object recognition algorithm can detect the objects correctly, and bounding-box is surrounding the objects with a tag written over them. Although this is just for showing visually, in reality, the visually impaired using the mechatronics kit would be able to listen about these objects and would be able to know more about their position.

For improving the position and distance of the objects, ultrasonic sensors give the feedback and could be detected by the vibration bands, which work on the bases

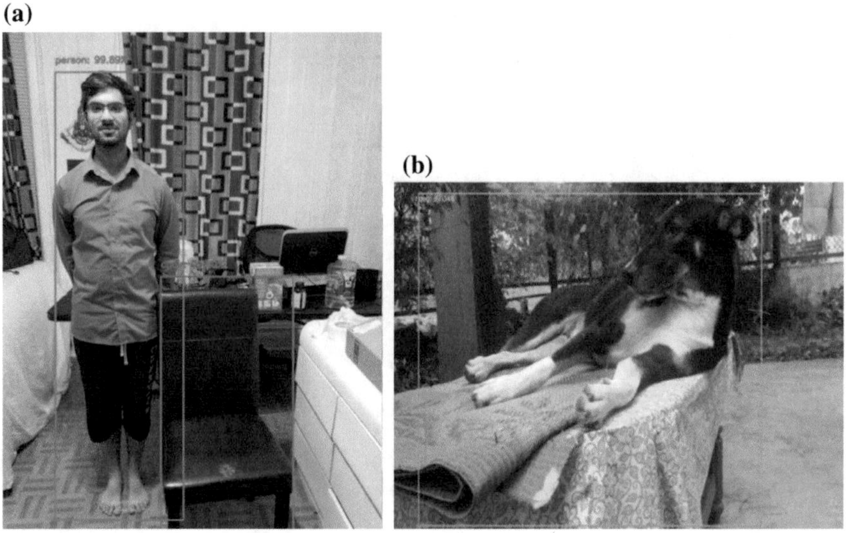

Fig. 7 **a** Person and a chair recognized. **b** Dog recognized in everyday surrounding

of frequency. More the frequency, more the vibration. Therefore, if the distance between the user and object decreases, the frequency will increase, which will intensify the vibration or vice versa.

4.1 Obstacle Detection in the Environment

The above Table 2 and Fig. 8 show the correlation between distance and the vibration frequency with vibration type which is provided as a feedback to the user. It can be inferred from the table that as the distance closes in on to the user, the vibration increases giving a warning to the user about the approaching object. Also as an added feature, the user can notify the device to stop vibration by providing a

Table 2 Obstacle detection with vibration indication for varying distance

Distance (m)	Time (s)	Vibration count	Vibration/sec	Vibration type
1	1	1	1	Continuous
2	1	2	2	Short
3	1	1	1	Long
5	2	1	0.5	Short
10	3	2	0.66	Short
15	3	1	0.33	Short

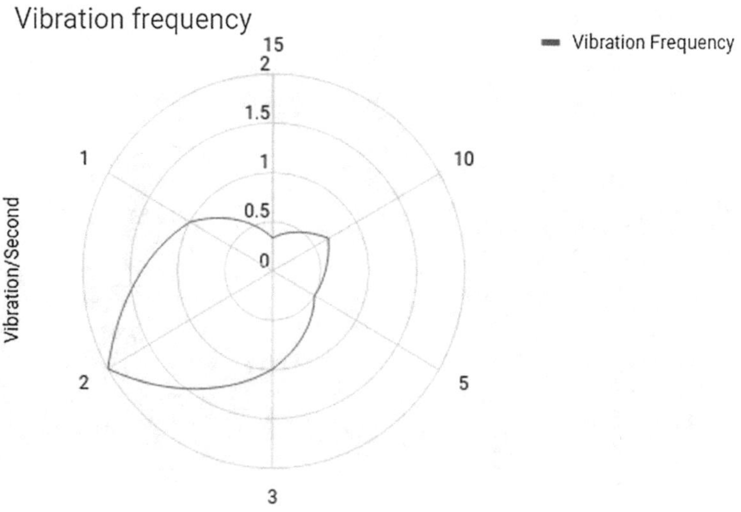

Fig. 8 Distance versus vibration

voice command "Stop Vibration", once the user is in contact with an object or the person. Even there is a difference in the vibration type from the object being in the farthest distance to closest distance. As the distance reduces, the vibration type changes too short to long and then to continuous mode. So there is a constant feedback to the user on the variation of the environment.

4.2 Validation from Visually Impaired People

Jason is 21 years old, lives in Brooklyn and would never leave the house without his smartphone. Because he is visually impaired, he relies on his smartphone for navigation, to read menus in restaurants and to recognize money denominations. But the speakers of smartphones can be very inefficient, after using our smart glass (Figs. 9, 10 and 11)

> "With the Smart Glass, I get all the information I need without standing out for day to day needs. And it's reliable and efficient, handy to use." Jason

> "I feel like I'll find myself using this product a lot at home because it will bring me back to do simple things quickly like when I had sight." Colin Watts

> "Your product would be handy when navigating around big campus buildings like Bobst and Kimmel; I would not bump into people as often. Also, I think it is able to read labels would be helpful as I don't have to ask for help when shopping for food." Emely

Fig. 9 Jason

Fig. 10 Colin Watts trying our designed smart glass prototype

Fig. 11 Emely

5 Application

People who get blindness due to old age will have trouble sustaining in doing chores. The product can be of an excellent use for old age people and can assist them in doing regular activities with much ease. Visual impairment due to accidents is a significant setback for people as they are not used to the blindness and can struggle for even small activities. Smart Glass can be of great assistance in guiding and navigating. Partial blindness may occur to some people who get affected by eye diseases like cataract, glaucoma, etc. may also use this device for assistance.

6 Conclusion and Future Discussion

Smart devices are thought as for future, but this paper tries to bring that in the present and got success to some extent. Smart Glass could be considered better than the products in the market regarding cost-effectiveness and functionality. The results show how it could be converted to a product from the prototype with minor modification in design. The device can also be addressed to a large population of visually impaired including the vision loss due to accidents, or any diseases.

As a future improvement, we will be implementing the same concept using a smartphone, so that we can avoid the use of GSM, GPS and Raspberry Pi zero modules. Also, we will be implementing the voice command in an advanced way using any of the available platforms like, i.e. google assist, Siri, Cortana, Bixby, Alexa. Also, the overall design and casing of the smart glass will be improved to achieve a very compact form.

References

1. Nguyen, T.H., Nguyen, T.H., Le, T.L., Tran, T.T.H., Vuillerme, N., Vuong, T.P.: A wearable assistive device for the blind using tongue-placed electrotactile display: design and verification. In: 2013 International Conference on Control, Automation and Information Sciences (ICCAIS), pp. 42–47. IEEE (2013)
2. The BuzzClip: IMerciv. Wearable Assistive Technology. www.imerciv.com/index.shtml
3. Dakopoulos, D., Bourbakis, N.G.: Wearable obstacle avoidance electronic travel aids for blind: a survey. IEEE Trans. Syst. Man Cybern. Part C (Appl. Rev.) **40**(1), 25–35 (2010)
4. Ren, S., He, K., Girshick, R., Sun, J.: Faster R-CNN: towards real-time object detection with region proposal networks. In: Advances in Neural Information Processing Systems, pp. 91–99 (2015)
5. Redmon, J., Divvala, S., Girshick, R., Farhadi, A.: You only look once: unified, real-time object detection. In: Proceedings of the IEEE Conference on Computer Vision and Pattern Recognition, pp. 779–788 (2016)
6. Liu, W., Anguelov, D., Erhan, D., Szegedy, C., Reed, S., Fu, C.-Y., Berg, A.C.: Ssd: single shot multibox detector. In: European Conference on Computer Vision, pp. 21–37. Springer, Cham (2016)
7. Howard, A.G., Zhu, M., Chen, B., Kalenichenko, D., Wang, W., Weyand, T., Andreetto, M., Adam, H.: Mobilenets: efficient convolutional neural networks for mobile vision applications (2017). arXiv:1704.04861
8. Ross, D.A.: Implementing assistive technology on wearable computers. IEEE Intell. Syst. **16** (3), 47–53 (2001)

A Proactive Robot Tutor Based on Emotional Intelligence

Siva Leela Krishna Chand Gudi, Suman Ojha, Sidra, Benjamin Johnston and Mary-Anne Williams

Abstract In recent years, social robots are playing a vital role in various aspects of acting as a companion, assisting in regular tasks, health, interaction, teaching, etc. Coming to the case of robot tutor, the actions of the robot are limited. It may not fully understand the emotions of the student. It may continue to give lecture even though the user is bored or left away from the robot. This situation makes a user feel that robot cannot supersede a human being because it is not in a position to understand emotions. To overcome this issue, in this paper, we present an Emotional Classification System (ECS) where the robot adapts to the mood of the user and behaves accordingly by becoming proactive. It works based on the emotion tracked by the robot using its emotional intelligence. A robot as a sign language tutor scenario is considered to assist speech and hearing impairment people for validating our model. Real-time implementations and analysis are further discussed by considering Pepper robot as a platform.

Keywords Social robots · Emotions · ECS · Sign language · AUSLAN

S. L. K. C. Gudi (✉) · S. Ojha · Sidra · B. Johnston · M.-A. Williams
The Magic Lab, Centre for Artificial Intelligence, University of Technology Sydney, 15 Broadway, Ultimo 2007, Australia
e-mail: SivaLeelaKrishnaChand.Gudi@student.uts.edu.au; 12733580@student.uts.edu.au
URL: http://www.themagiclab.org

S. Ojha
e-mail: Suman.Ojha@student.uts.edu.au

Sidra
e-mail: Sidra@student.uts.edu.au

B. Johnston
e-mail: Benjamin.Johnston@uts.edu.au

M.-A. Williams
e-mail: Mary-Anne.Williams@uts.edu.au

J.-H. Kim et al. (eds.), *Robot Intelligence Technology and Applications 5*, Advances in Intelligent Systems and Computing 751,
https://doi.org/10.1007/978-3-319-78452-6_11

1 Introduction and Related Work

The presence of social robots amongst human beings is drastically increasing in recent years mainly in the application area of service robotics. Service robotics include telepresence, cleaning, entertainment, security, tutors, health, personal assistants, etc [1]. They are made to emulate an individual life by replacing them to do their current jobs. To further develop human-robot interaction, many robotic platforms are proposed but not limited to Nao, Erica, Pepper [2–4]. Based on research, a child without social interaction or a partner have problems in developing skills [5, 6]. Another study says, one to one tutoring gives higher knowledge for a kid than group education [7]. Due to this reason, researchers proposed robots as a tutor to become a study partner so that a student can improve his skills [8]. Few researchers claim robot instructors increase cognitive learning gains of children [9]. Robot instructors include teaching the English language in schools [10], storyteller [11, 12], sign language teachers [13] and the list goes on.

Shortly, social robots are in a position to supersede human tutors. But to achieve this goal, a lot of considerations must be taken into account. Some researchers claim robots can negatively affect child learning due to its social behavior [14]. This says that robot should be in a position to improve its social interaction with kids. The primary concern is that robot should be proactive by understanding the feelings of the user and try to interact based on his mood. This can make the user feel robot as a general human tutor. Currently, robot trainer has various issues. Even though they are shown in different advertisements/showcases/exhibitions reacting to people on some occasions, they are limited to do only the particular task. It cannot go beyond its limits. There is a long way for the robot to be proactive and this is an emerging research area. The expectations of people are very high, but in reality, the robot still needs an operator/programmer to do individual tasks. For example, if a user is sad, angry or left the class, the robot without emotional intelligence will continue interacting with the user irrespective of the actions showed. This calls for the need of robot tutors to be proactive. We developed an Emotional Classification System (ECS) where the robot takes action independently based on mood perception of the individual it is interacting with. To validate our model, a robot as a sign language tutor is considered, which can communicate with people having speech and hearing impairment. As part of our test, Australian Sign Language (AUSLAN) [15] was used being the first research group to try on a humanoid.

In this paper, we discuss about our robot platform Pepper, the proposed model of ECS, working methodology using a flowchart, real-time implementation results and a survey comparing ECS with general tutoring. Finally, conclusion and future scope are discussed.

2 Proposed Model

We have chosen Pepper, a social humanoid robot from SoftBank Robotics [16] to test our Emotional Classification System (ECS) in a real-time scenario. It is designed to be human-friendly in the day to day person's life. It includes 10.1 in. tablet on its chest to make interaction intuitively by being 1.2 m (4 ft.) tall with a weight of 28 kg running based on NAOqi OS. Its wheelbase consists of various sensors ranging from Sonar, Bumper, Laser, and Gyro sensors to make it a safer robot while interacting with kids. Due to its features of being a friendly robot, we chose Pepper robot as our platform and made it be a teaching assistant based on ECS.

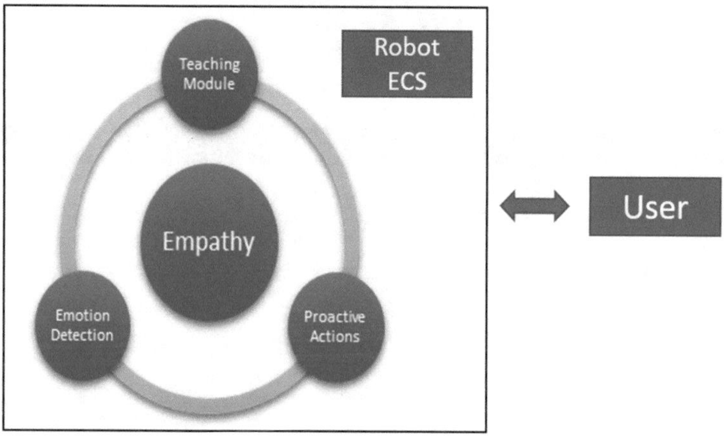

Fig. 1 Proposed Model with ECS

Our empathy model consists of a teaching module where we considered robot is teaching deaf and mute student various gestures of AUSLAN. The emotion of a user is continuously monitored, and proactive actions are chosen based on the user mood. Empathy model is as shown in Fig. 1. For our experiment, we considered few sign language gestures made by pepper and four of them are listed in Scenario and Results section for further discussion. There are few limitations by the robot because it cannot express gestures which are shown by fingers due to the lack of movement. So, gestures with a movement of hands along with wrist movement are considered for analysis. When pepper robot is expressing its gesture, then the meaning of the particular gesture is also shown on its tablet for better human-robot interaction.

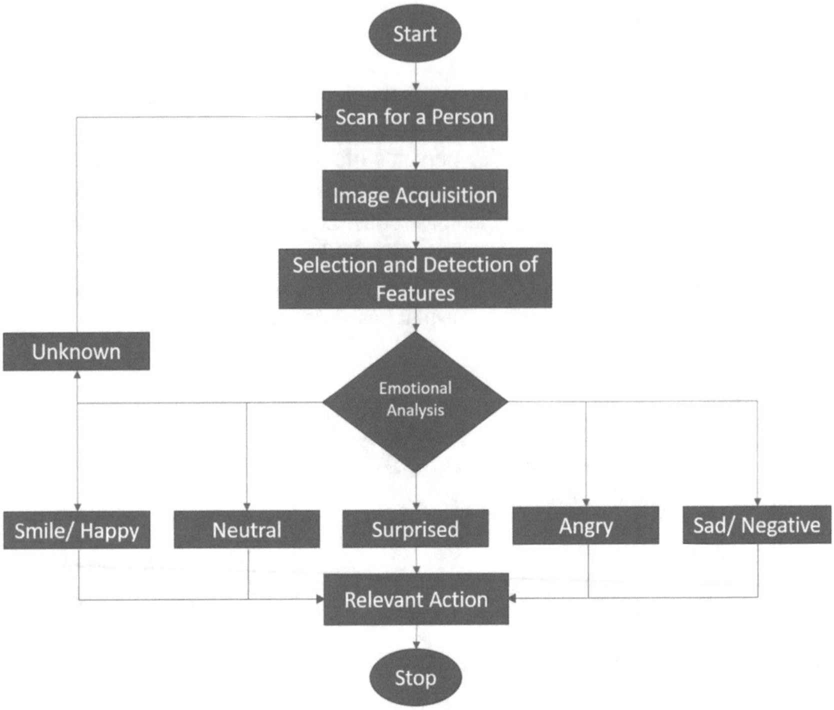

Fig. 2 Flowchart mechanism of ECS

3 Flowchart Mechanism of ECS

After feeding our robot with emotional intelligence, it will be looking for a person to recognize and start tracking user's movement in addition to giving a lecture simultaneously. Internally, after acquiring the image, the features will be selected and detected for further emotional analysis using NaoQi libraries. Considered emotions for detection are Smile/Happy, Neutral, Angry, Sad/Negative and unknown when the robot cannot trace a feeling. After analyzing, relevant actions are considered based on the recognized emotion. Flowchart of our working model is as shown in Fig. 2. In addition to ECS/emotion detection, pepper robot has an ability to show emotions during interaction with the user. It can generate them autonomously with the information from its camera, accelerometer, touch sensor and other sensors. This makes human-robot interaction in a better way as emotions are recognized from the robot as well as the user. NaoQi software library is used to control the ECS of pepper robot. It provides services, network access, and remote calling of methods for faster computation. Navigation modules are disabled while the interaction is activated during the mechanism of ECS.

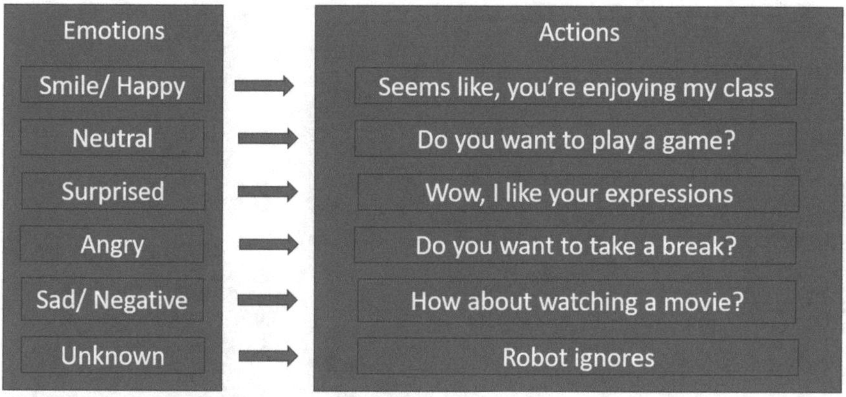

Fig. 3 Actions based on the perceived emotions

For the robot to be proactive, actions are considered entirely based on collected user interests by surveying 30 people. We received various responses from the people on what they wish the robot would do based on their emotional feeling. Finally, considering an average and rating of individuals intention, we have chosen the following tasks for the robot to perform by being a tutor. For example, if robot detects the user as happy and smiling then it responds by saying Seems like, you are enjoying my class. While if an angry mode is detected then it responds Do you want to take a break?. Other considered actions based on the perceived emotions are as shown in Fig. 3. These activities are limited to the scenario of deaf and mute. They could be different from other schema or the same could be applied. If the case is recognized as unknown, the robot will randomly choose the actions when the time limit is in between 5 to 10 min. It can trigger either at any stage between the stipulated time. Meanwhile, to make it more interactive sometimes robot choose the action randomly making the user feel like the robot is proactive. Moreover, if a student leaves robot then it inquiries about the problem.

4 Scenario and Results

In our scenario, the robot can interact with the deaf and hearing impairment person teaching the AUSLAN language. By being a tutor, robot continuously monitors the emotions of the user and decides autonomously on what to do if in case the user is not interested. We considered five different emotions and an unknown situation for the experiment. Considered emotions and concerning actions are as shown in Fig. 3.

Next, the robot is programmed to perform these proactive actions in real time. Interactions can be seen in the results (Fig. 4) where the robot is teaching (1) whereabouts, location, where (2) good, well, fine (3) next, demotion, demote (4) who, whom, whose, someone, whoever, whomever. The same scenario can be applied in

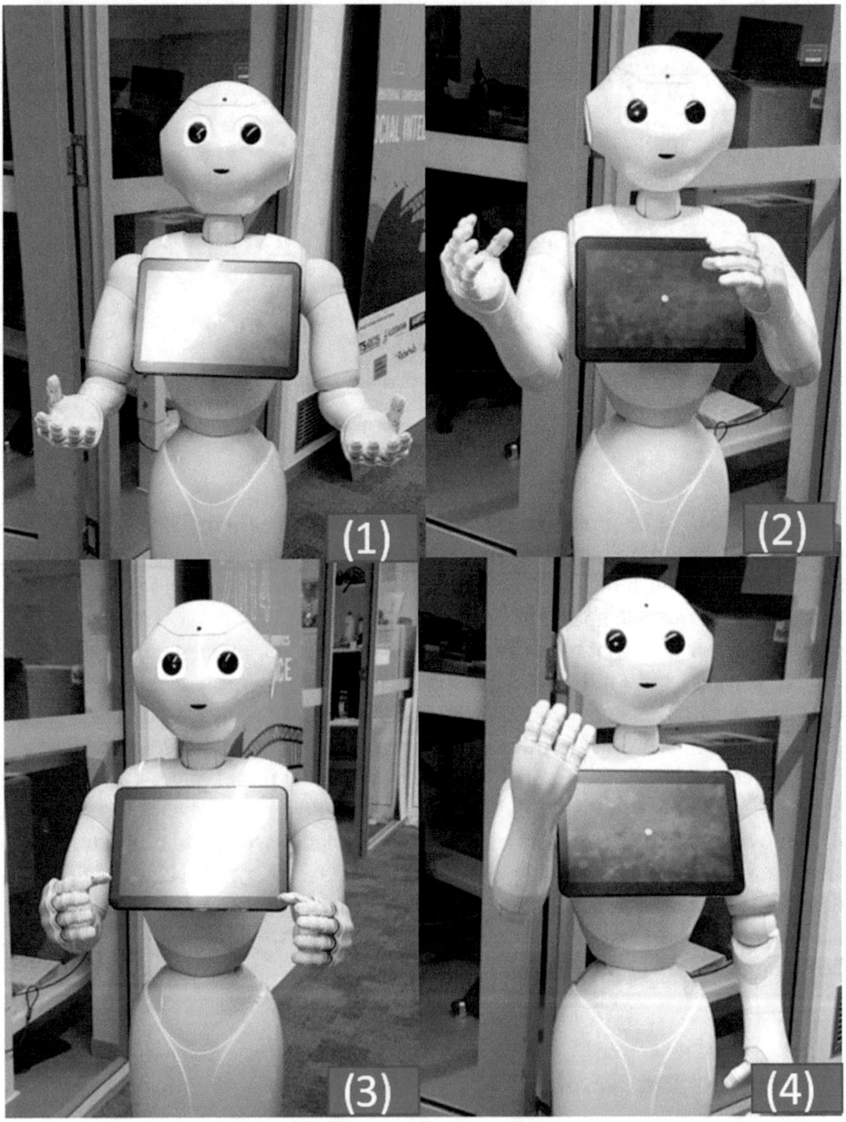

Fig. 4 (1) Whereabouts, location, where; (2) good, well, fine; (3) next, demotion, demote; (4) who, whom, whose, someone, whoever, whomever

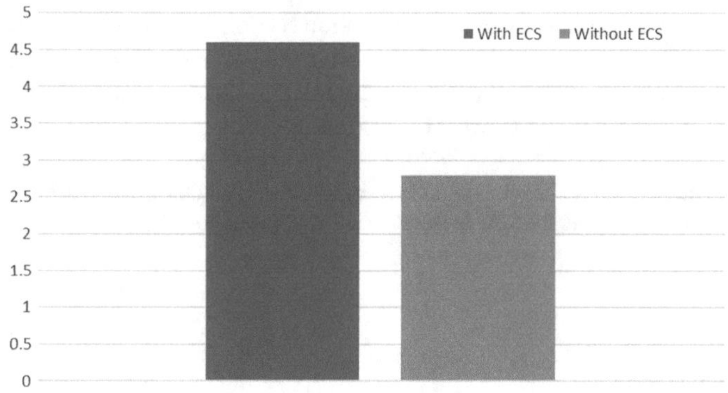

Fig. 5 Comparison between ECS and general system

some other cases as a robot being a storyteller or technical teaching subjects for kids etc.

Later, a survey is done by asking the participants to give a rating between the ECS and a system without ECS. It is as shown in Fig. 5 with respect to a rating scale of 1–5. ECS got a score of 4.6 while the general scheme got 2.8 which shows that our method performed better than the general scheme. User's felt more excited with ECS because the robot can understand a user's mind based on emotional intelligence. On the other hand, people felt happy by seeing the pepper expressing its emotions towards the user, but the recognition of emotion from user's end is missing which made it to have a low score compared to ECS.

5 Conclusion

The identification rate of emotions by the robot is less in some cases and need to be improved for better results. Most of the participants claimed that they like interacting with the pepper robot along with our ECS module and it seems like robot knew about them beforehand by performing like a human being understanding their emotions. On the other hand, pepper got difficulty in identifying people exact feelings sometimes. The reasons could be the environmental conditions like lighting while in some issues, pepper cannot detect emotion quickly because everyone has their own facial expression. Based on our survey, we wish to say that there should be some more improvement in the emotion recognition system which makes ECS more accurate. We look forward to enhancing our modules for better accuracy in future.

By utilizing our Emotional Classification System (ECS), we conclude that we made the robot proactive by thinking itself to take its decisions autonomously depending on the user's mood or feeling. It can impact the user by increasing his learning experience. In future, we wish to make the robot understand the emotions/feelings based on the signal of voice received. To make our model more independent, we plan to use our Ethical Emotion Generation System (EEGS) [17] system where robot works using the appraisals that it received. Also, the sign language gestures are going to be utilized to make human-robot interaction closer to a user by considering animated gestures for every word said by a robot.

Acknowledgements This research is supported by an Australian Government Research Training Program Scholarship. We are thankful to the University of Technology Sydney and ARC Discovery Project scheme.

References

1. Alonso, I.G.: Service Robotics. Service Robotics Within the Digital Home. Springer, Netherlands (2011)
2. Hugel, V., Blazevic, P., Kilner, C., Monceaux, J., Lafourcade, P., Marnier, B., Serre, J., Gouaillier, D., Maisonnier, B.: Mechatronic design of NAO humanoid. In: International Conference on Robotics and Automation, pp. 769–774. IEEE (2009)
3. Alami, R., Gestranius, O., Lemon, O., Niemel, M., Odobez, J.M., Foster, M.E., Pandey, A.K.: The mummer project: engaging human-robot interaction in real-world public spaces. In: International Conference on Social Robotics, pp. 753–763. Springer International Publishing (2016)
4. Milhorat, P., Lala, D., Zhao, T., Inoue, K., Kawahara, T.: Talking with Erica, an autonomous android. In: SIGDIAL Conference, pp. 212–215 (2016)
5. VanLehn, K.: The relative effectiveness of human tutoring, intelligent tutoring systems, and other tutoring systems. In: Educational Psychologist, pp. 197–221 (2011)
6. Kuhl, P.K.: Social mechanisms in early language acquisition: understanding integrated brain systems supporting language. In: Oxford Handbooks (2011)
7. Bloom, B.S.: The 2 sigma problem: the search for methods of group instruction as effective as one-to-one tutoring. Educ. Res. 4–16 (1984)
8. Han, J.: Robot-aided learning and r-learning services. In: Human-Robot Interaction. InTech (2010)
9. Spaulding, S., Toneva, M., Leyzberg, D., Scassellati, B.: The physical presence of a robot tutor increases cognitive learning gains. In: Cognitive Science Society (2012)
10. Alemi, M., Meghdari, A., Ghazisaedy, M.: Employing humanoid robots for teaching English language in Iranian Junior High-Schools. Int. J. Hum. Robot. **11**, 1450022 (2014)
11. Park, H.W., Lee, J.J., Gelsomini, M., Breazeal, C.: Engaging children as a storyteller: backchanneling models for social robots. In: International Conference on Human-Robot Interaction, pp. 407–407. ACM (2017)
12. Martinho, C., Paradeda, R.B., Paiva, A.: Persuasion based on personality traits: using a social robot as storyteller. In: International Conference on Human-Robot Interaction, pp. 367–368. ACM (2017)
13. Kose, H., Akalin, N., Uluer, P.: Socially interactive robotic platforms as sign language tutors. Int. J. Hum. Robot. **11**, 1450003 (2014)
14. Baxter, P., Kennedy, J., Belpaeme, T.: The robot who tried too hard: social behaviour of a robot tutor can negatively affect child learning. In: International Conference on Human-Robot Interaction, pp. 67–74. ACM (2015)
15. Johnston, T., Schembri, A.: Australian Sign Language (Auslan): An Introduction to Sign Language Linguistics. Cambridge University Press (2007)
16. Pereira, T., Connell, J., Perera, V., Veloso, M.: Setting up pepper for autonomous navigation and personalized interaction with users (2017)
17. Ojha, S., Williams, M.A.: Ethically-guided emotional responses for social robots: should I be angry? In: International Conference on Social Robotics, pp. 233–242. Springer International Publishing (2016)

Reusability Quality Metrics for Agent-Based Robot Systems

Cailen Robertson, Ryoma Ohira, Jun Jo and Bela Stantic

Abstract Programming for robots is generally problem specific and components are not easily reused. Recently there has been push for robotics programming to integrate software engineering principles into the design and development to improve the reusability, however currently no metrics have been proposed to measure this quality in robotics. This paper proposes the use of reusability metrics from Component-Based Software Engineering (CBSE) and Service-Oriented Architecture (SOA) to measure reusability metrics, and finds that they are applicable to modular, agent-based robotics systems through a case study of an example system.

1 Introduction

As a field, robotics programming has mainly focused on developing control systems for task-specific projects and their performance in that specific application. The actual process of designing software architecture and developing the control systems in an efficient manner takes a lower priority compared to the speed, accuracy and efficacy of the end product. As a solution to this issue, research in robotic control system design has advocated for the adoption of software engineering principles as other mainstream fields of computer science have done [1]. Object-oriented programming (OOP) has been proposed and implemented for

C. Robertson · R. Ohira · J. Jo (✉) · B. Stantic
School of Information Communication and Technology, Griffith University, Queensland, QLD, Australia
e-mail: j.jo@griffith.edu.au

C. Robertson
e-mail: cailen.robertson@griffith.edu.au

R. Ohira
e-mail: r.ohira@griffith.edu.au

B. Stantic
e-mail: b.stantic@griffith.edu.au

J.-H. Kim et al. (eds.), *Robot Intelligence Technology and Applications 5*, Advances in Intelligent Systems and Computing 751,
https://doi.org/10.1007/978-3-319-78452-6_12

robotics [1, 2]. Player/Stage was created and improved upon to attempt to provide an OOP platform for developing and reusing of code between projects [3]. Component-based architecture has been proposed as an improvement on OOP practices with more reusability and design abstraction [4–6]. Recent research has also proposed agent-based systems as a new suitable application of software engineering principles to robotics [7]. These architecture abstractions do not currently feature agreed upon metrics for measurement of the reusability quality of an application, which in comparison of different project designs and architectures. This paper proposes the adaptation and use of software engineering reusability metrics from Component-Based Software Engineering (CBSE) and Service-Oriented Architecture (SOA) to measure these qualities. This paper also investigates for their applicability in modular, agent-based robotics systems proposed in [7, 8].

Software engineering principles regarding reusability have been subject to extensive investigation and analysis in software development paradigms such as OOP and CBSE. Being a key component to object-oriented approaches to software development, many significant metrics emerged for measuring the reusability [9]. CBSE was then later introduced as a logical evolution on OOP that further developed on the foundation of reusability. However, due to the lack of standards in interoperability of components, obstacles to improving the reusability of components still remained [10].

Service-Oriented Architecture (SOA) is a natural evolution of the modular approach to software engineering where common features are published as services that can be reused [11]. SOA adopts a number of principles from CBSE such as information hiding, modularisation and reusability but implements a higher level of abstraction where loosely coupled systems are provided as services. As a service is made available to a group of service consumers, as opposed to a single consumer, reusability is a key metric in evaluating the quality of such services [12].

Sections 2 and 3 describe the approaches for CBSE and SOA and how their metrics are determined. Section 4 is a case study showing the application of the two metric systems to an example agent-based system.

2 Component-Based Software Engineering

CBSE is a modular approach to software architecture with a focus on reusability. The difference between CBSE and OOP can be seen in the level of abstraction where OOP can only be measured at the source code where as CBSE metrics can begin to consider the component, which could contain a number of classes. Furthermore, the metrics for CBSE measure different aspects to OOP such as the interface methods and interoperability of components as opposed to the relationships and responsibilities of classes such as coupling between classes and lack of cohesion [13, 14]. Brugali further investigates the application of CBSE design paradigms to robotic control system development and overview potential methods of implementation [4, 5]. While the study identified the benefits of CBSE, it did not propose or identify any metrics for quantifying the reusability of the software components.

A fully reusable component in CBSE architecture is said to be designed without featuring implicit assumptions about the setting, scope, robot structure or kinematic characteristics. Three aspects of reusability are defined: quality, technical reusability and functional reusability.

Two of the standard systems of reusability measurement in CBSE are the Fenton and Melton metric, and the Dharma metric [13, 15, 16]. This section investigates how these metrics are applicable to robotic control systems.

2.1 Fenton and Melton Software Metric

Fenton and Melton [17] proposed to measure the degree in which two components are coupled with the following metric:

$$C(x, y) = i + \frac{n}{n+1} \tag{1}$$

where n is defined as the number of interconnections between the components x and y, i being the level of highest coupling type between the two components. The type of coupling is further defined in the following Table 1.

While the Myers Coupling Levels proposes a classification system for the degree of coupling, the Fenton and Melton metric directly quantifies the degree to which two components are coupled. With a given level of coupling, an increasing number of interconnections between two components results in the Fenton and Melton metric to approach the next level of coupling.

Table 1 Fenton and melton modified definitions for myers coupling levels

Coupling type	Coupling level	Modified definitions between components
Content	5	Component x refers to the internals of component y, i.e. it changes data or alters a statement in y
Common	4	Components x and y refer to the same global data
Control	3	Component x passes a control parameter to y
Stamp	2	Component x passes a record type variable as a parameter to y
Data	1	Components x and y communicate by parameters, each of which is either a single data item or a homogenous structure that does not incorporate a control element
No coupling	0	Components x and y have no communication, i.e. are totally independent

2.2 Dharma Coupling Metric

Dharma [16] proposed a metric to measure the coupling inherent to a component *C*. This can be seen in Eq. (2).

$$C = \frac{1}{(i_1 + q_6 i_2 + u_1 + q_7 u_2 + g_1 + q_8 g_2 + w + r)} \tag{2}$$

where,

q_6, q_7 and q_8 are constants, usually a heuristic estimate with a value of 2,

i_1 is the quantity of data parameters,

i_2 is the quantity of control parameters,

u_1 is the quantity of output data parameters, and

u_2 is the quantity of output control parameters.

Global coupling is represented with g_1 being the number of global variables used and g_2 being the number of global control parameters used.

Environmental coupling is represented with w being the number of other components called from component C and r being the number of components that call component C with a minimum value of 1.

Where the Fenton and Melton metric considers all interconnections to have the same level of complexity, the Dharma metric considers the effects of local and global interconnections to have different effects on the coupling of two components. This enables the metric to calculate the coupling value of each component individually.

3 Service-Oriented Architecture

Parallels can be drawn between the current proposals in agent-based control systems and the evolution of development principles in software engineering. With the many similarities in agent-based control systems to CBSE, the number of legacy systems and devices in robotics poses a number of problems the same problems faced by CBSE. While SOA was introduced to solve these problems in enterprise software systems, the same architecture could be applied to robotics. SOA allows for control systems to be encapsulated as a service for reuse.

3.1 Service-Oriented Device Architecture (SODA)

As development paradigms evolved from OOP to CBSE and then to SOA, Service-Oriented Device Architecture (SODA) was proposed as extension of SOA that encapsulates devices in order to publish them as a service [18]. As physical devices are published as services, this is particularly relevant to agent-based control systems in robotics. SODA enables the integration of hardware and software services and consumers in order to fulfill its business goals. For example, an effector may be coupled with its controller within a service which can be accessed by a central controller on the same service bus. However, the integration of devices into SOA introduces another set of unique problems regarding how a device is encapsulated, decorated and published. Mauro identifies seven design problems in incorporating devices into SOA and proposes five candidate patterns to address these issues [19]. Mauro then proposes a compound pattern consisting of a number of SOA patterns to address these problems [20]. These include patterns from both Erl and Mauro [20, 21]:

- Service Encapsulation (Erl)
- Legacy Wrapper (Erl)
- Dynamical Adapter (Mauro)
- Autopublishing (Mauro)

While coupling is the degree to which a service relies on another service, a higher level of abstraction is needed for SOA and SODA. As SODA is an extension of SOA, existing metrics are applicable. At this level of abstraction, reusability of services is a result of the degree in which a service is dependent on another service, also known as modularity. Feuerlicht and Lozina demonstrate that reducing coupling improves service evolution by reducing the interdependencies between published services [22]. While Kazemi et al. propose a number of metrics for measuring the modularity of a service [15], Choi and Kim proposes a general metric to measure the reusability of service providers. These cover five quality attributes identified as:

- Business Commonality: A high level of business commonality results in the functionality and non-functionality of the service is commonly used by the service-consumers.
- Modularity: the degree to which a service provider is able to operate independently of other service.
- Adaptability: the capability of a service provider to provide services to different service consumers.
- Standard Conformance: in order to provide a universal service, service providers should conform to standards accepted by industry.
- Discoverability: as a service is added or removed from the domain, it should be easily found and identified by service consumers.

The following section investigates the components of the SOA reusability metric proposed by troy and Kim for its applicability to agent-based control systems.

3.2 Business Commonality Metrics

Business commonality measures how a service aligns to the requirements of its business domain. In order to do this, the Functional Commonality (*FC*) and Non-Functional Commonality (*NFC*) of the operations of a given service must be first calculated.

$$FC_{Op^i} = \frac{Num_{ConsumersRequiringFRofOp^i}}{Num_{TotalConsumers}} \tag{3}$$

where the numerator represents the number of consumers requiring the functionality of the ith operation by the service. The denominator represents the total number of consumers. The non-functional commonness can quantified by determining the value of NFC_{Op^i} in the following metric:

$$NFC_{Op^i} = \frac{Num_{ConsumersSatisfiedByNFRofOp^i}}{Num_{ConsumersRequiringFRofOp^i}} \tag{4}$$

where the numerator represents the number of consumers requiring the non-functionality of the service operation and the denominator is the number of consumers requiring the functionality of the operation. From these, the Business Commonality (*BCM*) metric can be calculated with the following:

$$BCM = \frac{\sum_{i=1}^{n} \left(FC_{Op^i} \times NFC_{Op^i}\right)}{n} \tag{5}$$

where n is the number of service operations of the given service.

As a ratio, the *BCM* metric values range from 0 to 1 where a lower value indicates a lower commonality of service.

3.3 Modularity Metrics

Kim and Choi identify that, in SOA, modularity is the functional cohesion of a service at level of abstraction higher than CBSE metrics. This is measured by the Modularity (*MD*) metric:

$$MD = 1 - \left(\frac{Num_{SRVOpWithDependency}}{Num_{TotalSRVOp}}\right) \quad (6)$$

where $Num_{SRVOpWithDependency}$ represents the number of service operations that are dependent on other services and $Num_{TotalSRVOp}$ representing the total number of service operations in the service. With the value range between 0 and 1, a service with a value of 1 indicates complete self-sufficiency.

3.4 Adaptability Metrics

The adaptability of a service depends upon how well a service is able to satisfy the needs of its consumers. This can be calculated with the Adaptability (*AD*) metric.

$$AD = \frac{V}{C} \quad (7)$$

where *V* represents the number of consumers satisfied by variant points in a service and *C* represents the total number of consumers. A variant point can be considered to be an adapter that allowed the service to provide a variation on its interface as a service provider. Again, with a value range of 0–1, an *AD* value of 1 indicates that the service is able to provide its services to all consumers requiring the service.

3.5 Standard Conformance Metrics

As there are often widely accepted industry standards for given business domains, these must be taken into consideration when discussing the reusability of a service provider. The Standard Conformity (*SC*) metric measures the degree to which a service will adhere to these standards.

$$SC = \frac{W_{MandatoryStd} \times C_{MandatoryStd}}{N_{MandatoryStd}} + \frac{W_{OptionalStd} \times C_{OptionalStd}}{N_{OptionalStd}} \quad (8)$$

where,

$W_{MandatoryStd}$ is the weight of the mandatory standards,

$C_{MandatoryStd}$ is the number of mandatory standards that the service conforms to,

$N_{MandatoryStd}$ is the total number of mandatory standards relevant to the business domain,

$W_{OptionalStd}$ is the weight of the optional standards,

$C_{OptionalStd}$ is the number of optional standards that the service conforms to, and

$N_{OptionalStd}$ is the total number of optional standards relevant to the business.

The sum of the two weights must equal 1 and acts to balance the importance of the mandatory standards to the optional standards. As such, the value of the *SC* ranges from 0 to 1 with a value of 1 indicating that all relevant standards are being conformed to.

3.6 Discoverability Metrics

The discoverability of a service is critical to its reusability as a service must be found and correctly identified in order to be consumed. In order for a service to be discoverable, it depends on descriptions in terms of syntactic and semantic elements. Syntax is measured by the Syntactic Completeness of Service Specification (*SynCSS*) metric.

$$SynCSS = \frac{D_{Syn}}{E_{Syn}} \tag{9}$$

where D_{Syn} is the number of well described syntactic elements of the service that is exposed to its consumers and E_{Syn} is the total number of syntactic elements. Similarity, semantics is measured by the Semantic Completeness of Service Specification (*SemCSS*) metric.

$$SemCSS = \frac{D_{Sem}}{E_{Sem}} \tag{10}$$

where D_{Sem} is the number of well described semantic elements and E_{Sem} is the total number of semantic elements of a given service. These two metrics can be used to calculate the overall Discoverability (*DC*) of a service.

$$DC = W_{Syn} \times SynCSS + W_{Sem} \times SemCSS \tag{11}$$

where W_{Syn} is the weight of the *SynCSS* and W_{Sem} is the weight of the *SemCSS*. The *DC* value ranges from 0 to 1 where a higher value indicates a greater level of discoverability.

3.7 Reusability Metric

Choi and Kim propose a Reusability (*RE*) metric which encompasses the previously defined metrics to measure the overall reusability of a given service provider.

$$RE = BCM\ (MD \times W_{MD} + AD \times W_{AD} + SC \times W_{SC} + DC \times W_{DC}) \tag{12}$$

where *W* is the corresponding weight for each of the predefined metrics. The value for *RE* ranges from 0 to 1 with a higher value indicating a greater level of reusability for the given service.

4 Case Study

After defining the metrics, their application to proposed system designs is investigated. The case study examines the agent-based drone system put forward by Jo et al. [7]. The system is examined at the system architecture level and the component level.

4.1 Drone-Based Building Inspection System

Jo et al. proposed an agent-based control architecture for use in autonomous unmanned aerial vehicles (UAV) for operation in close proximity to man-made structures. Telemetric data from camera and infrared sensors is transmitted by the control system to an off-board processing agent for classification and recording. An emergency override control was included as an additional module which upon activation changed the UAV from autonomous flight mode to a manual flight mode operated by a human controller with the emergency controller. The proposed system is implemented as a modular architecture (Fig. 1) with the UAV control, camera, vision analysis, machine learning and human operation systems separated into different modules that communicate with each other through a shared event manager.

The component diagram for the UAV module (Fig. 2) displays each of the individual components that comprise the module. The communication between components is shown through the interfaces connected to components within the module. The CBSE metrics are applied to this diagram to determine the reusability of the components of this module.

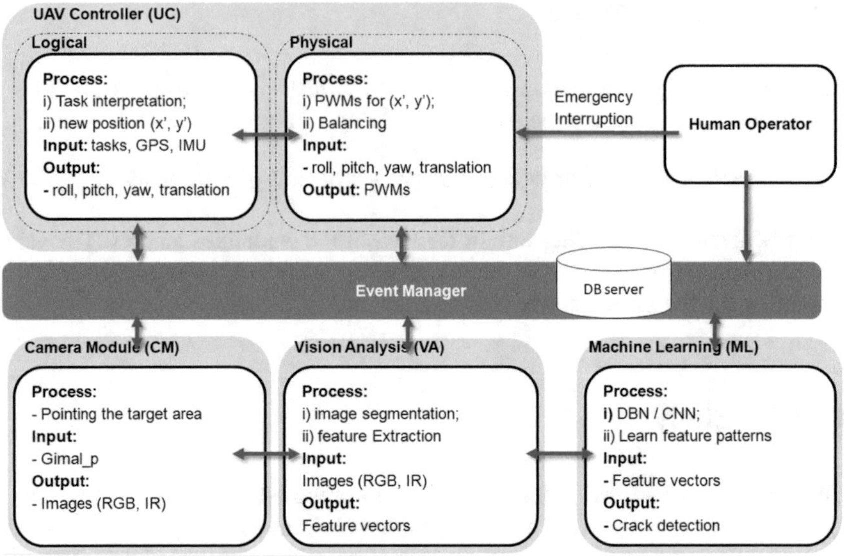

Fig. 1 Proposed agent-based control system [7]

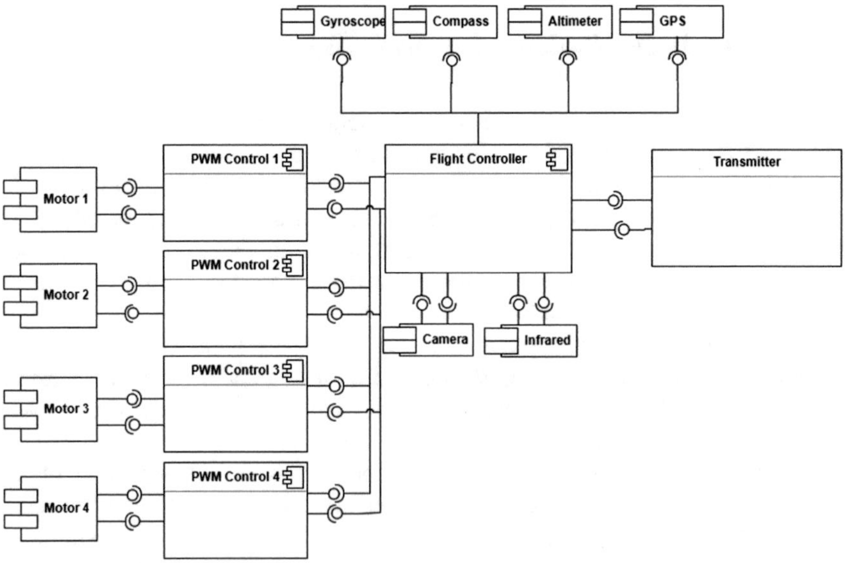

Fig. 2 Component design for UAV controller

4.2 Metric Measurement

In order to quantify the reusability of the proposed system, SOA metrics are applied to the control system level design and the CBSE metrics are applied to the component level system design.

Standard Conformance and Discoverability metrics were not specified in the design scope originally proposed by Jo et al. and it is difficult to precisely quantify to these metrics. Table 2 extrapolates the information to determine these two metrics from the proposal's design documents that are available. The reusability metric for the two agents shows that there is a major difference between reusability of the vision analysis agent and the machine learning agent. This difference is due to the direct contribution of an agent's service operations to the business goals. As the training function of the machine learning operation does not directly contribute to the goals, it can be seen as having a lower overall business commonality as compared to the vision analysis agent. Both agents scored 0 on the modularity metric due to their dependencies on external services. This shows that the agents are in fact highly coupled and cannot be reused without significant effort. Improvements to the agent's modularity would result in greater reusability in future projects.

The CBSE metric is determined for both the individual components and as a whole module using both the Fenton and Melton and Dharma metrics, Tables 3 and 4. For this case study the UAV flight module component design was examined due to the high level of information on its components being available as opposed to other modules in the case study.

In Table 4 the Dharma is seen to be more granular and takes into consideration the type of interfaces that a component features. However, the overall results of the Dharma and the Fenton and Melton metrics in Table 5 shows that the total and standard deviation of the two metrics is quite similar as a final result.

From these results it is seen that sensors such as the Gyroscope and Compass are relatively easy to reuse and replace, due to their single one-way interface connection. Components such as the PWMs are harder to replace as they feature interfaces with the motors as well as the flight controller. Finally, it is clear that the FC module is highly integrated with the other modules and cannot be replaced without substantial effort.

Table 2 SOA reusability metrics

Metric	Vision analysis	Machine learning
Business commonality	1.00	0.50
Modularity	0.00	0.00
Adaptability	1.00	1.00
Standard conformance	1.00	1.00
Discoverability	0.81	0.81
Reusability	0.70	0.35

Table 3 CBSE metrics: Fenton and Melton

Components		Inputs		Metric
x	*y*	*i*	*n*	
Motor 1	PWM 1	2	2	2.67
Motor 2	PWM 2	2	2	2.67
Motor 3	PWM 3	2	2	2.67
Motor 4	PWM 4	2	2	2.67
PWM 1	FC	2	2	2.67
PWM 2	FC	2	2	2.67
PWM 3	FC	2	2	2.67
PWM 4	FC	2	2	2.67
Gyroscope	FC		1	0.5
Compass	FC		1	0.5
Altimeter	FC		1	0.5
GPS	FC		1	0.5
Camera	FC		1	0.5
Infrared	FC		1	0.5
Transmitter	FC	3	3	3.75

Table 4 CBSE metrics: Dharma

Components		Input interfaces		Output interfaces		Dharma metric
x	y	Data	Control	Data	Control	
Motor 1	PWM 1		1	1		3
Motor 2	PWM 2		1	1		3
Motor 3	PWM 3		1	1		3
Motor 4	PWM 4		1	1		3
PWM 1	FC		1	1		3
PWM 2	FC		1	1		3
PWM 3	FC		1	1		3
PWM 4	FC		1	1		3
Gyroscope	FC	1				1
Compass	FC	1				1
Altimeter	FC	1				1
GPS	FC	1				1
Camera	FC	1				1
Infrared	FC	1				1
Transmitter	FC	1	1	1		4

Table 5 CBSE metric final results

Coupling	Fenton and Melton	Dharma metric
Motors	2.67	3
PWM	5.33	6
FC	17.42	22
Gyro	0.50	1
Compass	0.50	1
Altimeter	0.50	1
GPS	0.50	1
Camera	0.50	1
Infrared	0.50	1
Transmitter	3.75	4
Sum	**32.17**	**41**
Average	**3.22**	**4.1**
Standard deviation	**5.01**	**6.19**

5 Conclusion

The case study demonstrates that the metrics discussed in this paper are useful in measuring the reusability of the robotic control system at a component level and at a system architecture level of abstraction. CBSE metrics, such as the Dharma metric and the Fenton and Melton metric, measure the reusability of the individual components that comprise modules. SOA metrics can be used at the system architecture level to determine the modulatory and reusability of the system as separate services within a domain. Current increasing interest in agent-based robotic control systems will require agreed upon metrics for the measurement and comparisons of different systems from a software architectural point of view. In this paper, the SOA metrics covered are compatible for such purpose. These metrics will ensure that the development of robotic control systems are reusable, future proof, and help cultivate recommended design and development practices. This will directly contribute to improving the quality of robotic applications. As research questions surrounding the reusability of components in robotic systems are similar to those from software engineering, this paper proposes future investigation into SOA its applications into the fields of robotics.

References

1. Zieliński, C.: Object-oriented robot programming. Robotica **15**(1), 41–48 (1997)
2. Miller, D.J., Lennox, R.C.: An object-oriented environment for robot system architectures. IEEE Control Syst. **11**(2), 14–23 (1991)
3. Collett, T.H., MacDonald, B.A., Gerkey, B.P.: Player 2.0: toward a practical robot programming framework. In: Proceedings of the Australasian Conference on Robotics and Automation, p. 145 (2005)

4. Brugali, D., Scandurra, P.: Component-based robotic engineering (part I). IEEE Robot. Automat. Mag. **16**(4), 84–96 (2009)
5. Brugali, D., Shakhimardanov, A.: Component-based robotic engineering (part II). IEEE Robot. Automat. Mag. **17**(1), 100–112 (2010)
6. Brooks, A., Kaupp, T., Makarenko, A., Williams, S., Oreback, A.: Towards component-based robotics. In: International Conference on Intelligent Robots and Systems (2005)
7. Jo, J., Jadidi, Z., Stantic, B.: A drone-based building inspection system using software agents. In: Intelligent Distributed Computing XI. IDC 2017. Studies in Computational Intelligence (2017)
8. Zielinski, C., Winiarski, T., Kornuta, T.: Agent-based structures of robot systems. In: Trends in Advanced Intelligent Control, Optimization and Automation. KKA 2017. Advances in Intelligent Systems and Computing
9. Khoshkbarforoushha, A., Jamshidi, P., Gholami, M.F., Wang, L., Ranjan, R.: Metrics for BPEL process reusability analysis in a workflow system. IEEE Syst. J. **10**(1), 36–45 (2016)
10. Vitharana, P.: Risks and challenges of component-based software development. Commun. ACM **46**(8), 67–72 (2003)
11. Erl, T.: Service-Oriented Architecture: Concepts. Prentice Hall, Technology and Design (2005)
12. Choi, S.W., Kim, S.D.: A quality model for evaluating reusability of services in SOA. In: 2008 10th IEEE Conference on E-Commerce Technology and the Fifth IEEE Conference on Enterprise Computing, E-Commerce and E-Services (2008)
13. Chidamber, S.R., Kemerer, C.F.: A metrics suite for object-oriented design. IEEE Trans. Softw. Eng. **20**(6), 476–493 (1994)
14. Washizaki, H., Yamamoto, H., Fukazawa, Y.: A metrics suite for measuring reusability of software components. In: Proceedings of Ninth International Software Metrics Symposium (2003)
15. Kazemi, A., Rostampour, A., Azizkandi, A.N., Haghighi, H., Shams, F.: A metric suite for measuring service modularity. In: 2011 CSI International Symposium on Computer Science and Software Engineering (CSSE) (2011)
16. Dhama, H.: Quantitative models of cohesion and coupling in software. J. Syst. Softw. **29**(1), 65–74 (1995)
17. Fenton, N., Melton, A.: Deriving structurally based software measures. J. Syst. Softw. **12**(3), 177–187 (1990)
18. de Deugd, S., Carroll, R., Kelly, K., Millett, B., Ricker, J.: SODA: service-oriented device architecture. IEEE Pervas. Comput. **5**(3), 94–96 (2006)
19. Mauro, C., Leimeister, J.M., Krcmar, H.: Service-oriented device integration-an analysis of SOA design patterns. In: 2010 43rd Hawaii International Conference on System Sciences (HICSS) (2010)
20. Mauro, C., Sunyaev, A., Leimeister, J.M., Krcmar, H.: Standardized device services-a design pattern for service-oriented integration of medical devices. In: 2010 43rd Hawaii International Conference on System Sciences (HICSS) (2010)
21. Erl, T.: SOA Design Patterns, Pearsons Education (2008)
22. Feuerlicht, G., Lozina, J.: Understanding service reusability. In: International Conference Systems Integration, Prague (2007)
23. Biggs, G., MacDonald, B.: A survey of robot programming systems. In; Australasian conference on robotics and automation (2003)

Part II
Autonomous Robot Navigation

SVM-Based Fault Type Classification Method for Navigation of Formation Control Systems

Sang-Hyeon Kim, Lebsework Negash and Han-Lim Choi

Abstract In this paper, we propose a fault type classification algorithm for a networked multi-robot formation control. Both actuator and sensor faults of a robot are considered as node fault on the networked system. The Support Vector Machine (SVM) based classification scheme is proposed in order to classify the fault type accurately. Basically, the graph-theoretic approach is used for modeling the multi-agent communication and to generate the formation control law. A numerical simulation is presented to confirm the performance of proposed fault type classification method.

Keywords Fault type classification · Networked multi-robot/agent · Formation control · Support vector machine (SVM) · Graph Theory

1 Introduction

The safety of mobile agent systems is growing more important in recent years due to the increasing demand for unmanned vehicles for a variety of tasks. Both unmanned ground and aerial robots are finding their way into wide application areas ranging from civilian to military applications. In order to ensure safe and reliable behavior of these unmanned systems, their ability to detect and classify a fault must be of primary importance. Therefore, unmanned systems in a coordinated mission must be able to respond appropriately to a fault in one of the agents. In this paper, we

S.-H. Kim · L. Negash · H.-L. Choi (✉)
Department of Aerospace Engineering, Korean Advanced Institute of Science and Technology, Daejeon, Korea
e-mail: hanlimc@kaist.ac.kr

L. Negash
e-mail: lebsework@kaist.ac.kr

S.-H. Kim
e-mail: k3special@kaist.ac.kr
URL: http://ae.kaist.ac.kr

J.-H. Kim et al. (eds.), *Robot Intelligence Technology and Applications 5*,
Advances in Intelligent Systems and Computing 751,
https://doi.org/10.1007/978-3-319-78452-6_13

study a method for fault detection and propose a fault type classification algorithm for the multi-agent mission.

A multi-agent mission such as surveillance mission relied on the cooperative control of the multi-agent system and their interaction with the surroundings. Fax and Murray [1] presented a vehicle cooperative network that performs a shared task. Their method involves Nyquist criterion that uses the graph Laplacian eigenvalues to prove the stability of the formation. Olfati-Saber et al. [2] provided a theoretical framework for the analysis of consensus algorithms for multi-agent networked systems which is based on tools from matrix theory, algebraic graph theory, and control theory. There are various ways to detect and classify faults in a system. For fault detection, observer-based approaches were studied for a networked power system fault detection [3]. Hwang et al. [4] attempted to comprehensively cover the various model-based fault detection methods in many control applications. Kwon et al. [5] evaluated the vulnerability level of a given system and develop secure system design methodologies. Shames et al. [11] proposed unknown input observer (UIO) based distributed fault detection and isolation in a network of power system nodes seeking to reach consensus. Negash et al. [6] also used UIO for cyber attack detection in Unmanned Arial Vehicles (UAV) under formation flight. For fault classification, Alsafasfeh et al. [7] presented a framework to detect, classify, and localize faults in an electric power transmission system by utilizing Principal Component Analysis (PCA) methods. Dash et al. [8] presented an approach for the protection of Thyristor-Controlled Series Compensator (TCSC) line using Support Vector Machine (SVM). Silva et al. [9] proposed a novel method for transmission line fault detection and classification by the oscillographic record analysis. For the oscillographic record analysis, Wavelet Transform (WT) and Artificial Neural Networks (ANNs) are used.

The main contribution of this paper is the fault detection and fault type classification of a possible system fault in a networked system of multi-agent in a formation control setup using Kalman filter and Support Vector Machine (SVM). In our previous work [10], we mainly considered the fault detection and isolation scheme. However, in order to properly cope with a system fault, figuring out what kind of fault has occurred in the system is very essential. In another word, before fault treatment sequence such as faulty node isolation, the fault characteristic should be recognized.

The rest of the paper is organized as follows. The graph theory based formation control algorithm is presented in Sect. 2. Section 3 describes a formal definition of a system fault in the formation, Kalman filter based fault detection, and SVM based fault type classification scheme. Simulation results of fault detection and fault type classification on the formation control setup are presented in Sect. 4. Section 5 concludes the paper.

2 Formation Control Algorithm

A coordinated motion of unmanned vehicles is necessary for most of the applications of coordinated task such as surveillance, finding and rescue missions. In this section,

N networked unmanned vehicles are considered which coordinate among themselves to achieve a predefined formation setup. Graph theory is used as a tool for modeling the network of a multi-agent system and consensus-based formation control law. From here on multi-robot and multi-agent used interchangeably.

2.1 Networked Multi-Agent Model

Graph theory is a powerful mathematical tool to model a network topology of multi-agent system. The interaction of agents is represented using a undirected graph $\mathcal{G} = (\mathcal{V}, \mathcal{E})$, where $\mathcal{V}$ is the node set, $\mathcal{V} = \{v_1, \ldots, v_N\}$ and $\mathcal{E} \subseteq \mathcal{V} \times \mathcal{V}$ is the edge set of the graph. Every node represents an agent and the edges correspond to the inter-vehicle communication. An adjacent matrix $\mathcal{A} \in R^{N \times N}$ is a matrix encoding of the adjacency relationship in the graph $\mathcal{G}$ with an element $a_{i,j} = 1$ if $(v_i, v_j) \in \mathcal{E}$ and $a_{i,j} = 0$ otherwise. In the undirected graph case, $\mathcal{A}$ is equal to $\mathcal{A}^T$. The neighbors of the ith agent is denoted by a set $N_i = \{j \in \mathcal{V} : a_{i,j} \neq 0\}$. The indegree matrix of $\mathcal{G}$ is a diagonal matrix $\mathcal{D}$ with diagonal entries $d_{i,j} = |N_i|$, where $|N_i|$ denotes the number of neighbors of node i.

The Laplacian of a graph $\mathcal{G}$ defined as

$$\mathcal{L} = \mathcal{D} - \mathcal{A}. \tag{1}$$

In this paper, we suppose that agents have a linear dynamics and each node i has double integrator.

$$\dot{\mathrm{x}}_i = \mathrm{A}\mathrm{x}_i + \mathrm{B}u_i + \mathcal{F}_{a,i}, \quad i = 1, \ldots, N \tag{2}$$

$$\mathrm{y}_i = \mathrm{H}\mathrm{x}_i + \mathcal{F}_{s,i} \tag{3}$$

where the entries of $\mathrm{x}_i = [x_i, v_i]^T \in \mathbf{R}^{2m}$ represents the state variable, m is the dimension of state, and u_i represents control input, $\mathcal{F}_{a,i}$ and $\mathcal{F}_{s,i}$ are the actuator and sensor faults respectively of the ith agent. In addition, in this paper, the constant velocity model with arbitrary control input and the global position measurement model are considered, thus the matrix A, B, and H are defined as:

$$\mathrm{A} = \begin{bmatrix} 0_{m \times m} \; I_{m \times m} 0_{m \times m} \; 0_{m \times m} \end{bmatrix} \tag{4}$$

$$\mathrm{B} = \begin{bmatrix} 0_{m \times m} \\ I_{m \times m} \end{bmatrix} \tag{5}$$

$$\mathrm{H} = \begin{bmatrix} I_{m \times m} \; 0_{m \times m} \end{bmatrix}. \tag{6}$$

2.2 Formation Control Law

For representing a formation control law, a triangular formation of N agents is shown in Fig. 1, where N taken here to be five. In order to formulate the formation control law, the control input u_i should be a function of the relative state of vehicles (agents).

The collective dynamics of agents following dynamics (2) and measurements (3) can be written as

$$\dot{X} = \mathrm{A}_{aug}X + \mathrm{B}_{aug}U \tag{7}$$

$$Y = \mathrm{H}_{aug}X. \tag{8}$$

where $\mathrm{A}_{aug} = I_{N\times N} \otimes \mathrm{A}$ and $\mathrm{H}_{aug} = I_{N\times N} \otimes \mathrm{H}$ are the augmented dynamic equation matrix and augmented measurement equation matrix, respectively. $I_{N\times N}$ is the identity matrix whose size is $N \times N$ and $\otimes$ represents the Kronecker product. X is the augmented state vector, $X = [\mathrm{x}_1^T\ \mathrm{x}_2^T\ \cdots\ \mathrm{x}_N^T]^T$, U is the augmented control input of agents and Y is the augmented measurement vector.

Applying feedback control for the position and velocity consensus, the resulting augmented control input U is given as follows:

$$U = -KH_{xv}(X - X_d) \tag{9}$$

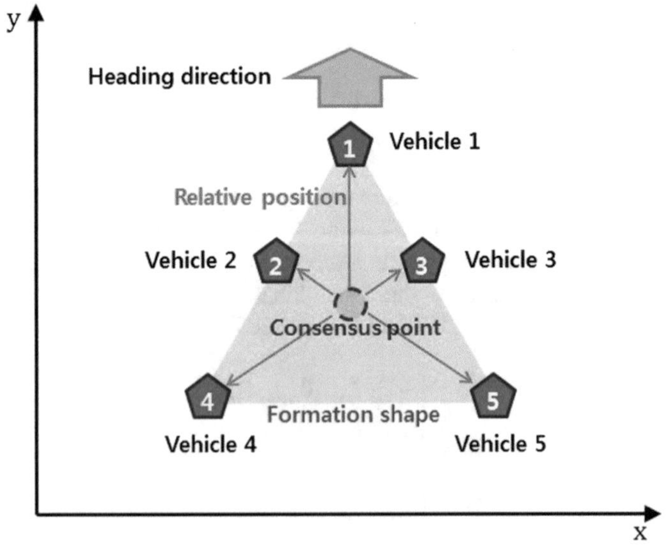

Fig. 1 Unmanned vehicles in the formation, $N = 5$

where

$$K = \begin{bmatrix} \kappa_1 \mathcal{L} & \kappa_2 \mathcal{L} + \kappa_3 I_{N\times N} \end{bmatrix} \otimes I_{m\times m} \tag{10}$$

$$H_{xv} = \begin{bmatrix} I_{N\times N} \otimes H_x \\ I_{N\times N} \otimes H_v \end{bmatrix} \tag{11}$$

$$H_x = \begin{bmatrix} I_{m\times m} & 0_{m\times m} \end{bmatrix} \tag{12}$$

$$H_v = \begin{bmatrix} 0_{m\times m} & I_{m\times m} \end{bmatrix}. \tag{13}$$

Moreover, X_d is the augmented desired state vector and $\kappa_1, \kappa_2, \kappa_3$ are feedback gains.

In the original consensus case where $U = -KH_{xv}X$, all agents move to one identical point (consensus point) therefore the formation shape is not formed. In order to form the predetermined formation shape, we apply the augmented desired state vector $X_d = [\mathrm{x}_{d,1}^T \; \mathrm{x}_{d,2}^T \; \cdots \; \mathrm{x}_{d,N}^T]^T$.

In order to derive the control gain matrix K, the Proportional-Derivative (PD) control law is used [11]. The PD control law for each agent i is described as follows:

$$u_i = -\kappa_3 v_i + \sum_{j\in N_i} \left[\kappa_1(x_j - x_i) + \kappa_2(v_j - v_i)\right] \tag{14}$$

where $N_i = \{j \in \mathcal{V} : \{i,j\} \in \mathcal{E}\}$ is the neighborhood node set of node i.

3 Support Vector Machine Based Fault Classification with Fault Detection

In this section, we consider a possible fault on the actuator or the sensor of each agent in the formation. In order to trigger the fault type classification sequence, it is necessary to confirm that current condition of formation control system is normal or not. After the fault detection, Support Vector Machine (SVM) based classifier can figure out the fault types. Fault Detection and Classification (FDC) scheme for the formation control system is illustrated in Fig. 2.

3.1 Fault Detection

In general, fault detection methods utilize the concept of redundancy, which can be either a hardware redundancy or analytical redundancy as illustrated in Fig. 3. The basic concept of hardware redundancy is to compare duplicative signals generated by various hardware. On the other hand, analytical redundancy uses a mathematical model of the system together with some estimation techniques for fault detection. Changes in the input and output behaviors of a process lead to changes of the output

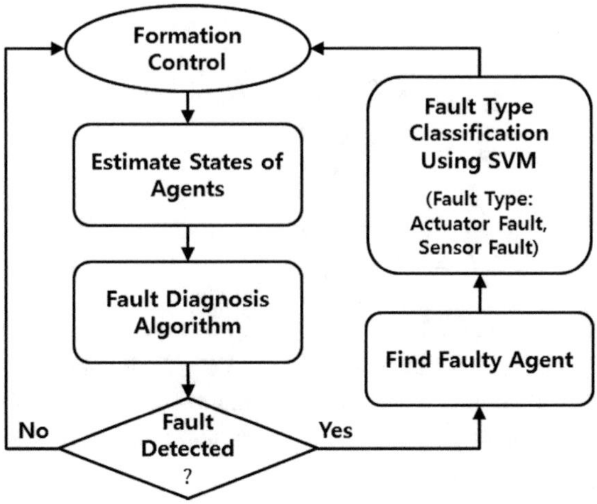

Fig. 2 Fault detection and classification scheme

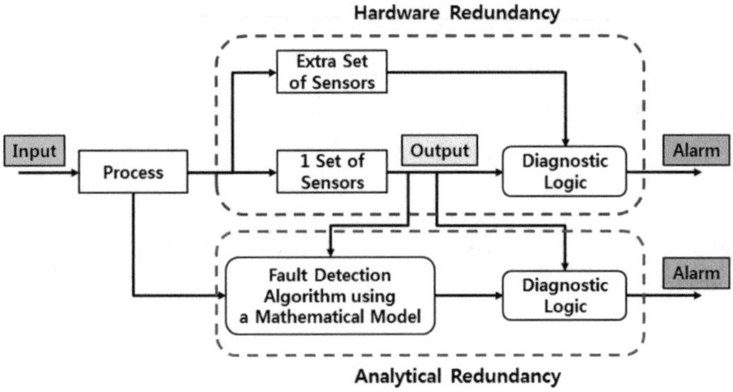

Fig. 3 Illustration of the concepts of hardware redundancy and analytical redundancy for fault detection [4]

error and state variables. Therefore, a mathematical model based fault detection technique can be applicable by using residuals of estimation [12]. This paper focuses on the analytical redundancy approach for fault detection using Kalman filter.

One of the most common fault detection algorithm using the residuals generated by Kalman filter is considered [5]. The residual of agent i is defined as:

$$\mathbf{r}_{i,k} := \mathrm{y}_{i,k} - \hat{\mathrm{y}}_{i,k|k-1}. \tag{15}$$

Without fault, the residual has a zero-mean Gaussian distribution with a covariance matrix $\mathcal{P}_{\mathbf{r}_i} = H_i \mathbf{P}_{i,k|k-1} H_i^T + R_{i,k}^T$, where $\mathbf{P}_i$ and $\mathbf{R}_i$ are the ith state estimation error and measurement noise covariance matrices respectively. Therefore, fault can be diagnosed by testing the following two statistical hypotheses:

$$\begin{cases} \mathcal{H}_0 : \mathbf{r}_{i,k} \sim \mathcal{N}(\mathbf{0}, \mathcal{P}_{\mathbf{r}_i}) \\ \mathcal{H}_1 : \mathbf{r}_{i,k} \nsim \mathcal{N}(\mathbf{0}, \mathcal{P}_{\mathbf{r}_i}) \end{cases} \tag{16}$$

where $\mathcal{N}(\cdot, \cdot)$ is the probability density function of the Gaussian random variable with mean and covariance. In this paper, we consider the above hypothesis test by checking the 'power' of residuals, $\mathbf{r}_{i,k}^T \mathcal{P}_{\mathbf{r}_i}^{-1} \mathbf{r}_{i,k}$. This is the Compound Scalar Testing (CST) which is described as follows:

$$\begin{cases} \text{Accept } \mathcal{H}_0 \text{ if } \mathbf{r}_{i,k}^T \mathcal{P}_{\mathbf{r}_i}^{-1} \mathbf{r}_{i,k} \le h \\ \text{Accept } \mathcal{H}_1 \text{ if } \mathbf{r}_{i,k}^T \mathcal{P}_{\mathbf{r}_i}^{-1} \mathbf{r}_{i,k} > h \end{cases} \tag{17}$$

where h is a threshold value for the fault detection and it is chosen to be greater than $\mathrm{E}[\mathbf{r}_k^T \mathcal{P}_{\mathbf{r}}^{-1} \mathbf{r}_k]$ to avoid the high false alarm rate. If $\mathcal{H}_0$ is accepted, the fault detection algorithm declares there is no fault in agent i, otherwise, $\mathcal{H}_1$ accepted, declares there is a fault in agent i.

3.2 Fault Type Learning and Classification

In this paper, the SVM-based fault type classifier which has two sequences such as, fault type learning sequence and fault type classification sequence, is proposed and these are illustrated in Fig. 4. In the fault type classifier algorithm, SVM is taken as a core part of this algorithm.

The Support Vector Machine (SVM) is one of supervised machine learning algorithms for two-group classification problems [13]. SVM conceptually implements the following idea. Input vectors are non-linearly mapped to a very high dimensional feature space. In this feature space, a linear decision surface is constructed (Fig. 5).

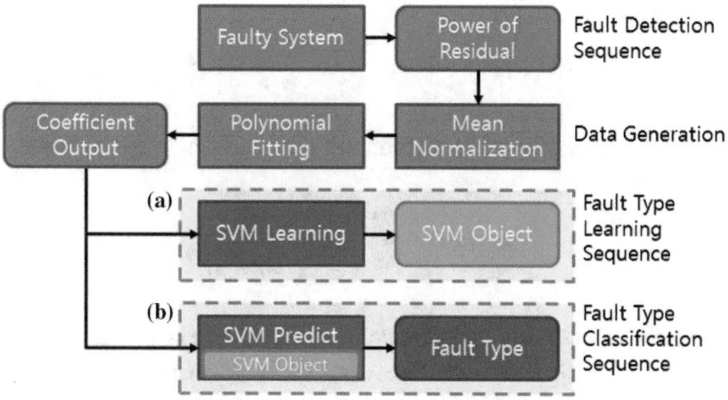

Fig. 4 Fault type (a) learning and (b) classification sequences

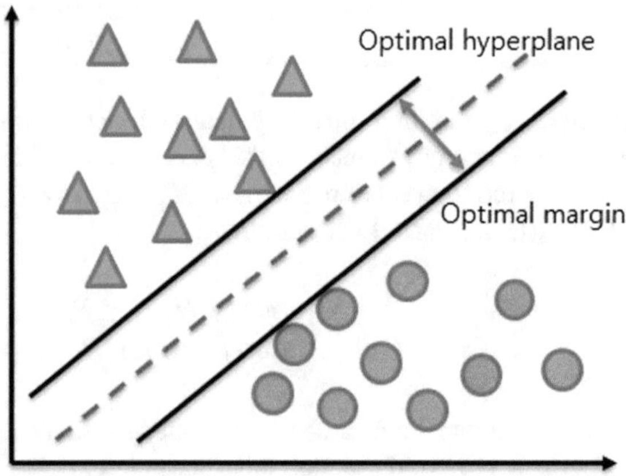

Fig. 5 SVM concept: Optimal hyperplane (decision surface) separates particles with the maximum margin [13]

The set of categorized (labeled) learning patterns:

$$(y_1, \mathbf{x}_1), \ldots, (y_K, \mathbf{x}_K), \quad y_k \in \{-1, 1\} \tag{18}$$

is said to be linearly separable if there exists a vector **w** and a scalar b such that the inequalities are valid for all elements of the training set:

$$\mathbf{w} \cdot \mathbf{x}_k + b \geq 1 \quad \text{if} \quad y_k = 1 \tag{19}$$

$$\mathbf{w} \cdot \mathbf{x}_k + b \leq -1 \ \text{if} \ \ y_k = -1. \tag{20}$$

The optimal hyperplane:

$$\mathbf{w}_0 \cdot \mathbf{x} + b_0 = 0 \tag{21}$$

is the unique one which separates the learning set with a maximal margin. It determines the direction $\frac{\mathbf{w}}{|\mathbf{w}|}$ where the distance between the projections of the training vectors of two different classes is maximal [13]. The distance $\rho(\mathbf{w}, b)$ is given by

$$\rho(\mathbf{w}, b) = \min_{\{x:y=1\}} \frac{\mathbf{x}.\mathbf{w}}{|\mathbf{w}|} - \max_{\{x:y=-1\}} \frac{\mathbf{x}.\mathbf{w}}{|\mathbf{w}|} \tag{22}$$

The hyperplane, $(\mathbf{w}_0, b_0)$ is the arguments that maximize the distance (22). The maximum distance is

$$\rho(\mathbf{w}_0, b_0) = \frac{2}{|\mathbf{w}_0|} = \frac{2}{\sqrt{\mathbf{w}_0 \cdot \mathbf{w}_0}}. \tag{23}$$

This means that the optimal hyperplane is the unique one that minimizes $\mathbf{w} \cdot \mathbf{w}$ under the constraints (19, 20). Therefore, constructing an optimal hyperplane can be formulated as a quadratic programming problem [13].

Fault type learning and classification sequences of the SVM-based classifier are illustrated in Fig. 4. The input signal of faulty system is power of residual, $\mathbf{r}_{i,k}^T \mathcal{P}_{\mathbf{r}_i}^{-1} \mathbf{r}_{i,k}$

which is introduced in fault detection algorithm. In order to extract features of fault type signal, in this paper, the polynomial regression method is used. In statistics, polynomial regression is one of a linear regression which is used for describing nonlinear phenomena by fitting a nonlinear relationship between the value of x and the corresponding conditional mean of y [14]. The general polynomial equation, $p(x)$ of degree n_p is given as follows:

$$p(x) = \sum_{i=1}^{n_p} c_{i+1} x^{n_p - i} \tag{24}$$

where c_i is the coefficient of the polynomial equation. In this paper, we use the polynomial equation of degree 20. Thus, the number of feature outputs (coefficients) is 21.

The outputs of the power of residual value have different magnitude scales according to the fault types or faulty system conditions. Therefore, the original outputs should be adjusted to the common scale. For that reason, in this fault type classifier, mean normalization step:

$$\tilde{y}_k = K \frac{y_k}{\sum_{k=1}^{K} y_k}. \tag{25}$$

is applied before polynomial fitting step. $\tilde{y}_k$ is the normalized y_k. As a result, we can get the normalized coefficients of the polynomial function and these are treated as features for SVM based learning or classification. In Fig. 4, SVM object means an output of SVM based learning algorithm and it is used for estimating the fault type.

Moreover, for generating the learning or test data set (i.e., from the power of residual of a faulty system), when a fault is detected, the fault type classification algorithm automatically saves up to 10 s of data from 5 s before, and during 5 s after fault detection. In other words, the time length of learning or test data is 10 s.

4 Simulation

4.1 Simulation Setup

For representing the formation control scenario, five agents in regular triangle formation are considered as illustrated in Fig. 1. It is supposed that all agents start a movement from an arbitrary position and then follow the predetermined path.

Agents are controlled to maintain the regular triangle formation and heading angle of formation shape equal to the predetermined path angle using fully connected communication network. In practical systems, the agents have actuator saturation which is related to the motor or engine power, therefore, we set up the control input limit as $|U_i| \leq 10\ [\mathrm{m}/s^2]$.

In this simulation, it is assumed that each agent uses GPS for navigation. GPS is one of the absolute sensors which takes information from the environment outside of agents. GPS model is defined as:

$$y_{GPSi,k} = \begin{bmatrix} x_{i,k} \\ y_{i,k} \end{bmatrix} + v_{GPS,k} \tag{26}$$

where $v_{GPS} \sim \mathcal{N}(0, (10\,\text{m})^2)$ is the GPS measurement noise.

The sampling time of simulations is 0.1 s and the end time of simulations is 120 s.

4.2 Simulation Results

In this subsection, the simulation results of different fault scenarios are presented. There are three possible different scenarios which are: the system with no fault, the system with an actuator fault or the system with a sensor fault. It is assumed that each fault occurs in agent 2.

(1) Fault Detection The result of 'no-fault' case shows the expected normal formation trajectories which are shown in Fig. 6 with the power of residual (Fig. 7) which doesn't show a change. In fault occurrence scenario, the agent 2 is suffering from an actuator fault (Fig. 8) and sensor fault (Fig. 9). In both cases, the UAVs were not able to maintain the formation flight. In this fault detection test scenario, the fault occurrence time is 30 s and the power of residual threshold value, h, for the fault detection is 30. As a result, the fault detection time of actuator fault is around 45 s and that of sensor fault is around 30 s (Fig. 10).

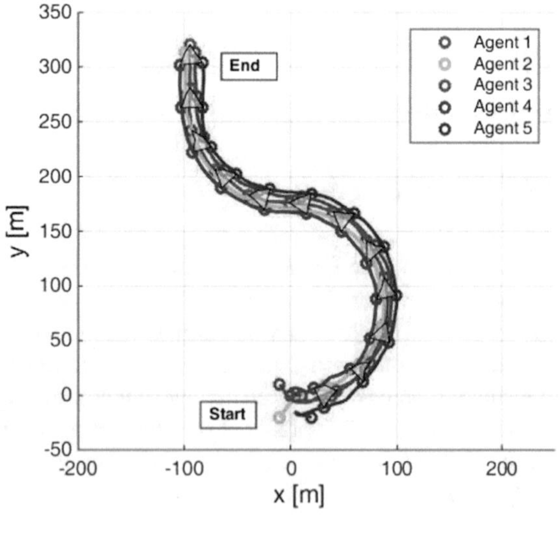

Fig. 6 No fault Case: Formation trajectory, yellow triangles represent the formation shape

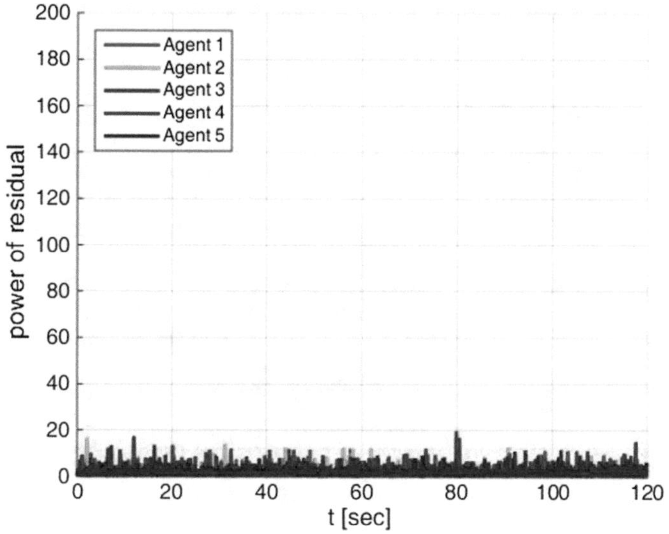

Fig. 7 No fault Case: Power of residual

Fig. 8 Actuator fault Case: Formation trajectory

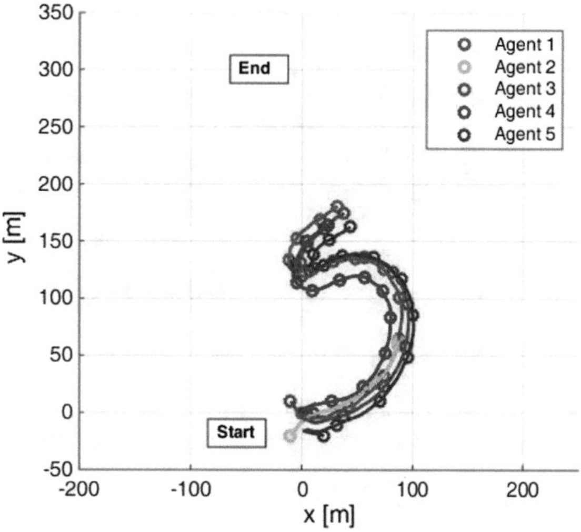

The power of residual for actuator fault is a little slower to build up as compared to the sensor fault one as it involves an integrating dynamics of the agent 2. As it is shown in Fig. 10, the FDI system detects the fault in the network and indicates Agent 2 is at fault.

(2) Fault Type Learning In this paper, for learning fault type, we set 60 scenarios of different fault occurrence time varying between 40–45 s and generate the polynomial regression (curve fitting) data of time length of 10 s with sampling time 0.1 s.

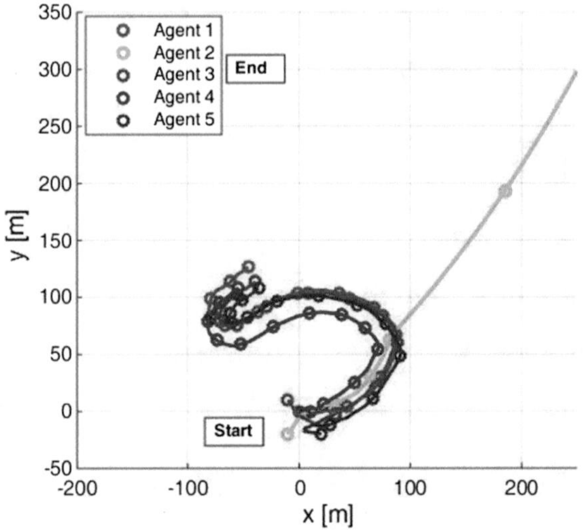

Fig. 9 Sensor fault Case: Formation trajectory

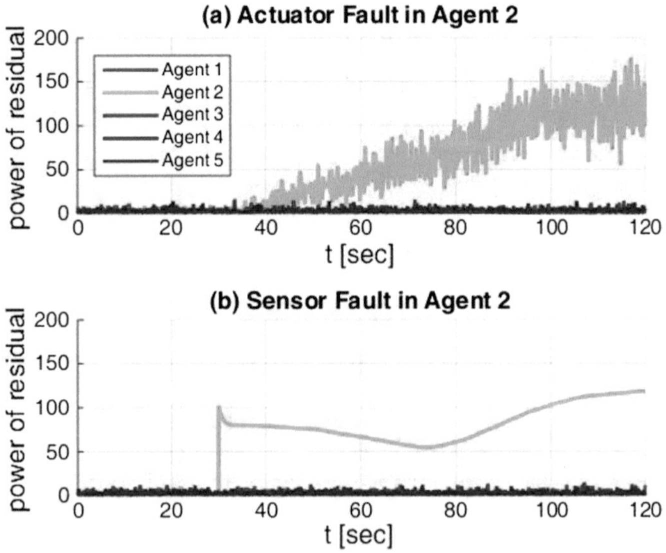

Fig. 10 Fault Case: Power of residual, (a) Actuator fault (b) Sensor fault

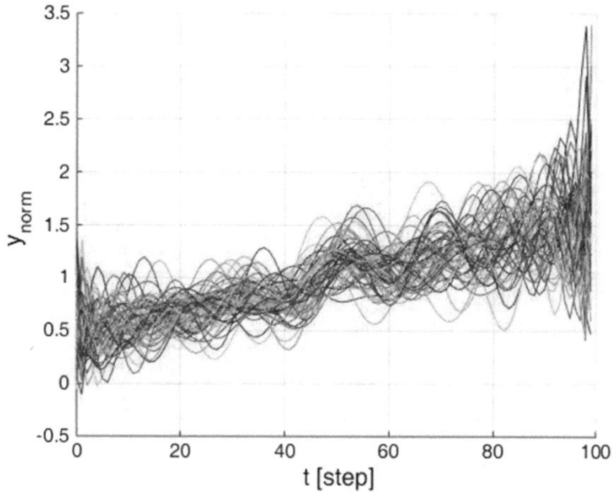

Fig. 11 Learning data set of normalized curve fitting value : Actuator fault

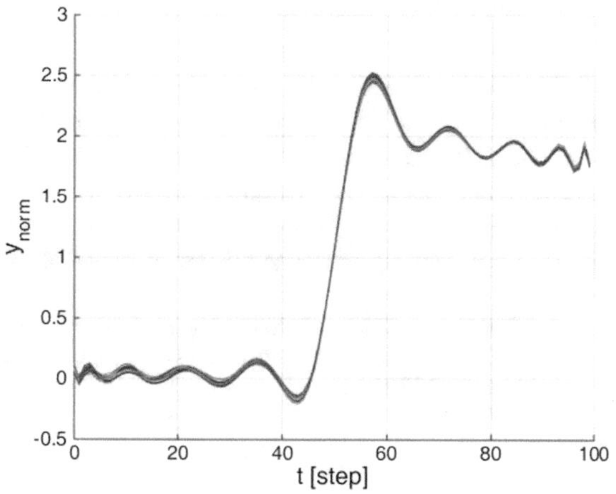

Fig. 12 Learning data set of normalized curve fitting value : Sensor fault

Therefore, the time step length of learning data is 100-time steps and the final learning data set of normalized curve fitting data is shown in Figs. 11 and 12. By using these learning data, the SVM object for fault type classifier is obtained.

(3) Fault Type Classification To investigate the performance of the proposed fault type classifier, 1,000 Monte-Carlo Simulation (MCS) runs are conducted and the associated average mean fault detection times and classification accuracies are computed for eight fault occurrence scenarios. The result of numerical simulations is shown in Table 1. The fault detection time of actuator fault is about 13 s later than that of a sensor fault. The performance of fault type classification accuracy is about 98 percent for actuator fault and about 99 percent for sensor fault. As a result, the proposed fault type classifier is confirmed that it has reasonable performance.

Table 1 Fault type classification: 1,000 Monte-Carlo simulation runs for each of the fault scenarios

Fault type	Occurrence time [s]	Mean detection time [s]	Classification accuracy [%]
Actuator fault	30	43.65	97.8
	40	53.61	98.2
	50	63.71	98.5
	60	73.48	98.0
Sensor fault	30	30.10	100
	40	40.10	100
	50	50.10	99.9
	60	60.09	99.9

5 Conclusion

In this paper, we considered a possible system fault in a networked multi-agent formation control. For fault type classification, the Support Vector Machine (SVM) based fault classification scheme is proposed with fault detection scheme based on the compound scalar testing algorithm. We confirm that the proposed algorithm is not only able to detect the system fault but also accurately classify the fault type. Simulation results demonstrate the performance of the proposed algorithm.

Acknowledgements This work was supported by the ICT R&D program of MSIP/IITP. [R-20150223-000167, Development of High Reliable Communications and Security SW for Various Unmanned Vehicles].

References

1. Fax, J.A.: Information flow and cooperative control of vehicle formations. IEEE Trans. Autom. Control **49**(9), 1465–1476 (2004)
2. Olfati-Saber, R., Fax, J.A., Murray, R.M.: Consensus and cooperation in networked multi-agent systems. Proc. IEEE **95**(1), 215–233 (2007)
3. Aldeen, M., Crusca, F.: Observer-based fault detection and identification scheme for power systems. IEEE Proc Gener. Transm. Distrib. **153**(1), 71–79 (2006)
4. Hwang, I., et al.: A survey of fault detection, isolation, and reconfiguration methods. IEEE Trans. Control Syst. Technol. **18**(3), 636–653 (2010)
5. Kwon, C., Liu, W., Hwang, I.: Security analysis for cyber-physical systems against stealthy deception attacks. American Control Conference (ACC), 2013. IEEE (2013)
6. Negash, L., Kim, S.-H., Choi, H.-L.: Distributed unknown-input-observers for cyber attack detection and isolation in formation flying UAVs (2017), arXiv:1701.06325
7. Alsafasfeh, Q.H., Abdel-Qader, I., Harb, A.M.: Fault classification and localization in power systems using fault signatures and principal components analysis. Energy Power Eng. **4**(06), 506 (2012)

8. Dash, P.K., Samantaray, S.R., Panda, G.: Fault classification and section identification of an advanced series-compensated transmission line using support vector machine. IEEE Trans. Power Deliv. **22**(1), 67–73 (2007)
9. Silva, K.M., Souza, B.A.: Fault detection and classification in transmission lines based on wavelet transform and ANN. IEEE Trans. Power Deliv. **21**(4), 2058–2063 (2006)
10. Kim, Sang-Hyeon, Negash, Lebsework, Choi, Han-Lim: Cubature Kalman filter based fault detection and isolation for formation control of multi-UAVs. IFAC-PapersOnLine **49**(15), 63–68 (2016)
11. Shames, I., et al.: Distributed fault detection for interconnected second-order systems. Automatica **47**(12), 2757–2764 (2011)
12. Isermann, R.: Model-based fault-detection and diagnosisstatus and applications. Annu. Rev. control **29**(1), 71–85 (2005)
13. Cortes, C., Vapnik, V.: Support-vector networks. Mach. Learn. **20**(3), 273–297 (1995)
14. Beck, J.V., Arnold, K.J.: Parameter Estimation in Engineering and Science. James Beck (1977)

Indoor Magnetic Pose Graph SLAM with Robust Back-End

Jongdae Jung, Jinwoo Choi, Taekjun Oh and Hyun Myung

Abstract In this paper, a method of solving a simultaneous localization and mapping (SLAM) problem is proposed by employing pose graph optimization and indoor magnetic field measurements. The objective of pose graph optimization is to estimate the robot trajectory from the constraints of relative pose measurements. Since the magnetic field in indoor environments is stable in a temporal domain and sufficiently varying in a spatial domain, these characteristics can be exploited to generate the constraints in pose graphs. In this paper two types of constraints are designed, one is for local heading correction and the other for loop closing. For the loop closing constraint, sequence-based matching is employed rather than a single measurement-based one to mitigate the ambiguity of magnetic measurements. To improve the loop closure detection we further employed existing robust back-end methods proposed by other researchers. Experimental results show that the proposed SLAM system with only wheel encoders and a single magnetometer offers comparable results with a reference-level SLAM system in terms of robot trajectory, thereby validating the feasibility of applying magnetic constraints to the indoor pose graph SLAM.

Keywords SLAM · Magnetic field · Pose graph optimization · Robust back-end

J. Jung (✉) · J. Choi
Marine Robotics Lab., Korea Research Institute of Ships and Ocean Engineering, 32 Yuseong-daero 1312 beong-gil, Yuseong-gu Daejeon 34103, Korea
e-mail: jdjung@kriso.re.kr

J. Choi
e-mail: jwchoi@kriso.re.kr

T. Oh · H. Myung
Urban Robotics Lab., KAIST, 291 Daehak-ro,, Yuseong-gu Daejeon 34141, Korea
e-mail: buljaga@kaist.ac.kr

H. Myung
e-mail: hmyung@kaist.ac.kr

J.-H. Kim et al. (eds.), *Robot Intelligence Technology and Applications 5*,
Advances in Intelligent Systems and Computing 751,
https://doi.org/10.1007/978-3-319-78452-6_14

1 Introduction

Simultaneous localization and mapping (SLAM) is one of the key technologies for autonomous mobile robot navigation [1]. For indoor environments, commonly used sensors for SLAM are cameras, laser scanners, ultrasonic ranging beacons, etc. Although these sensor modalities can provide reasonable performance in SLAM, additional and ambient signal sources can extend robot's operation area and further improve the performance. Among the various signals of opportunity, magnetic field is what we are trying to utilize in this paper. Indoor magnetic field is mainly contributed by geomagnetism, where the geomagnetic field is measured with distortion caused by ferromagnetic objects in indoors [2, 3]. It is shown that the distortion is sufficiently stable over time [4] and can be used for navigational purposes [5–7]. Moreover, magnetic field can be measured with high resolution and accuracy with cheap sensors.

Generally, pose graphs consist of nodes and edges, which represent robot poses and the constraints between nodes by relative pose measurements, respectively [8, 9]. Using magnetic field measurements, two types of constraints can be designed which can improve estimation of (i) rotational motion of the robot and (ii) loop closures [10]. For loop closing constraints, a sequence of magnetic measurements can be used. In this way we mitigate the ambiguity and orientation-dependency problems in the magnetic field-based loop closing. Further improvement on loop closing can be done by employing existing robust back-end methods [11, 12]. In the following sections we explain the formulation details of magnetic pose graph SLAM with robust back-end and its experimental results.

1.1 Basic Formulation

Let us define a pose graph $\mathbf{x}$ as the collection of $\mathbf{SE}(2)$ robot poses as follows:

$$\mathbf{x} = [\mathbf{x}_1^{\mathrm{T}}, \ldots, \mathbf{x}_n^{\mathrm{T}}]^{\mathrm{T}} \tag{1}$$

where the i-th $\mathbf{SE}(2)$ robot pose is defined as $\mathbf{x}_i = [x_i, y_i, \theta_i]^{\mathrm{T}}$ and n is the number of poses. Given k-th measurement $\mathbf{z}_k$, the residual $\mathbf{r}_k$ is calculated as the difference between the actual and predicted observations as follows (Gaussian noise is assumed for all measurements):

$$\mathbf{r}_k(\mathbf{x}) = \mathbf{z}_k - h_k(\mathbf{x}) \tag{2}$$

where h_k is an observation model for the k-th measurement. With an assumption of independence between observations, a target cost function $E(\mathbf{x})$ can be defined as a weighted sum of the residuals:

$$E(\mathbf{x}) = \sum_k \mathbf{r}_k(\mathbf{x})^{\mathrm{T}} \boldsymbol{\Lambda}_k \mathbf{r}_k(\mathbf{x}) \tag{3}$$

where $\mathbf{\Lambda}_k$ is an information matrix for the k-th measurement. The optimal configuration of the pose graph $\mathbf{x}^*$ is then obtained by minimizing (3) and this can be represented as follows:

$$\mathbf{x}^* = \arg\min_{\mathbf{x}} \sum_k \mathbf{r}_k(\mathbf{x})^{\mathrm{T}} \mathbf{\Lambda}_k \mathbf{r}_k(\mathbf{x}) \tag{4}$$

In terms of probabilistic inference, the solution $\mathbf{x}^*$ is a realization of the maximum a posteriori estimation of the robot poses given all observations.

1.2 Design of Constraints

Since the calculation of the residuals and Jacobians is dependent on the constraint types, the cost function can be decomposed into the corresponding constraint terms. Assuming ground vehicles, pose graphs generally have two types of constraints—odometric and loop closing constraints. The cost function $E(\mathbf{x})$ then can be described as

$$E(\mathbf{x}) = E_{\mathrm{odm}}(\mathbf{x}) + E_{\mathrm{LC}}(\mathbf{x}) \tag{5}$$

where $E_{\mathrm{odm}}(\mathbf{x})$ and $E_{\mathrm{LC}}(\mathbf{x})$ are the cost caused by the odometric and loop closing constraints, respectively. Each cost term can be further described as

$$E_{\mathrm{odm}}(\mathbf{x}) = \sum_{\{i,j\}\in\mathcal{P}} \mathbf{r}_{ij}^{\mathrm{odm}}(\mathbf{x})^{\mathrm{T}} \mathbf{\Lambda}_{ij}^{\mathrm{odm}} \mathbf{r}_{ij}^{\mathrm{odm}}(\mathbf{x}) \tag{6}$$

$$E_{\mathrm{LC}}(\mathbf{x}) = \sum_{\{i,j\}\in\mathcal{Q}} \mathbf{r}_{ij}^{\mathrm{LC}}(\mathbf{x})^{\mathrm{T}} \mathbf{\Lambda}_{ij}^{\mathrm{LC}} \mathbf{r}_{ij}^{\mathrm{LC}}(\mathbf{x}) \tag{7}$$

where $\mathcal{P}$ and $\mathcal{Q}$ are the sets of constrained pose pairs by odometric and loop closing constraints, respectively, and $\mathbf{\Lambda}_{ij}^{\mathrm{odm}}$ and $\mathbf{\Lambda}_{ij}^{\mathrm{LC}}$ are information matrices for odometric and loop closing measurements, respectively. The residuals $\mathbf{r}_{ij}^{\mathrm{odm}}$ and $\mathbf{r}_{ij}^{\mathrm{LC}}$ can be defined as

$$\mathbf{r}_{ij}^{\mathrm{odm}}(\mathbf{x}) = \mathbf{z}_{ij}^{\mathrm{odm}} - h_{ij}^{\mathrm{odm}}(\mathbf{x}) \tag{8}$$

$$\mathbf{r}_{ij}^{\mathrm{LC}}(\mathbf{x}) = \mathbf{z}_{ij}^{\mathrm{LC}} - h_{ij}^{\mathrm{LC}}(\mathbf{x}) \tag{9}$$

where $\mathbf{z}_{ij}^{\mathrm{odm}}$ and $\mathbf{z}_{ij}^{\mathrm{LC}}$ are the relative pose measurements from odometry and loop closures, respectively, and $h_{ij}^{\mathrm{odm}}(\mathbf{x})$ and $h_{ij}^{\mathrm{LC}}(\mathbf{x})$ are the corresponding observation models.

As for the observation models, standard inverse composition of two poses [8] is used to describe both odometric and loop closing measurements.

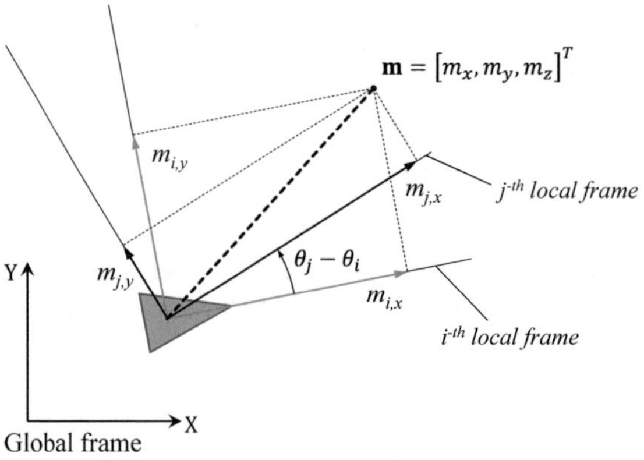

Fig. 1 The magnetic pivoting (MP) constraint. When the robot orientation changes from θ_i to θ_j without any translational motion, the global measurement of the magnetic field should be maintained. $\mathbf{m}$ is a vector of the magnetic measurement represented in a global frame and $\mathbf{m}_i$ and $\mathbf{m}_j$ are the vectors describing the same magnetic measurement based on the i-th and j-th frame, respectively

2 Magnetic Field Constraints

2.1 Magnetic Pivoting

Additional constraints can be made by exploiting the measurement characteristics of the magnetic fields under certain conditions. Assuming a robot runs on a 2D plane, the magnetic field vector represented in a global coordinate frame should be stationary when the robot rotates without any translational displacement. Figure 1 illustrates this situation. Since this constraint constrains the robot's rotational motion, let us call it a magnetic pivoting (MP) constraint. The MP constraint is added to the pose graph if the following condition is satisfied:

$$\begin{aligned} \sqrt{(x_j - x_i)^2 + (y_j - y_i)^2} &\leq T_{\mathrm{t}}^{\mathrm{MP}} \\ |\theta_j - \theta_i| &\geq T_{\mathrm{r}}^{\mathrm{MP}} \end{aligned} \tag{10}$$

where i, j are successive pose indices, and $T_{\mathrm{t}}^{\mathrm{MP}}$ and $T_{\mathrm{r}}^{\mathrm{MP}}$ are thresholds for translational and rotational motions, respectively, for magnetic pivoting.

The cost function E now can be written as the sum of three cost terms:

$$E(\mathbf{x}) = E_{\mathrm{odm}}(\mathbf{x}) + E_{\mathrm{LC}}(\mathbf{x}) + E_{\mathrm{MP}}(\mathbf{x}). \tag{11}$$

The cost $E_{\mathrm{MP}}(\mathbf{x})$ by MP constraints can be further described as

$$E_{\mathrm{MP}}(\mathbf{x}) = \sum_{\{i,j\}\in\mathcal{R}} \mathbf{r}_{ij}^{\mathrm{MP}}(\mathbf{x})^{\mathrm{T}} \boldsymbol{\Lambda}_{ij}^{\mathrm{MP}} \mathbf{r}_{ij}^{\mathrm{MP}}(\mathbf{x}) \tag{12}$$

where $\mathcal{R}$ is a set of constrained pose pairs by MP constraints, and $\mathbf{\Lambda}_{ij}^{\mathrm{MP}}$ is an information matrix for MP measurement. The residual $\mathbf{r}_{ij}^{\mathrm{MP}}$ is defined as

$$\mathbf{r}_{ij}^{\mathrm{MP}}(\mathbf{x}) = \mathbf{z}_{ij}^{\mathrm{MP}} - h_{ij}^{\mathrm{MP}}(\mathbf{x}). \tag{13}$$

Based on the relations described in Fig. 1, the MP measurement $\mathbf{z}_{ij}^{\mathrm{MP}}$ is treated as zero, and the observation model $h_{ij}^{\mathrm{MP}}(\mathbf{x})$ is defined as

$$h_{ij}^{\mathrm{MP}}(\mathbf{x}) = \mathbf{R}(\theta_j - \theta_i)\mathbf{m}_j - \mathbf{m}_i \quad (i < j) \tag{14}$$

where $\mathbf{m}_i = [m_{i,x}, m_{i,y}, m_{i,z}]^{\mathrm{T}}$ is the vector describing a magnetic measurement $\mathbf{m}$ based on the i-th frame. Since this observation model is the function of the robot's orientation, we can expect the heading corrections by the MP constraints. The information matrix $\mathbf{\Lambda}_{ij}^{\mathrm{MP}}$ is then calculated as

$$\mathbf{\Lambda}_{ij}^{\mathrm{MP}} = \begin{bmatrix} \sigma_B^2 & 0 & 0 \\ 0 & \sigma_B^2 & 0 \\ 0 & 0 & \sigma_B^2 \end{bmatrix}^{-1} \tag{15}$$

where σ_B is the standard deviation of the magnetic field measurement.

2.2 Sequence-Based Loop Closing

Since similarity between single magnetic measurements is too significant, a sequence of measurement is employed and this can greatly reduce the magnetic ambiguity. It is assumed that the robot's path includes linear segments due to the structured indoor environments. Whenever a pre-specified number of robot poses are made along a linear path, all the measurements of the three-component magnetic field are grouped into a single sequence vector $\mathbf{s}_i$. The corresponding pose nodes are also grouped as a super node and their indices are stored and managed. The reason for requiring a linear motion is that in this way we can restrain the magnetic fluctuation occurring by the robot's orientation change, which enhances the matching performance. The linear motion is defined by the following conditions:

$$\begin{aligned} \max\{\sigma(x_i, \ldots, x_{i+N_s-1}), \sigma(y_i, \ldots, y_{i+N_s-1})\} &\geq T_{\mathrm{t}}^{\mathrm{s}} \\ \sigma(\theta_i, \ldots, \theta_{i+N_s-1}) &\leq T_{\mathrm{r}}^{\mathrm{s}} \end{aligned} \tag{16}$$

where $\sigma(\cdot)$ is a function calculating the standard deviation, N_s is the number of poses in a sequence, and $T_{\mathrm{t}}^{\mathrm{s}}$ and $T_{\mathrm{r}}^{\mathrm{s}}$ are thresholds for the translational and rotational motions, respectively, for sequence generation.

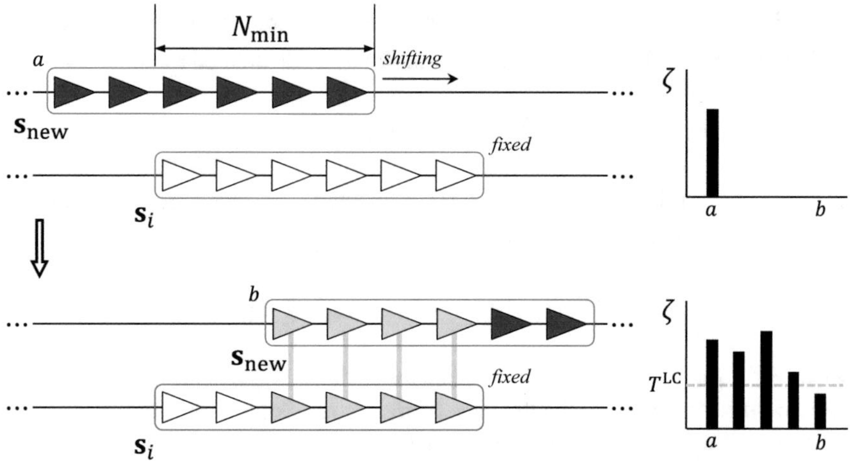

Fig. 2 Sequence-based matching. $\mathbf{s}_{\text{new}}$ are the newly generated sequence and $\mathbf{s}_i$ is an existing sequence to be matched. Shifting of the sequence is started from the point defined by N_{min}, and the similarity ζ of the overlapped sub-nodes are evaluated. When the ζ becomes below certain threshold T^{LC}, corresponding sub-nodes are constrained for loop closing

Whenever a new sequence is generated, a sequence-based matching is performed to detect a loop closing. Basically, a line by line matching is performed, shifting one line segment while the other is fixed (see Fig. 2 for illustration). To reduce the occurence of false positive matchings, at least N_{min} sub-nodes are required to be included in the matching. Since the matching can be done in the reverse direction, the reverse case is also examined using the reversed sequence.

During the matching procedure, the matching score is evaluated using the similarity measure ζ defined as follows:

$$\zeta = \frac{||\mathbf{s}_i - \mathbf{s}_{\text{new}}||_2}{\sqrt{\sigma(\mathbf{s}_i)\sigma(\mathbf{s}_{\text{new}})N_{\text{matched}}}} \tag{17}$$

where N_{matched} is the number of matched sub-nodes and $\mathbf{s}_{\text{new}}$ is the newly generated sequence vector.

When ζ is lower than a threshold T^{LC}, it can be assured that a loop closing has occurred and we check if the matching is done in the reverse direction. In the case of reverse matching, we set $\mathbf{z}_{ij}^{\text{LC}} = \pi$, and in the other case $\mathbf{z}_{ij}^{\text{LC}} = 0$. The corresponding sub-nodes are then added into a set of LC-constrained pose pairs Q and are optimized.

3 Robust Back-End

Now robust optimization techniques are applied to the proposed magnetic pose graph SLAM. This extension relieves the strict tuning of loop closure detection threshold and potentially occurring false positives. In this paper two types of robust back-ends are examined: (i) switchable constraint [11] and (ii) max-mixture [12] methods.

3.1 Switchable Constraint

In this approach, additional switch variable $s_{ij} \in \mathbf{s}$ is assigned to each LC constraint and this handles whether switch on or off the pertinent LC constraint. The formulation of LC residual (Eq. 9) is then modified to take into account the switch variable as follows:

$$\mathbf{r}_{ij}^{\mathrm{LC}}(\mathbf{x}) = s_{ij}(\mathbf{z}_{ij}^{\mathrm{LC}} - h_{ij}^{\mathrm{LC}}(\mathbf{x})). \tag{18}$$

And another cost term $E_{\mathrm{SP}}(\mathbf{s})$ related with switch value's prior is added into overall cost $E(\mathbf{x})$ (Eq. 11). $E_{\mathrm{SP}}(\mathbf{s})$ is formulated as follows:

$$E_{\mathrm{SP}}(\mathbf{s}) = \sum_{\{i,j\}\in\mathcal{P}} \frac{||s_{ij} - \gamma_{ij}||^2}{\sigma_{\mathrm{SP}}^2}. \tag{19}$$

where γ_{ij} is an initial value for switch variable and σ_{SP}^2 is a variance for this prior.

3.2 Max-Mixture

In max-mixture back-end, probability of each LC measurement is represented by a mixture of Gaussian distributions. Specifically, a significantly flat Gaussian component is employed to represent a null hypothesis of loop closure. In this way we can impose large covariances (i.e. small Λ_{null}) on those LCs with large residuals. And by using max-mixture instead of sum-mixture, we can maintain the original form of least-square formulations for graph optimization.

4 Experiments

4.1 Experimental Setup

In order to verify the proposed method, a set of experiments are conducted in real indoor environments. Figure 3a shows the Pioneer 3-AT robot used in the exper-

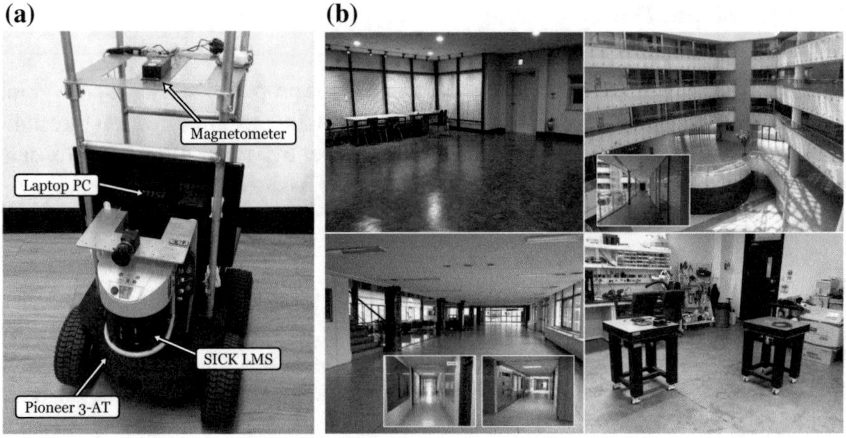

Fig. 3 **a** The robot used in the experiments. **b** SLAM test environments at KAIST, Daejeon, Korea. (From left to right, row-wise) W16 Building 3F, KI Building, W1-2 Building 1F, and Laboratory

Table 1 Obtained dataset summary

Dataset	Poses	Coverage (Approx.)	Features
W16	144	3×3 m	Repetitive small loops
KI 3F	1631	45×56 m	A few large loops
W1-2 #1	1211	42×10 m	Many small loops
W1-2 #2	839	40×50 m	Reverse loops
Laboratory	463	4×3 m	Many smooth curves

iments. It is equipped with 4 wheel encoders, one magnetometer (Honeywell's HMR2300) and one laser range finder (SICK LMS511). The laser range finder is merely used to obtain reference paths by employing a grid-based SLAM algorithm called GMapping [13]. The robot runs in four different indoor environments, shown in Fig. 3b. All the computations including reference path generation and sensor data acquisition are done in an Intel i7-4700MQ based laptop, equipped with 8 GB RAM. Table 1 summarizes the obtained datasets. Gathered sensor data are processed offline and in an incremental fashion to generate the pose graph SLAM results.

All the parameters used in the evaluation is summarized in Table 2. In the table, T_{LC} values are set to allow false positives. And $r_{(\cdot)}$ represents the ratio of noise covariances used in $\Lambda_{(\cdot)}$ to those used in Λ_{odm}.

Table 2 Parameters used in the evaluation of robust back-ends

Dataset	Front end			Switchable back-end			Max-mixture back-end		
	N_s	N_{min}	T_{LC}	r_{LC}	σ^2_{SP}	*epoch*	r_{LC}	r_{null}	*epoch*
W16	7	4	0.4	10^2	1	50	1	10^7	10
KI 3F	15	9	0.4	10^6	10	100	10^2	10^7	10
W1-2 #1	15	9	0.15	10^6	10	100+	1	10^9	15
W1-2 #2	20	12	0.15	10^6	10	50	1	10^7	10
Laboratory	7	4	0.4	10^2	10	100	1	10^7	10

4.2 Results

Figure 4 shows comparison between robust and non-robust optimization procedures. In the first column of figures, every detected loop closings are shown in thin grey lines. These loop closing constraints are generated by the sequence-based magnetic matching described in Sect. 2.2. Not all of these constraints are correct ones and some of them are false positives due to the magnetic ambiguity and measurement noise. When the graph is optimized, non-robust back-end fails in estimation due to the strong impact of the false magnetic constratins. The graph reaches to the state that cannot be recoverd even if more correct constraints are added. In contrast, both switchable and max-mixture back-ends overcome the false constraints and give correctly optimized robot paths. This means that even if magnetic field is slightly changed along the path, robust back-end can handle this and the robot can proceed to further explore the environment. It also greatly reduces efforts required in selection of front-end parameters. Finding appropriate values for $r_{(\cdot)}$ still requires additional efforts and this should be relieved in the future works.

5 Conclusion

In this paper, a solution to indoor SLAM problems is realized in a framework of pose graph optimization with the aid of the magnetic field anomalies. Two types of constraints for local heading corrections and global loop closings, respectively, were designed using the magnetic field measurements. A loop closing algorithm was designed based on the sequence of magnetic measurement and was applied to the pose graph optimization. Additional robust back-end methods are applied to improve the robustness of optimization. The experimental results showed that the proposed SLAM system with only wheel encoders and a single magnetometer offers comparable results with a reference-level laser-based SLAM system in terms of robot trajectory, thereby validating the feasibility of the magnetic field-based constraints.

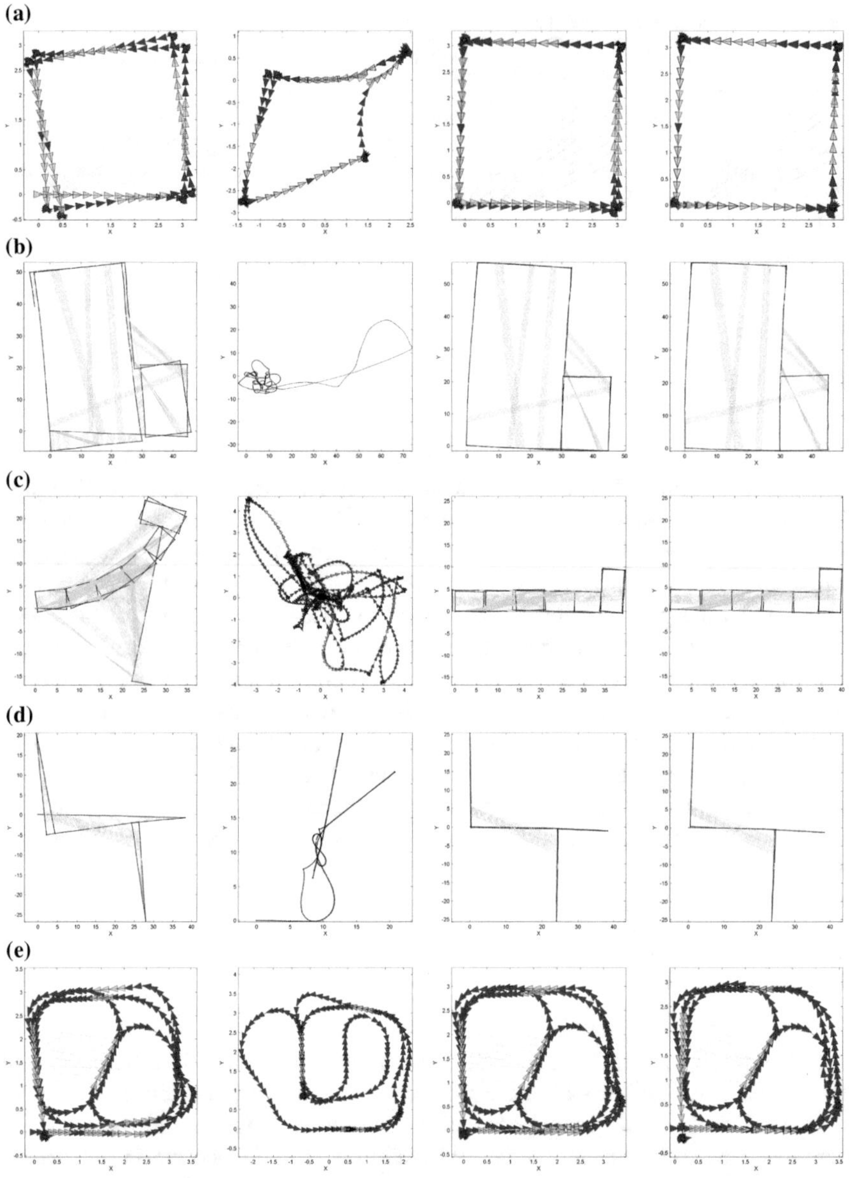

Fig. 4 From left to right: initial graph, non-robust optimization results, switchable back-end results, and max-mixture back-end results. **a** W16 dataset. **b** KI 3F dataset. **c** W1-2 1 F dataset #1. **d** W1-2 1 F dataset #2. **e** Laboratory dataset

Acknowledgements This research was supported by a grant from Endowment Project of "Development of fundamental technologies on underwater environmental recognition for long-range navigation and intelligent autonomous underwater navigation" funded by Korea Research Institute of Ships and Ocean Engineering(PES9390).

References

1. Thrun, S., Burgard, W., Fox, D.: Probabilistic Robotics. The MIT press, Cambridge, MA (2005)
2. Gozick, B., Subbu, K., Dantu, R., Maeshiro, T.: Magnetic maps for indoor navigation. IEEE Trans. Instrum. Meas. **60**(12), 3883–3891 (2011)
3. Burnett, J., Yaping, P.D.: Mitigation of extremley low frequency magnetic fields from electrical installations in high-rise buildings. Build. Environ. **37**, 769–775 (2002)
4. Angermann, M., Frassl, M., Doniecy, M., Julianyz, B., Robertson, P.: Characterization of the indoor magnetic field for applications in localization and mapping. In: Proceedings of IEEE International Conference on Indoor Positioning and Indoor Navigation (IPIN2012), Sydney, NSW pp. 1–9 (Nov. 2012)
5. Haverinen, J., Kemppainen, A.: A geomagnetic field based positioning technique for underground mines. In: Proceedings of IEEE International Symposium on Robotic and Sensors Environments (ROSE2011), Montreal, QC pp. 7–12 (Sept. 2011)
6. Vallivaara, I., Haverinen, J., Kemppainen, A., Roning, J.: Magenetic field-based SLAM method for solving the localization problem in mobile robot floor-cleaning task. In: Proceedings of IEEE International Conference on Advanced Robotics (ICAR2011), Tallinn, Estonia pp. 198–203 (June 2011)
7. Frassl, M., Angermann, M., Lichtenstern, M., Robertson, P., Julian, B., Doniec, M.: Magnetic maps of indoor environments for precise localization of legged and non-legged locomotion. In: Proceedings of IEEE/RSJ International Conference on Intelligent Robots and Systems (IROS2013), Tokyo, Japan pp. 913–920 (Nov. 2013)
8. Fernandez-Madrigeal, J., Claraco, J.L.: Simultaneous Localization and Mapping for Mobile Robots : Introduction and Methods. Information Science Reference, Hershey, PA (2013)
9. Kuemmerle, R., Grisetti, G., Strasdat, H., Konolige, K., Burgard, W.: G20: a general framework for graph optimization. In: Proceedings of IEEE International Conference on Robotics and Automation (ICRA2011), Shanghai, China pp. 3607–3613 (May 2011)
10. Jung, J., Oh, T., Myung, H.: Magnetic field constraints and sequence-based matching for indoor pose graph slam. Robot. Autonom. Syst. **70**, 92–105 (2015)
11. Sünderhauf, N., Protzel, P.: Switchable constraints for robust pose graph SLAM. In: Proceedings of IEEE/RSJ International Conference on Intelligent Robots and Systems (IROS2012), Vilamoura, Portugal pp. 1879–1884 (Oct. 2012)
12. Olson, E., Agarwal, P.: Inference on networks of mixtures for robust robot mapping. Int. J. Robot. Res. **32**(7), 826–840 (2013)
13. Grisetti, G., Stachniss, C., Burgard, W.: Improved techniques for grid mapping with rao-blackwellized particle filters. IEEE Trans. Robot. **23**(1), 34–46 (2007)

Multi-robot SLAM: An Overview and Quantitative Evaluation of MRGS ROS Framework for MR-SLAM

Mahmoud A. Abdulgalil, Mahmoud M. Nasr, Mohamed H. Elalfy, Alaa Khamis and Fakhri Karray

Abstract In recent years, multi-robot systems (MRS) have received attention from researchers in academia, government laboratories and industry. This research activity has borne fruit in tackling some of the challenging problems that are still open. One is multi-robot simultaneous localization and mapping (MR-SLAM). This paper provides an overview of the latest trends in tackling the problem of (MR-SLAM) focusing on Robot Operating System (ROS) enabled package designed to solve this problem through enabling the robots to share their maps and merge them over a WiFi network. This package had some out-of-date dependencies and worked with some packages that no longer exist. The package has been modified to handle these dependencies. The C-SLAM package was then tested with 2 robots using Gmapping and Hector SLAM packages. Quantitative metrics were used to evaluate the accuracy of the generated maps by comparing it to the ground truth map, including Map Score and Occupied/Free cells ratio.

Keywords Multi-robot SLAM · Map merging · Data association · SLAM
Occupancy grid map · ROS · C-SLAM

M. A. Abdulgalil · M. M. Nasr · M. H. Elalfy
Aerospace Engineering, Zewail City of Science and Technology, Giza, Egypt
e-mail: s-mahmoud.abdelgalil@zewailcity.edu.eg

M. M. Nasr
e-mail: s-mahmoud.nasr@zewailcity.edu.eg

M. H. Elalfy
e-mail: s-mohamed.elalfy@zewailcity.edu.eg

A. Khamis (✉) · F. Karray
Centre for Pattern Analysis and Machine Intelligence (CPAMI), University of Waterloo, Waterloo, ON, Canada
e-mail: akhamis@pami.uwaterloo.ca

F. Karray
e-mail: karray@uwaterloo.ca

J.-H. Kim et al. (eds.), *Robot Intelligence Technology and Applications 5*,
Advances in Intelligent Systems and Computing 751,
https://doi.org/10.1007/978-3-319-78452-6_15

1 Introduction

A key requirement for autonomous vehicles is situation awareness in indoor/outdoor, structured/unstructured and static/dynamic environments. Wrong beliefs about the vehicles state or environment state leads to wrong actions by the vehicle. In her description of a theoretical model of situation awareness [5], Endsley defined situation awareness as the perception of elements in the environment within a volume of time and space, the comprehension of their meaning, and the projection of their status in the near future. This model encompasses three main processes, namely, perception, comprehension and projection. Perception provides an awareness of multiple situational elements (objects, events, people, systems, environmental factors) and their current states (locations, conditions, modes, actions). Comprehension produces an understanding of the overall meaning of the perceived elements—how they fit together as a whole, what kind of situation it is, what it means in terms of ones mission goals. Projection produces an awareness of the likely evolution of the situation; its possible/probable future states and events. Simultaneous localization and mapping (SLAM) is a technology enabler for perception as the first level of situation awareness required for vehicle's autonomy.

Generally speaking, robot's behaviors can be classified into individual or *i*-level behaviors and group or *g*-level behaviors. A challenging problem in robotics is how to design *i*-level behaviors for a given desired *g*-level behavior. Though SLAM as *i*-level behavior is increasingly being considered a solved problem, Multi-robot SLAM (MR-SLAM) as *g*-level behavior still has room for extensive research to find stable mature solutions given the fact that collective behavior is not simply the sum of each participants behavior, as others emerge at the society level [12].

In this paper, an overview of the algorithms for solving mutli-robot SLAM problem is presented. A comprehensive and detailed analysis of each and every algorithm is not feasible due to the wide area covered by the scope of MR-SLAM problem and the huge variation in the way the problem is approached. However, an overview of the algorithms and a proposed taxonomy is presented. A ROS package to enable MR-SLAM is reviewed and modified to enable multiple robots sharing their maps and merge them over a WiFi network. The package had some out-of-date dependencies and worked with some packages that no longer exist. We were able to update the dependencies and change the package so that it works again. The C-SLAM package was then tested with 2 robots using Gmapping SLAM packages.

The remainder of the paper is organized as follows: Sect. 2 describes simultaneous localization and mapping (SLAM) problem followed by discussing multi-robot SLAM problem from different perspectives in Sect. 3. The modified ROS package for MR-SLAM is presented in Sect. 4. Test scenarios and experimental results are reported in Sect. 5. Finally, conclusion and future work are summarized in Sect. 6.

2 Simultaneous Localization and Mapping

Simultaneous Localization and Mapping (SLAM) is an essential feature of any autonomous vehicle. This feature allows the vehicle to explore an unknown environment, potentially with an unknown initial position, and be able to navigate in the environment while localizing itself with respect to its surroundings and building a model for the environment (an abstract form of a map). For an autonomous mobile robot, SLAM is the first step upon which other functionalities depend such as motion-planning and task allocation since all of these tasks require that the robot is able to localize itself and/or has a model of the environment. The difficulty of SLAM lies in the fact that a map is needed to localize the robot and the robot needs to localize itself first to build a map, in other words, it is another version of the 'chicken or the egg' causality dilemma. Other critical aspects are the uncertainties, elaborated in [40]. Single-robot SLAM in dynamic environments is still an active area of research. In such an environment, where the scene is not static, an additional dimension is added to problem, requiring estimation of the moving objects in the environment. Relevant work is provided in [16, 20].

2.1 Mathematical Formulation

The SLAM problem as stated above, and from the name, is nothing but a combination between the localization problem and the mapping problem. Stated in mathematical terms, the localization problem is the estimation of the state belief trajectory of the system from time 1 to time t, denoted as $x_{1:t}$, represented as a probability density function, given all of the observations of the surroundings of the environment $z_{1:t}$ and the control inputs issued to the robot(s) $u_{1:t-1}$,

$$p(x_{1:t}|z_{1:t}, u_{1:t-1})$$

When extended to SLAM, the above problem becomes the estimation of the state trajectory of the robot along with the map, denoted as m

$$p(x_{1:t}, m|z_{1:t}, u_{1:t-1}) \tag{1}$$

In some missions, however, and real-life situations, a single robot will not be able to fully meet the objectives at all or in the specified time requirements. This calls for a need to use a team of cooperating robots to perform such tasks. Nevertheless, the extension from a single-robot SLAM to a multi-robot SLAM is not straight-forward and in general requires additional layers of processing to be added to the algorithms to allow efficient collaboration between individual robots in serving the global goal of the mission.

2.2 Online Versus Offline

Offline SLAM is the the full estimation of both the trajectory of the robot since the beginning of time and the map, in mathematical terms this is equivalent to estimating the probability density function (pdf):

$$p(x_{1:t}, m|z_{1:t}, u_{1:t-1})$$

A simplified version of this problem is the online SLAM problem, stated as the problem of estimating the state belief at the current time only and the most recent map given the most recent control input, most recent measurement and the previous state belief. In mathematical terms, this is equivalent to estimating the pdf:

$$p(x_t, m|x_{t-1}, z_t, u_{t-1})$$

2.3 SLAM Approaches

There are two main commonly-used SLAM approaches, namely, *Smoothing-based SLAM* [6] and *Filtering-based SLAM*. The following subsections provide more details.

Smoothing SLAM Despite the fact that there is not much literature tackling this approach, there are some promising efforts done. One of these algorithms is *Incremental Smoothing and Mapping (iSAM)* [13]. In this approach, the SLAM problem is formulated into a least squares optimization problem via MAP. The resulting probabilistic framework is similar to Kalman Filter framework in some aspects. The algorithm is able to solve the full SLAM problem (offline SLAM), not just the online SLAM problem.

Filtering SLAM

a **Extended Kalman Filter**: Extended Kalman Filter is one of the most widely-used solutions to address the problem of localization in SLAM. It is a powerful solution due its ease of use and low computational cost. Despite the convenience of this algorithm, it is still an approximation of the real process since it can not work with non-Gaussian noise models, and thus in the presence of non-Gaussian noise, this filter fails to converge. Additionally, linearizing highly nonlinear systems may cause inconsistency. Detailed analysis of convergence and consistency of the EKF based SLAM is provided in [37].

b **Particle Filter**: A particle filter is another variant of the Bayes filter that uses random sampling instead of a closed-form for representing the estimated belief. This approach is computationally expensive because it requires generating a big number of samples (particles) to obtain acceptable results. Nevertheless, it is able to tackle problems that EKF can't tackle such as the case of the presence of non-

Gaussian noise and highly non-linear systems. An efficient implementation of Particle filter-based SLAM is introduced in [10]. This approach is based on a variant of the particle filter called *Rao-Blackwellized Particle Filters (RBPF)*, which greatly enhances algorithm performance by utilizing available gaussian substructures in the model.

Other Algorithms Other algorithms exist for state belief estimation, using other approaches such as Maximum Likelihood Estimation (MLE) methods. A comparative study of MLE algorithms is provided in [33]. Other approaches exploit Neural-networks based methods such as the work in [18, 34].

2.4 Map Representation

Different kinds of maps serve under different kinds of sensors used, the level of detail needed and computational power available. The 6 most common map types, in order of popularity are [35]:

a **Occupancy Grid Maps**: Occupancy grid maps (OGM) divide the map into cells which form a grid each with a binary random variable that shows whether or not the cell is occupied. An advantage of OGM is that it can work with dynamically changing environments due to moving objects and allow accurate modeling of it nevertheless.

b **Feature Maps**: Feature maps use distinctive features of the environment, each marked in the global map by its position. These features must be static and distinguishable from the rest of the environment to act as a landmark for the mapping process. This kind of mapping makes localization efficient but the process of data association is not an easy task.

c **Topological Maps** Using a series of nodes and arcs, topological maps are able to represent a compact form of the map showing only the abstract model. Compared to metric/grid maps, topological maps require less storage and computation time but they are harder to construct and may not valid for map matching and may suffer from perceptual aliasing in recognizing identical place.

d **Semantic Maps** A map similar to topological map is the semantic map. It is also an abstract map which makes it memory and computationally efficient. However, unlike topological maps, semantic maps include more details about the object and places and so it needs sufficient training before being used for high level planning.

e **Appearance Maps** Whenever a vision system is employed on the robot for SLAM, appearance maps are great candidate. In appearance maps, similar images are connected to each other and the weight of the edge between them represents the similarity between the images. Appearance maps are also extra useful when no odometer data can be obtained from the robot.

f **Hybrid Maps** Sometimes, a combination of more than one map representation, to include the advantages or properties of each map representation is required.

This merge of mapping techniques results in hybrid maps. Although this may increase the complexity and the amount of map processing, it makes handling inconsistencies and loop-closure much easier and more efficient.

3 Multi-robot SLAM

Multi-robot SLAM (MR-SLAM) is an extension of regular single-Robot SLAM exploiting several collaborating robots to map the unknown environment. However, this extension is not straight-forward since each robot maps the environment on its own using its local reference frame, the reference frames need to be related and a transformation between them has to be found if the maps are to be merged into a global map. Using MR-SLAM, the environment can be mapped much faster and the redundancy in the data reduces the uncertainty in the system state belief. In addition, if heterogeneous sensors are used on the separate robots, the map can include much more features than a single map created by a single robot. Needless to mention that this also makes the system robust to failure of one or more of the robots (specially if data processing is decentralized). The following subsections provide an overview of some of the main aspects of the MR-SLAM problem.

3.1 Centralization Versus Decentralization

In MRS, the question regarding where data processing occurs is the main question of which an answer dictates the rest of the aspects of the MR-SLAM problem. In general, data-processing can happen in on of two common fashions, *centralized* and *decentralized* data processing. In the centralized approach data is processed on a central agent of the collaborating group of robots, or on an external agent. The central unit is responsible for information fusion of all the robots, providing feedback to individual robots or sharing the fused information with them to better enhance their self-localization [7, 26]. On the other hand, decentralized systems usually consist of several interconnected agents that are managed by a coordination architecture, which makes these systems able to cope easily and efficiently with increasing number of agents. Many examples of this approach are found in literature, some are the work in [2, 8, 17, 39]. Despite the relative robustness of this configuration as compared to the normal centralized configuration, the decentralized configuration has its potential draw-backs such as large bandwidth requirements and network traffic.

Depending on the adopted approach of data processing, consistent assumptions about communication between robotic group individuals and also the type of the information shared between them are made.

3.2 Communication and Information Sharing

Another question mark regarding any proposed MR-SLAM algorithm is how the communication problem between robots is handled. All wireless communication systems have limited bandwidth, latency and coverage area. Many forms of communication configurations are proposed in the literature. Some assume, implicitly, the availability of a communication channel all the time or at least periodically such as the work in [26, 36]. Others, require no communication channel at all, but assume that the robots rendezvous together in order to calculate the relative transform between their co-ordinate systems, and combine the maps into a global mapsuch as [1, 9, 41]. Each of the approaches has its advantages and disadvantages. The first being that in the case of time-critical or task-critical missions, rendezvous might be non-feasible and thus the proposed approaches that require robot rendezvous are rendered as poor for the mission requirements. However, in the presence of network congestion and/or limited communication range, the assumption that a communication channel exists and that bandwidth permits data exchange is no longer valid, which renders the first approach as poor according to the requirements of the mission.

Another thing to consider is the type of information shared between the agents [4]. The robots can either share unprocessed *measurements* which is then expected to be processed by a central unit or on other robots, or *belief* states that is already processed. A merge between the two algorithms, to create a hybrid algorithm which automatically switches between the two according to the network state can found in [4].

3.3 Map Fusion

Finally, the last aspect to be considered regarding MR-SLAM is the map fusion process. Information fusion is one of the greatest challenges faced in the MR-SLAM problem. Robots need to merge the local maps they independently created in order to have a global unified map representing the surroundings. Several approaches has been proposed in the literature to solve the problem of map merging.

Some of the methods that already exist are optimization-based such as the work in [15], where a particle swarm optimization algorithm is adopted to solve the problem of map-merging. The experimental results show that the proposed algorithm enhances map merging accuracy. Nevertheless, since the problem is formulated as a particle optimization problem, it is generally going to require more processing time that other methods, and may not be feasible for real-time constraints.

In [1], a technique in which map fusion is dependent on rendezvous measurements was used. This requires that the two robots be in the line of sight of each other which enables them to measure the relative poses between them using techniques like visual detection in which detectable landmarks are mounted on robots. These measurements are then used to calculate a transform from the local coordinates of one robot to the other one. Once this transform is obtained, it turns into a trivial matter to merge the information obtained by each local map. The most challenging problem of this technique is the lack of accuracy of the relative pose measurement and the requirement that the robots be in line of sight.

Another approach proposed by [19] tackled the problem in a different way in which robots carry a camera that creates a stream of images. Map merging in that case is done by looking for the mutual locations that exist in both images from which relative poses could be deduced. Fox et al. proposed, in [8], a Graph-Based merging technique that combines graphical representations obtained by different robots. The process of merging is dependent on relative pose measurements and collocation which a location visited by more than one robot.

On the other hand, other approaches exist such as the one presented in [38], which provides an opportunistic approach for map merging. Each individual robot solves a local SLAM problem and stores a historical queue containing its history (sensor readings and state belief), and when a peer map is received, the robot issues a separate thread of execution that runs on the CPU, to localize the robot in the externally received map. Despite the algorithm efficacy and simplicity, it requires storage of past states of the robot, which grow linearly with time, thus leading to potential memory problems. Other common approaches for map matching include feature matching [25, 29] in which the relative transformation between the maps frames of references is done by matching features in both maps and inferring a transformation between them.

Also, inter-robot observation is used [1], where robots rendezvous and localize each other in their respective local maps, then communicate their respective relative location, which then becomes the transformation between their co-ordinate frames, and map merging is performed using that transformation. A comparative study for map merging techniques is provided in [31].

Another technique used by [3] only deals with occupancy grids which are commonly produced by SLAM implementations will be used in our implementation.

3.4 Mutli-robot SLAM Implementations

Several implementations to MR-SLAM problem are proposed in literature. Combining several approaches from the above aspects of MR-SLAM mentioned in the previous sections, a solution to the MR-SLAM problem is formed. Below, a few implementations along with some of their advantages and disadvantages are mentioned.

- **Extended Kalman Filter (EKF)**: EKF is the most widely used of the solutions in cooperative SLAM. It is a powerful solution yet it is worth to mention that it is a filter and not a smoothing algorithm, therefore the linearization of nonlinear systems may cause estimator inconsistency and uncertainty in pose. In [14], EKF was used to fuse the environment states of the vehicles where the poses of the detected vehicles are represented by a single system. This approach takes into account several constraints and provides proof that the system still yields good output. Reference [32] presented a new approach in which a single estimator, in the form of a Kalman filter, processes the available positioning information from all the members of the team and produces a pose estimate for every one of them. This single Kalman filter was divided into smaller communicating filters which are responsible of processing sensor data collected by its own robot.

- **Particle Filter (PF)**: Unlike the Kalman filter, although both are derived from Bayes filter, a particle filter is a non-parametric filter since it does not assume a certain distribution of noise the motion. In [21] FastSLAM is presented; a new solution which is powerful for mapping and localization. It works by decomposing the SLAM problem into a robot localization problem and a collection of landmark estimation problems that are conditioned on the robot pose estimate using a modified particle filter. Also, in [22], FastSLAM 2.0, a modified version of FastSLAM, is described. Convergence of this algorithm is validated for linear SLAM problems. FastSLAM 2.0 performs full and online SLAM, decreases logarithmic complexity in number of features, removed the need for parametrization, and also proved to be effective for loop closure. On the downside, it gets us back to the problem of computational load.
- **Maximum Likelihood Estimation (MLE)**: Given a set of observations, MLE can be used to find the parameters that maximize the likelihood of finding the set of observations provides. MLE serves as an alternative to particle filter when it comes to dealing with non-linearities in systems [11]. This approach is advantageous as it can deal with dynamic environments since it does not require the robots to be stationary, making it useful in missions of search and rescue. However, it is vital to keep in mind that MLE ignores prior probability, which is important for robot localization.
- **Maximum a Posteriori (MAP) Estimation**: [27] presents a distributed Maximum A Posteriori (MAP) estimator for multi-robot Cooperative Localization (CL) which -unlike centralized approaches- harnesses the computational and storage resources of all robots in the team to reduce the processing requirements. The MAP estimator works with non-linear system and improves the accuracy of the robots' pose estimates, acting as a smoother and a filter. Despite this great advantage, the computational complexity of MAP grows over time since all the previous states and observed measurements are involved in the optimization, making it unsuitable for real-time applications.

4 mrgs ROS Stack

The mrgs stack [36] is a combination of five modules incorporating four ROS packages forming a modular ROS stack. The stack depends on the maps being provided to it using external SLAM packages. The occupancy grids provided enters via the *Map Dam* node which crops the excess empty space and attaches the local pose of

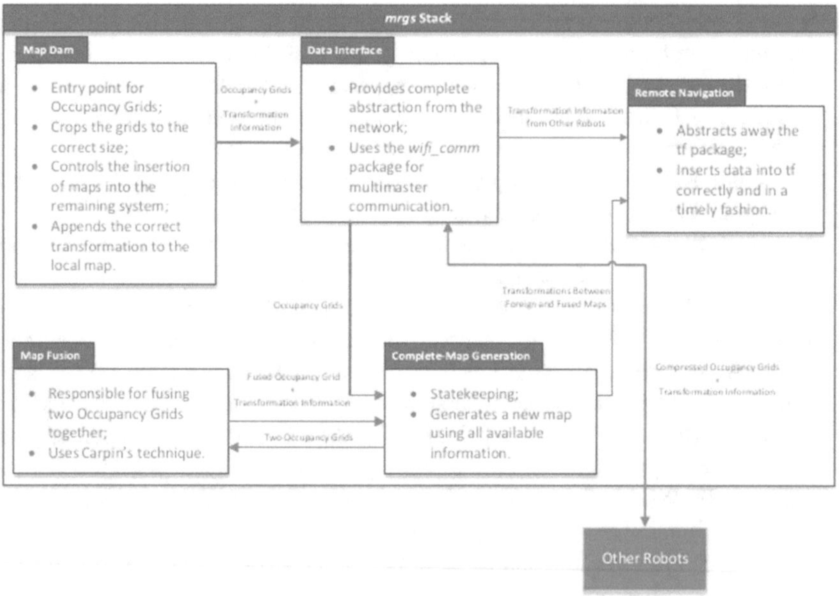

Fig. 1 Overview of the mrgs stack system design and data flow [36]

the robot that provided the map to it, and then sends the map to the *Data Interface* node. The *Data Interface* node compresses the map size, includes the robot's MAC address as an attachment and sends the map to the *Complete-Map Generation* node while receiving other maps simultaneously. The *Data Interface* node relies on the *wifi_comm* package to send the messages thus eliminating the need of rendezvous to share the maps. The *Complete-Map Generation* node's main task is to provide a single, global map of the environment through the use of the *Map Fusion* node which fuses the occupancy grid maps in pairs. The *Complete-Map Generation* node also stores the generated occupancy grid maps into tree-like structure. Finally the *Remote Navigation* node handles the transformations through taking care of the requirements of the *tf* package. Figure 1 shows the data flow and system design as described in the above lines.

4.1 SLAM Packages Used

Gmapping Gmapping[1] is the most widely used SLAM package in ROS due to the fact that it is the most one being updated. It is a laser-based algorithm that depends on odometry information as well as laser scans to construct occupancy grid maps. Gmapping is based on Rao-Blackwellized Particle filter and works by using the pose

[1] http://wiki.ros.org/gmapping.

which is continuously updated by the motion model and the estimated odometry. The weights of the particles are updated according to which samples gave more effective results [30]. The formula used to find how well each particle represents the posterior probability is as follows:

$$N_{eff} = \frac{1}{\sum_{i=1}^{N} (\tilde{\omega}^i)^2} \tag{2}$$

where $\tilde{\omega}^i$ is the normalized weight of the particle i. Through using the mentioned method and utilizing RBPF, it is possible to obtain results of good accuracy using a low number of particles. This directly means that the computational requirements are lowered without sacrificing performance.

4.2 Map Fusion

After each robot generates its own map using one of the mentioned techniques, all the generated maps require to be aligned together and merged to a single, global map. This is done by several techniques. Map merging is done on two maps at a time. The mentioned discussion is restricted to occupancy grid maps as they are type used in our ROS implementation. Once two maps are given that need to be aligned and fused together, the general approach is to try rotations that may result in correct alignment of the maps and check the generated map for consistency. Generally speaking, the only thing required, to merge two given maps M1 and M2 together is to check every possible rotational and translational transform and check if it gives a good match. This transformation (T) is simply a combination of rotation (ψ) followed by a translation along the x and y axis of magnitude (Δ_x and Δ_y):

$$T(\Delta_x, \Delta_y, \psi) = \begin{bmatrix} cos\psi & -sin\psi & \Delta_x \\ sin\psi & cos\psi & \Delta_y \\ 0 & 0 & 1 \end{bmatrix} \tag{3}$$

After the application of this transform on a map, a new map is obtained M' where $M = TM$ given that M is the old map. A large number of transformations could be applied. However, it is obvious that such an approach is both computationally expensive and time inefficient, thus a metric should be formulated to decide which transformation is better. This is done by defining an acceptance index ω; the higher the value, the higher the agreement is. It is given by:

$$\omega(M_1, M_2) = \frac{agr(M_1, M_2)}{agr(M_1, M_2) + dis(M_1, M_2)} \tag{4}$$

where:

- M_1: Map 1—The original map

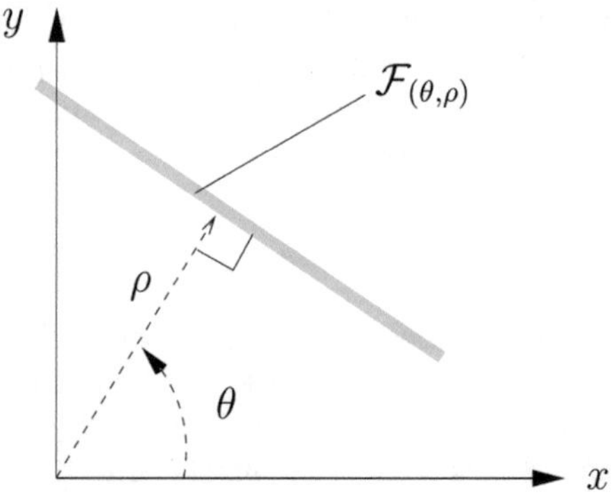

Fig. 2 Hough transform line $\mathcal{F}(\rho, \theta)$ representation

- M_2: Map 2—The second map (after rotation and translation)
- *agr*: The agreement between the maps (number of cells in M_1 and M_2 that match each other)
- *dis*: The disagreement between the maps (cells assumed to be the same but contain different values for occupancy).

Let T_1 and T_2 are two transformations that are applied to overlap M_1 and M_2, T_1 is preferred if

$$\omega(M_1, T_1M_2) > \omega(M_1, T_2M_2)$$

The goal is to determine the transformation that maximizes acceptance index. The overall transformation is then done on two separate steps the first is to determine the convenient rotation then the translation can be deduced. In this context, the map merging/fusion can be discussed as an optimizations problem.

The method followed by [36] in map merging/fusion is called *The Hough Transform*. The Hough Transform is a method that works on detecting parametric shapes such as lines, ellipses, or circles. This approach utilizes the algorithm called *The Hough Spectrum* which works on a spectra-based method to determine the appropriate rotations to align the maps. The rotations given by the algorithm then need to be compared to find the most promising transformation. Lines could be represented in polar coordinates with the following formula:

$$\rho = xcos\theta + ysin\theta$$

where (as shown in Fig. 2):

- ρ: the distance of the line from the origin
- θ: the angle between the x axis and the normal from the line to the origin.

Using Discretized Hough transform (DHT) the map grid could be divided into a grid of accumulators with n_θ columns and n_ρ rows which together forms a matrix called $\mathcal{H}$. Each entry of (ρ,θ) in this matrix represents a line in the occupancy grid. Making use of cross circular relations between the two spectra generated for each map the needed rotation is defined then the translations could be deduced. Hough Spectrum then could be defined in terms of $\mathcal{H}$:

$$\mathcal{H}S(k) = \sum_{i=1}^{n_\rho} \mathcal{H}(i,k)^2 \qquad 1 \leq k \leq n_\theta \tag{5}$$

4.3 Technical Overview of ROS Implementation

Our contribution in ROS is mainly in fixing dependencies issues in the work by [36]. The system provided in their work, provides highly modular infrastructure for multi-robot SLAM in which the module responsible for map merging is fully decoupled from the rest of the system infrastructure. However, this infrastructure was built on very old ROS APIs, thus it was not possible to utilize the power of the stack because the system lacked dependencies. We managed to fix these dependencies by modifying and building from source the relevant packages.

The following is a technical overview of the details associated with this contribution

a ***foreign-relay***: This package provides APIs that allow multi-master communication. The underlying system architecture of ROS is based on a remote procedure call protocol named XMLRPC which utilizes XML for encoding its procedure calls and uses HTTP for transferring these calls over compatible networks. The central node in a normal ROS network is the master node. The master node API provides ways to communicate with the master through XMLRPC protocol. *foreign relay* uses the same API to register processes that are on different nodes on the same wifi network with local ROS masters, allowing for communication between different masters running on different machines.

b ***wifi_comm***: This package is built upon *foreign relay*'s API, and it is used explicitly in the data interface node to decouple system internal processing from external communication between ROS masters using OLSR (Optimized Link State Routing Protocol) which is a type of routing protocols employed in computer networks that is optimized for mobile and ad-hoc networks. Exploiting this nature of OLSR communication framework, it was feasible to develop a framework suitable for a network of multiple robots in a mission with no prior preparations or infrastructure. Such a network configuration is not node-critical, in other words, losing a node is not critical to the overall performance of the network. In addition. adding a new node to the network is not hard.

c ***mrgs_stack***: Finally, the package itself, which was dependent on the packages *wifi-comm* and *foreign-relay* didn't work out of the box. It used an old build system, thus we had to re-write the build configuration files.

All these packages were outdated, meaning that they were no longer compatible with newer versions of ROS. We had to fix the dependency-issues, and rewrite their build configurations to be able to use them with *ROS Kinetic* version.

5 Experimental Results

5.1 Experiment Setup

The package was run on two separate machines; the first was an i7-core 2.4 GHz laptop with 16 GB RAM running Ubuntu 16.04 and ROS Kinetic version and the other was an i7-core 2.1 GHz laptop with 8 GB RAM running the same version of Linux and ROS. The simulation environment was the known Willow Garage world available in Gazebo models. Two separate robots were spawned (each on a separate machine), in the same gazebo world, running GMapping package for SLAM. The used sensor was a simulated LIDAR sensor with 360° FoV and 30 m range. Odometry values were obtained from Gazebo to ensure that they are correct, in order to test the effectiveness of the matching algorithm only. The maps where then compared using several algorithms mentioned shortly using MATLAB 2014a.

5.2 Evaluation Metrics

Until recently, quantitative measures were not used in map comparisons in robotics and instead human qualitative analysis was the one to decide if a map is a good representation of the environment. However, to be able to compare the map to the ground truth map, we need a quantitative measure.

1—Map Score: In [24], a comparison measure called the Map Score is introduced which aims to come up with a value that indicates how much a generated map matched another. This method is based on comparing the map on a cell-by-cell basis [28] and therefore first needs the two maps to be registered and have the same position and orientation. To test our generated maps to the ground truth, the generated map was registered and an example of the result can be seen in Fig. 3.

Map Score is defined in according to the following equation:

$$MapScore = \sum_{m_{X,Y} \in M, n_{X,Y} \in N} (m_{X,Y} - n_{X,Y})^2 \tag{6}$$

where:

- $m_{X,Y}$ is the value of the cell at position (X,Y) in map M
- $n_{X,Y}$ is the value of the cell at position (X,Y) in map N

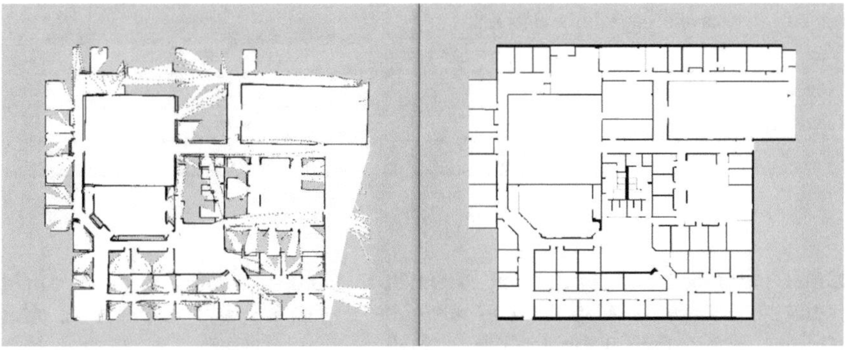

Fig. 3 Generated map using mrgs package (left) versus ground truth map (right)

2—Occupied/Free Cell Ratio: The map was also tested using a very intuitive method known as the occupied/free cells ratio. This is defined as the ratio of occupied and free cells in the generated map to the occupied and free cells in the ground truth map respectively. Therefore, the closer the value is to 1, the better the results. The equations to find the ratios are:

$$OccupiedCellsRatio = \frac{\sum(cell_{testmap,occ,true})}{\sum(cell_{referencemap,occ})} \tag{7}$$

$$FreeCellsRatio = \frac{\sum(cell_{testmap,free,true})}{\sum(cell_{referencemap,free})} \tag{8}$$

5.3 *Results and Discussion*

We can give a qualitative measure as was mentioned in [23, 28] to say that the map generated gives a good representation of the environment. However, this is still not enough when it comes to certain applications in which accuracy is critical, thus there must be the need for the quantitative metrics. As the Map score decreases, the better the match between the maps is. It can be seen intuitively though, that the score will greatly depend on the size of the maps being used (number of pixels). To account for this, a normalized map score was obtained [28] by generating the worst Map score possible through creating a new map with the occupied cells marked as empty and vice versa and the unexplored cells marked as either of them (both will give same result).

After implementing the algorithm in MATLAB, the final results obtained from robot 1 and from robot 2 for the map score was 26795. Given that we have 246950 pixels, we get an average difference of around 0.11, resulting in a great approximation

Table 1 Results of occupied/free cells ratio

Map	Occupied cells ratio	Free cells ratio
Final local map by robot 1	0.7415	0.4904
Final local map by robot 2	0.7079	0.3434
Final merged map	0.9362	0.8237

of the environment and ground truth. The map matching algorithm, however, can be improved to make the Map Score even smaller so that it would be useful for other missions such as path planning. The normalized score was also seen to drop from 0.16 to 0.12 between the final local map and the merged map for each robot, thus showing a 22% increase in the efficiency of the map. The results for Occupied/Free Cell ratios are shown in Table 1 for the final local map and the merged map using the package. An increase in both ratios can be seen therefore demonstrating the great use of MR-SLAM and the effectiveness of the package as well.

6 Conclusion

We summarize and provide an overview of the SLAM approaches. We also shed a spotlight on some of the available ROS packages that are able to successfully solve the SLAM problem. In addition, a ROS package in charge of allowing multi-robots to communicate over a WiFi network and build a global map together was out-of-date and not working. We were able to update the *mrgs* package to work with the currently available SLAM packages. We also evaluated the accuracy of the maps generated compared to ground truth using the map score metric and occupied/free cell ratio metrics. In literature, few work is done on machine learning and evolutionary algorithms approaches for map merging in MR-SLAM problem. Our future work will target implementing genetic algorithm-based map matching algorithm and also deep learning based algorithms. We will work on extending the capabilities of the mrgs stack framework to incorporate the mentioned approaches. A more detailed quantitative study is also to be performed on the different approaches for map merging that could compare the accuracy of the map merging algorithms as a function of the parameters involved such as the overlap between individual maps. We shall also seek verifying the quantitative evaluation on real-world data or hardware.

References

1. Andersson, L.A.A., Nygards, J.: On multi-robot map fusion by inter-robot observations. In: 12th International Conference on Information Fusion, FUSION'09, pp. 1712–1721 (2009)

2. Bresson, G., Aufrre, R., Chapuis, R.: Consistent multi-robot decentralized slam with unknown initial positions. In: 16th International Conference on Information Fusion (FUSION), pp. 372–379 (2013)
3. Carpin, S.: Fast and accurate map merging for multi-robot systems. Auton. Robot. **25**(3), 305–316 (2008). https://doi.org/10.1007/s10514-008-9097-4
4. Chang, C.K., Chang, C.H., Wang, C.C.: Communication adaptive multi-robot simultaneous localization and tracking via hybrid measurement and belief sharing. In: 2014 IEEE International Conference on Robotics and Automation (ICRA), pp. 5016–5023 (2014). https://doi.org/10.1109/ICRA.2014.6907594
5. Endsley, M.: Toward a theory of situation awareness in dynamic systems. Hum. Factors **37**(1), 32–64 (1995). https://doi.org/10.1518/001872095779049543
6. Fernandez-Madrigal, J.-A., Blanco Claraco, J.L.: Simultaneous Localization and Mapping for Mobile Robots, 1st edn. Information Science Reference (2013)
7. Forster, C., Lynen, S., Kneip, L., Scaramuzza, D.: Collaborative monocular slam with multiple micro aerial vehicles. In: IEEE/RSJ International Conference on Intelligent Robots and Systems, pp. 3962–3970 (2013). https://doi.org/10.1109/IROS.2013.6696923
8. Fox, D., Ko, J., Konolige, K., Limketkai, B., Schulz, D., Stewart, B.: Distributed multi-robot exploration and mapping. Proc. IEEE **94**(7), 1325–1339 (2006). https://doi.org/10.1109/JPROC.2006.876927
9. Gong, C., Tully, S., Kantor, G., Choset, H.: Multi-agent deterministic graph mapping via robot rendezvous. In: IEEE International Conference on Robotics and Automation (ICRA), pp. 1278–1283 (2012). https://doi.org/10.1109/ICRA.2012.6225274
10. Grisetti, G., Stachniss, C., Burgard, W.: Improved techniques for grid mapping with rao-blackwellized particle filters. IEEE Trans. Robot. **23**(1), 34–46 (2007). https://doi.org/10.1109/TRO.2006.889486
11. Howard, A., Matark, M.J., Sukhatme, G.S.: Localization for mobile robot teams using maximum likelihood estimation. In: IEEE/RSJ International Conference on Intelligent Robots and Systems, vol. 1, pp. 434–439 (2002). https://doi.org/10.1109/IRDS.2002.1041428
12. Jacques, P., Deneubourg, J.L.: From Individual to Collective Behavior in Social Insects, 1st edn. Birkhäuser Verlag, Basel (1987)
13. Kaess, M., Ranganathan, A., Dellaert, F.: Isam: incremental smoothing and mapping. IEEE Trans. Robot. **24**(6), 1365–1378 (2008). https://doi.org/10.1109/TRO.2008.2006706
14. Karam, N., Chausse, F., Aufrere, R., Chapuis, R.: Localization of a group of communicating vehicles by state exchange. In: IEEE/RSJ International Conference on Intelligent Robots and Systems, pp. 519–524 (2006). https://doi.org/10.1109/IROS.2006.282028
15. Lee, H.C., Roh, B.S., Lee, B.H.: Multi-hypothesis map merging with sinogram-based pso for multi-robot systems. Electron. Lett. **52**(14), 1213–1214 (2016). https://doi.org/10.1049/el.2016.1041
16. Lee, S., Cho, H., Yoon, K.J., Lee, J.: Mapping of Incremental Dynamic Environment Using Rao-Blackwellized Particle Filter, pp. 715–724. Springer, Berlin (2013). https://doi.org/10.1007/978-3-642-33926-4_68
17. Leung, K.Y.K., Barfoot, T.D., Liu, H.H.T.: Distributed and decentralized cooperative simultaneous localization and mapping for dynamic and sparse robot networks. In: IEEE International Conference on Robotics and Automation (ICRA), pp. 3841–3847 (2011). https://doi.org/10.1109/ICRA.2011.5979783
18. Li, Q.L., Song, Y., Hou, Z.G.: Neural network based fastslam for autonomous robots in unknown environments. Neurocomputing **165**, 99–110 (2015). https://doi.org/10.1016/j.neucom.2014.06.095, http://www.sciencedirect.com/science/article/pii/S0925231215004312
19. Li, Z., Riaz, S., Chellali, R.: Visual place recognition for multi-robots maps merging, pp. 1–6. IEEE (2012)
20. Miller, I., Campbell, M.: Rao-blackwellized particle filtering for mapping dynamic environments. In: Proceedings of the IEEE International Conference on Robotics and Automation, pp. 3862–3869 (2007). https://doi.org/10.1109/ROBOT.2007.364071

21. Montemerlo, M., Thrun, S., Koller, D., Wegbreit, B.: Fastslam: A factored solution to the simultaneous localization and mapping problem. In: Eighteenth National Conference on Artificial Intelligence, pp. 593–598. American Association for Artificial Intelligence, Menlo Park, CA, USA (2002). http://dl.acm.org/citation.cfm?id=777092.777184
22. Montemerlo, M., Thrun, S., Koller, D., Wegbreit, B.: Fastslam 2.0: An improved particle filtering algorithm for simultaneous localization and mapping that provably converges. In: Proceedings of the International Joint Conference on Artificial Intelligence (IJCAI), pp. 1151–1156 (2003)
23. Moravec, H., Elfes, A.: High resolution maps from wide angle sonar. In: Proceedings of the IEEE International Conference on Robotics and Automation, vol. 2, pp. 116–121 (1985). https://doi.org/10.1109/ROBOT.1985.1087316
24. Moravec, M.: Robot Evidence grids (1996)
25. un Nabi Jafri, S.R., Ahmed, W., Ashraf, Z., Chellali, R.: Multi robot slam for features based environment modelling. In: IEEE International Conference on Mechatronics and Automation, pp. 711–716 (2014). https://doi.org/10.1109/ICMA.2014.6885784
26. un Nabi Jafri, S.R., Li, Z., Chandio, A.A., Chellali, R.: Laser only feature based multi robot slam. In: 12th International Conference on Control Automation Robotics Vision (ICARCV), pp. 1012–1017 (2012). https://doi.org/10.1109/ICARCV.2012.6485296
27. Nerurkar, E.D., Roumeliotis, S.I., Martinelli, A.: Distributed maximum a posteriori estimation for multi-robot cooperative localization. In: IEEE International Conference on Robotics and Automation, ICRA'09, pp. 1402–1409 (2009). https://doi.org/10.1109/ROBOT.2009.5152398
28. OSullivan, S.: An empirical evaluation of map building methodologies in mobile robotics using the feature prediction sonar noise filter and metric grid map benchmarking suite. Master's thesis, University of Limerick (2003)
29. Park, J., Sinclair, A.J., Sherrill, R.E., Doucette, E.A., Curtis, J.W.: Map merging of rotated, corrupted, and different scale maps using rectangular features. In: IEEE/ION Position, Location and Navigation Symposium (PLANS), pp. 535–543 (2016). https://doi.org/10.1109/PLANS.2016.7479743
30. Pedro, J.: Smokenav simultaneous localization and mapping in reduced visibility scenarios. Master's thesis, University of Coimbra (2013)
31. Romero, V.A., Costa, O.L.V.: Map merging strategies for multi-robot fastslam: a comparative survey. In: Latin American Robotics Symposium and Intelligent Robotic Meeting (LARS), pp. 61–66 (2010). https://doi.org/10.1109/LARS.2010.20
32. Roumeliotis, S.I., Bekey, G.: Distributed multirobot localization. IEEE Trans. Robot. Autom. **18**(5), 781–795 (2002). https://doi.org/10.1109/tra.2002.803461
33. Rybski, P.E., Roumeliotis, S.I., Gini, M., Papanikolopoulos, N.: A comparison of maximum likelihood methods for appearance-based minimalistic slam. In: Proceedings of the IEEE International Conference on Robotics and Automation, ICRA'04, vol. 2, pp. 1777–1782 (2004).https://doi.org/10.1109/ROBOT.2004.1308081
34. Saeedi, S., Paull, L., Trentini, M., Li, H.: Neural network-based multiple robot simultaneous localization and mapping. IEEE Trans. Neural Netw. **22**(12), 2376–2387 (2011). https://doi.org/10.1109/TNN.2011.2176541
35. Saeedi, S., Trentini, M., Seto, M., Li, H.: Multiple-robot simultaneous localization and mapping: a review. J. Field Robot. **33**(1), 3–46 (2015). https://doi.org/10.1002/rob.21620
36. dos Santos, G.: A cooperative slam framework with efficient information sharing over mobile ad hoc networks. Master's thesis, University of Coimbra (2014)
37. Shoudong Huang, G.D.: Convergence and consistency analysis for extended kalman filter based slam. IEEE Trans. Robot. **23**(5), 1036–1049 (2007). https://doi.org/10.1109/tro.2007.903811
38. Stipes, J., Hawthorne, R., Scheidt, D., Pacifico, D.: Cooperative localization and mapping. In: IEEE International Conference on Networking, Sensing and Control, pp. 596–601 (2006). https://doi.org/10.1109/ICNSC.2006.1673213
39. Swinnerton, J., Brimble, R.: Autonomous self-localization and mapping agents. In: 7th International Conference on Information Fusion, vol. 2 (2005), pp. 7. https://doi.org/10.1109/ICIF.2005.1591990

40. Thrun, S., Burgard, W., Fox, D.: Probabilistic Robotics, 1st edn. MIT press, Cambridge (2005)
41. Zhou, X.S., Roumeliotis, S.I.: Multi-robot slam with unknown initial correspondence: The robot rendezvous case. In: IEEE/RSJ International Conference on Intelligent Robots and Systems, pp. 1785–1792 (2006). https://doi.org/10.1109/IROS.2006.282219

A Reinforcement Learning Technique for Autonomous Back Driving Control of a Vehicle

H. M. Kim, X. Xu and S. Jung

Abstract In this paper, a reinforcement learning technique is applied for a vehicle to learn and copy the trace of the forward driving path so that the vehicle can drive the path back to the initial position autonomously. Reinforcement learning algorithm is used to correct the trajectory tracking control where there are both lateral and longitudinal slips of the vehicle. Simulation studies of following the forward path a vehicle passed backward are conducted.

1 Introduction

Autonomous driving control of a vehicle becomes an important topic in the era of Industry 4.0. Based on the sensor information, data cloud is formed and intelligent decision tools are used to determine the optimal solutions for autonomous control of the trajectory following task for a vehicle. Successful driving performances depend upon accurate and rich sensor data to drive the optimal solutions by intelligent decision makers [1].

It is very difficult for the vehicle to achieve the completely autonomous driving control without human intervention at this stage since the safety issue of the human driver is the first priority. A terrible disaster of a Tesla car driver has been reported due to the minor sensing problem in an autonomous driving mode [2]. This accident warns auto-industries how dangerous small faults in sensing to a driver's life are.

H. M. Kim · S. Jung (✉)
Department of Mechatronics Engineering, Intelligent Systems and Emotional Engineering (ISEE) Lab, Chungnam National University, Daejeon 34134, Korea
e-mail: jungs@cnu.ac.kr

H. M. Kim
e-mail: khm9527@naver.com

X. Xu
College of Mechatronics and Automation, National University of Defense Technology Changsha, Changsha, China
e-mail: xinxu@nudt.edu.cn

J.-H. Kim et al. (eds.), *Robot Intelligence Technology and Applications 5*,
Advances in Intelligent Systems and Computing 751,
https://doi.org/10.1007/978-3-319-78452-6_16

To overcome those faults, many attempts for autonomous driving performances have been made by auto industries. Their partially autonomous driving control tasks, in other words, autonomous driving control tasks with a human involvement have been successfully performed.

Related research on partial autonomous driving tasks such as an automatic parking task of a mobile robot using fuzzy logic [3] and an autonomous driving of a vehicle on highway using a reinforcement learning method [4].

In the framework of the partial autonomous driving control, in this paper, the autonomous back driving control task of the vehicle is presented. When a driver can meet the situation where the road is blocked and is narrow, the driver has to drive back. However, back driving in the narrow road is more difficult than forward driving especially for the beginners and the elderly.

Therefore, in this paper, the autonomous back driving control algorithm is developed. The vehicle is required to follow the previous path where it has been moved back autonomously to arrive the starting location. This autonomous technique helps drivers to move back in the narrow pathway with ease when the front way is blocked. The vehicle has to memorize the previous path for a certain distance. Although the vehicle remembers the path it followed by storing the positional data, the returning path may not be exact due to the slip on the road conditions. The deviation from the stored path should be compensated.

The reinforcement learning algorithm is applied to learn the path under the slip conditions of both the longitudinal and lateral directions [5]. Simulation studies of the autonomous back driving control task are performed.

2 Vehicle Dynamics

A vehicle can be modeled as a bicycle, which is shown in Fig. 1 [6]. X axis is the longitudinal direction and Y axis is the lateral direction. In the bicycle model in Fig. 1, the steering angle is only present on the front wheels. The steering angle δ is used as an input to the system. In Fig. 1, C_f and C_r are cornering stiffnesses, I_z is the moment of inertia, m is a mass, l_f and l_r are the distances from the COG to the center of wheels, δ is a steering angle. F_{yr} and F_{yf} are lateral forces, α_f and α_r are slip angles, $\dot{\psi}$ is the yaw rate, V_x is the longitudinal velocity, and V_y is the lateral velocity

The dynamics equation for the y-axis is obtained as

$$ma_y = F_{xf} sin\delta + F_{yf} cos\delta + F_{yr} \quad (1)$$

where a_y is the lateral acceleration. The acceleration in the COG for the y-axis is

$$a_y = \dot{V}_y + V_x\dot{\psi} \quad (2)$$

Substituting Eq. (2) into Eq. (1) and assuming that the steering angle δ is very small such as $(cos\delta \approx 1, sin\delta \approx \delta)$ yields the following Eq. (3).

Fig. 1 Bicycle model of the vehicle

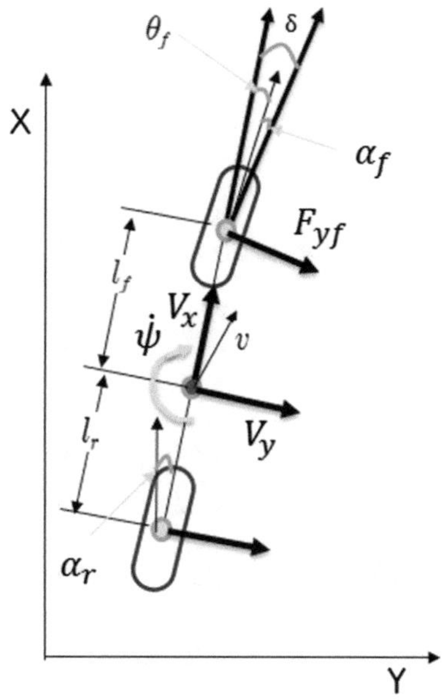

$$m(\dot{V}_y + V_x\dot{\psi}) = F_{yf} + F_{yr} \tag{3}$$

The equation for the z-axis is obtained as

$$I_z\ddot{\psi} = l_fF_{yf} - l_rF_{yr} \tag{4}$$

The slip angles of the front and rear wheels are expressed by Eqs. (5) and (6).

$$\alpha_f = \delta - \theta_f \tag{5}$$

$$\alpha_r = -\theta_r \tag{6}$$

In Fig. 1, $\tan\theta_f$ and $\tan\theta_r$ are expressed by Eqs. (7) and (8).

$$\tan\theta_f = \frac{V_y + l_f\dot{\psi}}{V_x} \tag{7}$$

$$\tan\theta_r = \frac{V_y - l_r\dot{\psi}}{V_x} \tag{8}$$

Assuming that θ_f and θ_r are very small, the lateral forces are described as the following equations.

$$\mathrm{F_{yf}} = 2\mathrm{C_f}\alpha_\mathrm{f} = 2\mathrm{C_f}\left(\delta - \frac{V_y + l_f\dot{\psi}}{V_x}\right) \tag{9}$$

$$\mathrm{F_{yr}} = 2\mathrm{C_r}\alpha_\mathrm{r} = -2\mathrm{C_f}\frac{V_y - l_r\dot{\psi}}{V_x} \tag{10}$$

Substituting (9) and (10) into (1) and (4) yields the dynamic equations as

$$\dot{V}_y = \frac{2C_{\alpha f}\delta}{m} - \frac{2(C_{\alpha f} + C_{\alpha r})}{mV_x}V_y - \left(V_X + \frac{2(C_{\alpha f}l_f - C_{\alpha r}l_r)}{mV_x}\right)\dot{\psi} \tag{11}$$

$$\ddot{\psi} = \frac{l_f 2C_{\alpha f}\delta}{I_z} - \frac{2(C_{\alpha f}l_f - C_{\alpha r}l_r)}{I_Z V_x}V_y - \frac{2\left(C_{\alpha f}l_f^2 + C_{\alpha r}l_r^2\right)}{I_Z V_x}\dot{\psi} \tag{12}$$

Equations (11) and (12) can be expressed in a state space form.

$$\begin{bmatrix} \dot{V}_y \\ \ddot{\psi} \end{bmatrix} = \begin{bmatrix} \frac{-2(C_{\alpha f} + C_{\alpha r})}{mV_x} & -V_X - \frac{2(C_{\alpha f}l_f - C_{\alpha r}l_r)}{mV_x} \\ \frac{-2(C_{\alpha f}l_f - C_{\alpha r}l_r)}{I_Z V_x} & -\frac{2(C_{\alpha f}l_f^2 + C_{\alpha r}l_r^2)}{I_Z V_x} \end{bmatrix} \begin{bmatrix} V_y \\ \dot{\psi} \end{bmatrix} + \begin{bmatrix} \frac{2C_{\alpha f}}{m} \\ \frac{l_f 2C_{\alpha f}}{I_z} \end{bmatrix}\delta \tag{13}$$

In the longitudinal model of the vehicle, the longitudinal slip is implemented using the Magic Formula graph. Applying the Magic Formula to the Newton's Law and summarizing $\dot{V}_x$ yields the following equation [8].

$$V_x = V(1 - S_x) \tag{14}$$

where V is the value for the given input speed and S_x is longitudinal slip ratio.

The magic formula graph of the slip and longitudinal force are shown in Fig. 2. The graph of the magic formula in Fig. 2 is [8].

$$\mathrm{F_x} = F_z \cdot \mathrm{D} \cdot \sin(\mathrm{C} \cdot \arctan(\mathrm{B} \cdot S_x - \mathrm{E} \cdot (\mathrm{B} \cdot S_x - \arctan(B \cdot S_x)))) \tag{15}$$

Figure 2 is a magic formula under dry road surface condition and the values used for each parameter are listed in Table 1.

The specifications of the vehicle model used in the simulation are listed in Table 2.

The difference between the forward model and the backward model is implemented by changing the following two variables.

1. Change the cornering stiffness coefficient of the front tire and the cornering stiffness coefficient of the rear tire ($C_{\alpha f} \rightarrow C_{\alpha r}$, $C_{\alpha r} \rightarrow C_{\alpha f}$).
2. The distance from the center of gravity (COG) of the vehicle to the center of the front tires and the distance from COG of the vehicle to the center of the rear tire ($l_f \rightarrow l_r, l_r \rightarrow l_f$).

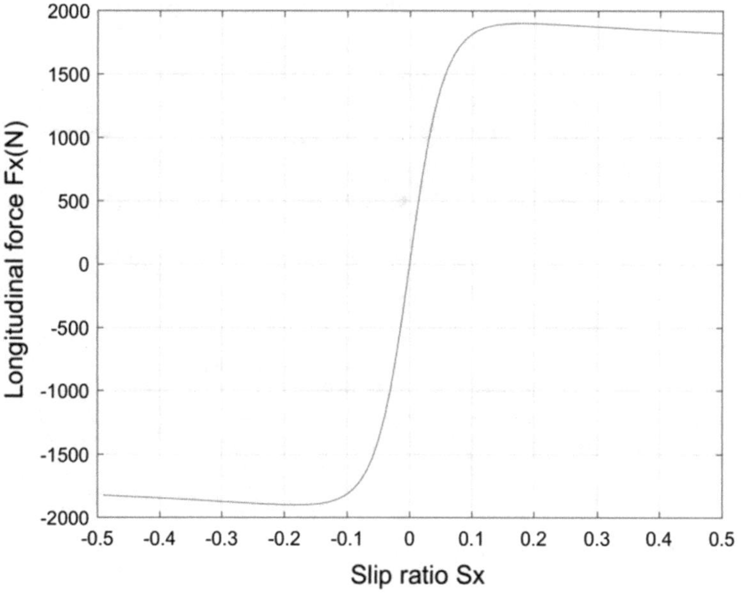

Fig. 2 Longitudinal force vs slip magic formula

Table 1 Values are typical sets of constant magic formula [8]

	B	C	D	E
DRY	10	1.9	1	0.97

Table 2 Specifications of the vehicle model [9]

Descriptions	Variables	Values
Mass	m	1900 (kg)
Moment of inertia body	I_Z	2900 (kg m^2)
Distance from center to tire	l_f, l_r	1.36, 1.44 (m)
Cornering stiffness coefficient	$C_{\alpha f}, C_{\alpha r}$	80000, 90000

3 Reinforcement Learning Algorithm

The Q learning algorithm is used, which is a commonly known as a reinforcement learning algorithm. Figure 3 shows the block diagram of implementing the reinforcement learning algorithm.

The reinforcement learning evaluates the behavior when an action, a_t is taken in a given environmental state, s, and stores the evaluation content as a Q function. The learning is gradually conducted to find the action, a_t that maximizes the Q function.

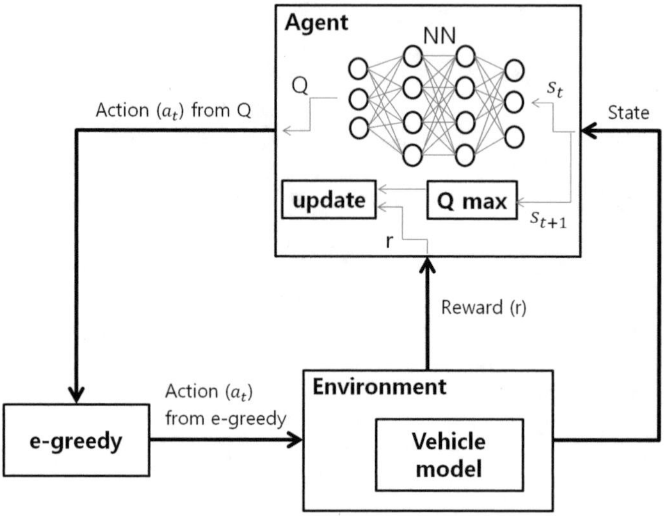

Fig. 3 Reinforcement learning structure

States (s) are the vehicle states when taking an action. When an action occurs, the state of the agent and the environment to be changed must match well. The states of the reinforcement learning framework contains the current states of the vehicle and its surroundings. It consists of a steering angle, a forward and backward distance errors, a running speed, and a longitudinal force.

Action (a) is chosen by an agent. When the selected action is performed, the states of the whole system are changed and a series of choices is founded. Selectable actions consist of six driving control modes such as a deceleration left turn, an acceleration left turn, a deceleration, an acceleration, a deceleration right turn and an acceleration right turn.

Reward (r) is the signal that the agent is expected to achieve. The reward signal is very important to achieve the goal. In our system, the distance error of the vehicle is considered as a reward. Therefore, the reward signal as a sum of the error signals is kept small.

The e-greedy process is to avoid falling into the local minimum and finding better action.

The cost function to be minimized is given as

$$J = \left(Q(s_t, a_t) - r + \gamma\, max\, \hat{Q}(s_{t+1}, a)\right)^2 \tag{16}$$

γ is the discount factor such that $\gamma = 0.1$. The neural net structure has an input-hidden–hidden-output layer of 4-100-100-6.

4 Simulation Results

The simulation was performed using the MATLAB SIMULINK program. The input conditions are the steering angle data and the speed data when the vehicle travels forward. When there is no additional control, the pre-learning data and the additional control are used.

Figure 4 shows the input conditions of the initial steering angle and the longitudinal force. The input conditions of the initial vehicle were implemented using an arbitrary model of the longitudinal torque and steering angle.

Figure 5 shows the simulation results of before and after the learning. Figure 5a shows the path trajectory of the forward trajectory, the trajectory without learning, and the trajectory after learning. Figure 5b shows the forward and backward tracking errors of without learning and after learning. We see that there is a small

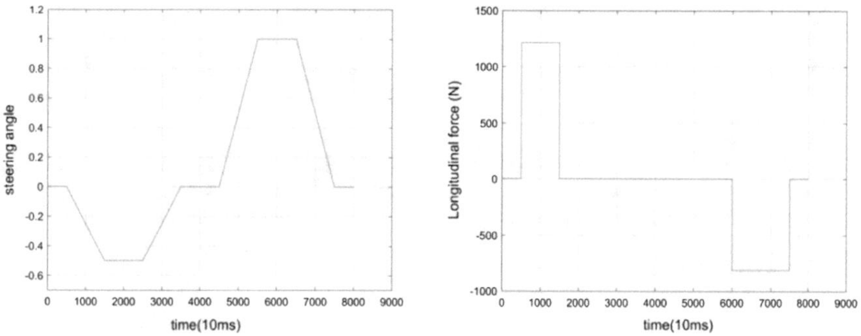

(a) Steering angle (b) Longitudinal force

Fig. 4 Input conditions

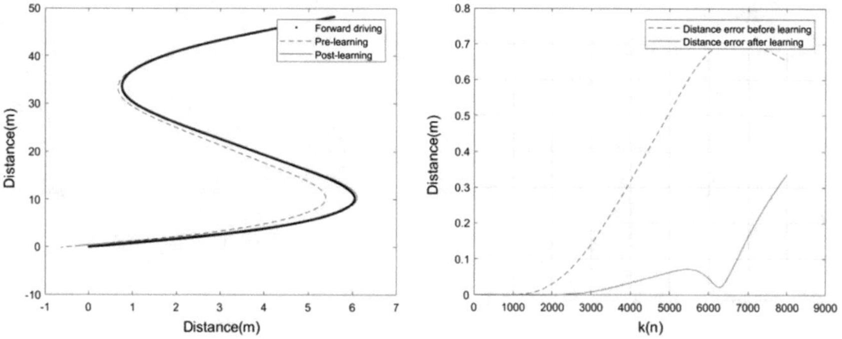

(a) Path following (b) Tracking Error

Fig. 5 Simulation results for tested learning

deviation between the forward and backward driving paths. Those deviations are improved by the learning.

Based on the learned parameters, the longitudinal values are used equally and the tracking results are confirmed when only the steering input values are changed. Figure 6 shows the different steering input conditions for the simulation.

Other input condition values were intended to represent an increased form of the steering angle and more frequent steering angle control. Figure 7 shows that the average tracking error decreases from about 0.5 m to about 0.2 m. In Fig. 8, the average tracking error decreases from about 0.35 m to 0.08 m.

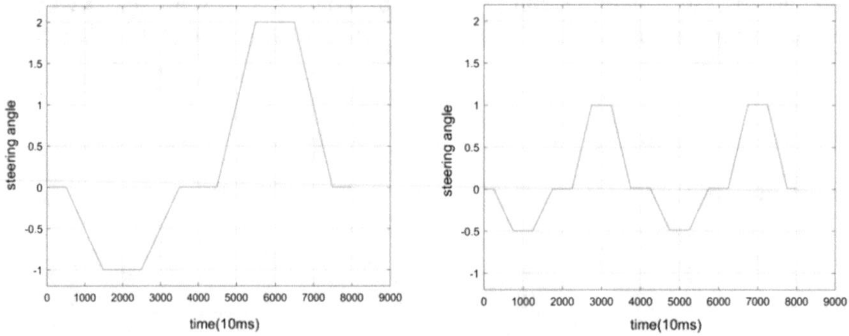

(a) Other steering angle condition 1

(b) Other steering angle condition 2

Fig. 6 Input conditions

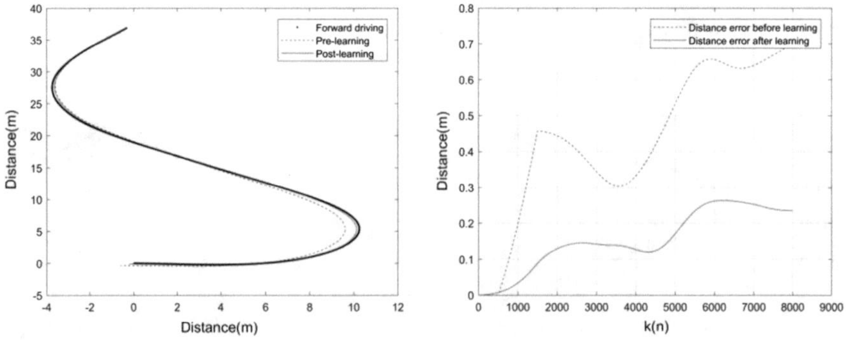

(a) Path following

(b) Tracking Error

Fig. 7 Simulation results for condition 1

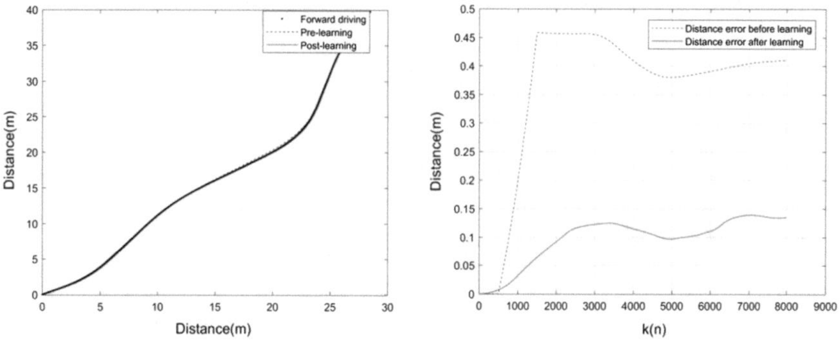

(a) Path following (b) Tracking Error

Fig. 8 Simulation results for condition 2

5 Conclusions

In this paper, the reinforcement learning technique is applied to an autonomous back driving control task of a vehicle under slip conditions of both lateral and longitudinal directions. Simulation results show that the back drive tracking errors have been reduced remarkably after the learning although there are some errors. Further research is required to minimize the deviation in the future along with experimental studies with the improved learning algorithms.

Acknowledgements This research has been supported by National Research Foundation of Korea (2016K2A9A2A06004776).

References

1. Google Car. https://waymo.com/tech/
2. Tesla. https://www.theverge.com/2016/6/30/12072408/tesla-autopilot-car-crash-death-autonomous-model-s
3. Li, T.H.S., Chang, S.J.: Autonomous fuzzy parking control. IEEE Trans. Syst. Man Cybernet. Part A Syst. Humans **33**(4), pp. 451–465 (2003)
4. Li, X., Xu, X., Zuo, L.: Reinforcement learning based overtaking decision-making for highway autonomous driving. In: Sixth International Conference on Intelligent Control and Information Processing, pp. 336–342 (2015)
5. Sutton, R.S., Barto, A.G.: Reinforcement Learning. The MIT Book (1998)
6. Rajamani, R.: Vehicle Dynamics and Control. Springer (2006)
7. Lee, C., Hedrick, K., Yi, K.: Real-time slip-based estimation of maximum tire-road friction coefficient. IEEE/ASME Trans. Mechatron. **9**(2), 454–458 (2004)
8. Longitudinal Magic Formula. https://kr.mathworks.com/help/physmod/sdl/ref/tirerodinteractionmagicformula.html
9. Schiehlen, W. O.: Dynamical Analysis of Vehicle Systems. Springer Wien New York (2007)

Rapid Coverage of Regions of Interest for Environmental Monitoring

Nantawat Pinkam, Abdullah Al Redwan Newaz, Sungmoon Jeong and Nak Young Chong

Abstract In this paper, we present a framework to solve the problem of rapidly determining regions of interest (ROIs) from an unknown intensity distribution, especially in radiation fields. The vast majority of existing literature on robotics area coverage does not report the identification of ROIs. In a radiation field, ROIs limit the range of exploration to mitigate the monitoring problem. However, considering the limited resources of Unmanned Aerial Vehicle (UAV) as a mobile measurement system, it is challenging to determine ROIs in unknown radiation fields. Given the target area, we attempt to plan a path that facilitates the localization of ROIs with a single UAV, while minimizing the exploration cost. To reduce the complexity of exploration of large scale environment, initially whole areas are adaptively decomposed by the hierarchical method based on Voronoi based subdivision. Once an informative decomposed sub area is selected by maximizing a utility function, the robot heuristically reaches to contaminated areas and then a boundary estimation algorithm is adopted to estimate the environmental boundaries. Finally, the detailed boundaries are approximated by ellipses, called the ROIs of the target area and whole procedures are iterated to sequentially cover the all areas. The simulation results demonstrate that our framework allows a single UAV to efficiently and explore a given target area to maximize the localization rate of ROIs.

Keywords Environmental monitoring · Regions of interest coverage
Energy-efficient path planning · UAV

N. Pinkam (✉) · S. Jeong · N. Y. Chong
School of Information Science, Japan Advanced Institute of Science and Technology, 1-1 Asahidai, Ishikawa, Nomi 923-1292, Japan
e-mail: pinkam.nantawat@jaist.ac.jp

S. Jeong
e-mail: jeongsm@jaist.ac.jp

N. Y. Chong
e-mail: nakyoung@jaist.ac.jp

A. A. R. Newaz
Graduate School of Information Science, Nagoya University, Furo-cho, Nagoya, Chikusa-ku 464-8603, Japan
e-mail: redwan.newaz@jaist.ac.jp

J.-H. Kim et al. (eds.), *Robot Intelligence Technology and Applications 5*,
Advances in Intelligent Systems and Computing 751,
https://doi.org/10.1007/978-3-319-78452-6_17

1 Introduction

In a large radiation field, it is important to localize Regions of Interest (ROIs) to monitor the radiation effects, to localize the hotspots, the sources, and so on. Recent advances in Unmanned Aerial Vehicle (UAV) offers the ability to access and navigate in unstructured or cluttered environments. Therefore, a single UAV equipped with dedicated sensors makes an attractive platform for such kind of tasks. However, it is difficult to monitor a large field with single UAV. In such situations, it becomes necessary to design a path planner that can localize the ROIs rapidly.

Radiation field monitoring has been commonly studied in robotics [6, 7]. The goal is to plan a path in which the robot can localize all the contaminated locations in a given target area. Since the contaminated locations could be spatially distributed throughout the target area, a search is needed to localize all of them. Thus, required tasks associated with the search inspire various methods in addressing the coverage problem. Spatial search techniques should be fitted according to the number of robots used for this application. In the case of multiple robots, the target area can be partitioned into smaller subregions to reduce the search space for each robot. The search strategy is exclusively benefited by the number of robots and the communication among them. However, in the case of a single robot exploration, the partitioning of the target area benefits neither the exploration cost, nor the accuracy.

The early survey on coverage algorithms was provided by Choset [3], where he classified the solution approaches either based on heuristic or cell decomposition. Heuristic methods explore the target area with predefined rules or a set of behaviors. The widely used heuristic methods are lawnmower pattern, raster scanning, inward spiral search, wall following, etc. Heuristic search is computationally less expensive, but cannot guarantee the optimal performance. On the other hand, in cell decomposition, the target area is decomposed into smaller areas. Galceran and Carries [5] provided a survey of an exact and uniform decomposition of the target area by a grid of equally spaced cells. Then, the coverage problem can be solved as the Traveling Salesman problem and is known to be NP-hard. Usually, in that case, a Hamiltonian path is determined using the spanning tree algorithm, which visits each cell exactly once. In recent year, a variant of Hamiltonian path utilized for the persistent coverage problem [10]. However, if there are obstacles in the target area, it is not possible to generate the Hamiltonian path in all the cases. The Boustrophedon cellular decomposition can then solve this problem for bounded planar environments with known obstacles [14]. The key idea is to construct a graph by decomposing the target area subject to obstacle positions and finding a minimal cost tour through all regions. In literature, we have seen an extension of that algorithm while respecting sensor feedback [1, 11, 12]. When unknown obstacles exist in the environment, the Morse decomposition used for determining critical points in the target area, and then incrementally construct the Reeb graph to solve the online coverage problem optimally [2]. Another way is to satisfy a temporal logic specification consisting of safety components in a partially unknown environment [8].

The majority of coverage planning work has been proposed for known environments [12, 15, 16]. Often these approaches are motivated to minimize the uncertainty metric of a given map. A common choice is to add an exploration to that location where the uncertainty metric such as entropy or mutual information is high. However, in many situations, a radiation map for the target area may not be *a priori* available. The problem can then be closely related relation to covering the entire target area for localizing the contaminated locations. Hence, complete coverage algorithms are often used. Even though complete coverage algorithms ensure the complete terrain visitation, they lack the opportunity to optimize the localization rate of contaminated locations.

Considering estimation on environmental boundaries instead of the complete coverage provides a useful abstraction that reduces the energy consumption [9, 13]. Here, the path planning problem consists of estimating boundary of contaminated areas that allow the robot to sense the ROIs. However, when the environment is unknown, it is hard to plan a path that identifies which areas are interesting and which are not. In conventional algorithms for the coverage planning with obstacles, the path is usually generated to cover the free space of the environment in an optimum fashion. In our problem, rather than avoiding ROIs, we want to identify locations and geometrical size of them rapidly. For example, when the robot opportunistically finds contaminated areas, firstly, it can expedite the boundary estimation process to determine ROIs, and then it can bypass exhaustively covering the entire regions. Determining ROIs in a radiation field allows us to prioritize the search area in such a way that minimizes the exploration of the robot.

In this work, motivated by a single UAV coverage, we investigate an additional component to the coverage problems by incorporating a localization rate factor for the radiation contaminated locations. Taking account of the localization rate factor which is important in a single UAV exploration, sometimes the target area is too large for the UAV to completely cover with limited exploration budget (maximum exploration time). Since it is also of the interest that the UAV is to localize all the contaminated locations as quickly as possible, the algorithm must behave as the complete coverage over long periods of operation. This problem might be thought of as target acquisition problems [4]. However, there is an important caveat. Target acquisition problems assumed that the robot equipped with a sensor that has a wide field of view, whereas in our problem, the robot sensor works in a point-wise fashion. Therefore, the robot needs to travel to a location to get a measurement.

In this paper, we discuss the online version of this problem, in which the coverage path of the robot is to be determined based on the information gain metric from the past exploration. To reduce the search space, we initially partition the target area in a random manner. Next, we update the partition size based on the size of ROIs. We propose an optimal path planner, which extends the complete coverage algorithm to reason about a localization rate factor. Under the assumption that there exist multiple ROIs in a given target area, the proposed algorithm can increase the localization rate of contaminated locations while guaranteeing a complete coverage path over long periods of operation.

The contributions of this work are as follows:

1. We have formulated the localization of ROIs which does not require *a priori* information at all.
2. Our algorithm can localize ROIs in a fast manner by minimizing the exploration of UAV.
3. The proposed algorithm is complete, which means all contaminated locations are identified for the long operation of UAV.
4. Focusing on the limited computational capabilities of the UAV, the proposed algorithm can robustly determine ROIs.
5. To best of our knowledge, this is the first approach that integrates the environmental boundary estimation problem to the area coverage problem.

To discuss the aforementioned topics, this paper is organized as follows: in Sect. 2, we describe the problem formulation; Sect. 3, we present the heuristic coverage algorithm based on adaptive hierarchical area decomposition. Section 4, we briefly explain generalization process of ROIs. Finally, in Sects. 5 and 6, we present simulation results and conclude our findings.

2 Problem Formulation

We are given a target area T, which contains radiation sources, strength of can be sensed by the robot. We assume that T can be decomposed into a regular grid with n cells. Let us denote this grid by G. Since radiation sources might be spatially distributed. Thus, G contains two type of cells, i.e., free cells and contaminated cells. Furthermore, nearby sources cumulatively affect the target area, resulting in a joint distribution of measurement attributes. Let us assume that each cell c is associated with a measurement attribute z. The robot is equipped with a sensor to make a pointwise measurement $z(t)$ at its position $x(t)$ at time t. The Regions of Interest (ROIs) in T are those cells $\mathbf{J} := \{c|z > 0\}$ where the robot finds $z > 0$. The contaminated areas are contiguous. Therefore, the robot can trace such areas by tracking only to the boundaries. Therefore, the definitions of the contaminated and the free cell are quantified through a binary probability value given by

$$p_c = \begin{cases} 0, & \text{if } z \approx 0 \\ 1, & \text{otherwise} \end{cases} \tag{1}$$

Figure 1 shows an example world map of size 50 × 50. Depending on the spatial locations of the radiation sources, measurement attributes are also spatially distributed throughout T. The dark blue cells are the cell where $p_c = 0$. The other colored cells represent the fact that measurement attributes are available such that $p_c = 1$. We can then find multiple ROIs while splitting $\mathbf{J}$ subject to spatial distances.

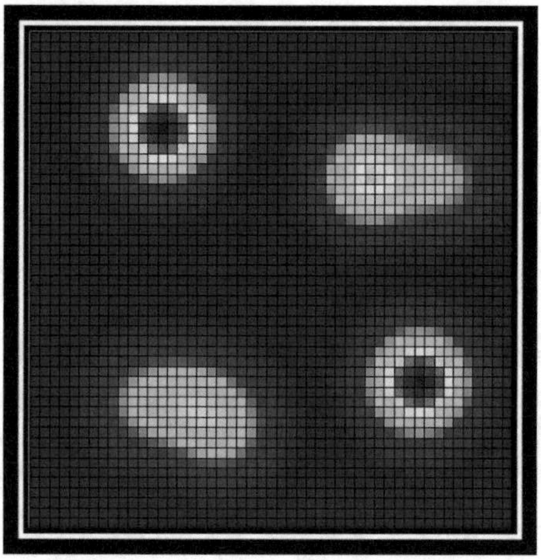

Fig. 1 The dark blue cells have no measurement attributes whereas other colored cells represent the measurement attributes

Definition 1 *Regions of Interest (ROIs):* A collection of cells corresponds to a set of contaminated locations in a given target area T, i.e. the set $\{\mathbf{J} \in T | p_c = 1\}$.

The global mission of the robot can be defined in two different ways, which implies two different objective functions as follows

- the minimum time to localize an ROI,
- the total time to localize all the ROIs in T.

Without loss of generality, we assume that the travel time is proportional to the travel distance. Therefore, firstly, we will use the boundary estimation technique that minimizes the robot's exploration to localize an ROI. Secondly, we will use the heuristic area coverage technique that ensures to localize all the ROIs in T. The total time is taken into account by summing up the boundary estimation paths and the heuristic area coverage paths.

Let us formally define these objective functions. First, starting from an initial cell, we denote the coverage path followed by the robot throughout the free cells by $\mathscr{P}$. We assume that, $|\mathbf{J}| << n$ i.e., the contaminated cells are far fewer than the number of free cells. We define the event $S_{\mathscr{P}}$ as the event that the robot reaches to any ROI which is not localized beforehand. The complete coverage path $\mathscr{P}$ can be then discretized by the presence of ROI. Therefore, the probability to find an ROI can be expressed as follows

$$E[S_{\mathscr{P}}] = \sum_{c \in \mathscr{P}} \left(1 - p_c\right). \tag{2}$$

Thus, the first objective is to find an online coverage path that minimizes $E[S_{\mathscr{P}}]$. Note that, in this objective, the heading of the path is not important, once the robot heuristically reaches any location of an ROI, the boundary tracking algorithm is followed to determine the ROI size.

For the second objective, we denote the sequence of newly discovered ROIs along the coverage path $\mathscr{P}$. if there exists k number of ROIs in T, we discretize $\mathscr{P}$ into a subset $Q = \{q_1.q_2, ..., q_k\}$. Since the travel time is proportional to the length of q_k, we want to find the minimum length paths in the set Q to localize all the ROIs. Therefore, the total events $C(\mathscr{P})$ that the robot is experienced to localize a finite set of ROIs given by

$$C(\mathscr{P}) = \sum_{q_k \in Q} S_{q_k} \ s.t. \ |Q| \leq |ROI|, \tag{3}$$

where $|Q|$ is the cardinality of set Q and $|ROI|$ is the number of ROIs are detected in T. If $|ROI|$ is *a priori* given, our focus is to find the minimum exploration time to achieve that number. We then derive the performance index of the robot from Eq. (3). A formal definition of the performance index as follows.

Definition 2 *Performance Index (PI):* The performance index of the robot is evaluated with respect to the minimum explored path to localize all of ROIs, i.e. $PI = \operatorname{argmin} C(\mathscr{P}) \ s.t. \ |Q| \leq |ROI|$.

Since we do not know the exact number of ROIs exists in T, it is not possible to stop the robot's exploration when all ROIs are localized. Then, the robot exploration can be terminated by exploration budget. Otherwise, the robot's task is to plan an online path through T such that every ROIs is rapidly localized while subject to complete area coverage.

3 Adaptive Hierarchical Area Decomposition and Coverage

Figure 2 shows the overall schematic of our proposed system. The algorithm we propose can be broken down into three steps. In the first step, *Adaptive Hierarchical Area Decomposition*, we adaptively partition the target area in hierarchical order to reduce the search space of the robot. We then find the subregions given by the partition using the *Finding subregions*. When the subregions are determined, we examine the utility to traverse each subregion that explained in the *Utility function design*. The subregion which has maximum utility, we plan a coverage path through the set of unvisited cells. The robot progresses through this path. If the robot notices an ROI along its path, it will drop exploring more and iterates whole steps. Otherwise, the whole steps iterated after traveling along the entire path.

3.1 Adaptive Hierarchical Area Decomposition

To reduce the computational complexity while navigating a large environment, the search space for the path planning needs to be at a tractable level. We argue that these objectives can be achieved by adaptive partitioning of the target area in hierarchical

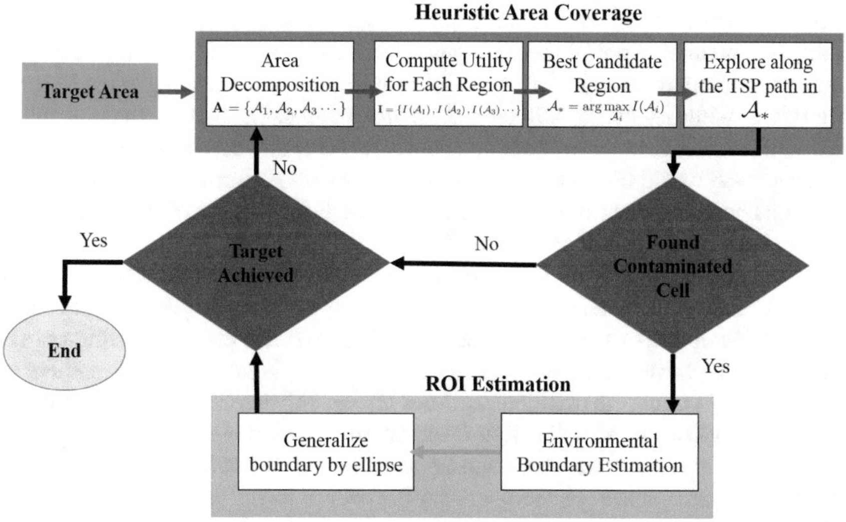

Fig. 2 System overview: the figure shows all the steps performed by the heuristic area coverage, and ROI estimation algorithms. Starting from an arbitrary location, the robot can iteratively localize the desired number of ROIs using this framework

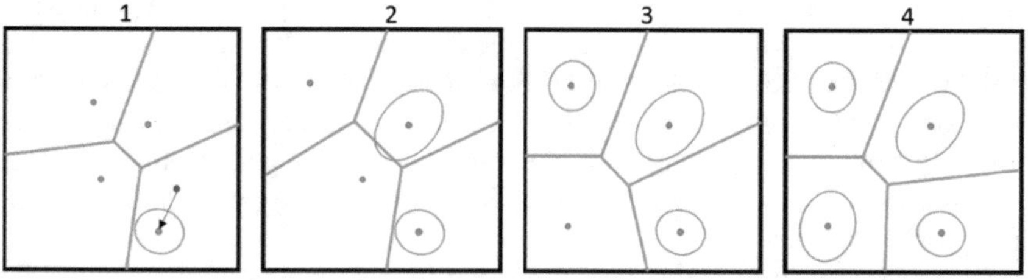

Fig. 3 VBS decomposition: starting with the random partitions, the partitions updated by the center position of ROIs. The algorithm iteratively approaches to optimal decomposition

order. Given the position of ROIs, the hierarchical order is determined by a local minimum distance with the respect to the robot's relative position. Therefore, we propose Voronoi-based partition in the sense of limiting the search space. Figure 3 shows the overall overview of each algorithm. With a given partition, our goal is to find an ROI through the limited exploration.

The Voronoi-based subdivision (VBS) uses the Voronoi-based approach to partition the target area. The main idea is to partition the area by representing the ROI centers as the Voronoi centroids. Since in our case the ROI centers are not apriori available, we have introduced a few changes to the original Voronoi-based partition algorithm. Firstly, it randomly partitions the target area using four random points inside the target area. Secondly, it leads the robot to the nearest centroid from its initial location. Finally, TSP algorithm generates the coverage path. The robot starts to explore along this route when a contaminated cell found; it switches to the boundary estimation planner. An ROI then computed from the estimated boundary. The robot finds a minimum route to ROI from its location either while traveling to the Voronoi centroid or while executing TSP path. Although these paths increase the probability of finding ROI, if there are no contaminated cells in the subregion, then the complete coverage path would be large because of traveling to the centroid. Note that in VBS, the initial search space limited by the random partition. The partition of the target area updated by the center position of the detected ROI.

In the second phase, VBS finds a coverage path that connects the desired number of ROIs. Finding such path is possible by iteratively updating the Voronoi centroids. The iterative updates of centroids lead VBS to generate an optimal partition of the search space. However, when the number of ROIs is greater than the number of random initial points, the partition centroids are not only iteratively updated but also incrementally constructed. The four basic operations of this decomposition are as follows. Firstly, we generate randomized incremental construction of partitions to reduce the search space. Secondly, the robot moves to the Voronoi centroid, and TSP algorithm creates a coverage path to explore the unexplored cells optimally of a given subdivision. Thirdly, when an ROI is determined, we terminate the exploration and update the Voronoi centroids. Finally, the region of each division is determined.

We demonstrate the Voronoi-based subdivision while the robot is covering its free space using an example depicted in Fig. 3. Voronoi Diagram is the partitioning method of a plane with n points into a specific subset of the plane such that each subset contains exactly one generating point. In typical Voronoi diagram, the set of generating points is *apriori* known. The Voronoi polygons are then constructed such that every point in a given polygon is closer to its generating point than to any other. However, in our case, we randomly initialize the generating points and iteratively update their positions.

The robot starts to cover the space in a vast cell by moving into the centroid of the current Voronoi region (red dot) which is located at the rightmost corner; the target area is shown as the black rectangle in Fig. 3. Then, the robot constructs a TSP path to cover the given region. Whenever the robot reaches the cell where $P_c = 1$, which is the unvisited location of a contaminated area, it finishes covering the centroid path or the TSP path. Since the contaminated area is unknown *a priori*, the robot follows the boundary tracking algorithm to cover it. The robot then constructs an ellipse over the estimated boundary to represent the ROI, shown as the orange ellipse in Fig. 3. At this point, it encounters the update of Voronoi centroid. The Voronoi centroid of the current region is replaced by the center point of the ellipse, shown as blue dots in Fig. 3. If there are more ROIs than the Voronoi centroids which are chosen initially, the overall Voronoi partitions are reconstructed with updated centroids. Note that the minimum number of subdivisions in this case is four, and the algorithm can also cover more than four subdivisions. The robot chooses the subdivision that maximizes the utility function and repeats the step described above as shown in Fig. 3. Since the Voronoi regions are connected, the robot is guaranteed to visit all the subdivisions in the target area, and thus completely cover the space.

3.2 Finding Subregions

At the end of the second phase, each algorithm finds the subregions based on its partition method. For this purpose, it begins by creating the graph $G = (V, E, B)$ induced from above mentioned methods. We represent the target area as a rectangular box B in G. The initial partitions are the edge set E that includes edges with infinite lengths. To find subregions Λ, firstly, we shorten each edge $e \in E$ subject to B. Let V be the set of vertices that includes three types of subsets such that $V = \{\{\psi_G\}, \{\psi_b\}, \{\psi_c\}\}$. Let ψ_G be the first subset of V that represents the vertices at the intersection between B and E. Also, let ψ_b be the set of vertices that represents the corner points of B, and let ψ_c be the centroid of ROIs. Once we trim the long edges, the new partition represented by E_ψ. Secondly, we find all the possible combination of edges on B and represent by E_b. The G is then updated by combining these two set of edges such that $E \leftarrow \{\{E_b\} \cup \{E_\psi\}\}$. Finally, we group all subregions Λ by finding the neighbor edges. Finding such a neighbors is straightforward. Given ψ_c, an anti-clockwise walk along the E can sort such neighbors.

3.3 Utility Function Design

In the third phase, each algorithm finds the best search space among all subdivisions of the target area. For this action, it computes the utility between each of subdivisions. The utility is designed to favor destinations which offer higher expected information gain. Throughout this work, we use an explored grid map, m, to model the environment. This map is a binary map where each cell represents visited or unvisited information. Let i be the index of each subdivision and the division of such a map satisfies the following equation

$$m = \sum_i m^{[i]}. \tag{4}$$

An action a_t generated at time step t is represented by a sequence of relative movements $a_t = \hat{u}_{t:T-1}$ which the robot has to carry out starting from its current position x_t. During the execution of a_t, if the robot finds a contaminated cell along its path, then it estimates an ROI in the map. Therefore, the explored trajectory of the robot indicates some of the cells in m as follows

$$x_{1:t} = \exists c \in m. \tag{5}$$

In the case when the robot finds an ROI in the map, we have to treat the ROI cells differently. We assumed that traveling inside an ROI is redundant, and want to avoid such a region. Therefore, the cells bounded by an ROI considered as similar as visited cells. Let d_t be the set that represents these cells as follows

$$d_t = \{\forall c \in ROI1, \forall c \in ROI2 \cdots\}. \tag{6}$$

Assuming that each cell c in m is independent of each other. Then the posterior entropy of m can be computed as follows

$$\begin{aligned} H(p(m|x_{1:t}, d_t)) = -\sum_{c \in m} p(c) \log p(c) + \\ (1 + p(c)) \log(1 - p(c)). \end{aligned} \tag{7}$$

Given a subdivision, since the robot does not know when it will find an ROI along its path, the coverage path should include all cells to compute the expected information gain. Thus, the entropy of target subdivision can write as follows

$$\begin{aligned} H(p(m^{[i]}|x^{[i]}_{t+1:T}, d_t, a_t)) = -\sum_{c \in m^{[i]}} p(c) \log p(c) + \\ (1 + p(c)) \log(1 - p(c)). \end{aligned} \tag{8}$$

To compute the information gain of a subdivision, we calculated the change in entropy caused by the integration of posterior and predicted prior into the robot's world model as follows

$$I(m^{[i]}, a_t) = H(p(m|x_{1:t}, d_t)) - H(p(m^{[i]}|x^{[i]}_{t+1:T}, d_t, a_t)). \tag{9}$$

After computing the expected information the utility for each action under consideration, we select the action a^*_t with highest expected information

$$a^*_t = \arg\max_{a_t} I(m^{[i]}, a_t). \tag{10}$$

There are some works in exploration and mapping problems that consider another quantity besides the information gain in Eq. (10). That is the cost to reach the subdivision. However, we observed that adding such a quantity with the utility function decreases the overall performance of both algorithms. Thus, every time the robot has to make the decision where to go next, it uses only information maximization metric to determine the action a^*_t.

4 Finding ROIs

We employ a boundary estimation algorithm to determine the ROIs by using the proposed exploration method. ROIs over the target area T are dependent on the boundary line estimated by environmental boundary algorithm. Memorizing a complex boundary is computationally expensive, therefore to obtain the tractable level of computation, we require the parametric estimation of the boundary.

Definition 3 *Boundary line:* The line is said to be boundary line if it represents the intersection between the contaminated area and non-contaminated areas.

Assume an contaminated area $\delta\mathscr{A}$ is a non-convex set where the continuous boundary is defined by a level set $\delta\mathscr{A}$ as follows

$$\delta\mathscr{A} = \left\{x \in \mathbb{R}^2 | z(x) = \beta\right\}, \tag{11}$$

where β is the measurement threshold.

Boundary algorithm ensures that an environmental boundary can be estimated by tracking the robot states such that $\delta\mathscr{A} = \{x_{1:t}\}$. When the exploration is terminated, this set $\delta\mathscr{A}$ can be used to estimate of the best fit to an ellipse. This generalization is done by the least squares criterion from the set $\delta\mathscr{A}$. We also consider the possible tilt of the ellipse from the conic ellipse representation as follows

$$ROI(\delta\mathscr{A}) = aS_x{}^2 + bS_xS_y + cS_y{}^2 + dS_x + eS_y + f = 0, \tag{12}$$

where $\{S_x, S_y\} \in \delta\mathscr{A}$ and a, b, c, d, e, f are the parameter for a second degree polynomial equation. After the estimation, the tilt is replaced by a rotation matrix from the ROI, and then the rest of parameters are extracted from the conic representation.

5 Simulation Results

To find the shortest coverage path, we perform 4 different experiments using MATLAB. We assume that the target area contains at most 5 ROIs. The performance of each algorithm was evaluated by the distance of coverage paths. To demonstrate the efficiency, we start localizing 2 out of 5 ROIs and conclude by 5 out of 5 ROIs. We also have analyzed the worst case performance and we present a statistical analysis of two algorithms from 20 trial runs. The performance of algorithms significantly varied from each other. In particular, we have observed a noticeable difference of the algorithms on localizing uniformly distributed random ROIs. It is noteworthy that to compute the efficiency, the ROIs shape should remain fixed for each algorithm, we then overlook the additional path cost required to estimate ROIs.

5.1 Finding Coverage Path that Connects the Desired Number of ROIs

We now consider the case of finding ROIs that meet the desired level of exploration. Therefore, we focus on the shortest coverage path for a given number of ROIs. We consider a 50×50 m grid area where 5 uniformly distributed random ROIs are located. Starting from an initial location (1, 1), the robot has to find the minimum

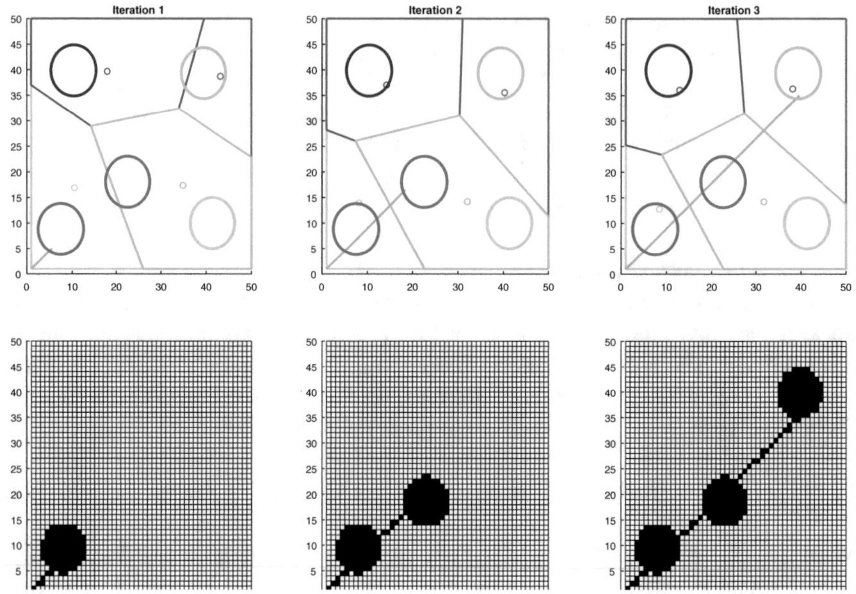

Fig. 4 The robot starts the coverage in cell (1,1) and detects any 3 ROIs out of 5. The shape of each ROI is elliptical and is represented in unique color. The lower grid map represents the coverage map. The measured cells are represented by black color. A cell is called to be measured if it is included either in coverage trajectory or it is bounded by the detected ROIs. VBS coverage paths on a sample map with uniformly distributed random ROIs. The dark green line in upper figure shows the coverage path, while the colored lines are the partition of the target area. The centroid of each region is represented by the same colored cycle. For a new region, the searching process is started from the centroid. The partitions are iteratively updated based on the true position of the center of ROIs

coverage path that connects the desired number of ROIs. The coverage path can be found by adjusting the cost to the inversely proportional to the unexplored area. In another word, the robot explores the mostly unexplored region first.

Figure 4 shows a toy example of VBS algorithm. In VBS, the initial search space is generated by randomly choosing 4 points bounded in the target area. We will call these points as the Voronoi centroids. The initial search space is then subdivided into four regions based on the Voronoi centroids. The robot moves the centroid of a Voronoi region first and exhaustively search for an ROI within that region. When an ROI is found whether traveling to the centroid or searching the entire subregion, the robot updates the Voronoi diagram. The robot avoids exploring the cells bounded by the ROI. These processes are iterated until the end of the mission. The VBS requires at least 3 points to partition the entire search space optimally. When there are less than 3 ROIs in total area and the robot has to localize all of them, the VBS performance is not stable.

5.2 Performance Comparison

We compare VBS algorithm to recursive quadratic subdivision (RQS) which follows a greedy approach, wherein each step it leads the robot to the nearest ROI to its current location that has not been covered yet. The three basic operations of RQS are as follows. Firstly, we generate a TSP path to explore the unexplored cells optimally. Secondly, when an ROI is determined, we terminate the exploration and decompose the area. Finally, the region of each division is determin.

Figure 5 shows a performance comparison. To access the long-term performance of each algorithm, we ran the same experiments for 20 times by gradually increase the target numbe. Figure 5 shows the results in area coverage percentage metric. We divided the given target area into three different regions—(1) explored by the robot (2) covered by the ROIs (3) remained unexplored. Our goal is to minimize

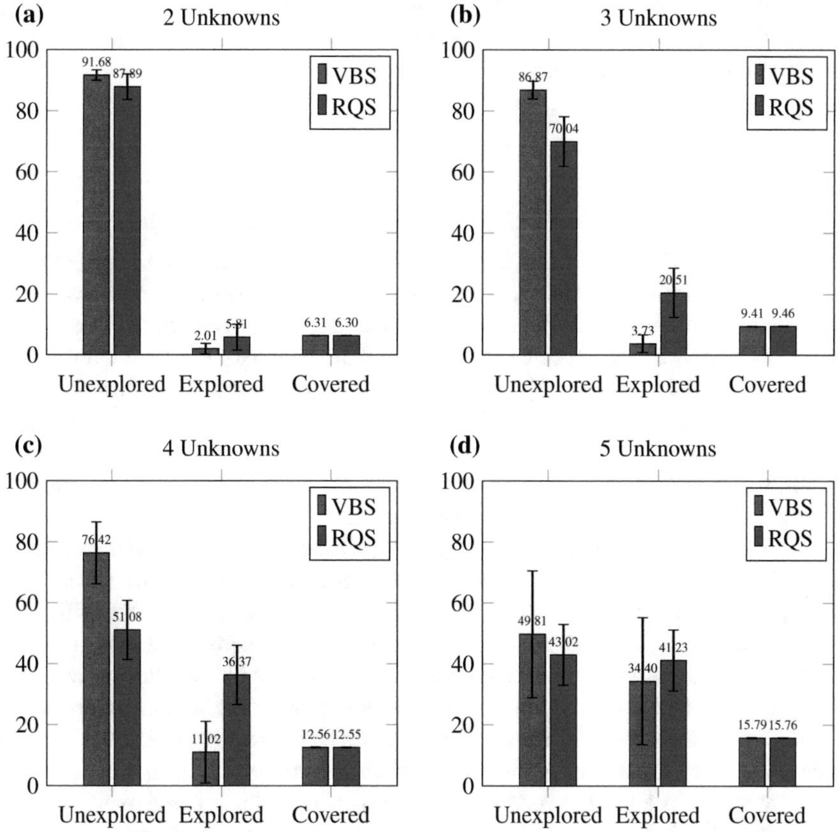

Fig. 5 Area coverage: every bar chart is generated from 20 trial runs of each algorithm. The performance is evaluated by comparing the size of following areas: unexplored, covered and explored area. The error bar of the bar chart represents the standard deviation of each area

the explored region as small as possible. To make a fair comparison, we use five uniformly distributed random ellipses and try to find the shortest path that connects 2, 3, 4, and 5 ROIs. For them, the covered regions by ROIs are 6, 9, 13, and 16 percentage of the target area. The unexplored region then determined by subtracting the covered and explored regions from the total area.

The reduction of search space directly influences of the explored areas. When the number target of ROIs is less than total ROIs existed in the target area, the robot dramatically reduces the amount of explored region. In worst case scenario, when the robot needs to localize all five ROIs, it requires traveling more locations to find the ROIs, resulting in higher exploration regions. However, the performance of each algorithm is not stable, and we use the error bar of the bar chart to represent their standard deviation (SD). For both algorithms, the SD increases with the increment of the number of target ROIs.

It is evident from the Fig. 5, the VBS always outperforms the RQS because of the optimal search space division strategy. Furthermore, when the number of target ROIs is less than the total number of ROIs, the VBS significantly reduces the explored region than the RQS. We reported the numeric performance comparison between the VBS and the RQS in Fig. 5.

6 Conclusion

In this paper, we have discussed the ROIs determining problem for a large environment and its various aspects. First, we have proposed a novel online framework to integrate the environmental boundary estimation and area coverage problems. Second, we theoretically analyze the properties of the boundary estimation algorithm which is deemed to best satisfy such conflicting requirements. Third, we proposed the adaptive area decomposition and search algorithm to localize the desired number of ROIs rapidly: VBS, which uses an optimal partitioning strategy for updating the search space. Fourth, we demonstrate these algorithms in a simulated environment, and statistically analyze their relative performance.

The simulation results show that, in general, VBS creates coverage path is shorter than the coverage path by RQS. VBS has clear benefit when handling fewer ROIs since it performs a global planning of the coverage according to the size of the target area. On the other hand, RQS plans only local best decomposition, resulting in overall poor performance. Both algorithms do not require to complete coverage of the target area and save a significant amount of redundant exploration. Comparing all the experiments, we have shown that, in general, required explored areas are less than unexplored areas. Furthermore, the robot does not need to visit the covered areas by ROIs. As a result, even in worst case scenarios, the required exploration to determine ROIs is always less than complete area coverage algorithms.

In future, we would like to extend the algorithms for multi-robot systems. We would also consider the problem associated with non-stationary environmental boundaries.

Acknowledgements The authors would like to thank Ministry of Education, Culture, Sport, Science and Technology (MEXT) - Japan, for the financial support through MEXT scholarship. In addition, this work was supported by the Industrial Convergence Core Technology Development Program (No. 10063172) funded by MOTIE, Korea.

References

1. Acar, E.U., Choset, H., Lee, J.Y.: Sensor-based coverage with extended range detectors. IEEE Trans. Robot. **22**, 189–198 (2006)
2. Acar, E.U., Choset, H., Zhang, Y., Schervish, M.: Path planning for robotic demining: robust sensor-based coverage of unstructured environments and probabilistic methods. Int. J. Robot. Res. **22**(7–8), 441–466 (2003)
3. Choset, H.: Coverage for robotics—a survey of recent results. Ann. Math. Artif. Intell. **31**(1–4), 113–126 (2001)
4. Dames, P., Kumar, V.: Autonomous localization of an unknown number of targets without data association using teams of mobile sensors. IEEE Trans. Automat. Sci. Eng. **12**, 850–864 (2015)
5. Galceran, E., Carreras, M.: A survey on coverage path planning for robotics. Robot. Autonom. Syst. **61**, 1258–1276 (2013)
6. Guillen-Climent, M.L., Zarco-Tejada, P.J., Berni, J.A.J., North, P.R.J., Villalobos, F.J.: Mapping radiation interception in row-structured orchards using 3d simulation and high-resolution airborne imagery acquired from a uav. Precision Agri. **13**, 473–500 (2012)
7. Han, J., Xu, Y., Di, L., Chen, Y.: Low-cost multi-uav technologies for contour mapping of nuclear radiation field. J. Intelligen. Robot. Syst. **70**, 401–410 (2013)
8. Lahijanian, M., Maly, M.R., Fried, D., Kavraki, L.E., Kress-gazit, H., Member, S., Vardi, M.Y.: Environ. Partial Satisfact. Guarant. **32**(3), 583–599 (2016)
9. Matveev, A.S., Hoy, M.C., Ovchinnikov, K., Anisimov, A., Savkin, A.V.: Robot navigation for monitoring unsteady environmental boundaries without field gradient estimation. Automatica **62**, 227–235 (2015)
10. Mitchell, D., Chakraborty, N., Sycara, K., Michael, N.: Multi-robot persistent coverage with stochastic task costs. In: IEEE International Conference on Intelligent Robots and Systems, pp. 3401–3406 (2015)
11. Paull, L., Seto, M., Li, H.: Area coverage planning that accounts for pose uncertainty with an AUV seabed surveying application. Proceedings of IEEE International Conference on Robotics and Automation pp. 6592–6599 (2014)
12. Paull, L., Thibault, C., Nagaty, A., Seto, M., Li, H.: Sensor-driven area coverage for an autonomous fixed-wing unmanned aerial vehicle. IEEE Trans. Cybernet. **44**, 1605–1618 (2014)
13. Soltero, D.E., Schwager, M., Rus, D.: Decentralized path planning for coverage tasks using gradient descent adaptive control. Int. J. Robot. Res. **33**, 401–425 (2014)
14. Strimel, G.P., Veloso, M.M.: Coverage planning with finite resources. In: IEEE International Conference on Intelligent Robots and Systems, pp. 2950–2956 (2014)
15. Xu, L.: Graph Planning for Environmental Coverage. Carnegie Mellon University, p. 135 Aug. 2011
16. Yehoshua, R., Agmon, N., Kaminka, G.A.: Robotic adversarial coverage of known environments. Int. J. Robot. Res. 1–26 (2016)

Part III
Intelligent Robot System Design

Embedded Drilling System Using Rotary-Percussion Drilling

Jongheon Kim, Jinkwang Kim and Hyun Myung

Abstract The drilling technology is widely used in various fields. This technology has been developed explosively by large companies. However, most of the conventional drilling systems require large equipment such as rig and mud systems. This makes it difficult for drilling in polar or space environments where large equipment is hard to enter or carry. To solve this problem, we propose an embedded drilling system which does not requires the rig, mud system, and pipes. In this paper, the concept of embedded drilling system will be proposed and we introduce a locomotion and rotary-percussion drilling mechanism of this system. Two simulations were performed to show efficiency of the proposed system compared toss the conventional rotary drilling system.

Keywords Drilling system · 1-DOF leg mechanism · Caterpillar gait
Rotary-percussion drill · Theo-jansen mechanism

1 Introduction

Drilling has been carried out in a way to collect resources and for geological survey, among which directional drilling has been developed as an efficient technique for the extraction of unconventional resources. However, in order to apply these techniques, huge drilling equipment such as a rig and mud circulation systems must be

---

J. Kim · J. Kim (✉) · H. Myung
Department of Civil and Environmental Engineering and Robotics Program,
KAIST (Korea Advanced Institute of Science and Technology),
Daejeon 305-701, Republic of Korea
e-mail: jinkwang@kaist.ac.kr

J. Kim
e-mail: rayn@kaist.ac.kr
URL: http://urobot.kaist.ac.kr

H. Myung
e-mail: hmyung@kaist.ac.kr

J.-H. Kim et al. (eds.), *Robot Intelligence Technology and Applications 5*,
Advances in Intelligent Systems and Computing 751,
https://doi.org/10.1007/978-3-319-78452-6_18

accompanied. Additionally, as the depth of the existing drilling system is deepened, more pipes required. According to this situation, these systems are difficult to apply in harsh environments such as Artic or planetary environment.

There had been several researches for self-burying robots for planetary drilling such as the planetary underground tool (PLUTO), self-turning screw mechanism (STSM), and lunar subsurface explorer robot; The PLUTO used hammer drilling method, STSM used rotary drilling with a screw, and lunar subsurface explorer robot used peristaltic movement and rotary drilling [1]. These robots are developed to install geophysical sensors or excavate soil under the lunar surface. However, these robots are designed to drill in straight path and cant build a winding, multi-wells like the conventional directional drilling systems.

Therefore, we propose a robot which can operate directional drilling without large equipment and additional pipes. This robot can be divided into three parts depending on the role which are the leg mechanisms, drilling and soil removal, and steering unit. The leg mechanisms are designed to supply weight-on-bit (WOB) for self-drilling and it replace the role of the rig and pipes. The drilling and soil removal rotates the bit and prevent the soil generated during drilling from influencing the steering unit. The steering unit generates the force to turn the heading of the robot in desired direction. In this paper, we will introduce embedded directional drilling robot focused on the leg mechanisms and drilling method. The leg mechanism uses peristaltic movement and walking simultaneously which is influenced by a caterpillar, and rotary-percussion drilling is applied for drilling method.

2 Locomotion Unit

2.1 Leg Mechanism

Walking machines on off-road terrain have advantages in mobility and energy efficiency compared to wheeled vehicles. Generally, three degrees of freedom (DOF) is adopted for each leg of the walking machine. Two DOF movements for the vertical movement and the horizontal movement and the other for turning motion. Among them, the turning motion is least used. Using two independent degrees of freedom on the feet, makes it easy to generate linear motion of the foot, which can increase the contact time with the ground. However, it requires seperate sensors, motors, and additional controls for efficient walking. Additionally, system gets more weight and less robustness [2].

The drilling system has some features that it can move only along the drilled path and it has a separate steering unit that provides directionality to the bit. With these characteristics and to improve mobility and energy efficiency, 1-DOF leg mechanism is applied for the locomotion unit of the proposed system. There are various 1-DOF leg mechanisms are used for legged robots. The most commonly used mechanisms were T-Hoekens mechanism and Theo-Jansen mechanism due to trajectory of

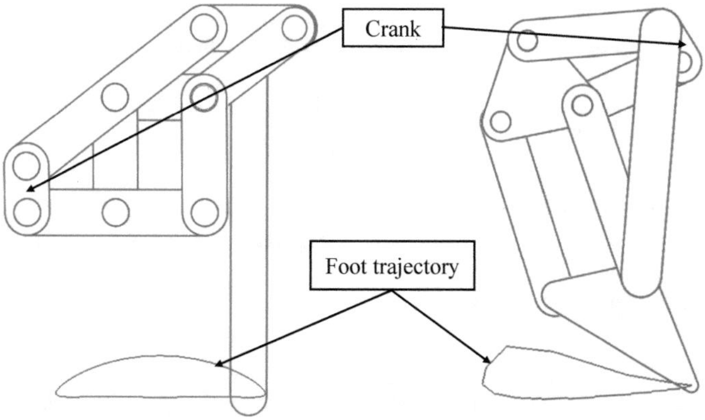

Fig. 1 T-Hoeken mechanism (left) and Theo-Jansen mechanism (right) with its foot trajectory

foot as shown in Fig. 1. Hoekens mechanism which looks like Greek letter λ, therefore it is also known as λ-mechanism. Both mechanisms are able to draw D-shaped foot trajectory with a single actuator. T-Hoeken's mechanism is translated version of Hoeken's mechanism to use it as a leg mechanism [3]. Although this mechanism can implement linear trajectory in the support phase easily, since the pipe shaped environment require at least three legs to hold its position, but it is difficult to apply this mechanism due to its workspace which is relatively large compared to other 1-DOF legs. Theo-Jansen mechanism is bioinspired mechanism and has three independent loops with eight links to show smooth foot trajectory [4]. Theo-Jansen mechanism is selected as the leg mechanism for the locomotion unit.

2.2 *Caterpillar Gait*

The borehole is an uneven space, and the WOB that pushes the bit during drilling is important. Therefore, the proposed locomotion mechanism is based on leg instead of wheel, because walking is stronger on off-load and slip than wheeled motion [5]. Trajectory of the foot can be divided into four-phase; the lower, support, lift, and return, as shown in Fig. 2. The down phase is the part where lower the leg from maximum height and the support phase is the part that moves with the foot in contact with the floor. And the lift phase is the part that raises the foot to maximum height, and the return phase shows the foot moves forward while lifted. Additionally, overcoming the obstacles in the path is easier with long step height.

Unlike earthworm, caterpillar uses both peristaltic movement and walking when it crawls. It has 8 pair of legs; three pair of thoracic legs and five pair of abdominal prolegs. When it becomes imago, the five pair of prolegs are degenerated, but these prolegs are more consistent than thoracic legs when the caterpillar crawls and a pair of prolegs in a segment moves in same phase [6]. To generate the peristaltic move-

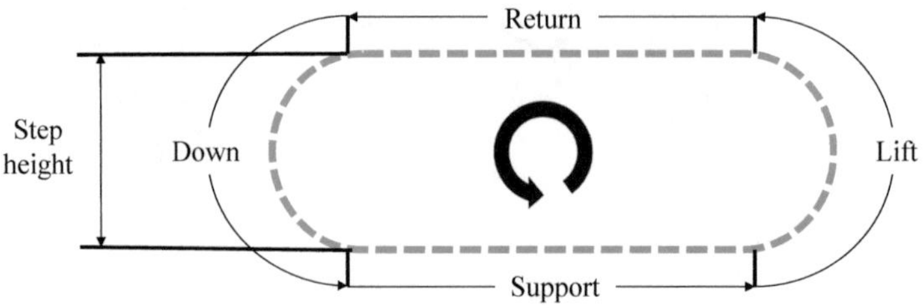

Fig. 2 The phase for foot trajectory

ment, least three segments are required and three legs are applied in each segment to support the system in pipe-shaped environment. There two-ways to apply the peristalitic movement, one is generate peristaltic wave by segment (all legs in a segment move in same walking phase) and the other is by each leg in each direction (all legs in a segment are in different walking phase). This was simulated with dynamic simulation. In case of all legs in a segment move in same walking phase showed better performance in speed and stability of the locomotion unit which is more similar to the movement of caterpillar.

3 Drilling Unit

3.1 Drilling Methods

There are three types of drilling which are the rotary drilling, hammer drilling, and rotary-percussion drilling. The rotary drilling is a basic drilling method that drills a hole in the work surface by using the rotation motion of the drill bit very similar to a general hand drill. It is possible to drill a deep hole but has some drawbacks such as the bit wear and low penetration rate. The hammer drilling is a drilling method that applies a shock in the axial direction of the drill bit and produces a hole in the work surface such as a lane or a cement through a linear motion instead of a rotation motion. In this method, however, the penetration rate changes depending on the material of the work surface. For example, this method shows high penetration rate in rocks such as granitic rocks, sandstones, and limestone rocks, but in the case of work surfaces such as soil, it shows low penetration rate. Rotary-percussion drilling is a combination of rotation and hammer drilling, which drills through rotary motion and linear motion of the drill bit. According to work of Franca [7], the penetration per revolution of the rotary-percussion drilling (d) is expressed as the sum of the penetration per revolution of rotary drilling (d_r) and hammer drilling (d_p) as follows:

$$d = d_r + d_p. \tag{1}$$

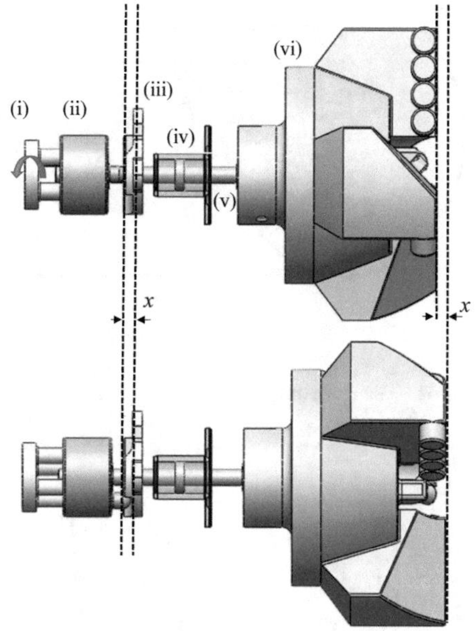

Fig. 3 The rotary-percussion drilling mechanism for embedded drilling system. Numbered components are ; (i) Motor connection and spring bracket (ii) Power transmission bracket with pin (iii) Stepped Bracket (Fixed) (iv) Linear bush (v) Transmission shaft (vi) Drill bit

In this way, not only does the penetration rate improve, but also it is possible to complement the disadvantages of each drilling method. Therefore, the rotary-percussion drilling is applied to drilling unit of the embedded drilling system.

3.2 Mechanism of the Drilling Unit

The designed drilling unit is able to generate rotational and linear motion to the drill bit with single actuator as shown in Fig. 3. The spring is located between (i) and (ii) to use elastic force for hammering movement. The pin in (ii) rotates on (iii), which has steps and slopes equivently around the bracket. Therefore, when the pin is on the stepped section, part (ii), (v), and (vi) moves forward with distance x by the elastic force from the spring. After this stroke, the pin follows the slope of (iii) to compress the spring.

4 Simulation

Two simulations are demonstrated to show the performance of the system. One is performed with the locomotion unit and rotary drilling and the other is performed with the locomotion unit and the rotary-percussion drilling. Figure 4 shows the environmental settings for these simulations. Leg mechanisms and the drill bit are activated

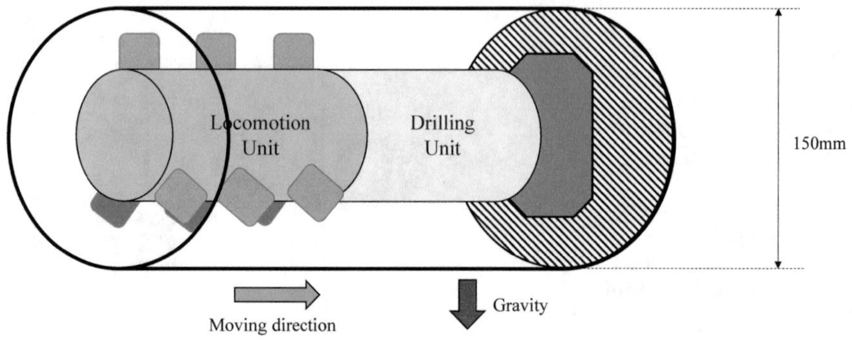

Fig. 4 The environmental settings for contact simulations

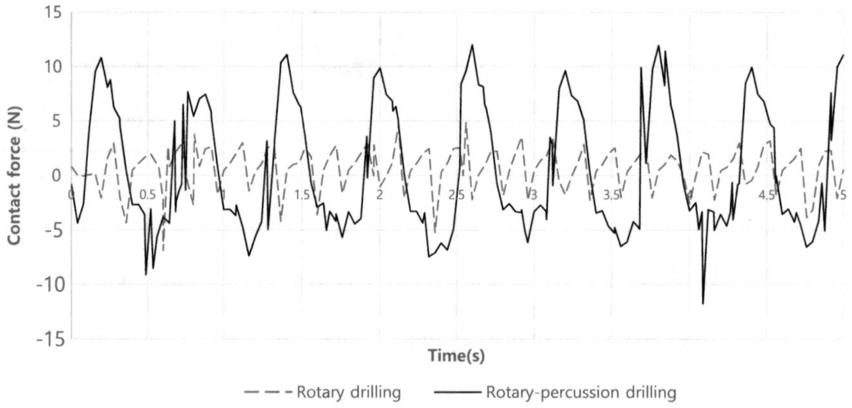

Fig. 5 The contact force between the bit and the wall from the simulation results

while simulation running and the contact force between the bit and the wall is measured to check the WOB of the drilling system for each simulation. Figure 5 shows the results from both simulations. The simulation with rotary-percussion drilling gained more contact force (average: 0.577 N, maximum: 11.993 N, minimum: −11.742 N) than the simulation with rotary drilling (average: 0.574 N, maximum: 4.832 N, minimum: −6.909 N). Although both maximum and minimum force is almost doubled, the average value is very similar. However, there is a bigger advantage to putting more force momentarily in order to drill the ground. And the negative forces must require additional research to improve bearing capacity by improving the leg mechanism.

5 Conclusion

The locomotion unit and the drilling unit of the embedded drilling system was proposed and drilling performance of the system was demonstrated by simulations. The comparison of the effects of the two structures through simulation was clear, but the drilling efficiency is affected by various factors in real world. Therefore, it is necessary to build an actual system and conduct experiments, and a method of increasing the efficiency of the locomotion unit is also required. In addition, the steering mechanism and localization algorithm for this system will be applied for directional drilling.

Acknowledgements This work is supported by the Technology Innovation Program (#10076532, Development of embedded directional drilling robot for drilling and exploration) funded by the Ministry of Trade, Industry & Energy (MOTIE, Korea). The students are supported by Korea Ministry of Land, Infrastructure and Transport (MOLIT) as U-City Master and Doctor Course Grant Program.

References

1. Ritcher, L., Coste, P., Gromov, V., Grzesik, A.: The mole with sampling mechanism (MSM)—technology development and payload of beagle 2 mars lander. In: 6th ESA Workshop on Advanced Space Technologies for Robotics and Automation (ASTRA) (2000)
2. Shieh, W.B., Tsai, L.W., Azarm, S.: Design and optimization of a one-degree-of freedom six-bar leg mechanism for a walking machine. J. Robotic Sys. **14**, 871–880 (1997)
3. Saha, S.K., Prasad, R., Mandal A.K.: Use of Hoeken's and pantograph mechanism for carpet scrapping operations. In: National Conference on Machines and Mechanisms (2003)
4. Patnaik, S.: Analysis of Theo Jansen mechanism (Strandbeest) and its comparative advantages over wheel based mine excavation system. IOSR J. Eng. **5**, 43–52 (2015)
5. Ghassaei, A., Choi, P., and Whitaker, D.: The Design and optimization of a crank-based leg mechanism. Pomona College Department of Physics and Astronomy (2011)
6. van Griethuijsen, L.I., Trimmer, B.A.: Kinematics of horizontal and vertical caterpillar crawling. J. Exp. Bio. **212**, 1455–1462 (2009)
7. Franca, L.F.P.: A bitrock interaction model for rotarypercussive drilling. J. Rock Mech. Mining Sci. **48**, 827–835 (2011)

Design and Development of Innovative Pet Feeding Robot

Aswath Suresh, Debrup Laha, Dhruv Gaba and Siddhant Bhambri

Abstract This paper describes a pet feeding robot which can be controlled from anywhere in the world. The robot features a differential drive mechanism, collision free distance sensing, wireless control, User-End GUI and Camera Vision. The robot hardware includes Arduino Microcontroller, Raspberry Pi 3, 6-DOF Robotic Arm, RF-SPI, HD Camera, Sabertooth Motor Driver and 4S Lipo Battery. Inspired from nature, a reflex mechanism has also been integrated into the robot design to minimize damage, by automated safety reflexes using ultrasonic distance sensor. The four-wheeled mechanism ensures that it can traverse stairs easily. The mechanism provides traction due to its body weight. The robot finds applications in feeding any kinds of pet from anywhere, remotely explore the house for any potential theft and deceive robbers to believe that someone is at home. It also includes personalize daily meal portions, stay connected with real time alerts, know your pet is alright when you are away, stores up to 7 lbs, keeps food fresh and takes cares of your pet's diet. The robot also enables a user to prevent a pet from eating a specific food while still allowing access to that food to other pets. All these features make the robot user friendly for users having more than one pet.

Keywords Camera vision · Drive mechanism · GUI · Pet
Raspberry pi 3

A. Suresh (✉) · D. Laha · D. Gaba
Department of Mechanical and Aerospace Engineering,
New York University, Brooklyn, USA
e-mail: as10616@nyu.edu

D. Laha
e-mail: dl3515@nyu.edu

D. Gaba
e-mail: dg3035@nyu.edu

S. Bhambri
Department of Electronics and Communication Engineering,
Bharati Vidyapeeth's College of Engineering, Delhi, India
e-mail: siddhantbhambri@gmail.com

J.-H. Kim et al. (eds.), *Robot Intelligence Technology and Applications 5*,
Advances in Intelligent Systems and Computing 751,
https://doi.org/10.1007/978-3-319-78452-6_19

1 Introduction

The pet feeding robot is in great need because pet keeping is a time-consuming responsibility and we want to provide convenience to users by helping them feed their pets easily and smartly from anywhere around the world.

Keeping pets takes many commitments. This includes giving them company, showing your concerns and of course, feeding them on time and in the right manner. However, not everyone is a pet expert; taking care of your pet's diet can be hard and time consuming. One of the top health concerns of pets are overeating and obesity. Especially at younger age, they are usually satisfied with however much is given to them. Many adult pets are fed unscientifically that later may cause short lifespan. Another problem of feeding pets is that users might not always be home regularly as they need to travel to another country for business or vacation. Being occupied by personal plans knowing that they still have a starving little fellow at home to be taken care of is always a concern that bothers users. The third concern that we want to deal with is the fact that there hasn't been any product in the market right now that is able to dispense food for pets monitored by its user in real-time. However, pets themselves might not necessarily recognize the potential health problems of eating the wrong food. There are products like Petnet, AutoPetFeeder, and Automation Pet Feeder [1–4] which can be scheduled to dispense food at certain interval of time, but it lacks real time monitoring and mobility. Therefore, we want to take care of the users' concern of feeding by building a phone/laptop controlled real-time semi-automatic pet feeder that can dispense the desired food as per the user by live camera feedback.

The autonomous features are the obstacle avoidance using the ultrasonic sensor which will enable the robot to avoid collision when something comes in its way. Also, the user has to take the robot to the pet to feed it because the pet could be anywhere. The reason why carrying food is better than the prior automatic pet feeding mechanisms, is that the pet may not go to the pet-feeder to eat food. Whereas, in our system we will carry food to the pet and make sure that it eats food, so as to take care of the diet of the pet.

Our solution not just only carries food to the robot, it also withdraws the appropriate amount of food from the storage for maintaining a healthy diet of the pet. This way we have full control over the diet of the pet, even if we are not at our home.

None the less, our robot will keep the pet engaged and will not allow it to feel lonely. Also, the robot is equipped with small speakers which could send the audio feed from the user to the pet, such that the pet feels that his owner is near it.

Apart from this, since we have a device which could give us live video feed of the house. It could easily help us to prevent thefts. As, it will act as a secondary watch dog for the house, and since it runs on electricity it could even permanently stay out and keep a watch out of our house, and whenever it senses someone is trespassing the user's property it will detect thief's presence and will make a video recording of the intruder. Also, the robot will secretly keep on moving in the house, which will make the thief that someone is in the house. And distract him/her, till the police comes.

Also, our design is made in a symmetric manner. Such that even if the robot topples, it could still function (for the basic design). Further, the design is made in such a manner that it could easily climb on stairs (coarse and large diameter tires), this allows the robot to freely move around in the house.

2 Mechanical Design

The four-wheeled mechanism ensures that it can traverse over a considerable height greater than the chassis height which could be as much as twice the diameter of the wheels. The advantage of this design is that it can still run even if it gets toppled allowing the robot to overcome obstacles and traverse over a highly-rugged terrain like the stairs. The robot gets traction only due to its body weight without having to compromise the strength of the chassis. A differential drive robot uses a simple locomotion system composed of four drive wheels. Drive wheels on left and right are paired, each pair is independently controlled, which is responsible for the robot's rotation and translation. A pure translation is reached when all drive wheels have the same velocity—magnitude and direction—, while a pure rotation appears when the velocities of both left and right-side drive wheels are the same in magnitude but with an opposed direction with respect to each side. Kinematics equations define the interaction between the control commands and the corresponding space state. Thus, in a differential drive locomotion robot, these equations will reflect the robot's position (x, y, θ) when the velocity of each drive wheel is controlled (V_{right}, V_{left}). Therefore, translational and rotational velocities can be expressed by means of the drive wheel velocities:

$$V = \frac{V_{right} + V_{left}}{2}$$

$$\omega = \frac{V_{right} - V_{left}}{L}$$

where L is a length of drive wheel axis. The kinematic equations of a differential drive locomotion robot can be expressed as:

$$\dot{x} = \frac{V_{right} + V_{left}}{2} \cdot \cos(\varnothing)$$

$$\dot{y} = \frac{V_{right} + V_{left}}{2} \cdot \sin(\varnothing)$$

$$\varnothing = \frac{V_{right} - V_{left}}{L}$$

2.1 Chassis

The entire chassis is comprised of a square tube pipe connected to four regular circular tube pipes on either ends by means of two custom-made T-joint pipe tube fittings, all of which are made of Polyvinylchloride (PVC). The four wheels of the robot are connected to the either ends of the four circular tube pipes. The schematic view of the chassis with the wheels is shown in Fig. 1.

2.2 Mechanism

The four-wheeled robot is symmetrical on either end. This ensures that the robot will still be in motion even if it topples while moving through a rugged surface. The Fig. 2 shows the prototype of the four-wheeled mechanism.

2.3 Finite Element Analysis

Finite Element Analysis was performed on the square tube pipe with a load of 35 N being equally distributed on the surface of the pipe. After analyzing, it is noticed that the maximum deformation in the pipe is 0.056 mm which can be allowed in the design range. We can also observe that the maximum von-Mises stress in the pipe is 0.54 MPa which is approximately 4500 times less than that Young's modulus of PVC. It is also noticed that the maximum principal stress is about 0.625 MPa. The results are shown in Figs. 3, 4 and 5.

Fig. 1 CAD model of the pet feeding robot

Fig. 2 Prototype

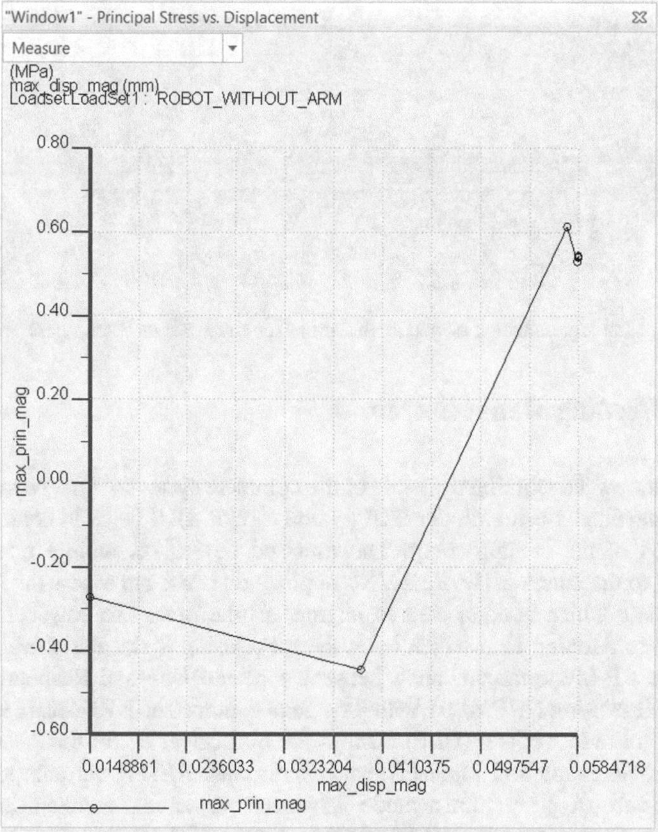

Fig. 3 Graphical representation of maximum principal stress versus maximum displacement

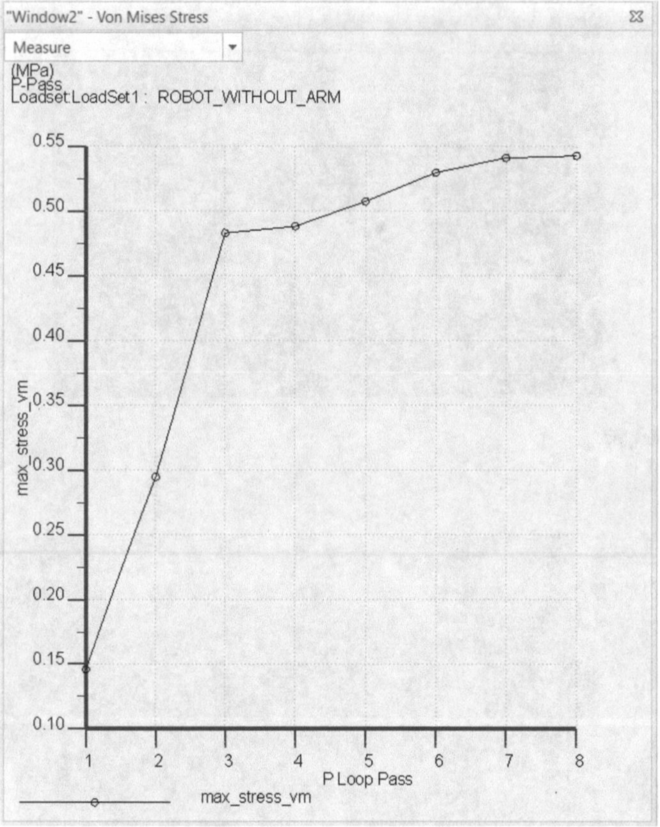

Fig. 4 Graphical representation of maximum von mises stress versus P loop pass

3 Pet Feeding Robot System

Figure 6 shows the working principle of the complete system. At the user-end there is a GUI developed using tkinter GUI python toolkit. GUI helps in controlling the movements of pet feeding robot. The user-end device i.e. mobile or laptop, is connected to the internet. The Real VNC application is used to access the Raspberry Pi 3 at home which is connected to internet all the time. The Raspberry Pi 3 is connected to Arduino Uno which is on the robot using Serial Peripheral Interface (SPI). The RF-SPI communication between Arduino Uno and Raspberry Pi 3 is made wireless using NRF24L01 with a range of 1 mile. The NRF24L01 is a highly integrated, ultra-low power (ULP) 2 Mbps RF transceiver IC for the 2.4 GHz ISM (Industrial, Scientific and Medical) band. With peak RX/TX currents lower than 14 mA, a sub μA power down mode, advanced power management, and a 1.9–3.6 V supply range, the NRF24L01 provides a true ULP solution enabling months to years of battery lifetime when running on coin cells or AA/AAA batteries.

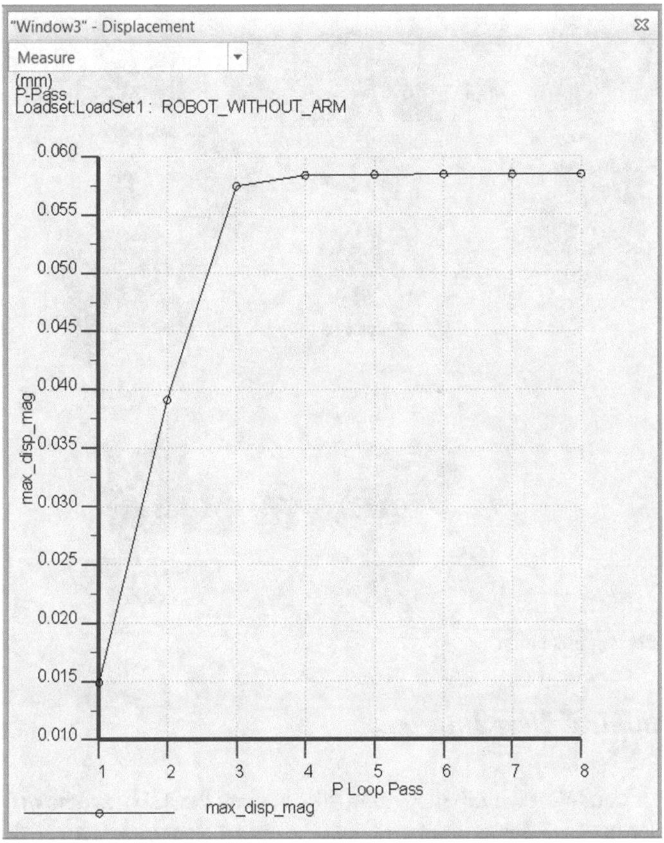

Fig. 5 Graphical representation of maximum displacement versus P loop pass

The Arduino gives commands to the motor using a motor driver called Sabertooth for the required robot movement. The Arduino cannot handle high voltage and high current requirement of the motors. So, we used high current (up to 60 A) and high voltage (up to 32 V) motor driver. When the button is pressed in the GUI at the user-end, a specific function is called in python program in the raspberry pi 3 based on which button is pressed. The function sends set of string value (Speed, Direction) to the Arduino through RF-SPI Communication. The camera feed from the HD IP Camera is always available at the user-end via internet. This system makes it very easy to find the pet and feed it on time.

Apart from this, the robot is equipped with a 6-DOF robotic arm which is meant for picking up the food from the rack and holding it carries the food till it reaches the pet, and then placing the food in front of the pet. The feature of dropping the food is pre-coded such that when the user comes near the pet, and the user just have to call the function of the dropping food. This way work of the user is reduced.

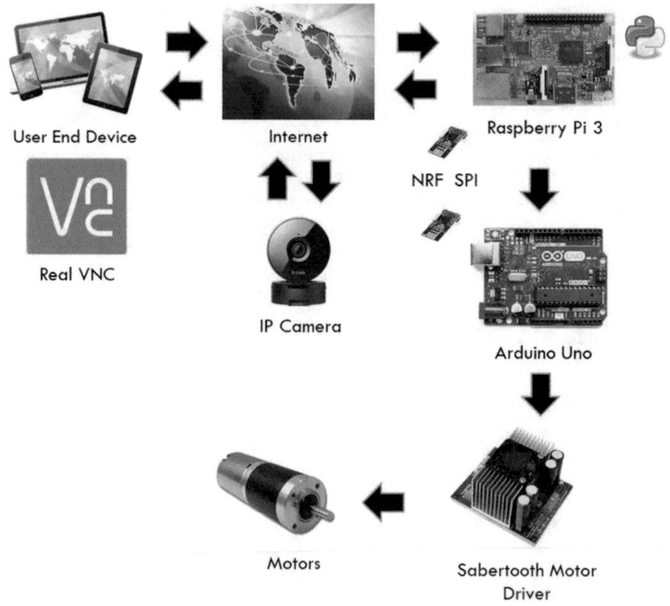

Fig. 6 System representation

3.1 Graphical User Interface

The Tkinter module is used in python to develop the GUI as shown in Fig. 7. Tkinter is Python's de facto standard GUI (Graphical User Interface) package. It is a thin object-oriented layer on top of Tcl/Tk. Tkinter is not the only GUI Programming toolkit for Python. It is however the most commonly used one. A window named 'Control Panel' is created with four buttons namely Forward, Backwards, Right and Left. If the 'FORWARD' button is pressed the robot will move forward, if 'RIGHT' is pressed it moves towards the right and so on. The GUI developed in python can be accessed using real VNC from anywhere in the world where internet is available. It is possible to make a GUI which is more user interactive.

3.2 Raspberry Pi-Arduino RF-SPI Communication

The Serial Peripheral Interface (SPI) bus was developed by Motorola to provide full-duplex synchronous serial communication between master and slave devices. The SPI bus is commonly used for communication with flash memory, sensors, real-time clocks (RTCs), analog-to-digital converters, and more. Standard SPI masters communicate with slaves using the serial clock (SCK), Master Out Slave In (MOSI), Master In Slave Out (MISO), and Slave Select (SS) lines. The wiring for

Fig. 7 GUI

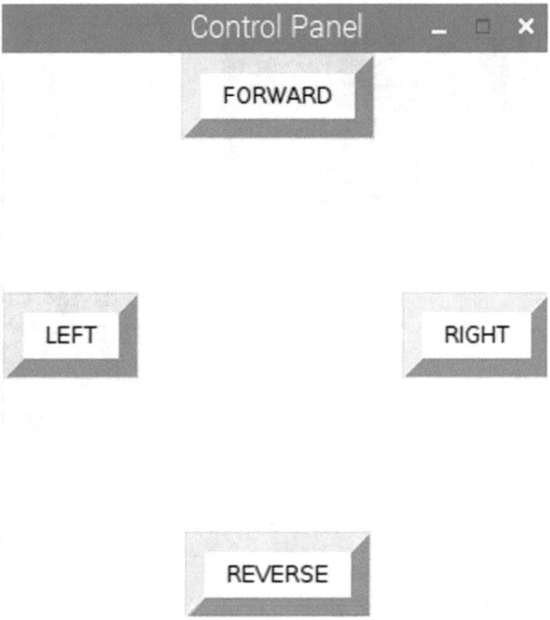

SPI communication between Arduino and Raspberry Pi along with NRF24L01 is done as shown in the Fig. 8.

3.3 Camera Feedback

The IP Camera on the pet feeding robot is connected to a wireless router using Wi-Fi Protected Setup (WPS). The router is always connected to the internet and the user can access the HD camera using the camera IP address over the internet cloud. The user will be receiving the live feedback of 1080 p live footage of what the pet feeding robot sees. This helps the user to navigate the robot to the required destination. The Fig. 9 shows the camera setup to get live feedback at the user end.

3.4 Mechanical Layer and Power Management

The pet robot uses four high torque 7 A motors for travelling in any sort of rugged terrain with ease. The four-wheel mechanism is developed in such a way that the robot can move even if it gets toppled. This feature allows it to climb up and down

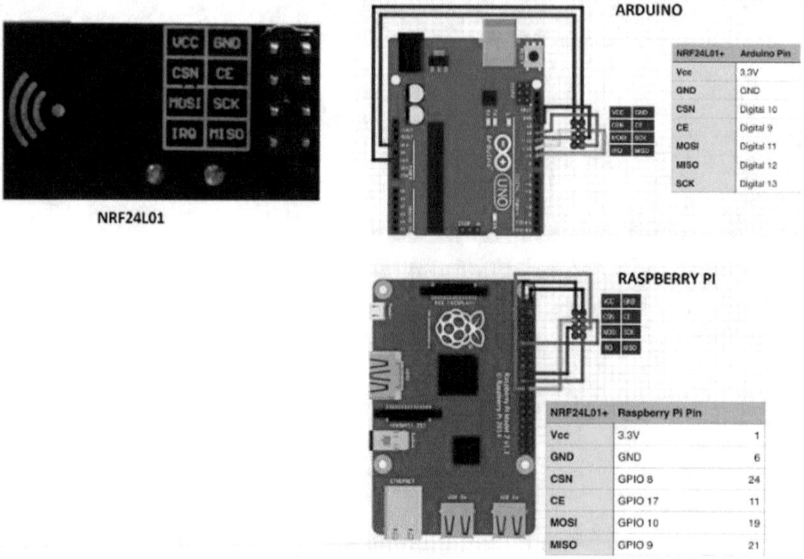

Fig. 8 SPI communication

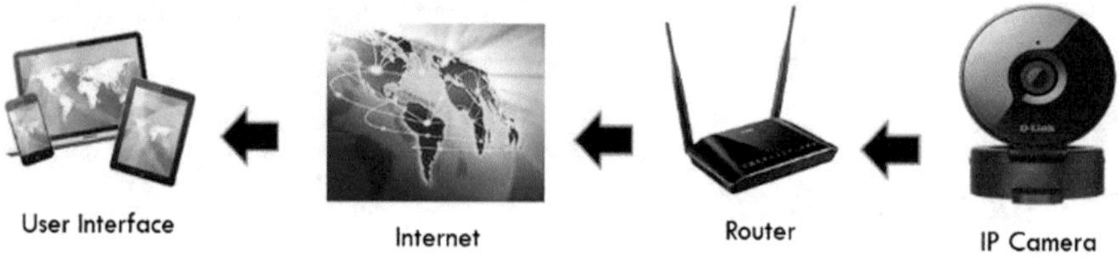

Fig. 9 Camera setup

the stairs without any issue. The signal received from Arduino to the Sabertooth motor driver drives the 28 A 12 V motor system in the required direction. The robot uses a 5000 mah 12 V high power lithium polymer batter. The Arduino and HD camera are powered from the 5 V regulated supply available from the sabretooth motor driver. Back current and short is taken care of using diode and protection circuit which makes the system totally safe. The Fig. 10 shows the power management and mechanical layer of the system.

Fig. 10 Mechanical and power management

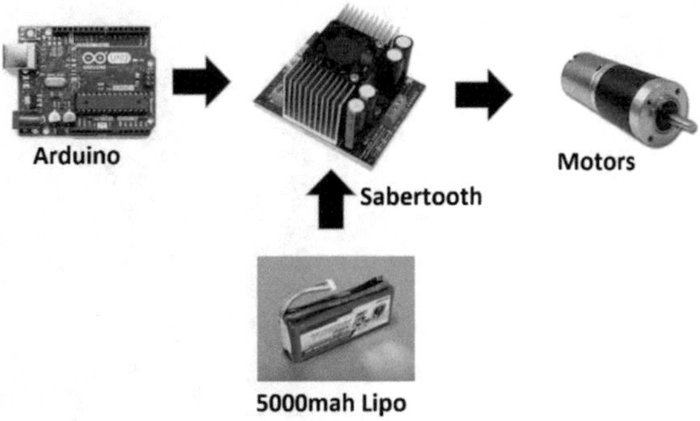

4 Result and Discussion

The robot was tested as a scenario of a person sitting at office and feeding his/her pet at home. The different scenarios are shown in Figs. 11, 12 and 13 till the result was achieved.

The Fig. 11 shows a person sitting in the office with the GUI of the pet feeding robot trying to feed his/her pet at home.

The Fig. 12 shows a pet feeding robot loaded with food at home, controlled from the office. Figure 13 shows that a robot is on its way to the target and the top left shows the camera feedback which the user sees while controlling the robot.

It is observed that the pet feeding robot at home is capable of doing the real-time pet feeding task without any difficulty from the office which is far away, and the performance is good. The four-wheeled mechanism for climbing up and down was tested and led to satisfactory result. The Fig. 14 shows results when the pet feeding robot was tested at the staircase.

Fig. 11 User sitting in office

Fig. 12 Pet feeding robot at home

Fig. 13 Pet feeding robot reaching target

Fig. 14 Pet feeding robot staircase test

5 Modification with Robotic Arm

This robotic arm is made of Aluminum and has a 6° of freedom inverse kinematic machine. The first of the six axes, is located at the base of the arm that allows for the rotational movement of the arm from left to right and allows the arm to spin to 180° from the central point. The second axis allows leaning of the lower arm of the robot to and fro. The third axis allows the upper arm to raise and lower which lets it access parts easily. The arm is equipped with a gripper turret with end effectors which are positioned with the fourth axis. This axis also helps in the rotation of the upper arm in a circular motion helping it move between horizontal and vertical orientations. The fifth axis allows for the pitch and yaw motion of the arm letting it to tilt up and down. The sixth axis allows for a twisting motion which in turn allows the arm to rotate freely in 360° in either clockwise or counter clockwise direction for positioning the end effectors as well as for controlling the parts.

After attaching the robotic arm, we required two cameras for optimal operation, one is placed towards the posterior end of the pet feeding robot and the other camera is placed on the end-effector gripper of the robotic arm. The second camera provides the user with a better view while operating the robotic arm.

The Fig. 15 is a CAD representation of the six DOF Robotic Arm of the pet feeding robot. The design shows three links and three joints, connected with the end effector-gripper to form the robotic arm.

The above Fig. 16 shows the CAD representation of the Robotic arm on the pet feeding robot. This robotic arm, is attached for loading the pet food from the storage

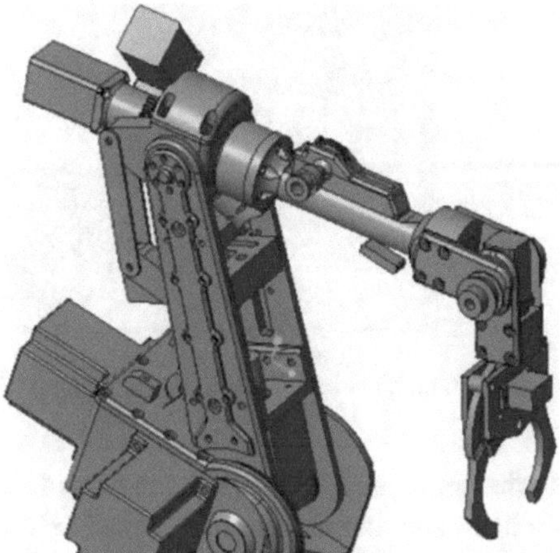

Fig. 15 6 DOF robotic arm

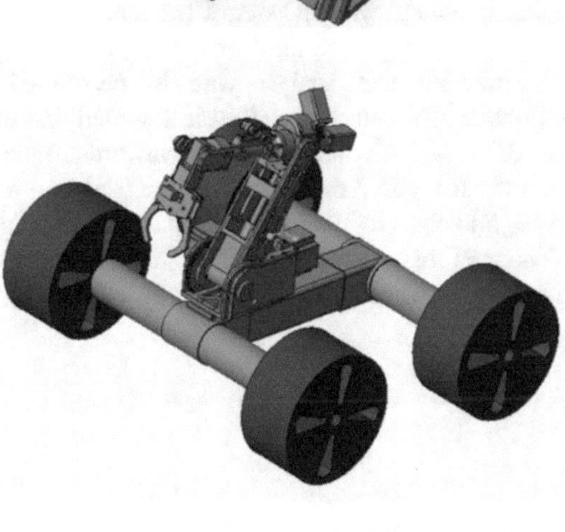

Fig. 16 CAD model of pet feeding robot with robotic arm

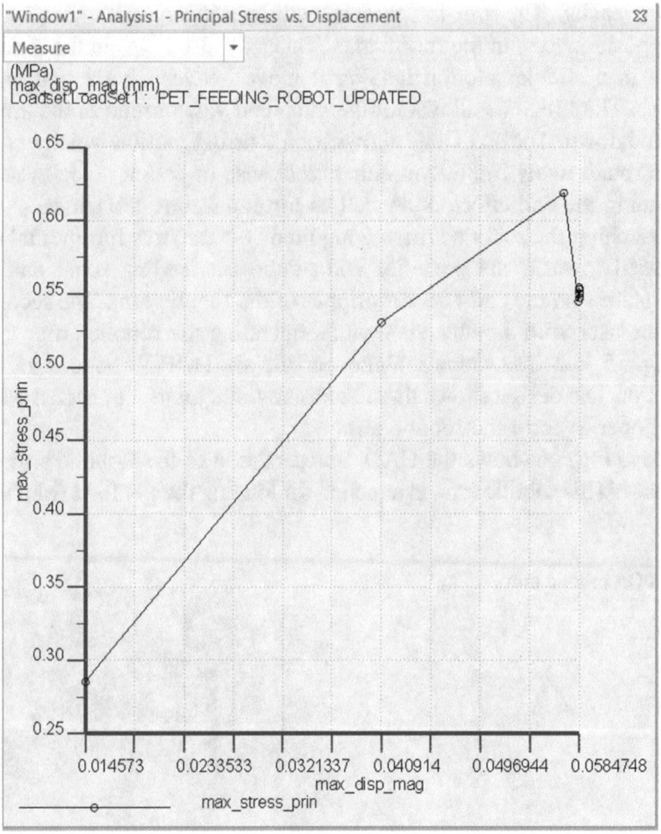

Fig. 17 Graphical representation of maximum principal stress versus maximum displacement

to the pet feeding platform on the robot. Thus, by our way we could keep the food contents safe in a storage area and place it on the platform, then the user will have to finally drive the robot to the pet for feeding it, via the mobile application (live video feedback). Since, we can control the amount of food given to the pet, we could avoid food wastage and limit overeating of food by the pet.

A Finite Element Analysis was also performed on the Pet Feeding Robot with the Robotic Arm attached to it. It is assumed that the weight of the Robotic Arm is about 35 N which will be acting downwards to the surface of the square tube with which the Robotic Arm is in direct contact. The results are shown in Figs. 17, 18 and 19. It is observed that that the maximum principal stress is about 0.62 MPa. It is also noticed that the curve defining the maximum Von Mises Stress as well as the maximum Displacement converge with increasing number of P Loop Passes. It is

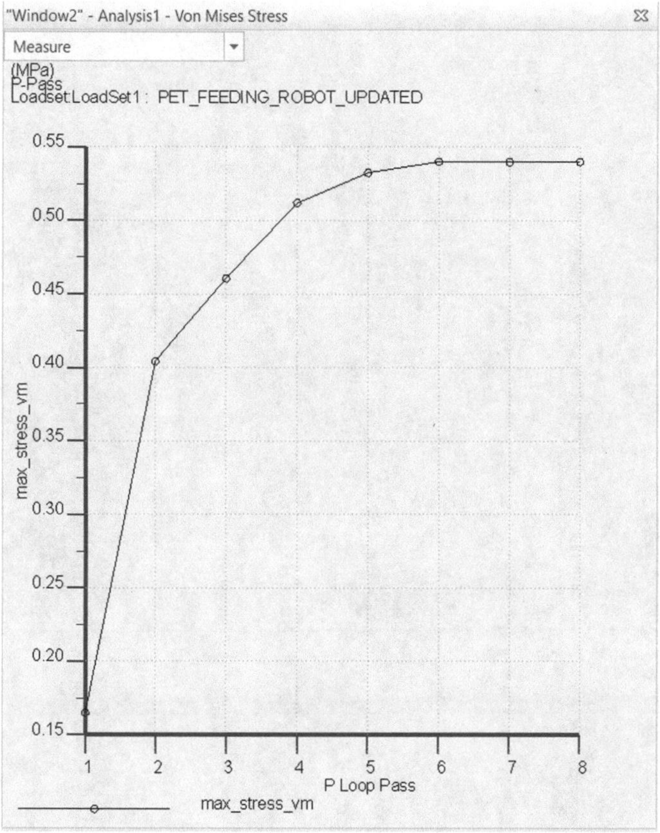

Fig. 18 Graphical representation of maximum Von Mises stress versus P loop pass

observed that the maximum Von Mises Stress is about 0.54 MPa and the maximum deformation is about 0.057 mm. It is noticed that the results are very similar to the FEA results done on the Pet Feeding Robot when the arm was not placed.

The above Fig. 20 illustrates the final prototype of the Pet Feeding Robot along with the 6 DOF Robotic arm. We have used high friction tires to allow our prototype to climb stairs easily, and attached our electronic system in such a manner to keep the center of gravity towards the mid-point of the prototype (considering that the Robotic arm is attached towards the other end of the prototype).

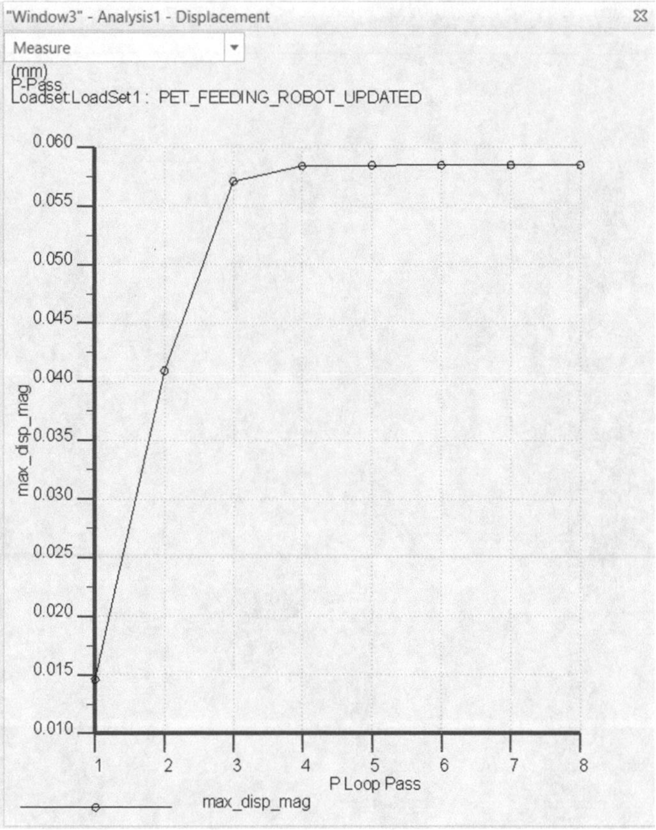

Fig. 19 Graphical representation of maximum displacement versus P loop pass

Fig. 20 Final prototype of pet feeding robot

6 Problems and Challenges

The only issue with this kind of a mobile design is that the pet might get scared of the robot and might attack. To avoid such a scenario, the user will have to introduce and make his/her pet familiar enough with the pet feeding robot. So that the pet should treat the robot like another individual of the house, and should not be afraid of the robot.

7 Conclusion and Future Discussion

Our team has done a lot of research, coding and testing in the field of pet feeding, which has made us feel that we have succeeded in our objective. Most of our design goals were achieved. We received a live feed from the camera mounted on the robot, over internet. We controlled the pet-feeder precisely through GUI of smart devices. The pet feeding robot was easily able to climb the stairs up and down. Also, the all-terrain vehicle design used for the prototype, set our product apart from any other product in the market. Moreover, the 6 DOF robotic arm is used to load the pet's food contents from the storage to the on-board platform. From this platform the pet (e.g. Dog), could easily eat the food. Thus, the application was successfully tested in a real-time environment and satisfactory results were achieved.

Acknowledgements The authors would like to thank Makerspace and NYU Tandon School of Engineering for providing support to carry out the research and experiments.

References

1. Heil, R., McCarthy, K., Rege, F., Rodriguez-Carlson, A.: The Smart Pet Feeder: A Proposal to Design and Build an Automated Pet Feeder Capable of Preventing One Pet from Eating Another Pet's Food, 30 Jan 2008
2. Zhao, Z., He, Z., Ling, F.: Automatic Pet Feeder Project, 10 Feb 2016
3. Kolmanovsky, I., McClamroch, N.H.: Developments in nonholonomic control problems. IEEE Control System Magazine **15**, 20–30 (1995)
4. Konolige, K.: Saphira Software Manual. SRI International (1998)
5. http://petnet.io/smartfeeder
6. http://www.autopetfeeder.com/

SignBot, Sign-Language Performing Robot, Based on Sequential Motion of Servo Motor Arrays

Rami Ali Al-Khulaidi, Nuril Hana Binti Abu Bakr, Norfatehah Mohammad Fauzi and Rini Akmeliawati

Abstract This paper presents the development of SignBot, a 3D printed robot which can perform Malaysian Sign Language (MSL). This work is the first attempt to eliminate the barriers of communication between hearing impaired individuals and the mainstream society as it is the first robot to perform MSL. The signing, in this work, can be done with two hands. The robot hands were developed with detailed finger joints. Micro servo motors were installed to allow for the signing motions for the relevant joints of selected letters, numbers as well as phrases for emergency cases. The sequential movements of the servo motor arrays are stored in the database to represent particular signs, and retrieved by the microcontroller based on the speech input detected, and finally executed accordingly by the servo motors to perform the sign.

1 Introduction

Based on the statistic from United Nations, the disabled people is around 10% among people on the earth. The number of people who live in Malaysia is about 28 million, so the number of disabled people in Malaysia is approximately 2.8 million people. Nevertheless, only 280 thousand disabled people registered at "Jabatan Kebajikan Masyarakat (JKM)" [1]. However, the gap of communication between this notable segment of the society and the healthy people is still unresolved as the researchers focusing on video databases and animation to deal with such issue [2, 3]. Luckily, in the recent years, humanoid robots have been implemented in

---

R. A. Al-Khulaidi · N. H. B. A. Bakr · N. M. Fauzi · R. Akmeliawati (✉)
Department of Mechatronics Engineering, International Islamic University Malaysia, Jalan Gombak, 53100 Kuala Lumpur, Malaysia
e-mail: rakmelia@ieee.org

R. A. Al-Khulaidi
e-mail: teemomory@gmail.com

J.-H. Kim et al. (eds.), *Robot Intelligence Technology and Applications 5*, Advances in Intelligent Systems and Computing 751,
https://doi.org/10.1007/978-3-319-78452-6_20

performing sign languages. Nino is the first full-sized humanoid robot, who can demonstrate sign language [4]. In [5], Nao-25 is a child-sized humanoid robot who can act as a sign language interactive game for hearing impaired children. In [6], a Humanoid Robotic Hand Demonstrates Sign Language motion needs five steps of realization of the robot hand that is capable of conveying feelings and sensitivities by finger motion. A robotic hand with the capability of spelling words using the manual alphabet was developed [7]. Additionally, there are some real works have been conducted recently including the speech system as an input for the robot. In other words, the robots perform the sign language in response to the uttered speech by ordinary speaking people. Jonathan Gatti et al. created an automated voice to sign language robotic hand translator. They took advantage of OpenSCAD software to design the robot hand which was printed by FDM technology [8].

However, there are some limitations in previous works such as the design and clear description of performance. The inadequate design of the Nao robot makes it insufficient to perform the sign language as it only has three fingers. Almost all conducted works of sign language robots lack the clarity in performance since they do not include a clear description of the movements of the robot nor did they assign the joints motion corresponding to targeted letters. Moreover, even the research that employs the speech system is still limited to the movement of fingers and neglecting the mobility of the wrist and the elbow. In addition, the most critical is that there are no works specifically in the Malaysian sign language robots.

In this paper, we present SignBot, Malaysian Sign Language Performing Robot developed by IIUM. The objective of the robot is to help the SHI community to communicate easily with the regular people by performing the MSL using its two hands based on the speech input from the ordinary people. The robot can perform selected letters and numbers. In addition to single characters, it can perform emergency phrases.

This paper is organized as follows. Section 2 gives a brief overview of the design and construction of the system including both the mechanical and electrical structures. The results and discussion are presented in Sect. 3. Finally, the conclusion is drawn in Sect. 4.

2 System Design and Structure

The proposed system consists of speech processing unit at the input side, a database of signs, microcontrollers, and actuators which comprises twelve servo motors; eight servos for the right hand and four servos for the left hand. The system uses a microphone to capture the speech signal, and the speech is processed by using Microsoft visual studio and translated into texts which will be checked by the microcontrollers whether the words exist in the database or not, and it will then retrieve the corresponding servo motors' sequential mechanism. The servo motors

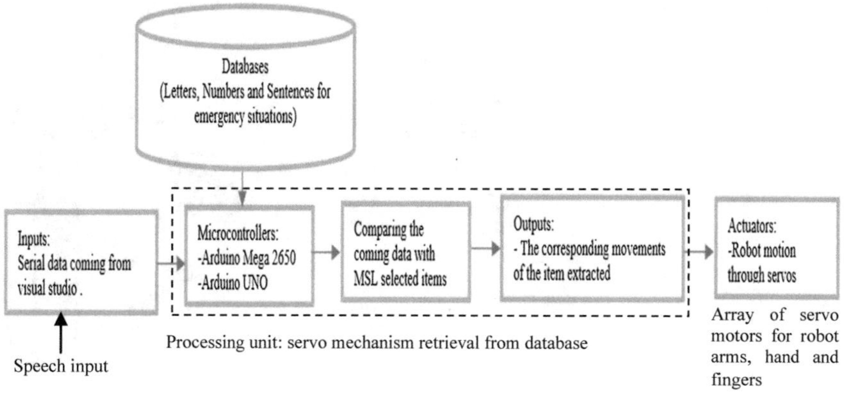

Fig. 1 Overall block diagram of the proposed system

are identified by number as later explained in more detail in Sect. 2.2. The movement of the servos representing the output stage. The overall process is shown in Fig. 1. At present stage, the database contains signs for letters, numbers and selected words that are often used in emergency cases, such as "tolong" (i.e., "help"), "sakit" (i.e., "painful"), etc.

2.1 Mechanical Design of SignBot

The physique of the system consists of two main parts: the body along with head and the hands. Each section is made up of different materials. The body and the head are made out of Perspex while the hand is printed by means of 3D printing. Additionally, we have designed the head and the body while the robot hands are implemented from an open source design downloaded from Thingiverse website [9]. Figures 2 and 3 show the mechanical design of the body and head, and Fig. 4 represents the STL 3D files of the hand.

2.2 Servo Mechanism

The system consists of two parts: the speech system as an input and the servo motors as an output. The speech system is developed in C# language in Microsoft visual studio environment. The recognized speech or word is sent to the Arduino as strings by means of serial communication. Each string is assigned to one sign representing the selected letters, numbers and/or the emergency words. If the sound of the letter A, for instance, is uttered and detected, the robot will perform the

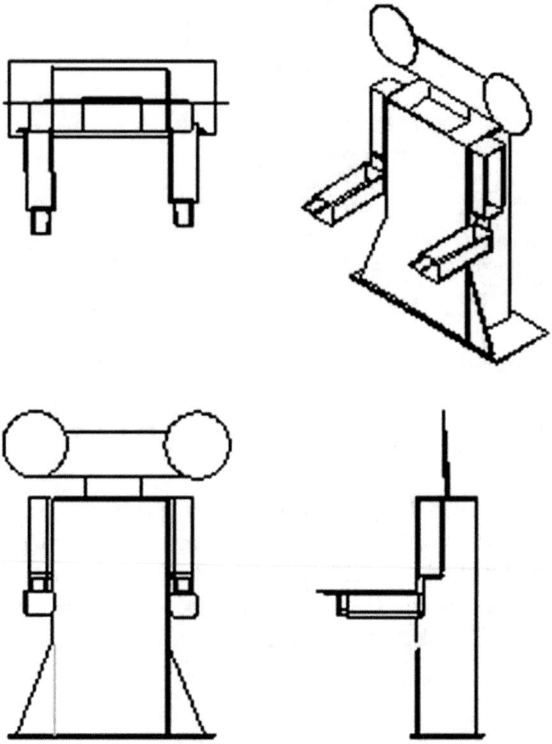

Fig. 2 Mechanical design

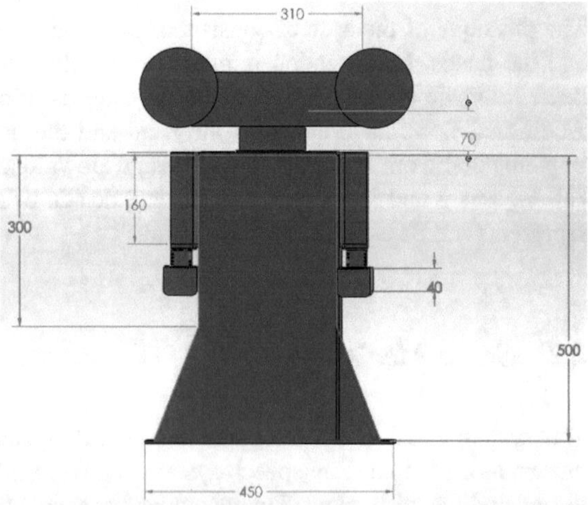

Fig. 3 The dimension of the robot

corresponding movements of the letter A in MSL. The actuation of the system is achieved by an array of twelve servo motors. The right side has eight servos while the left has four servos as shown in Fig. 5. Every single finger of the right hand has

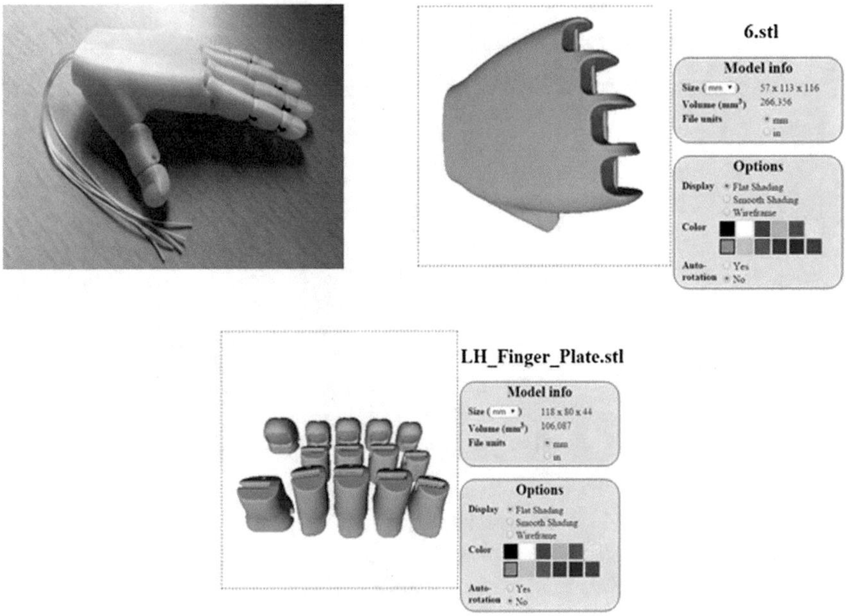

Fig. 4 Mechanical structure of the robot right hand [9]

Fig. 5 Servo distribution for both hands

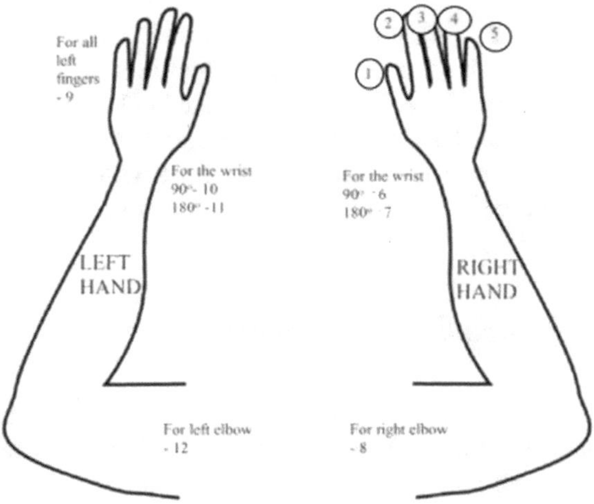

a servo motor, and the remaining three servos are for the pitch and roll movements of the wrist as well as flexion/extension movement of the elbow. The left side, on the other hand, has only four servos; one for all the fingers as we need the hand to form a fist when representing the words in emergency case. Similar to the right side, the roll and pitch movement of the wrist and the elbow have three servos

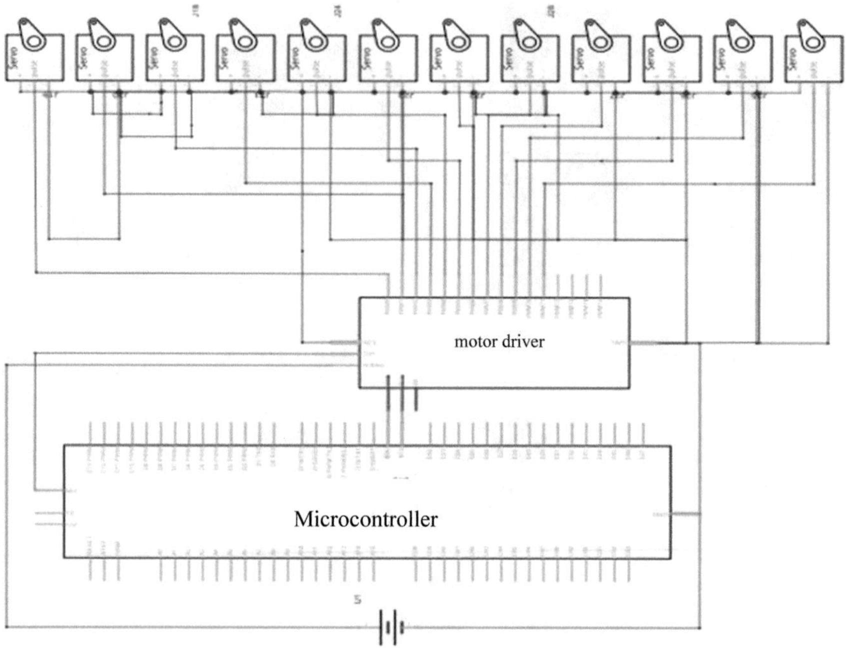

Fig. 6 Schematic circuit diagram for servo arrays

representing their movements. The left-hand does not need to move in the full degree of freedom as most of the signs in MSL hardly use the left hand. An array of four servo motors is sufficient for the left hand to perform selected essential words/phrases at this stage. The circuit diagram of the servos is shown in Fig. 6. Two microcontrollers are used; one for the servo motor arrays and one for controlling the wheel motion of the robot allowing it to move around.

2.3 *Robot Hands' Movements*

The robotic hands' motions were based on the standard MSL. Each item (letter, number or phrase) has been investigated intensely by assigning the servos to each joint and the sequence of motion. Tables 1, 2, 3 and 4 show samples of detailed description including the servos in action and the series of movements for letters, numbers and the emergency phrase. In other words, the database stores the sequential motions of the servo motor arrays for both hands that correspond to particular words/letters/numbers.

Table 1 Sample of word stored in database

Command	Awak (You)	Sakit (Sick)
Picture		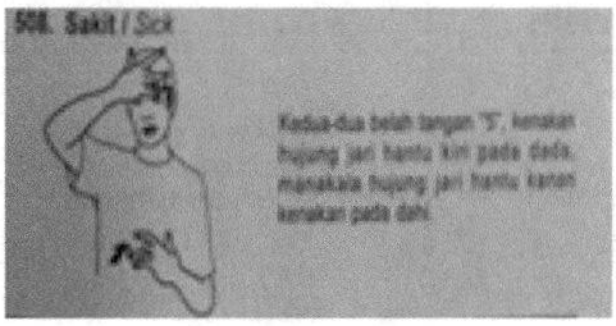
Number of hand? Which hand used? How many servo ? Which servo?	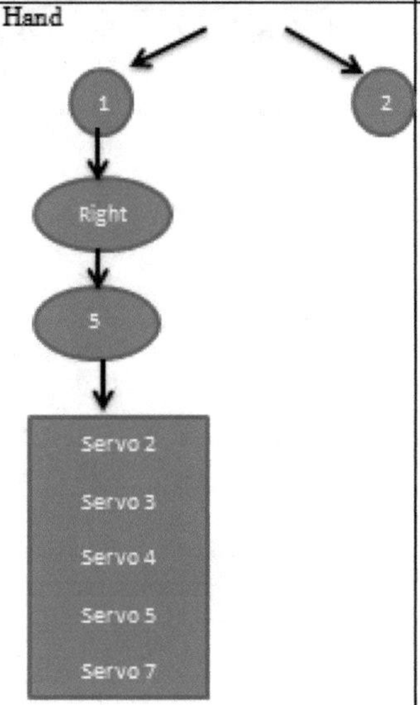	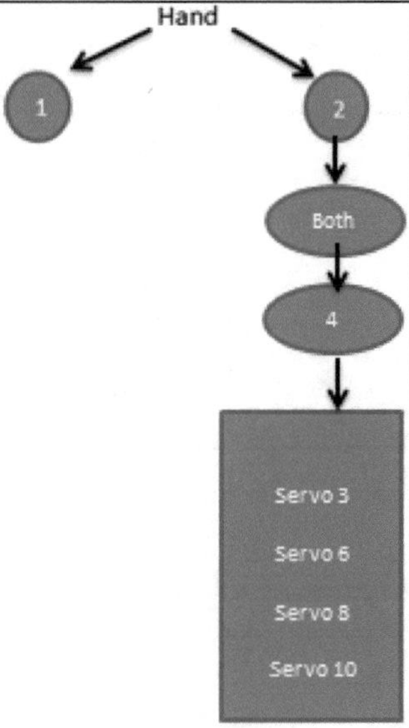

3 Result and Discussion

To test SignBot, one person started speaking any random letters and numbers of Malaysian Language (i.e., Bahasa Melayu). Consequently, when the uttered letter/number exists in the database, the microcontroller turned on the Green LED showing that the character has been identified. Then, accordingly, the command proceeds to the performance of the intended letter by executing its own code according to Tables 1, 2, 3 and 4 (in Sect. 2.3); after the assigned servos for each letter determined as shown in the tables, the corresponding move will be performed representing the letter in MSL as in the feature column of the tables. For example, when the speaker uttered the letter A, the servos number 2, 3, 4 and 5 were actuated in unison. The performance of the robot is depicted in Fig. 7. The robot showed the capability to perform seven alphabets (A, B, H, I, L, S, U), the numbers from 1 to 5,

Table 2 Selected MSL alphabetical signs

Letter	Features	Servo used	Sequences of servos
A		2, 3, 4, 5	Simultaneously 2, 3, 4, 5
B		1	1
H		1, 4, 5, 6, 7	Simultaneously 1, 4, 5 then 6 and 7
I		1, 2, 3, 4	Simultaneously 1, 2, 3, 4
L		3, 4, 5	Simultaneously 3, 4, 5
S		1, 2, 3, 4, 5	Simultaneously 1, 2, 3, 4, 5
U		1, 4, 5	Simultaneously 1, 4, 5

Table 3 Selected MSL numerical signs

Numbers	Features	Servo used	Sequences of servos
1		1, 3, 4, 5	Simultaneously 1, 3, 4 and 5
2		3, 4, 5	Simultaneously 3, 4 and 5
3		4, 5	Simultaneously 4 and 5
4		1	1
5	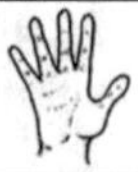	none	None

Table 4 Selected emergency phrases

Setences	Features	Servo used	Sequences of servos
(1) Can I help you? *Boleh saya tolong awak?*	Boleh / *Can* 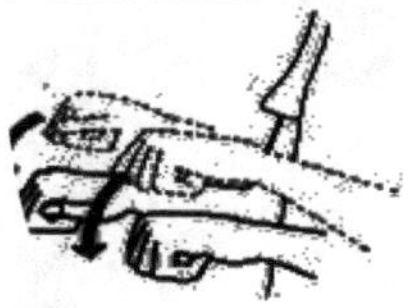	2, 3, 4, 5, 6	Simultaneously for both hands 2, 3, 4, 5, then 6
	aya / *I* 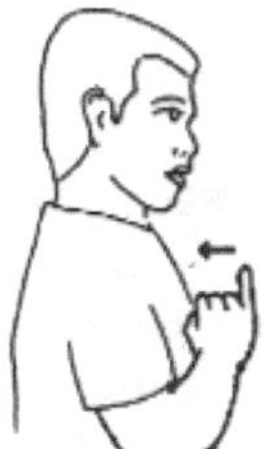	1, 2, 3, 4, 7	1, 2, 3, 4 then 7
	Tolong / *Help* (II) 	1, 2, 3, 4, 5, 6, 7, 8, 10, 11, 12	Simultaneously for both hands 1, 2, 3, 4, 5 then 6, 7, 8, 10 then 11, 12
	Anda / *You* 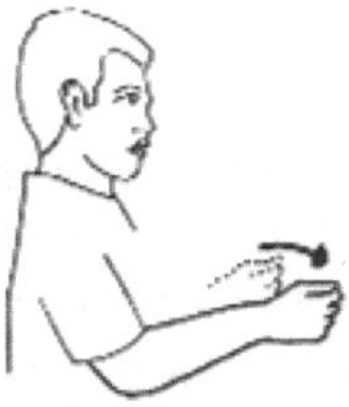	2, 3, 4, 5, 6, 7	2, 3, 4, 5 then 6, 7

and "*Boleh saya tolong awak*?" (Malay sentence that means "Can I help you?" in English).

SignBot is the first robot that can perform Malaysian sign language based on the sequential motion of the array of servo motors. The tests conducted in front of SHI

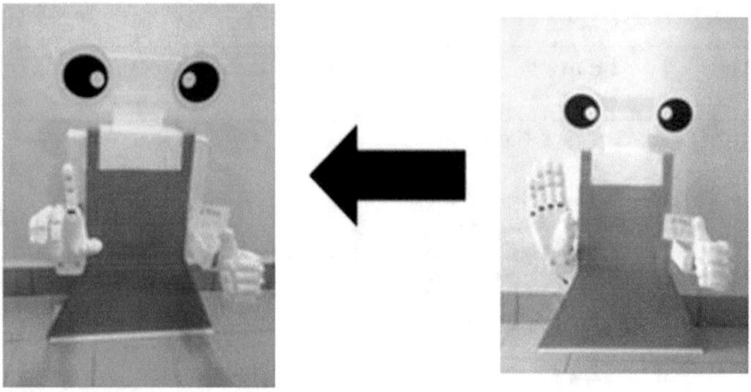

Fig. 7 Testing the performance of SignBot

individuals have demonstrated the capability of the robot to perform Malaysian sign language.

However, at present, SignBot has two limitations. First, the number of signs and phrases implemented in the database is limited at this stage. Second, the motion of the robot is only performed by the hand. Arm and elbow do not move in full degrees of freedom. Their restricted movements, however, do not make the signs perform less understood as most of the signs in MSL require less involvement of arm and elbow.

4 Conclusion

In this paper, we have presented the design of our proposed sign language performing robot, SignBot. The robot has been equipped with its servo motor database for some letters, numbers and emergency phrases selected from Malaysian Sign Language (MSL) as an attempt to help Malaysian speech/hearing-impaired community communicate with mainstream people. For hand signing, each of servo motors used is assigned with numbers, from 1 to 12, for reference. Although the robot has performed the targeted characters and phrases, it still needs some additional improvements to include other signs for words. However, SignBot is the first of the kind that can perform Malaysia sign language. Future work may enhance the quality of performance by implementing some adequate, dexterous design that will be able to employ more signs.

References

1. Yusuf, M.A.M.: Jangan Pinggir mahasiswa OKU, 15 November 2011. http://www.harakahdaily.net/index.php/letter/6529-jangan-pinggir-mahasiswa-oku. Accessed Saturday Apr 2015

2. Anuja, K., Suryapriya, S., Idicula, S.M.: design and development of a frame based MT system for english- to-ISL. In: Nature & Biologically Inspired Computing, 2009. NaBIC 2009. World Congress _on, Coimbatore (2009)
3. Su, A.: ASL Finger-spelling in VRML. http://www.csdl.tamu.edu/~su/asl/. Accessed 19 Aug 2016
4. Falconer, J.: Humanoid Robot Demonstrates Sign Language. 31 Oct 2013. https://spectrum.ieee.org/automaton/robotics/humanoids/ntu-taiwan-humanoid-sign-language
5. Kose, H., Yorganci, R.: Tale of a robot: humanoid robot assisted sign language tutoring. In: IEEE-RAS international conference on humanoid robots (2011)
6. Hoshino, K., Kawabuchi, I.: A humanoid robotic hand performing the sign language motions, Tsukuba, Japan (2003)
7. Jaffe, D.L.: Evolution of mechanical fingerspelling hands for people. J. Rehabil. Res. Dev. **31** (3), 236–244 (1994)
8. Gatti, J., Fonda, C., Tenze, L., Canessa, E.: Voice-controlled artificial handspeak system. International J. Artif. Intell. Appl. (IJAIA) **5**(1), 107–112 (2014)
9. Gyrobot: "Flexy-Hand," Thingiverse, 4 March 2014. https://www.thingiverse.com/thing:242639. Accessed Nov 2016

Isolated Bio-Regenerative System for Vertical Farming Through Robots in Space Explorations

Shivam Bhardwaj, Aswath Suresh, Ragini Arora, Shashank Vasishth, Debrup Laha, Dhruv Gaba and Siddhant Bhambri

Abstract For colonization in deep space, we need to explore the feasibility of a bioregenerative system in microgravity or artificial gravity environments. The process has various complexities form ranging to biological obstacles to engineering limitations of the spacecraft. The concentration of microbes in the confinements of a spacecraft can be fatal for the crew. In this paper, a solution to the elevated microbial levels by farming in an isolated chamber using robots is discussed. Manual control of the robot using a virtual reality environment are considered.

Keywords Colonization · Vertical farming · Bioregenerative
Robots

S. Bhardwaj · A. Suresh (✉) · D. Laha · D. Gaba
Department of Mechanical and Aerospace Engineering,
New York University, Brooklyn, USA
e-mail: as10616@nyu.edu

S. Bhardwaj
e-mail: sb6377@nyu.edu

D. Laha
e-mail: dl3515@nyu.edu

D. Gaba
e-mail: dg3035@nyu.edu

R. Arora · S. Vasishth · S. Bhambri
Department of Electronics and Communication Engineering,
Bharati Vidyapeeth's College of Engineering, Delhi, India
e-mail: ragini.arora15@gmail.com

S. Vasishth
e-mail: shashankvasisht@gmail.com

S. Bhambri
e-mail: siddhantbhambri@gmail.com

J.-H. Kim et al. (eds.), *Robot Intelligence Technology and Applications 5*,
Advances in Intelligent Systems and Computing 751,
https://doi.org/10.1007/978-3-319-78452-6_21

1 Introduction

For the survival of human species, it is now time to explore ways to make colonization in deep space practical [1]. One of the most prominent issue with space exploration is the duration of space travel. To make such an endeavour possible, we must take care of our metabolic needs without depending on earth for a re-supply. To do so, we need to explore various aspects of farming in space. A possible approach is vertical farming using a controlled environment [2]. In recent years, we have successfully grown plants like lettuce in the International Space Station which is a testament that with further prominent research, significant discoveries can be made to make this a viable reality.

In furtherance of creating a sustainable environment in space, we ought to develop a bio-regenerative system. By doing so, an unlimited supply of food, oxygen and water can be provided as the fundamental life particles are processed naturally in the form of composting. This is a possible solution to the rudimentary problem of increase in the initial launch payload of the spacecraft with the increase in travel time.

Farming in space has a lot of inherent challenges like microgravity and reduced air pressure. In our paper, we tried to address such an issue by associating growth with a bioregenerative system in the enclosed of a spacecraft. An inevitable part of a bioregenerative system is the microorganisms that break down the complex natural molecules to provide the tree with other fundamental resources required for it to grow. Apart from this, every plant has a natural microbial level. On earth, since the process is often done in an open space, the concentration of the microbes developed during the process of farming is moderate. However, in the enclosure of a spacecraft, the level of concentration of such organisms can be fatal for the astronauts.

To tackle this problem, a similar environment needs to be created in a sealed enclosure inside the spacecraft. This will create a safe, controlled sealed environment where the bioregenerative system [3] can thrive without causing any harm to the astronauts. Moreover, all the activities such as harvesting are performed using robots. To reduce the initial payload of the spacecraft at the time of launch, the robot's end effector will be equipped with a changeable gripper called as so that same set of robots can be used to perform a variety of tasks. In our paper, we have further discussed the autonomous and manual control robots to carry out the steps of farming as well as implementing the testing of the produce.

The manual control of the robot can be achieved by an augmented reality system which can map the input given by hand gestures of the astronaut and consequently perform the desired tasks using a robot manipulator. To add the autonomous features in the robots, we must exploit the scope of machine learning efficiently and mesh it with the bioregenerative system. Using computer vision and of a dedicated set of sensors such as leaf and humidity sensors, the system can be trained about farming of specific crops.

Growing a crop and collecting raw data for training the system takes a significant amount of time. In the paper, the contemporary methods of data collections are thoroughly discussed. The ways to sterilize the crops are extensively explored as well. Such systems should prove to be salutary for growing crops in a controlled environment, without affecting the produce as well as the astronauts on board.

1.1 Need for Exploring Extra-Terrestrial Life Support

One of the most prominent reasons for colonizing space or any other planet is solely, for the survival of our species. Human extinction is certain on planet earth; the reason may be anthropogenic, i.e. due to human themselves or due to some natural phenomenon. One of the most plausible human originated cause of extinction can be the technology itself. According to the equation [4].

$$\mathrm{I} = \mathrm{P} \times \mathrm{A} \times \mathrm{T} \tag{1}$$

where,

I Impact on Environment
P Population
A Affluence (Consumption of resources per person)
T Technology of resource utilization

The human impact on the environment is ever increasing, and with advancement in technology to procure the resources, we are fast moving towards to frivol away. Some human activities that can have an antagonistic effect on earth are over irrigation that leads to soil erosion, destruction of carbon filtering coral reefs due to excessive fishing, greenhouse gases produced due to extravagant meat production, improvident energy utilization harvested from non-renewable resources, nuclear waste generated during nuclear energy generation, and worst of all wars [5]. With the enhanced weapons systems, a global war in the present time will be devastating for human civilization.

Some most common non-anthropogenic global catastrophic risk can arise from an asteroid impact, natural climate change, geomagnetic reversal [6], a global pandemic and numerous cosmic threats.

1.2 Challenges of Extraterrestrial Life Support

According to a study conducted in NASA AMES Summer Study on Space Settlements and Industrialization [7] using non-terrestrial material, the whole aspect of surviving in space with minimal/no re-supply from earth can be taken as five broad challenges. These are

1. Creating of a bioregenerative Life support system.
2. Effect of microgravity on humans as well other biotic lifeforms in the above system.
3. Detection and analysis of space rocks that can be utilised as resources.
4. A practically unperishable source of energy for the propagation of the space vehicle as well as to support life in isolation of planet Earth.
5. Processes to transform non-terrestrial material, i.e. manufacturing of metals and other infrastructural materials.

In our paper, we are concentrating on the first aspect of human survival in space.

2 Challenges in Bioregenerative System in Space

2.1 Effect of Gravity on Such System

Gravity plays an inevitable role in a plant's growth. Plants have evolved a very sensitive system to detect the slightest change in the environment. Over millions of years, flora on earth has been accustomed to its gravity. Plants have developed a system in which the roots grow towards the gravity to absorb more nutrients from the soil. In a micro-gravity system, the growth of the plant is drastically hit, and this possesses a significant complication in space farming.

2.2 Elevated Microbial Concentration in a Closed System

Another major complication in the approach of the bioregenerative system is the microbes involved in the system. For such a system, the process of composting is critical. The raw food, the human waste and other natural and complex organic molecules must be broken down to obtain vital minerals again. This process is aided by specialised bacteria and other microbes.

On earth the since farming is done in an open area, the concentration of these microbes doesn't affect much on human health. But, inside a chamber in the space vehicle, the level of microbes can be lethal.

Apart from the bacteria required for composting, plants always carry some amount of natural microbial level. These microbes can be a cause of an epidemic in the system.

3 Proposed Solution for Such System

In this paper, it is intended that such a system is developed in an isolated chamber which can inhibit these bacteria from coming in contact with the astronauts. An augmentation to this solution is, allowing only robots in the chamber. Since robots are immune to any biological influences, they can work continuously.

3.1 Isolated Chamber

In this paper, a significant part is devoted to a proposed system with which we can accumulate the process of vertical farming in the spacecraft.

The following image is an exploded view of our proposal for augmenting the SpaceX Dragon Spacecraft to have an augmented chamber for the purpose of an isolated vertical farming environment.

The proposed system is not limited to the SpaceX Dragon spacecraft as the chamber is generic, and can act as a standalone augmentation in any spacecraft (Figs. 1 and 2).

The idea behind such a system is that humans have adapted towards the Coriolis forces over time of during evolution.

To create an artificial gravity environment, we have an option of rotating the chamber such that the centrifugal force results in an artificial gravity environment. But the problem with this approach is that this will result in medical issues like vertigo etc. for humans on board.

Therefore, the system described above has two concentric cylinders inside the main spacecraft. Both of the cylinders are rotating in the opposite direction with an angular velocity such that net angular momentum is zero.

The overall system is equipped with a gyroscopic sensor in the astronauts' chamber to detect any total rotation of the spacecraft. In case of such rotation, the

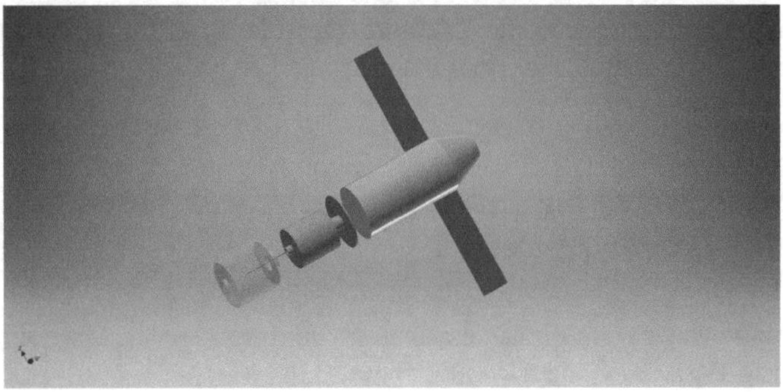

Fig. 1 This is an exploded view of the SpaceX Dragon Spacecraft with concentric cylindrical chambers

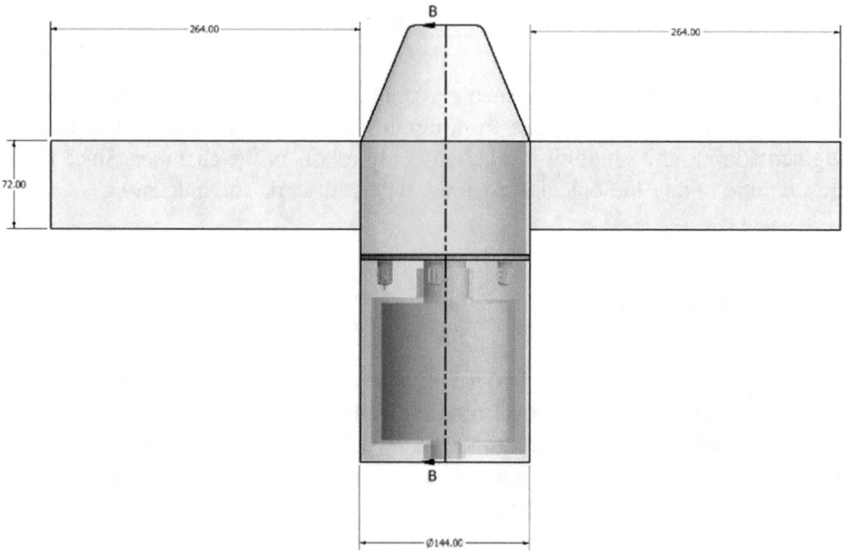

Fig. 2 Front view of the proposed spacecraft

rotational velocities of the cylinders are automatically adjusted to nullify the effect of the mutual rotation again.

$$L = r \times m \times v_{\perp} \quad (2)$$

where,

L Angular momentum
r Radius
m Mass
v Velocity

The mass of the outer cylinder is less than the inner cylinder to reduce the initial launch payload of the spacecraft. Therefore, the velocity of the outer shell will be significantly higher than the internal cylinder.

3.2 Dimensions of the Isolated Chamber with Respect to the Dragon Spacecraft

The schematic views of the isolated chamber with respect to the Dragon spacecraft are shown in Figs. 3, 4 and 5.

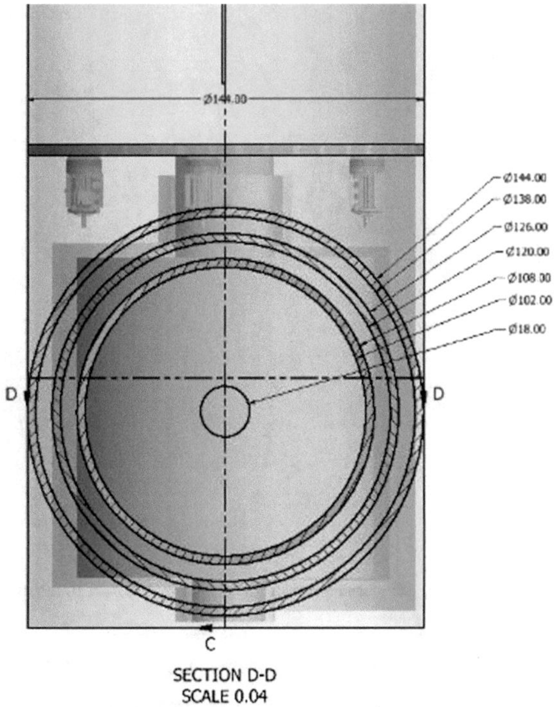

Fig. 3 The three concentric circles are the projection of the Spacecraft, the outer cylinder and the inner cylinder

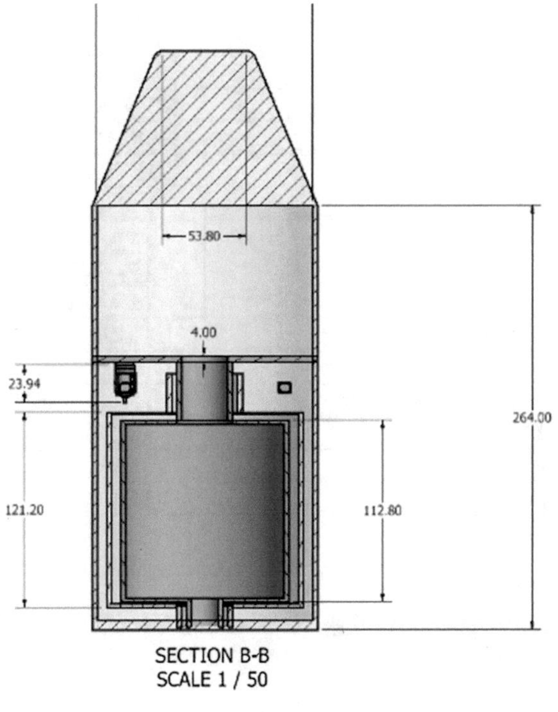

Fig. 4 This is the cross-section of the spacecraft at the scale 1/50

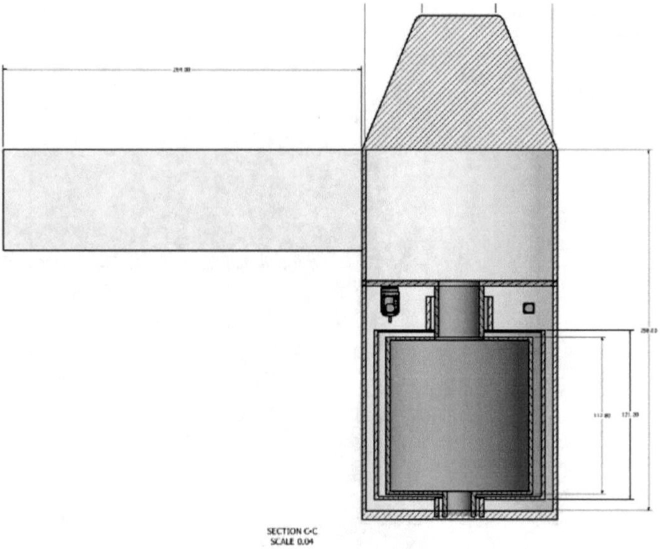

Fig. 5 This is the cross-section view of the capsule including the solar panel

3.3 *Stress Analysis of the Isolation Chamber During the Launch. Physical Properties of the Material*

Material	Steel, Alloy
Density	0.279264 lb mass/in^3
Mass	53693.2 lb mass
Area	1010.03 ft^2
Volume	111.266 ft^3

Center of gravity	x = 0.000000000000498396 ft
	y = 4.95686 ft
	z = 0 ft
Mass density	0.279264 lb mass/in^3
Yield strength	36259.4 psi
Ultimate tensile strength	58015.1 psi
Young's modulus	29732.7 ksi
Poisson's ratio	0.3 ul
Shear modulus	11435.7 ksi

Load

The magnitude of linear acceleration is taken considering the launch acceleration of a space shuttle.

Linear	Acceleration	
Magnitude		27.000 ft/s^2
Vector	X	0.000 ft/s^2
Vector	Y	−27.000 ft/s^2
Vector	Z	0.000 ft/s^2

Selected Body

The selected body is depicted in blue, and constraint body is represented in red. (Fig. 6).

Result

Name	Minimum	Maximum
Volume	192267 in^3	
Mass	53693.2 lb mass	
Reaction force (lbf)	0	45043.5
Von Mises stress (ksi)	0.00428436	6.59093
1st Principal stress (ksi)	−2.13312	7.95529
3rd Principal stress (ksi)	−7.82182	1.22482
Displacement	0 in	0.0519638 in
Safety factor (ul)	5.50141	15
Stress XX (ksi)	−6.9796	7.33874
Stress XY (ksi)	−2.68151	2.5519
Stress XZ (ksi)	−2.42353	2.43882
Stress YY (ksi)	−3.26476	2.63815
Stress YZ (ksi)	−2.98435	2.64803
Stress ZZ (ksi)	−6.97103	7.44338
X Displacement	−0.0016733 in	0.00166564 in
Y Displacement	−0.0519525 in	0.000112361 in
Z Displacement	−0.0016928 in	0.00165476 in
Equivalent strain (ul)	0.000000135363	0.00020254
1st Principal strain (ul)	−0.00000240395	0.000230885
3rd Principal strain (ul)	−0.000215018	0.000000996451
Strain XX (ul)	−0.00018062	0.000217896
Strain XY (ul)	−0.000117243	0.000111576
Strain XZ (ul)	−0.000105964	0.000106632
Strain YY (ul)	−0.0000906127	0.0000721264
Strain YZ (ul)	−0.000130484	0.000115779
Strain ZZ (ul)	−0.000181866	0.000220795

The stress and displacement analyses are shown in Figs. 7 and 8.

Fig. 6 Constraint body

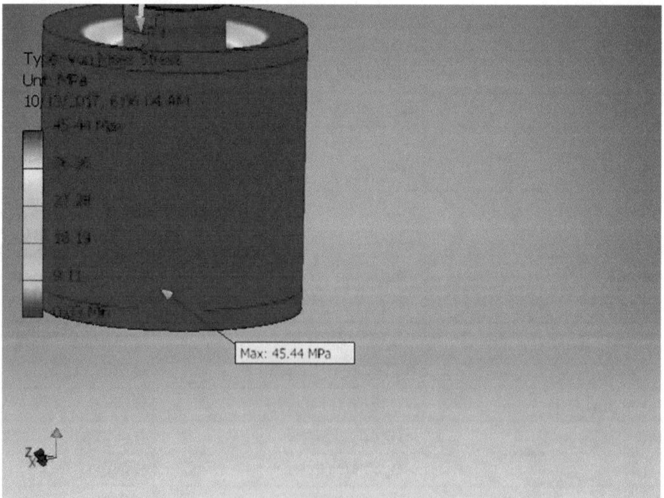

Fig. 7 Stress analysis

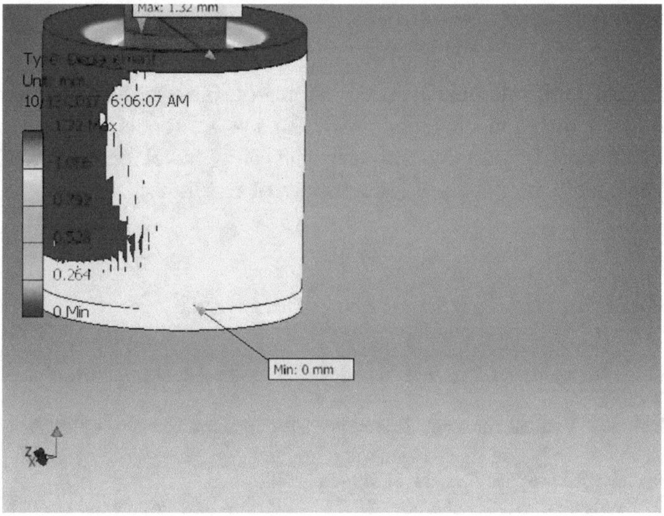

Fig. 8 Displacement due to properties of the material

3.4 Use of Robots

An augmentation to the isolated chamber in the spacecraft is to use robots for vertical farming.

In the present technological advances, humans have developed robotic solutions that can perform as intricated tasks as performing surgeries. The process is so efficient that it is called minimally invasive surgery.

Such an efficient system of robot manipulators can undertake the responsibility of maintaining the bioregenerative life support in the spacecraft.

The isolated chamber will contain multiple dexterous robots that can harvest the crops and can even repair the other robots if required, as the possibility of direct human intervention in the system is almost null.

4 Control of Robots in the Absence of Human Intervention

Robots can be manually controlled by the astronauts with the help of system specially designed for vertical farming in space [8].

The method comprises of a virtual reality environment in which the astronauts will be projected with the live feed of the chamber using a 360-degree camera. Using a combination of depth sensors and flex sensors the astronauts can give the control signals to the robot in the chamber.

The robots then, can respond to the signal and replicate the hand movement of the astronauts.

5 Conclusion

The problem of high microbial concentration in a bioregenerative system can be solved using an isolation chamber. With the use of robots, the system can be controlled from outside, and the astronauts can safely travel long distances in space without the dependency on earth for resupply of resources.

References

1. Monje, O.: Farming in space: environmental and biophysical concerns. Adv. Space Res. **31**(1), 151–167 (2003)
2. Al-Chalabi, M.: Vertical farming: Skyscraper sustainability. Sustain. Cities Soc. **18**, 74–77 (2015)
3. Silverstone, S.: Adv. Space Res. **31**(1), 69–75 (2003)
4. Lutz, W., Sanderson, W.C., Scherbov, S.: The End of World Population Growth in the 21st Century: New Challenges for Human Capital Formation & Sustainable Development, pp. 283–314. Earthscan Press, London
5. Gledistch, N.: Conflict and the Environment. Kluwer Academic Publishers (1997)
6. Plotnick, R.E.: Relationship between biological extinctions and geomagnetic reversals. Geology (1980)
7. Billingham, J.: SEE N79-32226 through N79-32241
8. Sawada, Hirokata, Ui, Kyoichi, Mori, Makoto, Yamamoto, Hiroshi, Hayashi, Ryoichi, Matunaga, Saburo, Ohkami, Yoshiaki: Micro-gravity experiment of a space robotic arm using parabolic flight. Adv. Robot. **18**(3), 247–267 (2004). https://doi.org/10.1163/156855304322972431

A Mobile Social Bar Table Based on a Retired Security Robot

David Claveau and Stormon Force

Abstract The design of a social robot that serves refreshments to party guests is presented. The robot has the physical form of a bar table on wheels. It has a simple three-state behavior: if it detects drinks and refreshments on its tabletop then it wanders the room in search of people by means of a simple phonotaxis approach. If the drinks and refreshments have been removed then it returns to a designated corner of the room for reloading. The robot serves as an interesting research platform for human-robot interaction. The design is based on a retired security robot that was popular in the 1990s. Its omnidirectional synchro-drive base serves the current design well as it can easily maneuver around people. The entire design and implementation occurred over approximately two months. The design process, implementation and preliminary test results are presented.

Keywords Social intelligence · Human-robot interaction · Education

1 Introduction

The total number of industrial and service robots operating in the world is likely to be over 20 million [1]. The number of retired robots is also likely to be in the millions. This represents a vast mechatronic resource for roboticists. Outdated controllers and sensors are usually not useful, however, the mechanical and electro-mechanical components, with a little work, can often be reused in new designs. By carefully tearing down these old robots, they can be creatively repurposed to meet new needs.

D. Claveau (✉) · S. Force
CSU Channel Islands, Camarillo, CA, USA
e-mail: david.claveau@csuci.edu

S. Force
e-mail: stormon.force211@myci.csuci.edu

J.-H. Kim et al. (eds.), *Robot Intelligence Technology and Applications 5*,
Advances in Intelligent Systems and Computing 751,
https://doi.org/10.1007/978-3-319-78452-6_22

There is currently a need for more robots that can safely interact with humans in social settings. These social robots need to be responsive to humans and support interactions that are natural and intuitive [2, 3]. General-purpose social robots are the subject of intense research efforts that involve multiple fields such as artificial intelligence and human-machine interaction. Special-purpose social robots, however, have been deployed and have become familiar as inter-office mail carriers, companions, and entertainers. Their design is made easier by limiting the social contexts they are expected to operate within. This paper describes the design and implementation of such a social robot. It documents the transformation of a retired security robot from the 1990s into a social robot that can serve as an interactive bar-table that serves refreshments to guests at parties and banquets. The design experience is offered as a contribution to future efforts in recycling and repurposing robots for social roles.

2 Before and After

Universities are frequently offered old and retired equipment for use in student projects. Our lab was fortunate to receive a Cybermotion Cyberguard robot [4] that a local company had used decades earlier for security. The robot, in its original form, is shown in Fig. 1a. It was typically used for automated safety patrol at warehouses and airports. It was 190.5 cm tall and 69 cm wide and weighed 295 Kg

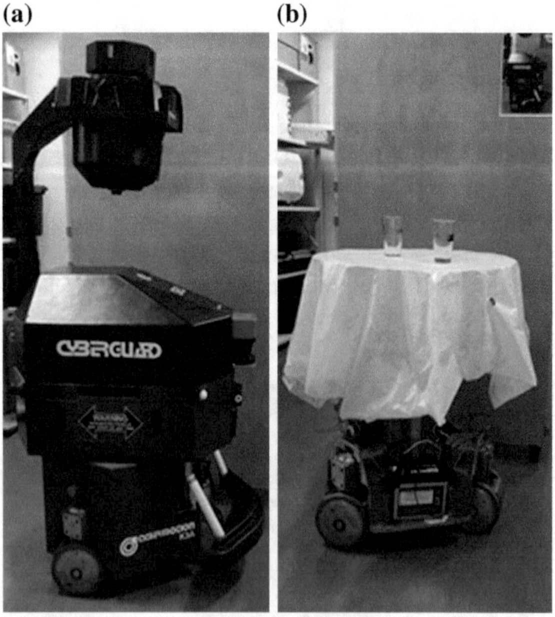

Fig. 1 **a** The retired security robot, **b** the social bar table

(650 lbs.). Its large, heavy body made it look more like an industrial robot than a service robot. It carried a wide range of sensors including a lidar, sonar, infrared, and a video camera that recorded to an onboard VHS tape deck. It made use of an innovative synchro-drive mechanism [5, 6] that allows the robot to rotate and move in any direction. No attempt was made to operate the robot in its original state as its control unit consisted of multiple Z80 processors and there were no available software resources to program it.

The robot was transformed into the mobile social bar table shown in Fig. 1b. Its purpose is to serve drinks and refreshments to party guests in a safe and efficient manner. As such it should look and behave like a standard bar table, except that if it is carrying refreshments it should seek out humans, and if it is empty it should return to its base to be replenished. It has a standard 30″ (76 cm) tabletop at a height of 40″ (102 cm). A minimalist approach was taken to keep the robot from being a spectacle. The goal is to have a piece of robotic furniture that interacts with humans in such a way that no one really pays much attention to it, but it serves its role of waiter in a safe and efficient manner.

A student in our lab conducted a careful teardown of the robot down to its base, as shown in Fig. 2a. This was the first step in the transformation. Following this a pedestal was constructed to fit over the central column and an acrylic tabletop was mounted, as shown in Fig. 2b, At this stage we could see the potential for a real social robot and proceeded to complete the transformation. In the following sections this robot 'makeover' is described in detail.

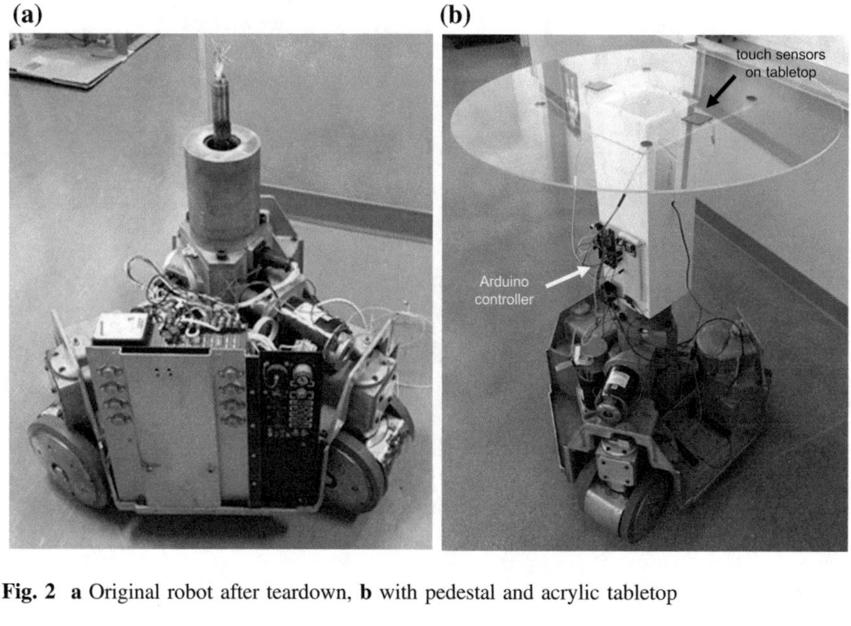

Fig. 2 **a** Original robot after teardown, **b** with pedestal and acrylic tabletop

3 How does a Bar Table Behave?

A social robot is an autonomous robot that interacts with humans in a safe, efficient and comfortable manner. Very often a social robot is an anthropomorphic robot, such as Snackbot [7, 8] and Robotinho [9]. They use speech and gestures to communicate in a human-like way. However, when humans see a robot that looks like them, they expect it to behave like them, raising expectations. When they see a bar table however, no one talks to it or expects it to talk back. They only expect it to have drinks, and so its behavior should be very simple. Fig. 3 shows a three-state behavior in which the robot is initially in the 'stop' state at its station. Refreshments are placed on the table and when the table is full, it transitions into the 'seek out talking people' state. After all of the drinks have been removed from the table by humans, it transitions into the 'return to base station' state. While in the two states in which it is moving it needs to avoid bumping into people. It should proceed in a slow manner when near people so that it can quickly stop if someone does make contact with it. Interaction is easy. The tabletop affords placing glasses onto it and taking glasses off of it. There is no need for speech, sounds, or flashing lights. In fact, any of those features would probably lead to less acceptance by humans.

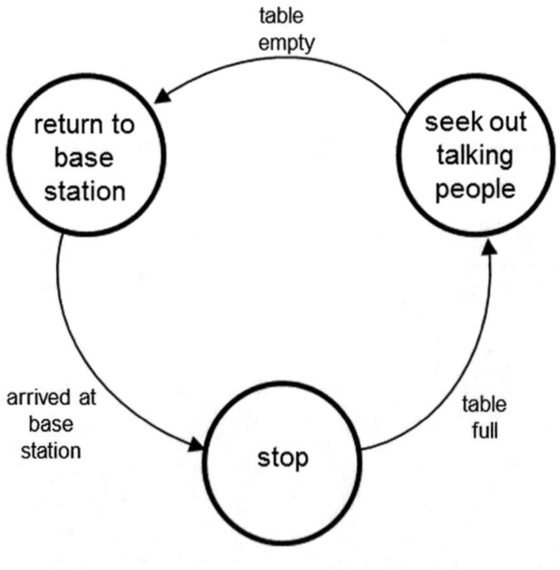

Fig. 3 Finite-state machine for robot behavior in a banquet room

4 Designed to Wait

The design of any interactive system should consider:

(i) who will be interacting with the system?
(ii) what will they be trying to do?
(iii) where will the interaction take place?

In this case it is expected that adults will be interacting with the robot. They will notice that it is nearby and they will remove drinks from its tabletop. The interaction will take place in a large rectangular banquet room.

Initially our goal was to create a robot that can demonstrate the basic behavior of Fig. 3. To that end, the robot has the simple design shown in the block diagram of Fig. 4. There is a single Arduino controller and it receives input from three types of sensors. There is an encoder on the central shaft of the robot to measure the current steering angle. There are touch sensors (force sensitive resistors) on the tabletop to detect the placement of glasses. There are microphones around the perimeter of the table to detect the sound of humans talking. The complete set of sensors are shown in Fig. 5. There are only two motors, one for steering and one for drive. The synchro-drive system steers all three wheels at once with one motor and drives all wheels at once with the other motor. Each wheel is actually composed of two wheels that can move in opposite directions when turning. The placement of the two motors is shown in Fig. 6.

The microphones were arranged as shown in Fig. 7a. Their 'lines-of-sight' are separated by 120° and their orientation with respect to the room remains fixed over time as the synchro-drive wheels rotate without rotating the robot. Fig. 7b shows how this might look in a banquet hall that has some people talking at one end. Depending on where the most human chatter is occurring, the robot will steer itself in that direction. It uses a simple vector addition of intensities from the

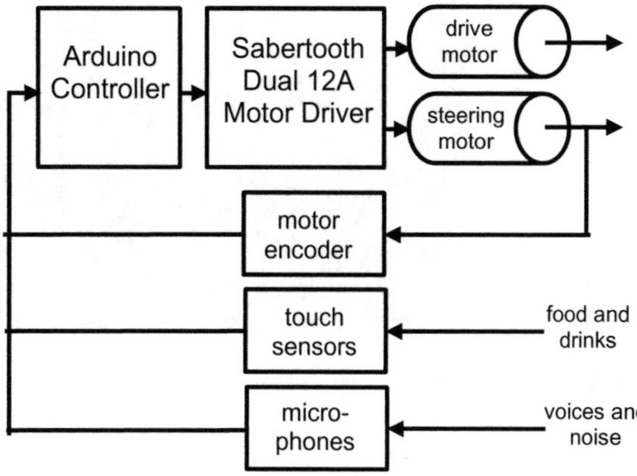

Fig. 4 Block diagram of the social bar table robot

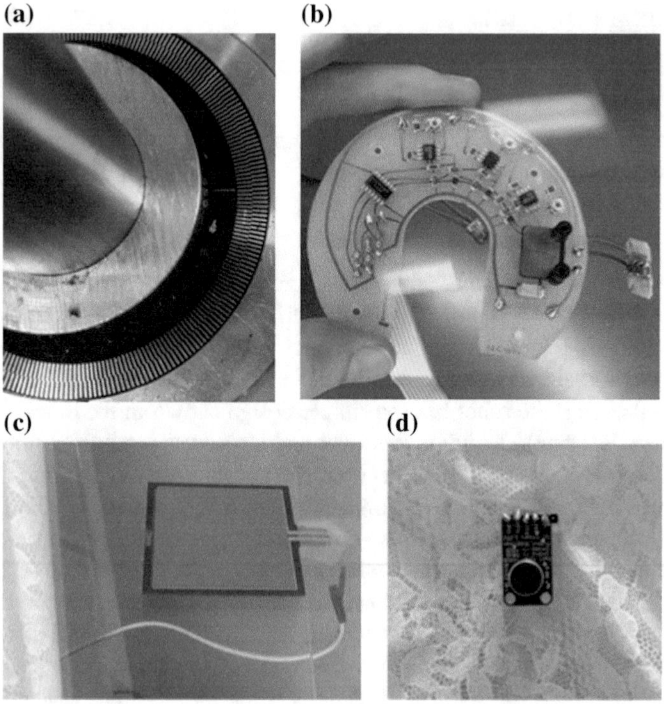

Fig. 5 Complete set of sensors used, **a** and **b** show the shaft encoder for steering, **c** shows the force-sensitive resistor used for the tabletop touch sensor, and **d** shows the microphone

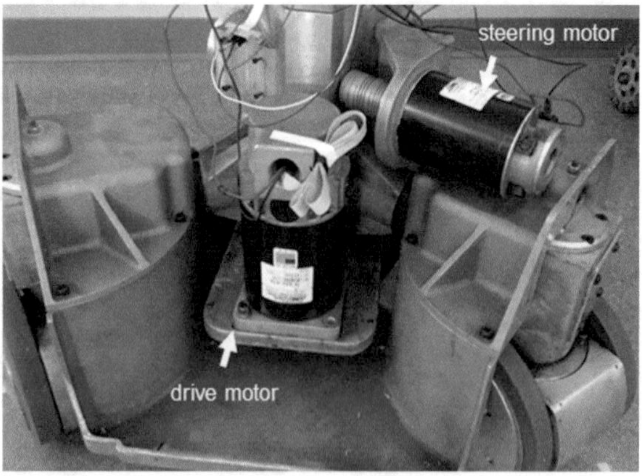

Fig. 6 Position of drive motor and steering motor

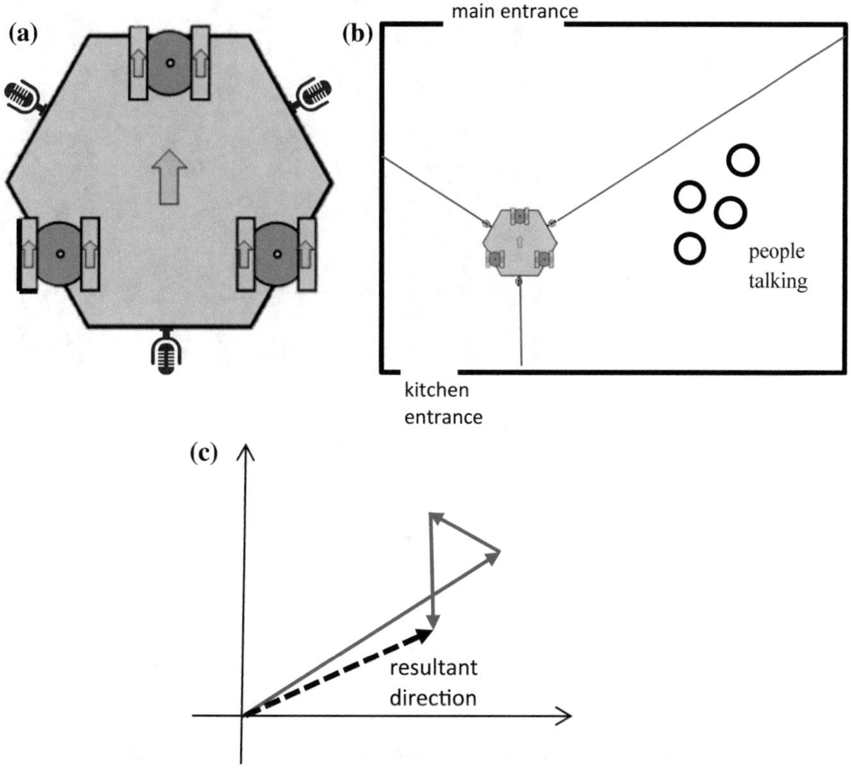

Fig. 7 **a** Arrangement of microphones on robot, **b** the directions of greatest sensitivity for the microphones in a room, and **c** the vector addition of input signals for positive phonotaxis

microphones, shown in Fig. 7c, to perform positive phonotaxis. The current implementation uses peak-to-peak values over 20 ms windows for the intensities from each microphone. A moving average of length eight is used to filter out spurious values. It is expected that more processing is required to filter the microphone data and remove frequencies that are not typically found in the 20 kHz band of human voices.

5 Preliminary Testing

An initial version of the mobile social bar table was tested in a large lobby in our building. Fig. 8 shows a frame from a video taken of our testing which is available online [10]. We were able to successfully demonstrate the desired behavior. When drinks were placed on the tabletop the robot began to move and was able to track a

Fig. 8 Testing the robot in a large open space similar to a banquet hall

human voice and stop when the drink was removed. Testing the robot was actually an enjoyable experience and this leads us to believe that such a robot will be well received by humans at a party. Much improvement needs to be done. We expect to add range sensors to avoid bumping into humans and obstacles. We also need to improve the audio processing to deal with various sounds that lead the robot astray.

6 Conclusion

This design experience has highlighted some valuable lessons in recycling retired robots and in repurposing them for social roles. By carefully tearing down the original robot we were able to see the possibilities for the new robot. By making use of the original robot's best features we were able to create a mobile platform that is very rugged and very maneuverable. By designing a special-purpose social robot we were able to keep the design simple. In less than eight weeks we were able to achieve a testable prototype that will serve as a research robot and a platform for future development.

References

1. http://www.futuretimeline.net/blog/2012/01/4.htm#.WbtJIz6GPcu
2. Breazeal, C.: Toward sociable robots. Robot. Auton. Syst. **42**, 167–175 (2003)
3. Feil-Seifer, D., Mataric, M.J.: Defining socially assistive robotics. In: Proceedings of the 2005 IEEE 9th International Conference on Rehabilitation Robotics, 28 June–1 July 2005
4. CyberGuard SR3/ESP High Mobility Security Robot with Enhanced Sensor Package 1354 8th Street, SW Roanoke, VA 24015, (800) 762-6848, www.cybermotion.com
5. Holland, J.M.: Cybermation, Inc., Roanoke, VA, Mobile Base for Robots and the Like, Patent Number: 4,573,548, Date of Patent, 4 Mar 1986
6. Fisher, D.E., Holland, J.M., Kennedy, K.F.: Cybermotion, Inc., Roanoke, Va., Mobile base having leg assemblies with two wheels, US 5609216 A, 11 Mar 1997

7. Lee, M.K., Forlizzi J., Kiesler S., Rybski, P., Antanitis, J., Savetsila, S.: Personalization in HRI: A longitudinal field experiment. In: Proceedings of HRI (2012)
8. Lee, M.K., Forlizzi, J., Rybski, P.E., Crabbe, F., Chung, W., Finkle, J., Glaser, E., Kiesler, S.: The Snackbot: documenting the design of a robot for long-term human-robot interaction. In: Proceedings of HRI 2009, pp. 7–14 (2009)
9. Nieuwenhuisen, M., Gaspers, J., Tischler, O., Behnke, S.: Intuitive multimodal interaction and predictable behavior for the museum tour guide robot Robotinho. In: 2010 10th IEEE-RAS International Conference on Humanoid Robots (Humanoids), pp. 653–658 (2010)
10. https://www.youtube.com/watch?v=g6urqa1HLFk

I Remember What You Did: A Behavioural Guide-Robot

Suman Ojha, S. L. K. Chand Gudi, Jonathan Vitale, Mary-Anne Williams and Benjamin Johnston

Abstract Robots are coming closer to human society following the birth of emerging field called *Social Robotics*. Social Robotics is a branch of robotics that specifically pertains to the design and development of robots that can be employed in human society for the welfare of mankind. The applications of social robots may range from household domains such as elderly and child care to educational domains like personal psychological training and tutoring. It is crucial to note that if such robots are intended to work closely with young children, it is extremely important to make sure that these robots teach not only the *facts* but also important social aspects like knowing *what is right and what is wrong*. It is because we do not want to produce a generation of kids that knows only the facts but not morality. In this paper, we present a mechanism used in our computational model (i.e EEGS) for social robots, in which emotions and behavioural response of the robot depends on how one has previously treated a robot. For example, if one has previously treated a robot in a good manner, it will respond accordingly while if one has previously mistreated the robot, it will make the person realise the issue. A robot with such a quality can be very useful in teaching good manners to the future generation of kids.

Keywords Social robots · Emotion · EEGS · Memory · Positivity
Response bias · Child behaviour

S. Ojha (✉) · S. L. K. C. Gudi · J. Vitale · M.-A. Williams · B. Johnston
Centre for Artificial Intelligence - The Magic Lab, University of Technology Sydney,
15 Broadway, Ultimo 2007, Australia
e-mail: Suman.Ojha@student.uts.edu.au
URL: http://www.themagiclab.org

S. L. K. C. Gudi
e-mail: SivaLeelaKrishnaChand.Gudi@student.uts.edu.au

J. Vitale
e-mail: Jonathan.Vitale@uts.edu.au

M.-A. Williams
e-mail: Mary-Anne.Williams@uts.edu.au

B. Johnston
e-mail: Benjamin.Johnston@uts.edu.au

J.-H. Kim et al. (eds.), *Robot Intelligence Technology and Applications 5*,
Advances in Intelligent Systems and Computing 751,
https://doi.org/10.1007/978-3-319-78452-6_23

1 Introduction

Social robots employed to work with young children like playing companion [1], carer [2] or tutoring expert [3] should also be able to teach them manners about what is right and what is wrong. Instead of just loading the robots with a bunch of facts, it is also crucially important to enable them to track the behaviour of children and provide feedback accordingly. As such, a robot should be endowed with an ability to remember the previous actions or behaviour of the child in subsequent interactions and adjust its own behavioural responses to teach the child that treating someone badly is not a nice thing and if one does that s/he may not expect good in return. While currently children are likely to treat robots like pieces of plastic or metal that are only meant to be played with and can even be mistreated without any impact. In such a case, if a robot is lifeless programmed hardware that only provides some predefined verbal responses or physical responses, children may end up enjoying treating them badly. In long run, this may become a habit and they might develop a natural tendency to mistreat people. Psychology literature suggests that children who are not adequately acknowledged of their bad behaviour end up becoming an ill-mannered person in adulthood [4]. This is only one example where applications of robotics may threaten the harmony of the human society. Therefore it is high time that we thought of bringing life to social robots and empowering them with an ability to track and remember the previous behaviour of the interacting child and bias its responses towards the child to make her/him realise what s/he did was desirable or not.

In this paper, we present how the mechanism of keeping track of previous action in our computational model i.e. Ethical Emotion Generation System—EEGS [5] causes the alteration in emotional and behavioural responses in social robots. We believe that such memory-biased responses can be useful for teaching valuable moral lesson to young children.

Remaining of the paper is organised as follows. In Sect. 2, we shall present the theoretical foundation and motivation behind our work. In Sect. 3, we shall present the mechanism by which our computational model EEGS is able to store the past events and retrieve those events to bias its emotional and behavioural responses towards the interacting person. In Sect. 4, we will present two different scenarios of interaction, where in one of them a child treats a robot in a nice way and in another misbehaves with the robot. We show the comparison of how EEGS handles those scenarios differently and provides an emotional feedback that can help reinforce the child's behaviour. Finally in Sect. 5, we shall briefly conclude the contribution of this paper.

2 Background and Motivation

Human beings are intelligent—both in terms of cognitive as well as emotional and behavioural aspects [6, 7]. Most of the times human behaviour and actions are guided by life experience. We tend to remember something similar that we experienced in the past in a situation where we have to make a decision—be it consciously or subconsciously. Consider a situation where you are purchasing a new car since your existing car is quite old. While doing so, you tend to remember the issues you had with the previous car and try to find the features that help you get rid of your problems in previous car. In the similar manner, when it comes to decision making in the context of social interaction, we tend to be biased by the previous actions of the person towards us or towards the person we love. For example, a simple greeting by a person you like a lot might give you a feeling of joy while the same action from the person you hate might trigger anger. Such emotional experience might then encourage you to respond or act accordingly. While being biased in our responses or decision based on previous memory may not sound justified sometimes, we argue that such a behaviour of an autonomous robot working closely with young children can be considered highly desirable.

As previously mentioned, such a mechanism can be effective in reminding young children of the bad actions or behaviour and encouraging them to be a nice and lovable person. In their work, Harris et al. [4] found that adult social reinforcement has significant effect on child behavioural development. They stressed that such a reinforcement can help in modifying problematic behaviour of the children. Moreover, Burgees et al. [8] argue that since behaviour is learned, a child whose misbehaviour is not reinforced by some form of feedback may develop criminal tendencies in long run. If we compare these situations in context of a child interacting with a robot, we can easily draw a thread of association. If a child is allowed to interact with a robot considering it as a lifeless piece of metal and plastic, then the lack of reinforcement may lead to the development of anti-social and violent behaviour in the same child as an adult. Therefore we propose a mechanism in which robot uses the past experience with an interacting child and biases its responses based on the previous actions of the child. This mechanism can be observed in Fig. 1. Whenever a child performs some action to the robot, the robot stores the event in its memory for future retrieval. If there are previous events associated with the child, the robot retrieves those events to bias the current response based on whether the past events had positive or negative impact on the robot. Finally, robot exhibits a response that helps the child realise if the previous actions of the child were acceptable or not. It should be noted that the interacting individual may also be an adult, yet this type of behavioural quality of a robot is more useful in shaping behaviour of children than that of adults. The mechanism discussed above is implemented in our computational model of emotion—EEGS [5], which shall be detailed in Sect. 3.

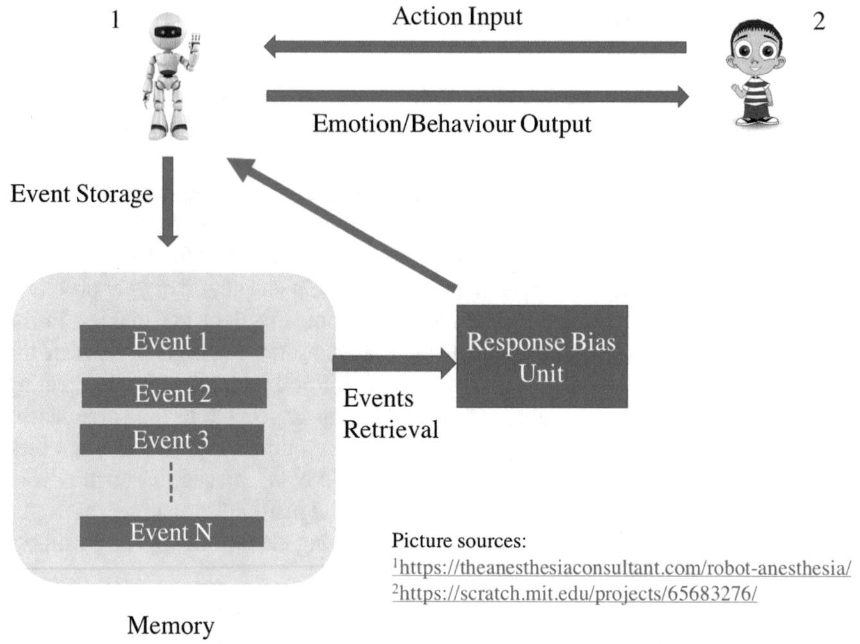

Fig. 1 A child interacting with a robot with an ability to bias its responses based on prior experience

3 Memory and Experience in EEGS

EEGS is a computational model of emotion [5, 9, 10] based on appraisal theory [11]. EEGS stores the sequence of actions performed by an interacting child and biases its actions based on how the child treated the robot in the past. In EEGS when an event happens, its attributes are stored in the following structure.

```
(<Person/Source>, <Action>, <Date/Time>, <Target>)
```

`Person/Source` indicates the person interacting with the robot. From the implementation viewpoint, the attribute `Person` can be a complex object encapsulating ID, age, sex and other features of the interacting individual. The choice of attributes can vary between applications of the intended system. `Action` indicates the action performed by the person to the `Target`. `Target` can be the robot itself or someone else the robot recognises. Similar to `Person`, `Target` can also be represented as a complex object of various features. `Date/Time` indicates the date and time of the action from `Source` to the `Target`. It can be argued that the event structure presented above is very simple and may not capture all the aspects that might be relevant to an event denoting an interaction between two individuals. Therefore, we also consider other aspects like familiarity of the robot about the interacting person, perception of the robot about the interacting person, familiarity and perception of the robot about the target in interaction (if the target of interaction is not the robot itself) and also the impact caused by the action of the interacting individual on the robot (a valenced number denoting the positive or negative effect of the event on the robot). With all this information related to each event in memory, EEGS is able to use this information to bias its emotional and behavioural responses in subsequent and future interactions with the same individual. In other words, EEGS is able to infer a positive or negative bias based on past history with the interacting individual. Our discussion in the following sections will revolve around an interacting child.

Suppose there are N events experienced by the robot with the child. If we denote an impact of the event on the robot as I_m and time (in days) since the event has happened as t, then the positive (or negative) bias of the robot towards the child (p_e) is given by the following formula.

$$p_e = \frac{\sum_{i=1}^{N} \frac{I_{m_i}}{t_i}}{N} \tag{1}$$

The reason for dividing the impact (I_m) by the time duration (t) is because the effect of an event which was experienced long ago should be less than the effect of a similar event which happened recently. The value of p_e is used to bias the emotional and behavioural responses of the robot towards the child. Mathematically, we can represent the response of the robot as the function of current action (a) as well as positive/negative bias (p_e) of the robot towards the child.

$$Response = \mathcal{F}(a, p_e) \tag{2}$$

Equation 2 denotes that response of the robot towards the child for the same action might be different depending on the bias value determined by the previous experience of the robot with the child. For example, consider a robot who has mostly been treated well by a child. If the child asks to play with the robot (Action = “Ask for playing”), the robot might express happiness or excitation on getting a chance to play with the child. However, if the robot has previous experience of foul play or other misbehaviour by the child, then it might express lack of interest or distress in response of the child’s offer to play. Such a response from the robot helps the child realise that his actions towards the robot in the past were desirable or not. This acts as a reinforcement to encourage the child to maintain good behaviour if he had treated the robot with good manner previously and to discourage him from doing something bad if he behaved with the robot in inappropriate way. In Sect. 4, we shall present an evaluation of the working of EEGS with the implementation of the mechanism of biasing the response based on prior interaction with the robot.

4 Evaluation

In the previous section, we presented the mechanism by which past events are stored in the memory of EEGS and how those events are considered to bias the response of the robot in the subsequent interactions with a child. In this section, we present a detailed analysis of the responses of EEGS in different situations of interactions making it possible for the child to realise what he did was right or wrong.

4.1 Experiment Design

In order to understand the working of EEGS, let us consider two separate scenarios. Suppose in the first scenario, a child treats the robot (running EEGS) with good manners and plays well and in the second scenario, a child does bad treatment to the robot instead of giving attention to playing. We will examine how these different treatments of a child towards the robot will make difference in robot's decision making and responses during the subsequent interaction from the child. Thus, we compare the differences in the responses of the robot for another play request from the child to the robot after each of the interaction scenarios. The interactions between the child and robot in each scenario was simulated in a computer environment. Each action from the child towards the robot was feed into the EEGS system as a valenced value in the range $[-1, 1]$.[1]

Scenario 1: Nice Treatment Scenario

It is a regular afternoon and Tom (a child of age around 5 years) wants to play ball with his companion robot. Tom requests the robot to play with him. Below are the sequence of interactions from Tom towards the robot[2] in the first scenario.

(i) Tom asks the robot to play with him.
(ii) Tom passes the ball towards robot with his hands.
(iii) Tom passes the ball towards robot with his hands.
(iv) Tom passes the ball towards robot with his hands.
(v) Tom kicks the ball towards robot.
(vi) Tom kicks the ball towards robot.
(vii) Tom passes the ball towards robot with his hands.

Figure 2 shows the dynamics of joy emotion exhibited by EEGS when Tom treats robot in a nice way and plays well. We can clearly see that intensity of joy emotion

[1]For more understanding of how the actions were extracted and assigned a valenced value, please refer to our previous paper [10].

[2]In this experiment, we are concerned about only the emotions/responses of the robot in reaction to the actions of Tom. Hence, we are enumerating only the actions of Tom towards the robot. These actions will be used to trigger the emotions of EEGS system in the simulated robot.

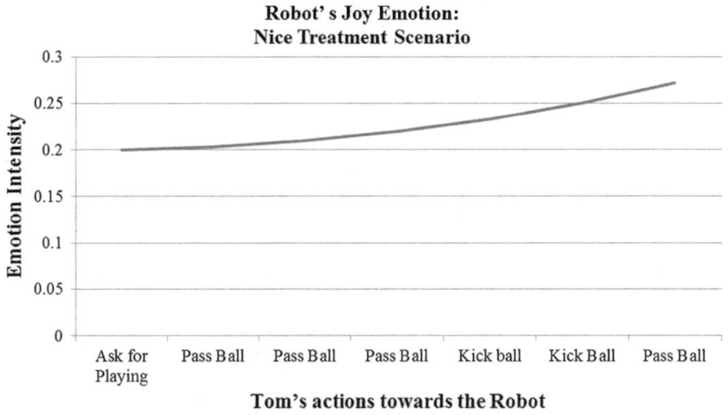

Fig. 2 Joy emotion dynamics of EEGS in nice treatment scenario

of the robot rises gradually on each ball pass from Tom. Expression of increased joy in such a situation provides an implicit reinforcement to Tom that his current actions are desirable.

Scenario 2: Bad Treatment Scenario

In the second scenario, Nick (also a child of age around 5) requests the robot to play ball with him. In this scenario, Nick misbehaves with the robot instead of playing in a good manner. Below are the sequence of actions from Nick towards the robot.

(i) Nick asks the robot to play with him.
(ii) Nick kicks the robot.
(iii) Nick kicks the robot.
(iv) Nick kicks the robot.
(v) Nick kicks the robot.
(vi) Nick kicks the robot.
(vii) Nick kicks the robot.

Figure 3 shows the dynamics of joy emotion of EEGS in the scenario of bad treatment by Nick. In contract to Fig. 2, even if the intensity of joy emotion was similar to the scenario of nice treatment for the action "Ask for playing" (i.e. 0.2 in both the scenarios), the intensity gradually decreases towards zero as soon as Nick starts kicking the robot instead of playing in a nice way. Such a response from the robot gives hint to Nick that the way he was behaving with the robot was not pleasing but rather undesirable. However, this may not be sufficient to reinforce a strong realisation of this misbehaviour. Therefore, we propose that the robot should make Nick recall about misbehaviour in the future interactions with the robot. Thus we ran another experiment to examine how the memory of event history can help EEGS to bias its responses to make the child realise about his actions (see below).

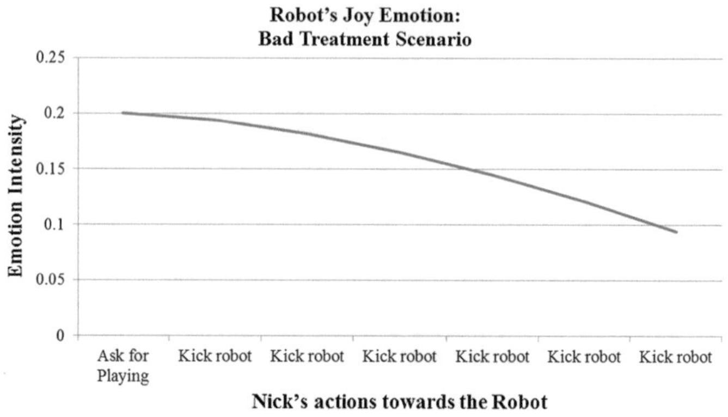

Fig. 3 Joy emotion dynamics of EEGS in bad treatment scenario

Post Nice/Bad Treatment Responses of the Robot

In order to examine the effectiveness of the events-biased (memory-biased) response of EEGS, we recorded the emotions of the robot in reaction to another play request from Tom and Nick respectively. Our hypothesis was that since Tom played with good manners and Nick misbehaved with the robot in the previous play, robot would express more positive emotions in response to the play request from Tom while the situation should be opposite in case of Nick because he misbehaved with the robot instead of playing properly.

Since EEGS provides the robot with an ability to remember the past actions of the interacting individual, robot kept all the actions of Tom and Nick in its memory. Next time when Tom and Nick asked the robot to play, the robot recalled their previous actions and exhibited different responses. For example, robot responds to Tom saying, "*Yes sure!! It was pleasure playing with you last time*" and responds to Nick saying "*I would play but I am little worried about how you might behave with me*". This phenomenon is depicted in Fig. 4. Comparing the bars of emotion intensities post the nice treatment scenario (Tom) and bad treatment scenario (Nick), we can clearly see that the response of the robot for the action "Ask for playing" from Tom is more positive than that of Nick. Table 1 summarises the actual difference in emotional response of EEGS for the request to play from Tom following the nice treatment scenario and same request from Nick following the bad treatment scenario. In case of Joy emotion, intensity is lower by 63% post the bad treatment scenario as compared to the intensity of post nice treatment scenario. Similarly, for Appreciation of the proposal to play is lower by 78% post the bad treatment scenario. Likewise, there is difference of 89 and 75% in case of Gratitude and Liking emotions respectively. These figures strongly suggest that EEGS has an ability to bias its emotional responses based on the previous actions of interacting child thereby giving them an opportunity to realise that they are not expected to behave in a bad manner with others (including the companion social robots).

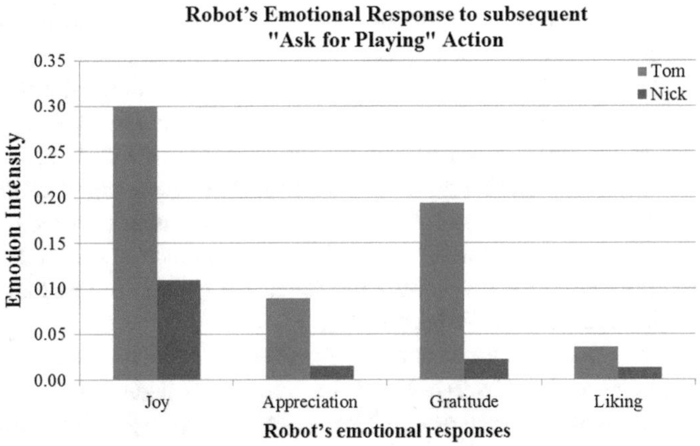

Fig. 4 Difference in emotional response of EEGS post the nice treatment and bad treatment scenarios

Table 1 Comparison of emotion intensities of various emotions post (i) Nice treatment scenario (Tom) and (ii) Bad treatment scenario (Nick)

Emotion	Post nice treatment	Post bad treatment	Percentage difference (%)
Joy	0.30	0.11	−63
Appreciation	0.09	0.02	−78
Gratitude	0.19	0.02	−89
Liking	0.04	0.01	−75

As previously mentioned, it is important to provide a feedback signal to a child to indicate if his/her behaviour is acceptable or not. Failure to do so may lead them to believe that they can do whatever they want and in the long run they may also develop violent and criminal tendency. The mechanism used in EEGS to remember the past actions of a child and recall the actions in order to tune the instantaneous responses to a child's actions may play a vital role in helping children develop socially acceptable behaviour.

5 Conclusion

Social robots working closely with young children as a companion or tutor should not only act as an entertainer or provide pre-defined facts to the children but also be able to teach them valuable moral behaviour. Previous studies suggest that adult reinforcement has major effect on a child's behavioural development. If robots are intended to be employed as companions or guardians of children, then this fact must also be taken into consideration. In other words, companion or tutoring robots can play a vital role in shaping a socially acceptable behavioural development in young children. Addressing this requirement, we employ our computational model of emotion (EEGS) with an ability to remember and retrieve the actions of a child from the past and bias its responses towards the child based on how the child has treated the robot in the past. Experimental results show that EEGS able to significantly tune its emotional responses post a series of good or bad actions and acknowledge the same to the child so that s/he can correct her/his behaviour in the future. We believe that this is a good direction for developing the social companion robots that will help develop a well-behaved generation of kids.

In our future work, we aim to conduct our experiments with physical robot and children and examine the effectiveness of the robot to teach good manners to young children. The study will have to be conducted for a long run—possibly a few months for the certainty of the outcome. While we anticipate to achieve similar results to the ones obtained in simulation, there might be some variability because of physical limitations of the robot such as vision and speech recognition—which are still far from perfect in robotic applications. Nonetheless, we believe that our work posits itself as an important pedestal in the advancement of the field of social robotics.

Acknowledgements This research is supported by an Australian Government Research Training Program Scholarship. We are thankful to the University of Technology Sydney; ARC Discovery Project scheme; and CBA-UTS Social Robotics Partnership.

References

1. Ziegler, A., Jones, A., Vu, C., Cross, M., Sinclair, K., Campbell, T.L.: Companion robot for personal interaction, 7 June 2011. US Patent 7,957,837
2. Osada, J., Ohnaka, S., Sato, M.: The scenario and design process of childcare robot, papero. In: Proceedings of the 2006 ACM SIGCHI international conference on Advances in computer entertainment technology, p. 80. ACM (2006)
3. Han, J.: Robot-aided learning and r-learning services. In: Human-robot Interaction. InTech (2010)
4. Harris, F.R., Wolf, M.M., Baer, D.M.: Effects of adult social reinforcement on child behavior. Young Child. pp. 8–17 (1964)
5. Ojha, S., Williams, M.-A.: Ethically-guided emotional responses for social robots: Should i be angry? In: International Conference on Social Robotics, pp. 233–242. Springer (2016)
6. Guilford, J.P.: The Nature of Human Intelligence. McGraw-Hill (1967)
7. Mayer, J.D., Roberts, R.D., Barsade, S.G.: Human abilities: Emotional intelligence. Annu. Rev. Psychol. **59**, 507–536 (2008)
8. Burgess, R.L., Akers, R.L.: A differential association-reinforcement theory of criminal behavior. Social Probl **14**(2), 128–147 (1966)
9. Ojha, S., Williams, M.-A.: Emotional appraisal: A computational perspective. In: Advances in Cognitive Systems (2017)
10. Ojha, S., Vitale, J., Williams, M.-A.: A domain-independent approach of cognitive appraisal augmented by higher cognitive layer of ethical reasoning. In: Annual Meeting of the Cognitive Science Society (2017)
11. Ortony, A., Clore, G.L., Collins, A.: The Cognitive Structure of Emotions. Cambridge University Press (1990)

Robot Supporting for Deaf and Less Hearing People

Nguyen Truong Thinh, Tuong Phuoc Tho, Truong Cong Toai and Le Thanh Ben

Abstract This paper discusses the development of a service robot for translating spoken language text into signed languages and vice versa The motivation for our study is the improvement of accessibility to public information announcements for deaf and less hearing people. The robot can translate Vietnamese sign language into speech and recognize Vietnamese/English speech to suitable gesture/sign language. The paper describes the use of service robot in a sign language machine translation system. Several sign language visualization methods were evaluated on the robot. In order to perform this study a machine translation service robot that uses display screen on robot as service-delivery device was developed as well as a 3D avatar. It was concluded that service robot are suitable service-delivery platforms for sign language machine translation systems.

1 Introduction

There are over one million deaf people in Vietnam. Approximately 400,000 of these are in school age and use Vietnamese Sign Language as their first and only language. Contrary to common belief, sign languages are not visual-gestural representations of spoken languages. They are rich fully-fledged languages of their own. Additionally, different countries have sign languages of their own. Each of these sign languages also experience regional variations within the country. This is akin

N. T. Thinh (✉) · T. P. Tho · T. C. Toai · L. T. Ben
Department of Mechatronics, Ho Chi Minh City University of Technology and Education, 70000 Ho Chi Minh City, Vietnam
e-mail: thinhnt@hcmute.edu.vn

T. P. Tho
e-mail: thotp@hcmute.edu.vn

T. C. Toai
e-mail: truongcongtoai110196@gmail.com

L. T. Ben
e-mail: benspktk14@gmail.com

J.-H. Kim et al. (eds.), *Robot Intelligence Technology and Applications 5*, Advances in Intelligent Systems and Computing 751,
https://doi.org/10.1007/978-3-319-78452-6_24

to dialect and idiolect variations in spoken languages. The objective of this study is contribution of active communication between deaf and hearing people and reduction of their communication barrier. As you known, deaf people need to communicate with hearing people on the very basic daily talks as a necessity of life. Besides, the deaf need to communicate with their hearing people on situations include public services, buying bus and train tickets, seeking directions and purchasing the goods. The inability to access these services effectively because of communication constraints is a source of stress for the deaf [1].

The solution for barrier between deaf and hearing people was the use of highly skilled and trained interpreters proven inadequate and inefficient. And of course, the services are also very costly. We believe that most deaf people in Vietnam cannot afford this service. Also, the use of interpreters is unsuitable in contexts in which confidentiality is vital such as when a deaf person seeks medical or psychological treatment. In such cases, the deaf person may not be keen to have a human interpreter present. The research carries out automated translation robot between sign/ gesture language and Vietnamese/English and vice versa. The given gesture/sign as input, the system aims to produce corresponding English audio as well as display on screen of robot, and given Vietnamese/English speech as input, the robot produce and render the corresponding sign language clip. The system of robot produced gesture recognition systems that can track hand-motions, hand-shape recognitions as well as facial recognition. Besides, the robot's system produced a animation avatar displaying on robot's screen based on synthetic sign language phrases [2]. The paper describes the use of mobile robot as service delivery devices in a gesture/sign language machine translation system.

2 Structures of Robot

The Robot is located in the public space to provide and guide information for customers as well as deaf and less-hearing people. It can move on smooth floor and avoid the statics and dynamics obstacles. Therefore, the design must ensure that the design of hardware and algorithms should make a high level decision. The robot was designed and set up the hardware being developed in parallel. The robot is designed 4 degrees of freedom. The main components of the robot are base, robot body and a touch screen on the head. Base of robot consists of 3 wheels: 2 wheel drive and a passive wheel. The volume under analysis intended robot about 30 kg including all mechanical parts, power systems, actuators, computers, screen… Actuators are DC motors. The maximum speed of robot for movement is 5 km/h.

The structure of controlling system is shown in Fig. 1. The main controller is based on the mini PC, that has the function of managing the entire operation of the system, gathering and processing key information, and linking the devices together. Here, the data obtained from the camera and microphone is processed by the CPU. Calculates appropriate actions such as displaying information to communicate with the user via the monitor and speakers, or controlling the robot's behavior through

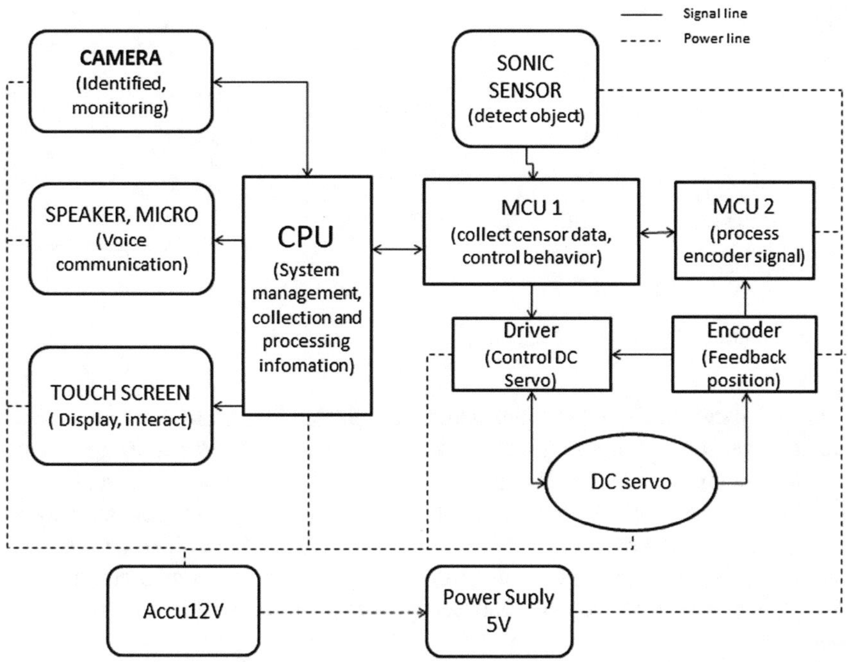

Fig. 1 Block diagram connecting devices across the robot

microcontrollers via the UART interface. The robot behavior control is performed and managed by a microcontroller (MCU).

This MCU receives the control signal from the mini PC and then analyzes it and sends the command to the actuator to perform (move the robot to a certain position, change the direction of the camera, etc.). In addition, the MCU also receives signals from the sensors to control the robot more intelligently, such as: using sonic sensors to dodge obstructions while moving, read encoder signals to move to the right position…

3 Communication of Robot for the Deaf People

The communication between deaf and hearing people (Fig. 2) based on service robot required to have two distinct translation tasks: Translation from spoken language to gesture/sign language; Translation from gesture/sign language to spoken language. The deaf people can not communicate with hearing people in directly, and require additional intermediary tool to be translation tool. Sign/Gesture-based tool carries out a gesture translation to deaf people message. The deaf person can stands opposite robot and camera attaching on robot's head. These captured images are then sent over the network to central processors of robot that

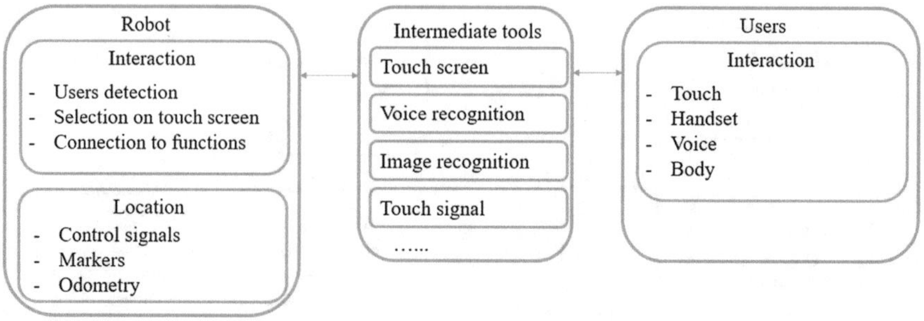

Fig. 2 Interaction between robot and human

recognize them. Such tool requires additional translation infrastructure in order for the deaf person to communicate with a hearing person. When using devices communicating between the different parties is indirect such as a text and speech based instant messenger by one party and a sign/gesture language application by the other party. The service robot's communication switches such as text to speech and vice versa, sign/gesture language to speech and vice versa. Gesture-based tool refers to specifically built to translate between gesture/sign and spoken language using translation tool.

In translating from gesture/sign to spoken language, gesture recognition algorithm is used to recognize gesture language from captured images. The output of this system is information of speech or message on screen. Conversely, in translation between hearing and deaf people, the input which is sentences of spoken language is translated into gesture information with animation avatar. The translation from spoken language to gesture/sign involves the following processes: capturing spoken speech input as audio; and converting it into text using speech recognition technology; translating the text into a gesture/sign transcription notation finally synthesizing like as clip on robot's screen. Capturing spoken speech is based on several microphones. Nowadays, many speech recognition technologies currently exist that can convert from spoken language audio to text, many of them open-source. On the other hand, translation from spoken language text to graphical user interface (GUI) that allows a hearing person to understand what the deaf person want to do from gesturing manual signs. The animation avatar is created using humanoid skeletons.

4 Technical Solution of Gesture Recognition

In order to translate sign/gesture language, deaf people stand opposite robot to recognition. The following block diagram summarizes our approach in Fig. 3. The input data will be recorded by camera on robot's head and sent to central processors of robot. Then robot will process data to determine subjects that standing opposite

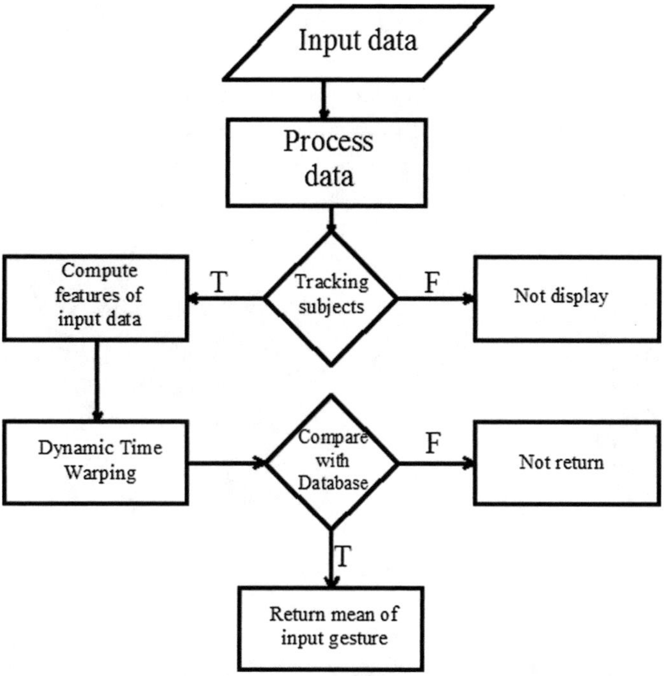

Fig. 3 Solution of gesture recognition

robot is human or not. If true, the robot will record and compute features of them gesture based on the characteristics of the skeleton.

Then we will have two sequences Q = {Q1, Q2, Q3, …, Qm} and C = {C1, C2, C3, …, Cm} where each sequence is features of skeleton in each frames (Q is data sequence and C is input sequence in real time). In order to determine similarities between the two sequences, we use Euclid method:

$$dist(Q, C) = \sqrt{\sum_{i=1}^{m} (Q_i - C_i)^2} \quad (1)$$

However, having a problem that is difference in time between sequence C and sequence Q. Euclid method can only compute exactly result when the two sequences have the same in time [3, 4]. Therefore, Dynamic Time Warping method is used to solve that problem. Two lines (Fig. 4) have the very same types of shape but different in time. In Fig. 2, if computing by Euclid method will have significant error although the two sequences have great similarities. And DTW method helps to increase accuracy. DTW method is used to compare the similarity between two sequences Q and C in real time, DTW algorithm will build a square matrix of size m, where:

$$T(i,j) = dist(i,j) + min[T(i,j-1), T(i-1,j), T(i-1,j-1)] \quad (2)$$

where dist(i, j) is computed by Euclid method. After having T matrix, determine minimum element of T and set it as optimal value in similarities between the two sequences.

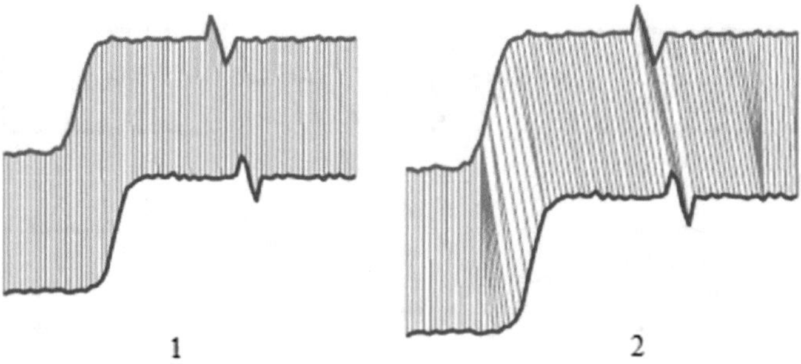

Fig. 4 Compute dynamic time warping method

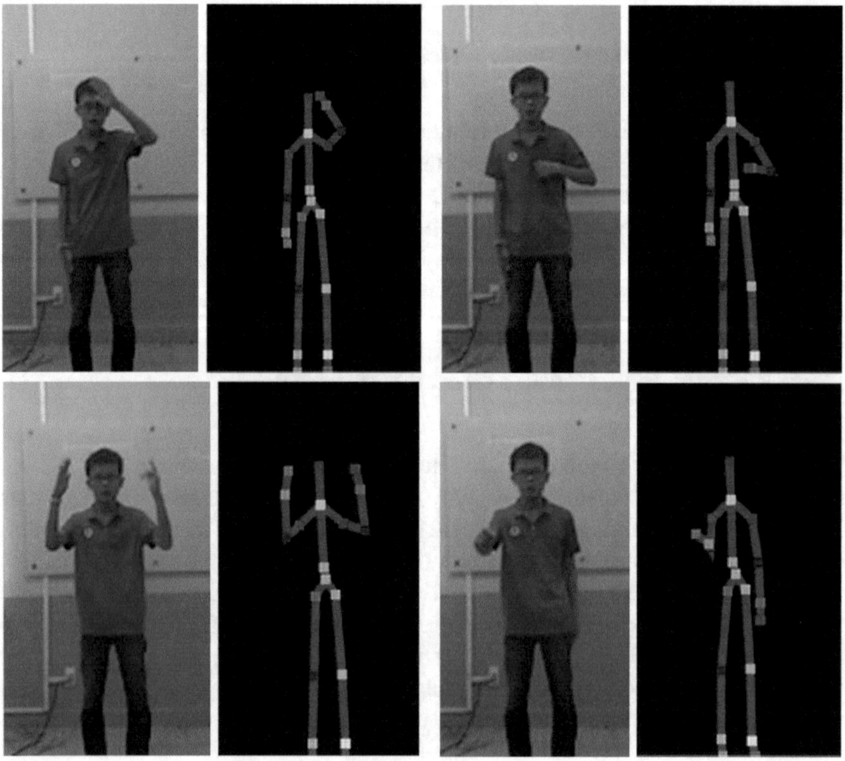

Fig. 5 Translate sign language results

The input sequence is compared with all sequences in database, each time will have a corresponding optimal value. Input sequence is compared with all sequences in database, each time will have a corresponding optimal value. Finally, comparing between the minimum value and given certain threshold (in order to reduce error). If below threshold, the optimal value is accepted, returning mean of input gesture. The study is developed into service robot for supporting deaf people.

With dynamic moving ability and friendly display, robot can easily communicate with deaf people. We translated Vietnamese Sign Language (VSL) with sentence: "Hello, Nice to meet you" (In Vietnamese). In grammar of VSL have not adjectives ...before adverbs and adjectives to add emphasis or exclamatory. The gesture in real life is recorded from the 1st to the 32nd frames of 640×480 frames. The Dynamic Time Warping algorithm doesn't care about how quickly the gesture is performed. The Robot is located in the public space to provide and guide information for customers as well as deaf and less-hearing people. We repeated several times with difference about speed of gesture from Robot (Fig. 5). The gesture was tracked and translated successfully with accuracy about than 90%. Besides that, we sent five clips sign language to center of deaf people to ensure that they can understand our database. The result is almost all deaf people can identify what we want to convey.

5 Conclusions

This paper presented a summary of the design and development of robot supporting for deaf and less hearing people. With dynamic moving ability and friendly display, robot can easily communicate with deaf people. The robot can reduce the cost of hiring or training sign language interpreters while ensuring quality of communication with deaf people. Besides, the working time of the interpreters can be limited but the robot can work 24/7 without complaint or get tired.

Acknowledgements This study was financially supported by Ho Chi Minh city University of Technology and Education, Vietnam.

References

1. Stokoe, W.C.: Sign language structure: an outline of the visual communication systems of the American deaf. Deaf Stud. Deaf Edu. (2005)
2. Hameed, A.: Information and communication technologies as a new learning tool for the deaf. Relation (2009)
3. Ghaziasgar, M.: The use of mobile phones as service-delivery devices in sign language machine translation system. University of the Western Cape (2010)
4. Escudeiro, P., Escudeiro, N., Reis, R., Barbosa, M., Bidarra, J., Gouveia, B.: Automatic Sign language translator model. Adv. Sci. Lett. (2014)

The Use of Machine Learning for Correlation Analysis of Sentiment and Weather Data

Hu Li, Zahra Jadidi, Jinyan Chen and Jun Jo

Abstract The development of the Internet of Things (IoT) drives us to confront, manage and analyse massive and complicated data generated from various sensors. Also, social media have rapidly become very popular and can be considered as important source of data. Twitter on average generates about 6,000 tweets every second, which total to over 500 million tweets per day. Facebook has over 2 billion monthly active users. The individual posts may be trivial, however, the accumulated big data can provide diverse valuable information, which can be also correlated with IoT and enable human sentiment identification of the environment changes. This work proposes a machine learning model for correlation analysis and prediction of weather conditions and social media posts. In experimental evaluation we demonstrate that the Big Data analysis approach and machine learning techniques can be used to analyse and predict sentiment of different weather conditions.

1 Introduction

With the rapid growth of the Internet and smartphone users, social media play important roles in our daily lives. Millions of 'connected' citizens share information through social media and the enormous amount of posts often turns into collective knowledge or valuable resources for the analysis and management of our environment. It is now recognised that such non-traditional approaches present

H. Li · Z. Jadidi · J. Chen · J. Jo (✉)
School of Information and Communication Technology, Griffith University,
Southport, QLD 4222, Australia
e-mail: j.jo@griffith.edu.au

H. Li
e-mail: hu.li@griffith.edu.au

Z. Jadidi
e-mail: z.jadidi@griffith.edu.au

J. Chen
e-mail: jinyan.chen@griffith.edu.au

J.-H. Kim et al. (eds.), *Robot Intelligence Technology and Applications 5*,
Advances in Intelligent Systems and Computing 751,
https://doi.org/10.1007/978-3-319-78452-6_25

inexpensive means for gathering rich, authentic, and unsolicited data on people's perceptions and experiences [1].

Big data is described as datasets that are diverse, volumes, complex and is not easy to handle by traditional database technologies. Compared with traditional datasets, big data typically includes masses of unstructured data, such as social media posts, structured data, data collected from various sensors, therefore it provides great potential for real-time analysis. Cloud computing usually provides big data with infrastructure of cost efficiency, elasticity, and easy management.

In the paradigm of IoT, we investigate how the public opinions from social media are related to the weather data in the context of human sentiments. This work builds on top of work related to sentiment analysis and environmental changes [2, 3]. Findings of this work will expand the extent of the IoT technology to predict human sentiments for different weather conditions.

This work relied on Twitter data posted from Cairns city in Queensland, Australia. In line with previous research this work was conducted on Hadoop cluster, running MongoDB and NoSQL databases where Twitter posts have been stored, and relational MySQL database to store IoT data. Data and processing were done in the distributed environment across nodes in a shared-nothing architecture. Utilization of artificial neural network has been proposed. A two-layer feed-forward neural network (with one hidden layer) was employed to predict sentiment pattern using non-linear regression [4]. A fitnet method was used in Matlab to simulate this fitting neural network [5]. A machine learning model was trained for the Twitter dataset and the accuracy of the prediction was tested.

2 Big Data

The amount of digital data is growing at an exponential rate producing data commonly referred as Big Data. Big Data is a term for datasets that are so large or complex that the traditional data processing methods are inadequate to deal with them. Big data challenges include issues related to capturing, storing, analysis, sharing, visualization, as well as information privacy [6]. Lately, the term "big data" tends to refer to the use of predictive analytics, user behavior analytics, and other advanced data analytics methods that extract value from data [7]. Defined challenges and opportunities brought about by increased data with a 3Vs model, i.e., the increase of Volume, Velocity, and Variety [8]. In the "3Vs" model, Volume means a massive amount of data; Velocity means the data collection and that the analysis must be rapidly and timely conducted; Variety indicates the various formats and types of data. A huge amount of valuable data can be generated by crawling the Internet, from forum posts, chatting records, and microblog messages. Such Internet data may be valueless individually, but through the analysis of accumulated data and their correlation, useful information can be identified. It is even possible to forecast users' behaviours and emotional moods. Traditional data management and analysis systems are based on the relational database management system

(RDBMS). However, such RDBMSs only apply to structured data, other than semi-structured or unstructured data. The traditional RDBMSs could not handle the huge volume and heterogeneity of big data [8]. Such data sources include sensors, videos, clickstreams, and other data sources. In the paradigm of IoT, sensors all over the world are collecting and transmitting data to be stored and processed in the cloud [6].

3 Sentiment Analysis on Social Media Data

Sentiment analysis is a new approach to text analysis, which aims at determining the opinion and subjectivity of reviewers. Sentiment relates to valence and reflects underlying emotions, broadly classified into positive, neutral and negative. Sentiment can be extracted from social media statements through the use of computational linguistics and natural language processing [9]. Specifically, in this work we relied on Twitter as a source of, because it is a relatively commonly used platform and some of the Twitter data through public API is freely available for analysts [10]. Twitter on average has about 6,000 tweets every second, which total to over 500 million tweets per day. These posts can provide valuable information of diverse topics. There are many sentiment analysis methods for English text presented in literature. The existing work on sentiment analysis can be classified into three popular groups based on methods employed [11]. Machine learning method, lexicon-based approach and rule-based approach.

Another classification is oriented more on the structure of the text: document level, sentence level or word/feature level classification. In general sentiment analysis of tweets has been coded into positive, negative or neutral [12]. To obtain higher accuracy it has been proposed to use classifiers that are trained with large volume of human annotated text [13]. Further improvements can be obtained over the time via deep (machine) learning. Accuracy of the computer-based classifier depends, and it has been reported from 41.9%, for a sample of Thai students based on tweets about Bangkok and Phuket, to 66% related to hotel reviews [14].

Considering that in this work we analyse sentiment of social media post we utilized Valence Aware Dictionary for Sentiment Reasoning (VADER) approach for sentiment analysis because it is purposely developed for sentiment analysis calculation of short text found in social media posts [7].

4 The Sentimental Analysis with Weather Data

For this research, the collected Twitter data is stored in a NoSQL MongoDB database, which is located on a cluster computer with a Hadoop architecture. Each Twitter data in the database contains Metadata, including the content of the tweet, language, location where account was opened, and potentially place from which the

Twitter message was sent, if GPS is enabled. We found that on average 15% of tweets have their GPS enabled.

Twitter users send tweets for a wide range of reasons and some of tweets refer to the topic. Therefore, it needs to be filtered to extract messages of interest. All keywords were extracted using a case insensitive search technique, and variations of the same word (e.g. 'dive', 'diving') were compiled as the same keyword. Tweets were analysed with regards to their positive or negative polarity.

As indicated above method adopted by this research is VADER rule-based approach with the text-level, which combines a general lexicon and a series of intensifiers, punctuation transformation, emoticons, and many other heuristics to compute sentiment polarity. The VADER sentiment lexicon is composed of more than 7000 items with associated sentiment intensity validated by humans. The sentiment score ranges from minus one (negative) to plus one (positive), with the zero being 'neutral'.

5 Twitter Dataset

This paper considers the impact of weather variables on users' mood during daytime. In this regard, the Twitter data are mapped with weather information in Cairns. The Bureau of Meteorology (http://www.bom.gov.au), which is an Australia national weather, climate, and water agency is used to find the relevant weather data. These data can also be obtained through APi provided as Open Data. The tweets mapped to the weather information are eventually labelled with hourly sentiment scores.

The Twitter dataset used in this paper consists of tweets posted between 8:00 a. m. to 5:00 p.m. (October 2016–October 2017) from Cairns in Australia. This study aims to find the correlation between weather conditions and sentiment scores. An artificial neural network (ANN) was used to predict sentiment values in new data. The ANN model was trained and evaluated with training and testing datasets respectively. 70% of the input data is used for training dataset and 30% is reserved for testing dataset. Table 1 shows the number of samples in each dataset.

Table 1 Samples in training and testing datasets

Dataset	Total samples	Training dataset (70%)	Testing dataset (30%)
Carins (Oct 2016–Oct 2017)	55,216	38,651	16,564

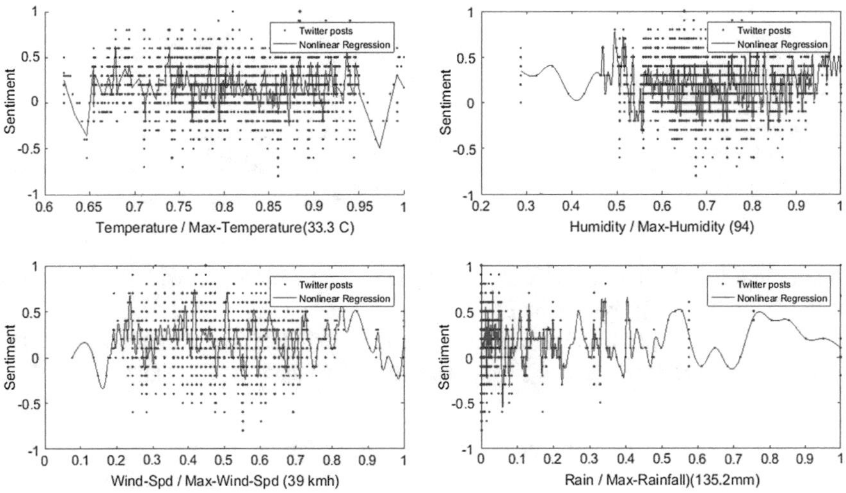

Fig. 1 Comparison of the sentiments

6 Sentiment Pattern Analysis

The Twitter dataset is very noisy and we need to remove noise to be able to analyze the sentiment patterns. A smoothing spline method is used to find the fitted curve for this noisy dataset. Figure 1 shows the relationship between temperature, humidity, wind, rainfall, and sentiment in our Twitter dataset collected from Cairns Australia.

The x-axes in all graphs were rescaled considering the maximum value for the variables in the dataset. This helps to convert x-axis to a dimensionless value in the [0 1] range and compare the impact of different variables on sentiment. According to the figure, the concerned tweets were in the range of 22–32 °C (temperature), 47–87% (humidity), 10–31 km/h (wind speed) and rain less than 14 mm. The averaged sentiment values for this ranges are positive. It shows that these ranges were desirable conditions for users and made them feel positive. Although there are some negative sentiments in these ranges, the averaged sentiment for each weather variable is positive.

7 Experimental Results

One of the objectives of this study was to forecast human sentiments under certain weather conditions using an artificial neural network. In this regard, a two-layer feed-forward neural network (with one hidden layer) is employed to predict sentiment pattern using non-linear regression [4]. A fitnet method was used in Matlab

Table 2 Parameters of NNP

Description	Values
Number of neurons in hidden layer	30
Number of input features	8
Training function	Levenberg-Marquardt (Trainlm)
Number of ephocs	5000

to simulate this fitting neural network [5]. The parameters of this neural network predictor (NNP) is provided in Table 2.

The NNP has [8 30 1] topology, and it is simulated in Matlab 2016a. For measuring the performance of NNP, Accuracy is defined as Eq. (1):

$$\text{Accuracy } (\%) = \frac{C_p}{C_p + W_p} * 100 \tag{1}$$

where C_p is the number of correct predictions and W_p shows the number of wrong predictions. The NNP method is trained with 5000 Ephocs. Afterwards, the trained method is separately evaluated with training dataset and testing dataset (new data). The results are as Table 3. In addition, Mean Square Error (MSE) is shown in Fig. 2. All experiments are repeated 10 times to achieve higher accuracy, and all provided results are averaged values of 10 trials. The NNP method could predict 95% of the sentiment values in new dataset.

NNP was trained with trainlm and the performance was compared with other training algorithms in Table 3. As the result shows, trainlm has the highest accuracy over 10 trials (95% correct prediction). In addition, MSE function shown in Fig. 2 is used to evaluate the performance of NNP. In this phase, 70% of input data is used for training dataset and 30% is for both validation and testing datasets. As Fig. 2 shows, MSEs in all data sets are about 0.02 after 3000 epochs. This also shows the high performance of NNP in predicting sentiments in new data.

Table 3 Comparison of the accuracy of FFNNP trained with different algorithms

Feed-forward training algorithms	Accuracy with training dataset (%)	Accuracy with testing dataset (%)
Trainlm	98	95
Traingd	97.5	93.6
Trainbr	96.8	94.8

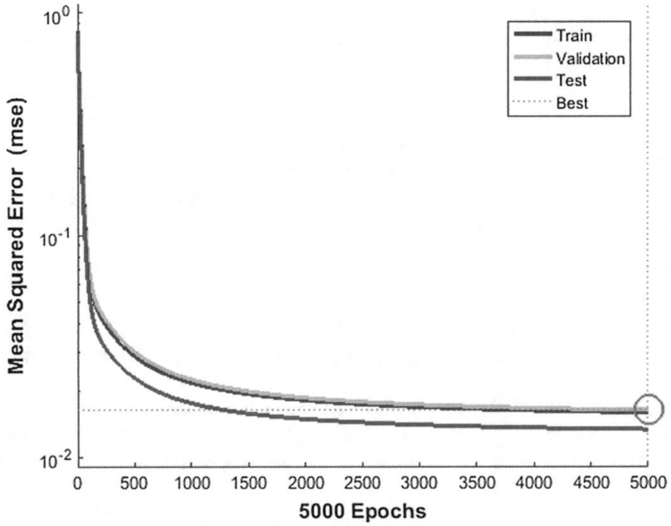

Fig. 2 Mean square error of FFNNP

8 Conclusion

This paper introduced Big Data concept to analyse and predict human sentiment depending on different weather conditions. Twitter social media post are used and we utilized VADER rule based sentiment analysis method to compute sentiment polarity. Additionally, BOM Open weather data have been captured. The correlation between the two has been investigated. A machine learning model was trained for the Twitter dataset and the accuracy of the prediction was tested. The analysis revealed that the Twitter-based sentiment analysis demonstrated a close relationship between different weather conditions and users' sentiments. The results show that the Big Data analysis and machine learning techniques can be used to analyse and predict sentiment to different weather conditions. It also revealed comfortable range of weather conditions to humans.

Acknowledgements This project was partly funded through a National Environment Science Program (NESP) fund, within the Tropical Water Quality Hub (Project No: 2.3.2).

References

1. O'Leary, D.: The use of social media in the supply chain: survey and extensions. Intell. Syst. Acc. Finance Manag. **18**(2e3), 121e144 (2011)
2. Becken, S., Stantic, B., Chen, J., Alaei, A.R., Connolly, R.M.: Monitoring the environment and human sentiment on the Great Barrier Reef: assessing the potential of collective sensing. J. Environ. Manag. **203**, 87–97 (2017)

3. Chen, J., Wang, S., Stantic, B.: Connecting social media data with observed hybrid data for environment monitoring. In: International Symposium on Intelligent and Distributed Computing, pp. 125–135 (2017)
4. Catalão, J.P.D.S., Mariano, S.J.P.S., Mendes, V.M.F., Ferreira, L.A.F.M.: Short-term electricity prices forecasting in a competitive market: a neural network approach. Electr. Power Syst. Res. **77**(10), 1297–1304 (2007)
5. https://au.mathworks.com/help/nnet/ref/fitnet.html
6. Chen, M., Mao, S., Liu, Y.: Big data: a survey. Mobile Netw. Appl. **19**(2), 171–209 (2014)
7. Laney, D.: 3-d data management: controlling data volume, velocity and variety. META Group Research Note (2001)
8. Stantic, B., Pokorný, J.: Opportunities in big data management and processing. In: DB & IS, pp.15–26 (2014)
9. Thelwall, M., Buckley, K., Paltoglou, G., Cai, D., Kappas, A.: Sentiment strength detection in short informal text. J. Am. Soc. Inf. Sci. **61**, 2544–2558 (2010). https://doi.org/10.1002/asi.21416
10. Chaffey, C.: Global social media research summary. Smart Insights (2016)
11. Collomb, A., Costea, C., Joyeux, D., Hasan, O., Brunie, L.: A study and comparison of sentiment analysis methods for reputation evaluation. Rapport de recherche RR-LIRIS-2014–002 (2014)
12. Philander, K., Zhong, Y.Y.: Twitter sentiment analysis: capturing sentiment from integrated resort tweets. Int. J. Hosp. Manag. **55**, 16–24 (2016)
13. Claster, W., Pardo, P., Cooper, M., Tajeddini, K.: Tourism, travel and tweets: algorithmic text analysis methodologies in tourism. Middle Eastern J. Manag. **1**(1), 81–100 (2013)
14. Kasper, W., Vela, M.: Sentiment analysis for hotel reviews. In: Proceedings of the Computational Linguistics-Applications Conference, pp. 45–52 (2011)

Design and Analysis of a Novel Sucked-Type Underactuated Hand with Multiple Grasping Modes

Nan Lin, Peichen Wu, Xiao Tan, Jinhua Zhu, Zhenjiang Guo, Xinbo Qi and Xiaoping Chen

Abstract The gripper of home service robot is required to be lightweight, appropriately-sized and compliant to grasp everyday objects. To meet the demand, an underactuated robot hand with suction cups is proposed which is capable of grasping objects of various shapes, sizes and materials. Using one actuator with tendon-driven mechanism, the inherent compliant hand can realize three grasping modes to adapt to different object shapes. Additionally, soft material applied on fingers makes the hand more flexible and controllable. Kinematic and mechanical analysis offer a theory evidence for design optimization and configuring proper grasping modes. Furthermore, in the experiment, robot hand is validated to have satisfactory grasping effect with three grasping modes.

Keywords Underactuacted hand · Sucked-type · Grasping mode

N. Lin · P. Wu · X. Tan · J. Zhu · Z. Guo · X. Qi · X. Chen (✉)
Department of Computer Science, University of Science and Technology of China Hefei, Anhui 230026, China
e-mail: xpchen@ustc.edu.cn

N. Lin
e-mail: fhln@mail.ustc.edu.cn

P. Wu
e-mail: wpc16@mail.ustc.edu.cn

X. Tan
e-mail: tx2015@mail.ustc.edu.cn

J. Zhu
e-mail: teslazhu@mail.ustc.edu.cn

Z. Guo
e-mail: gzj10416@mail.ustc.edu.cn

X. Qi
e-mail: hmbshn@mail.ustc.edu.cn

J.-H. Kim et al. (eds.), *Robot Intelligence Technology and Applications 5*,
Advances in Intelligent Systems and Computing 751,
https://doi.org/10.1007/978-3-319-78452-6_26

1 Introduction

In recent years, artificial intelligence and sensor technology have promoted the rapid development of home service robots. However, there is no doubt that grasping various objects is the essential capability to serve people better in the home environment. Unlike just facing specific products in the industrial environment, robots in the home environment need to handle various of objects with different sizes, materials and kinds. Thus, the gripper of the home service robot should have adaptive ability to grasp common objects as many as possible.

In order to achieve this goal, diverse grippers have been invented, which can be found in the literature. An intuitive idea is to mimic human hands, in which every degree of freedom can be controlled, such as UTAH/MIT hand [1], Robonaut [2], Gifu hand II [3], DLR-HIT-Hand [4]. Theoretically this kind of dexterous gripper with multiple fingers can pick up and manipulate the target objects what we do in daily life, but it meets trouble in practice. Using a large number of actuators and sensors results in extensive physical and computational complexity [5, 6]. Meanwhile, they're usually costly to manufacture (Fig. 1).

Another approach is making grippers passive compliant by utilizing flexible materials. They are prone to have a lower number of actuators than degrees of freedom. Pneumatic grippers based on soft materials have been widely studied in recent years [7–10]. By inflating the pneumatic actuators like pneumatic networks or PneuFlex, fingers will bend and adapt themselves to the contour of objects. So this kind of gripper is compliant, highly adaptable and has a good shape matching with the object, and its capability to perform dexterous grasping has got proved [11]. Pneumatic grippers are lightweight and pretty compliant. However, their grasping modes are simple. In most cases, they can only bend fingers to envelop objects. Another downside is that oversize of pumps/compressors are required, limiting its application in the human environment.

(a) (b)

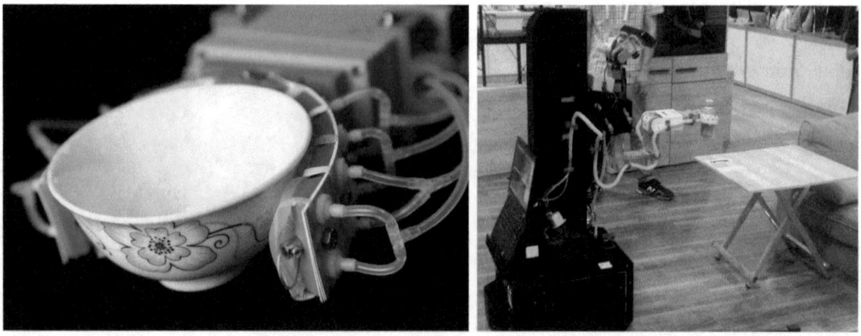

Fig. 1 In **a**, the gripper is grasping a ceramic bowl. In **b**, it won the best manipulation award in RoboCup2017@Home, open challenge

Additionally, there is a novel form of vacuum gripper which makes uses of granular jamming transition. It changes the fragile matter from a fluid to a solid-like state by vacuum pump [12, 13], and it is easy to control when grasping a rigid object with uneven surface. T. Takahashi et al. developed this kind of gripper by using suction cups array on the membrance to increase adhesive force [14]. Jamming gripper is suitable for handling small-sized objects, but grasping objects with large size or little curvature like paper is beyond its capability.

Between hard dexterous hands and soft ones, there exists an intermediate form. This kind of gripper usually possesses hard skeleton, but uses soft material in contact surface, joints or actuators. i-HY Hand, a compliant, underactuated three-fingers hand presented in [15] can grasp diverse objects owing to its flexure joints and relative hard links. The hand made by Shadow Robot Company [16] uses McKibben-type artificial muscles to realize compliance. They're able to manipulate the objects like complicated hard grippers but more compliant, however, the size of target objects is also limited in a small range. Recently, T. Nishimura et al. has invented a new underactuated gripper with three grasping modes, its microgripper-embedded fluid fingertip can pick up objects whether they are soft, rigid, deformable, fragile, big or heavy [17]. However, generally it needs an additional surface to support the object when grasping.

In this paper, we introduce a new kind of underactuated gripper driven by cable, which combines with suction cups. The gripper and total vacuum system are lightweight and proper size for using in home environment. By utilizing the suction force from suction cups and configuring the stiffness of fingers, three grasping modes are realized, makes the gripper capable to grasp a wide range of objects with different shapes and sizes. Meanwhile, Thanks to the flexibility of the finger material, it is easy to control and adaptable, guaranteeing the safety with human.

The rest of this paper is structured as follows: Sect. 2 describes the mechanical design in detail, and introduces the concept of three grasping modes. Section 3 contains kinematic and mechanical analysis of the fingers. Section 4 presents the experimental validation of model, suction cups and grasping modes, shows capability of the gripper to grasp diverse objects. Finally, Summary and conclusions are presented in Sect. 5.

2 Sucked-Type Underactuacted Hand

The gripper is expected to grasp various objects in the home environment. Common materials in regular life such as ceramics, glass, plastic, metal, organic matter are included for grasping. Two grasping strategies, gripping from the side and from above, are realized, so that the gripper will be more controllable when working in a complex environment. Moreover, the gripper must be compliant, which means that it must be self-adaptive in case of breaking objects or harming human.

2.1 Mechanical Structure

In this work, a low cost gripper is designed combining underactuated technology with suction cups. As shown in Fig. 2, the gripper consists of two flexible fingers, tendon-driven system, vacuum system and the base made from 3D printing. In order to increase friction force on contact surface, fingertips are covered with silica gel skin.

Schematic diagram of cable-driven system can be seen in Fig. 3. When spooler motor rotates, steel wire becomes shorter. Then the elastic deformation of the PVC board between two 3D printing parts happens, causing fingers bent and gripper closed. Adjust the thickness of joints by choosing the piece of PVC board, the stiffness of joints can be easily changed.

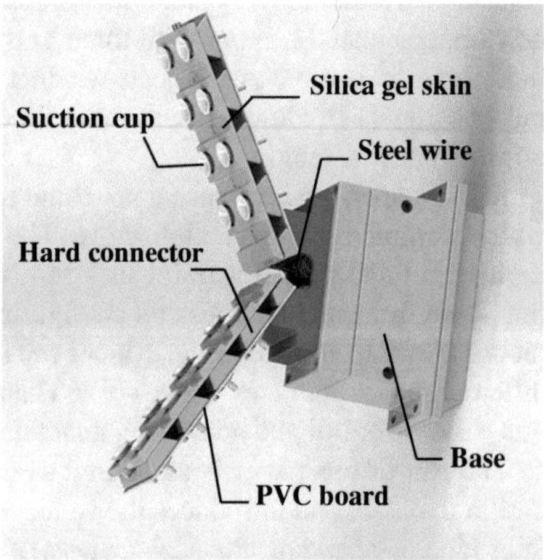

Fig. 2 CAD model of the gripper

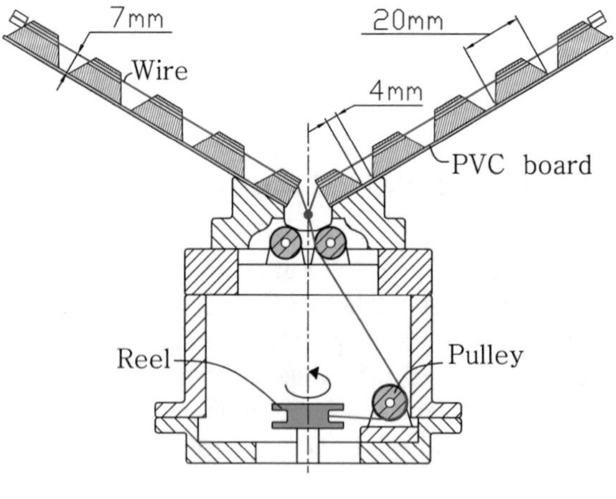

Fig. 3 Schematic diagram of the mechanism

We have designed two prototypes whose PVC board differs in thickness (i.e. have distinct configuration of stiffness), to test grasping effect. We call them single thickness finger (STF) and multiple thickness finger (MTF) which can be found in the contrast experiment.

2.2 Vacuum System

Figure 4 shows the working process of the vacuum system. Mini vacuum pump (a), connecting with suction cups through solenoid valve (b), can produce maximum 80 KPa negative pressure. Computer controls the working state of gas circuit. The suction cup is connected with mini vacuum pump if the solenoid valve is opened, otherwise its pressure is equal to the air pressure. Pressure sensor placed near the solenoid valve can detect the pressure of suction cup, obeying the Pascal's Principle. Two suction cups on the same hard connector are classified as a group, sharing one vacuum pump. The whole vacuum system, which located on the chassis of the robot and connected with suction cups by silicone tube, weighs only 2.5 kg, lighter than other pneumatic systems.

Suction cups on the finger can significantly enhance the grasping stability owing to their suction. For our gripper, the effective area of a suction cup is approximately 1.1 cm^2, then the maximum suction of a suction cup is 0.9 kg. So theoretically eight groups of suction cups can hold an 14 kg object. However, leakages always exist. In the experiment, one is defined as **effective suction cup** if its pressure reaches −40 KPa.

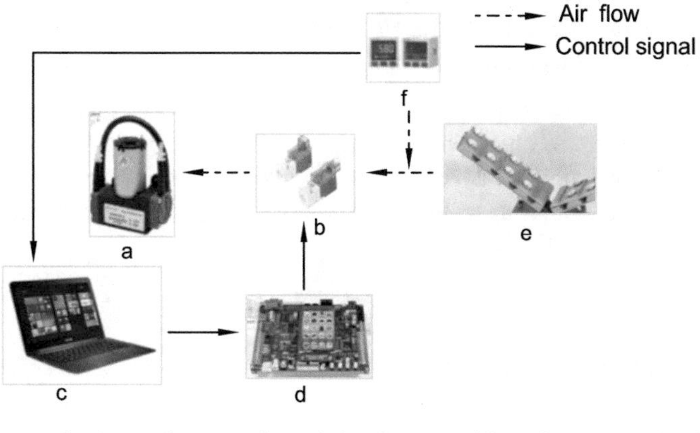

Fig. 4 Framework of control system: Control signal generated from the computational device (**c**) is converted to analog voltage signal by controller (**d**) and passed to solenoid valve (**b**) which can control the on-off of air flow; mini vacuum pumps (**a**) produce negative pressure through solenoid valve which is connected with suction cups (**e**); Pressure sensor (**f**) can detect the pressure of suction cups and feeds the information back to the computational device

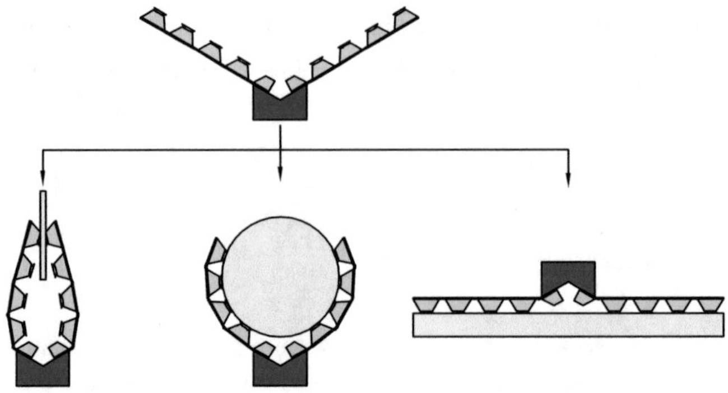

Fig. 5 Three grasping modes

2.3 *Three Grasping Modes*

The gripper has three grasping modes. Figure 5 illustrates the concept.

1. **Pinching mode** This mode is used to grasp small objects. At this time, base of finger has the largest curvature and only fingertips can touch targeted objects. Furthermore, only one group (even no) of suction cups on fingertips work.
2. **Enveloping mode** Objects with similar size of gripper and normal curvature are suitable for this mode. Owing to self adaptability of the gripper, at least two groups of suction cups contact with the object, improving the grasping capability.
3. **Unenveloping mode** This mode is used to grasp flat plates like paper, whose surface curvature is tiny. At this time, fingers don't bend and tendon is slightly relaxed so that the gripper can grasp plates by just sucking them.

The main factors of grasping mode choosing are shape and position of the object. In pinching mode, targeted object contacts gripper with fingertips and keeps away from base, while in enveloping mode objects are close to base. In unenveloping mode, fingers are straight and objects are also close to base. Further motion planning strategy is beyond the scope of this study.

3 Modeling and Analysis

3.1 *Kinematic Model*

For most underactuated grippers, when the actuator works, their phalanxes will rotate around the joint which usually has two types, one is link joint and the other is tendon joint. Differently, our gripper will bend the whole *tsoftpart* while grasping. Figure 6

Fig. 6 Kinematic analysis of single section

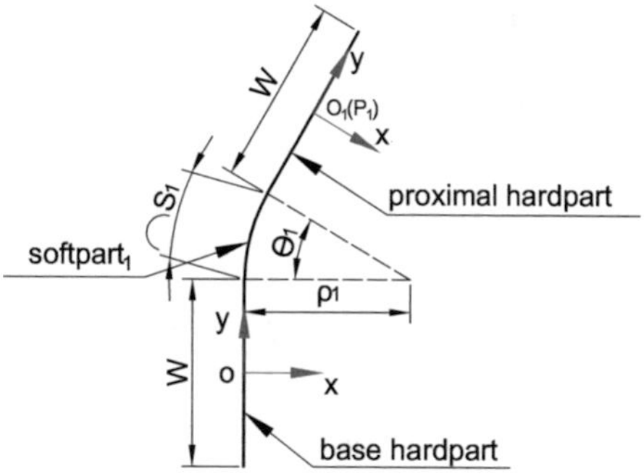

shows the sketch of the *tsoftpart*$_1$, the base *hardpart* and the proximal *hardpart*. The parameters in this picture are defined as: θ_1 is the proximal *hardpart*'s bend angle relative to the base *hardpart*'s; S_1 is the arc length between the base *hardpart* and the proximal *hardpart*; ρ_1 is the radius of arc; W is the width of *hardpart*; O is the origin of frame which is based on the base *hardpart*; O_1 is origin of frame which is based on the proximal *hardpart* (In frame O, O_1 is called P_1). For each frame, the X axis is the normal vector of the *hardpart* which point to palm and the Y axis is tangential of corresponding *tsoftpart* while the origin is the center of *hardpart*.

According to geometry, point P_1 locates at $(\rho_1 - \rho_1 \cdot \cos(\theta_1) + \frac{W}{2} \cdot \sin(\theta_1),\ \rho_1 \cdot \sin(\theta_1) + \frac{W}{2} + \frac{W}{2} \cdot \cos(\theta_1))^T$ in the frame O where $\rho_1 = \frac{S_1}{\theta_1}$. So, we can produce transformation from base *hardpart* to proximal *hardpart*.

$$T_{O_1}^{O} = \begin{pmatrix} \cos\theta & -\sin\theta & \frac{S_1}{\theta} - \frac{S_1}{\theta} \cdot \cos\theta + \frac{W}{2} \cdot \sin(\theta) \\ \sin\theta & \cos\theta & -\frac{S_1}{\theta} \cdot \sin\theta - \frac{W}{2} - \frac{W}{2} \cdot \cos(\theta) \\ 0 & 0 & 1 \end{pmatrix} \tag{1}$$

where $T_{O_1}^{O}$ means the homogeneous transformation matrix which transforms the frame O to O_1. In the frame O_i, the coordinate of *hardpart*$_{i+1}$'s origin can be obtained in the same way is $(\rho_i \cdot (1 - \cos(\theta_i)) + \frac{W}{2} \cdot \sin(\theta_i), \rho_i \cdot \sin(\theta_i) + \frac{W}{2} + \frac{W}{2} \cdot \cos(\theta_i))^T$ where $\rho_i = \frac{S_i}{\theta_i}$ and $i \in [1, 3]$. Premultiplying the $T_O^{O_i}$ that is the homogeneous transformation matrix from O_i to O, we can obtain the coordinate of *hardpart*$_{i+1}$'s origin P_{i+1}, as

$$P_{i+1} = T_O^{O_i} \cdot \begin{pmatrix} \rho_{i+1} \cdot (1 - \cos(\theta_i)) \\ \rho_{i+1} \cdot \sin(\theta_i) \\ 1 \end{pmatrix} \quad i \in [1, 3] \tag{2}$$

So, we can produce each P_i coordinate which is the center of each *hardpart* by (2).

In (2), S_1, S_2 and W are constant for the gripper. So, we can get a conclusion that, in the frame O, each *hardpart*'s coordinate can be calculated as long as we get each

θ. In other words, we can know the posture of gripper. While grasping, from the variation tendency of θ_i we could know which grasp mode happened. For example, if only θ_1 increases during grasp and other angles are almost invariant, we will know the pinching grasp has happened.

3.2 Mechanical Equilibrium

This part focuses on the mechanical model of finger, in order to find the relationship between finger posture and tendon pulling force or displacement without touching the objects. We assume that the maximum static friction force equals to the sliding frictional force. Friction force at both ends of hole is considered carefully while that inside is ignored. Suction cup is also be ignored because its existence has no influence on mechanical analysis.

Figure 7 shows the simplified model of bending fingers. β is the allowable flexural angle in design; S_i represents arc length of each *tsoftpart*$_i$; θ_i is radian of each *tsoftpart*$_i$ and α_i is the angle between friction force and wire; $T_{i,i+1}$ is the tensile stress of wire between *hardpart*$_i$ and *hardpart*$_{i+1}$; $f_{i-1,i}$ and $f_{i,i+1}$ are the friction forces at both ends of wire hole; M_i is the bending moment of *tsoftpart*$_i$; ρ_i is the radius of curvature mentioned above and E is the elasticity modulus of PVC board. Known from mechanics of materials:

$$\frac{1}{\rho_i} = \frac{M_i}{E \cdot I_i} \tag{3}$$

where I_i, the cross sectional moment of inertia, can be calculated by the following equation:

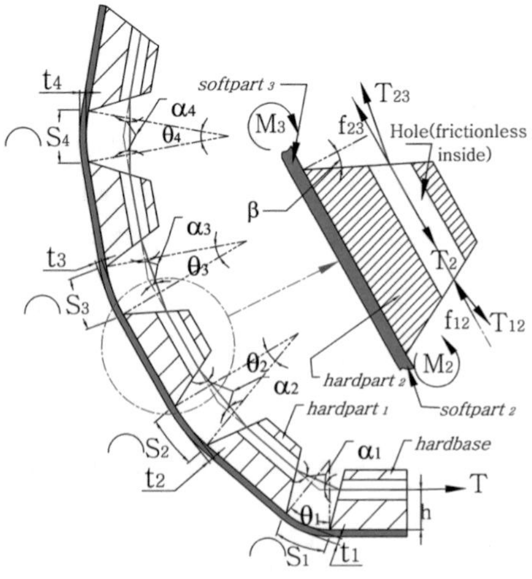

Fig. 7 Mechanical model of the finger

$$I_i = \frac{B \cdot t_i^3}{12} \tag{4}$$

where B and t_i are the width and thickness of cross section of PVC board respectively. The bend angle of each $softpart_i$ is:

$$\theta_i = \frac{S_i}{\rho_i} \tag{5}$$

Combine (3)–(5), one achieves

$$\theta_i = \frac{12 S_i M_i}{E B t_i^3} \tag{6}$$

Seen from the equation above, as the thickness rises, stiffness of *softpart* increases then θ sharply decreases. So stiffness can be easily configured by just adjusting the thickness of PVC board each section. After configuration, it improves the performance of the gripper, optimizes grasping mode, and finally gets the balance between pinching mode and enveloping mode. Comparison tests can be found in experiment part to support the claim.

Next we analyze one section of fingers under force balance. h is the distance between hole and PVC board. Wire receives the forces $f_{i-1,i}$ and $f_{i,i+1}$ at both ends of $hardpart_i$. Relatively, $hardpart_i$ will inevitably suffer the reaction of steel wire. Known from moment equilibrium,

$$M_{i+1} + f_{i,i+1} \cdot h + f_{i-1,i} \cdot h = M_i \tag{7}$$

Assuming that the coefficient of friction between wire and hole is μ, using friction formula,

$$f_{i,i+1} = T_{i,i+1} \cdot \sin\frac{\theta_{i+1}}{2} \cdot \mu \tag{8}$$

The tension of wire can be expressed by the following equation:

$$T_{i-1,i} \cos\frac{\theta_i}{2} = T_{i,i+1} \cos\frac{\theta_{i+1}}{2} + f_{i,i+1} + f_{i-1,i} \tag{9}$$

let $\xi_i = \frac{12 S_i}{E B H_i^3}$, yielding $\theta_i = \xi_i M_i$. Combine (7)–(9), one achieves

$$T_{i-1,i} = \frac{\cos\frac{\theta_{i+1}}{2} + \mu \sin\frac{\theta_{i+1}}{2}}{\cos\frac{\theta_i}{2} - \mu \sin\frac{\theta_i}{2}} T_{i,i+1} \tag{10}$$

$$M_{i+1} + \mu h T_{i,i+1} \left[\sin \frac{\xi_{i+1} M_{i+1}}{2} + \sin \frac{\xi_i M_i}{2} \frac{\cos \frac{\xi_{i+1} M_{i+1}}{2} + \mu \sin \frac{\xi_{i+1} M_{i+1}}{2}}{\cos \frac{\xi_i M_i}{2} - \mu \sin \frac{\xi_i M_i}{2}} \right] = M_i \quad (11)$$

Boundary condition is $T_0 = T$, in which T means the force that the actuator provides. Although none of analytical solution exists, qualitative analysis shows bending moment at the root of every finger is larger than other parts. If the thickness of PVC board at each section is the same, root of the finger will bend most. But here the key factor contributed to bending angle is the wire's tension. As a consequence, the degree of curve of each section differs indistinctively.

The wire displacement in each section can be inferred from geometrical relationship, shown in the following equation:

$$\delta_i = \left[S_i + 2h \tan \beta \right] - 2 \cos \alpha_i \left[\left(\frac{S_i}{\theta_i} - h \right) \tan \frac{\theta_i}{2} + h \tan \beta \right] \quad (12)$$

where the angle of wire in each section α_i equals $\frac{\theta_i}{2}$. the front part of the equation is the displacement of wire in nominal configuration and the latter part is that after bending. Then we know the whole displacement (i.e. displacement of steel wire attached to spooler motor):

$$\delta_{all} = \sum_{i=1}^{4} \delta_i \quad (13)$$

Finally, the displacement of wire is associated with the gesture of fingers by mechanical analysis.

4 Experiment

The gripper is mounted on a mobile robot, with a manipulator of five degrees of freedom. It is 200 mm long, 130 mm wide and 42 mm tall in nominal configuration, and weighs 150 g excluding actuator and vacuum system. As the PVC board is the weakest part of gripper, load capacity mainly depends on the joint's stiffness. In this work, we use PVC board 1 mm-thick per piece, and the load capacity of the gripper is about 2000 g.

4.1 Model Validation

Mechanical analysis shows that during gripper closing without object, the variation trend of each θ is different in STF and MTF. Here we choose the thickness ratio of PVC board at *softparts* from distal to proximal to be 1:1:1:1 for STF and 3:3:2:1 for

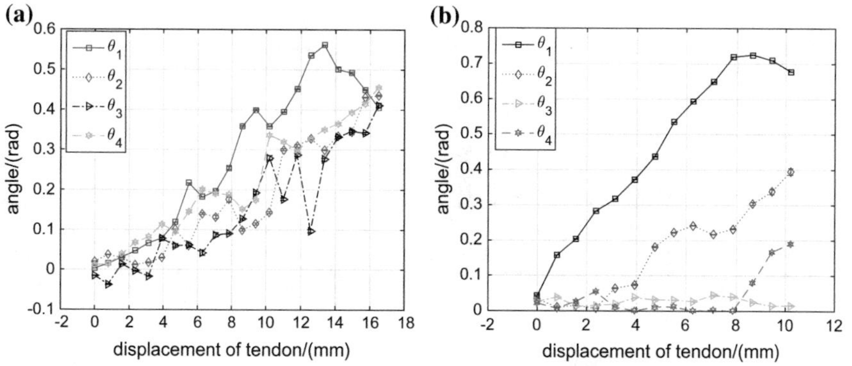

Fig. 8 The relationship between bend angles and displacement of tendon, **a** is for STF, **b** is for MTF

MTF. We use motion capture system (17W × 12) to measure the value of θ_i, trying to find out the relationship between θ_i and the displacement of tendon. Because of the symmetrical structure of the two fingers, we focus on only one of them. We carry out 10 repeated experiments on closing without object for each gripper prototype, and calculate the average value to describe the trend. As shown in Fig. 8a, when the finger is closing, all θs of STF go rising while the displacement of wire increases. At the same time, however, only θ_1 & θ_2 of MTF become larger while θ_3 & θ_4 maintain almost unchanged. It is the tendency of MTF's angle that makes pinching mode possible, as shown in Fig. 8b.

Thanks to its soft materials, the gripper won't break itself in the accident. When the fingertip touches each other, if the displacement of wire still increases, $\theta_{2,3,4}$ will rise, but θ_1 will be smaller. This phenomenon, which we called *fingertipcurving*, avoids to damage the finger structure. In Fig. 8a, b we see the two θ_1 reach the top and then go down, which supports our analysis well. The figures also show the standard deviation is small which means high repeated positioning accuracy of the gripper.

Another interesting thing we discovered from STF's data is the fluctuation of curve, especially θ_3. That may be caused by the mechanical coupling of finger sections and transition between sliding and static friction. When fingers are bending, if α_i increases, the force balance on the other *softparts* may be disturbed. In this circumstance, static friction in the hole will be transformed to sliding friction and θs on the other *softparts* will decrease. The model is not accurate enough to explain this phenomenon clearly at present and we will study on advanced optimization in future work.

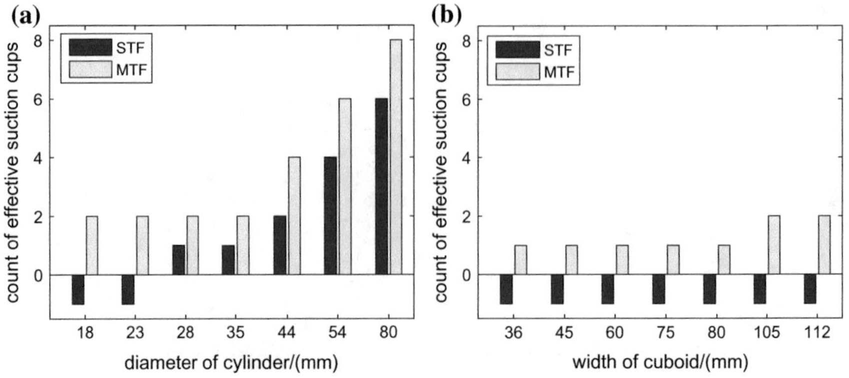

Fig. 9 Relationship between size of objects and count of effective suction cups. In **a** target object is cylinder whose diameter varies from 18 to 80 mm, and **b** is cuboid. negative means the gripper is unable to grasp the object

4.2 Effective Suction Cups

In this part, we use two gripper prototypes, STF and MTF, to grasp cylinders (PVC pipe) and cuboids (carton). The results demonstrate how different stiffness configuration influences effective suction cups (mentioned in vacuum system). The height of cylinders is 85 mm, and data of diameters can be found in Fig. 9a. The height and length of cuboids are 110 mm and 200 mm respectively and their width is shown in Fig. 9b. In the picture, negative means gripper can't grasp the object. Tests shows that the cylinders with small diameter and all cuboids can not be grasped by STF, while MTF can grasp them by fingertips in pinching mode. As for the cylinders with larger diameter, MTF has more effective suction cups than STF. It is because of the different motion trajectory between STF and MTF shown in Fig. 8, and obviously MTF performs better than STF in enveloping mode. In theory, effective suction cups should be odd because of symmetrical structure. But the position of objects may have some deviations in reality. Thus the count of effective suction cups becomes even when grasping slim objects. This phenomenon also shows high accuracy of position of the objects is required when in pinching grasping.

4.3 Grasping Modes Test

In this part, we present the test of MTF with three grasping modes to show how our robot hand generates adaptable grasp by exploiting the combination of underactuated mechanisms and suction cups. In Fig. 10, top row demonstrates the pinching mode, middle row adopts enveloping mode and bottom is unenveloping mode. It is evident that the gripper is capable to handle wide range of target objects, from coin to

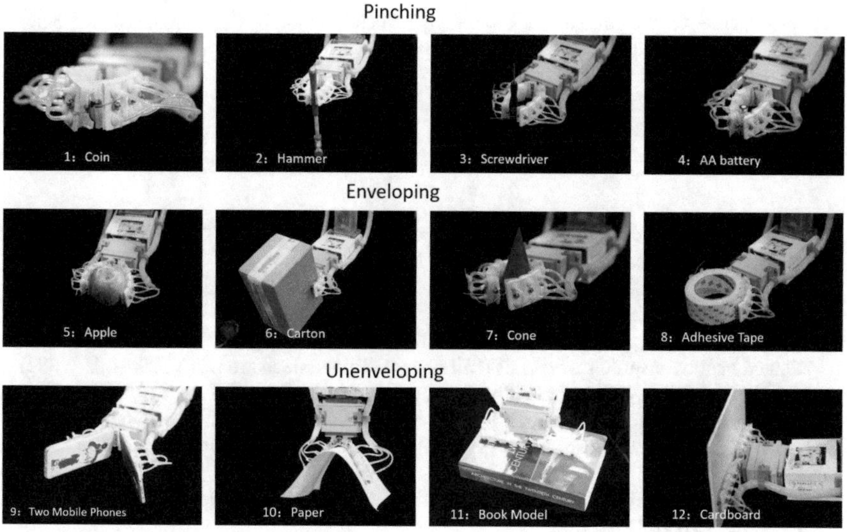

Fig. 10 MTF gripper grasps common objects with three grasping modes. From top to bottom row: pinching, enveloping and unenveloping mode

carton, with satisfactory grasping force (maximum 20 N). The ratio of grasping size between large objects (like adhesive tape, 220 mm) and small ones (like AA battery, 10 mm) is beyond 20, and flat objects such as mobile phone and cardboard also can be easily catched, which is almost impossible for other robotic grippers. The gripper has multiple grasping posture, not only from the side face but also from above (10, 11 in the picture). Moreover, we find that the suction cups would work effectively as long as they touch the object surface. In other words, it's not necessary to exert large normal pressure on suction cups to seal it up.

5 Conclusion

This paper has introduced a novel underactuated gripper which can grasp objects in everyday life. Because of inherent flexibility of materials and using suctoin cups, the gripper has great flexibility and strong maneuverability. It can operate in three grasping modes, i.e. pinching, enveloping and unenvoloping mode, then the objects of different sizes, materials and shapes can be grasped. We analyze the relationship between motion trajectory and the stiffness configuration of *softparts*, and comparative tests show better grasping results in MTF. However, there is still room to increase its load capacity by optimizing the design, and the effect of suctoin cups during grasping is not discussed enough. In future work, we will focus on increasing load capacity and analyzing suction cup's effect on grasp stability.

Acknowledgements We gratefully acknowledge financial support by the National Natural Science Foundation of China under grant 61573333.

References

1. Jacobsen, S.C., Wood, J.E., Knutti, D., Biggers, K.B.: The Utah/MIT dextrous hand: work in progress. Int. J. Robot. Res. **3**(4), 21–50 (1984)
2. Cipriani, C., Controzzi, M., Carrozza, M.C.: The smarthand transradial prosthesis. J. Neuroeng. Rehabil. **8**(1), 29 (2011)
3. Kawasaki, H., Komatsu, T., Uchiyama, K.: Dexterous anthropomorphic robot hand with distributed tactile sensor: Gifu hand II. IEEE/ASME Trans. Mechatron. **7**(3), 296–303 (2002)
4. Liu, H., Meusel, P., Seitz, N., Willberg, B., Hirzinger, G., Jin, M., Liu, Y., Wei, R., Xie, Z.: The modular multisensory DLR-HIT-Hand. Mech. Mach. Theory **42**(5), 612–625 (2007)
5. Shimoga, K.B.: Robot grasp synthesis algorithms: a survey. Int. J. Robot. Res. **15**(3), 230–266 (1996)
6. Saxena, A., Driemeyer, J., Ng, A.Y.: Robotic grasping of novel objects using vision. Int. J. Robot. Res. **27**(2), 157–173 (2008)
7. Ilievski, F., Mazzeo, A.D., Shepherd, R.F., Chen, X., Whitesides, G.M.: Soft robotics for chemists. Angew. Chem. **123**(8), 1930–1935 (2011)
8. Stokes, A.A., Shepherd, R.F., Morin, S.A., Ilievski, F., Whitesides, G.M.: A hybrid combining hard and soft robots. Soft Robot. **1**(1), 70–74 (2014)
9. Deimel, R., Brock, O.: A compliant hand based on a novel pneumatic actuator. In: 2013 IEEE International Conference on Robotics and Automation (ICRA), pp. 2047–2053. IEEE (2013)
10. Homberg, B.S., Katzschmann, R.K., Dogar, M.R., Rus, D.: Haptic identification of objects using a modular soft robotic gripper. In: 2015 IEEE/RSJ International Conference on Intelligent Robots and Systems (IROS), pp. 1698–1705. IEEE (2015)
11. Deimel, R., Brock, O.: A novel type of compliant and underactuated robotic hand for dexterous grasping. Int. J. Robot. Res. **35**(1–3), 161–185 (2016)
12. Brown, E., Rodenberg, N., Amend, J., Mozeika, A., Steltz, E., Zakin, M.R., Lipson, H., Jaeger, H.M.: Universal robotic gripper based on the jamming of granular material. Proc. Natl. Acad. Sci. **107**(44), 18809–18814 (2010)
13. Amend, J.R., Brown, E., Rodenberg, N., Jaeger, H.M., Lipson, H.: A positive pressure universal gripper based on the jamming of granular material. IEEE Trans. Robot. **28**(2), 341–350 (2012)
14. Takahashi, T., Suzuki, M., Aoyagi, S.: Octopus bioinspired vacuum gripper with micro bumps. In: 2016 IEEE 11th Annual International Conference on Nano/Micro Engineered and Molecular Systems (NEMS), pp. 508–511. IEEE (2016)
15. Odhner, L.U., Jentoft, L.P., Claffee, M.R., Corson, N., Tenzer, Y., Ma, R.R., Buehler, M., Kohout, R., Howe, R.D., Dollar, A.M.: A compliant, underactuated hand for robust manipulation. Int. J. Robot. Res. **33**(5), 736–752 (2014)
16. Kochan, A.: Shadow delivers first hand. Ind. Robot Int. J. **32**(1), 15–16 (2005)
17. Nishimura, T., Mizushima, K., Suzuki, Y., Tsuji, T., Watanabe, T.: Variable-grasping-mode underactuated soft gripper with environmental contact-based operation. IEEE Robot. Autom. Lett. **2**(2), 1164–1171 (2017)

Robotizing the Bio-inspiration

Ahmad Mahmood Tahir, Giovanna A. Naselli and Matteo Zoppi

Abstract Imitating natural living beings remains a perpetual curiosity of human beings. The pursuit of replicating biological systems led humans to develop contemporary machines—the robots with diverse range of shapes, sizes, capabilities and applications. Such systems may exhibit strength, control and operation sustainability; however, rigidness of the hard underlying mechanical structures is one of the major constraints in achieving compliance like that of natural organisms and species. This constraint is required to be softened to create biological duos with enhanced structural compliance. This demarcation has led to a new corridor to craft biological mockups, and has made doors opened to exploit new materials, novel design methodologies and innovative control techniques. This paper is an exclusive appraisal to bio-inspired state-of-the-art developments conferring their design specific importance. The methodical study and survey of corresponding structural designs, actuation techniques, sensors, and materials is potentially useful to demonstrate the novelty of bio-inspired robotic developments. Keywords: Soft robotics, bio-inspiration, novelty, flexibility, softness and compliance.

1 Introduction

Human beings have been evidently interested in developing bio-inspired systems like moving, walking and flying machines since ever. Such kinds of structures have been developed since ancient times. "The Pigeon" of Archytas [1–3] from Greek around 350 B.C propelled by steam (Fig. 1), may be considered as one of the primitive efforts of that journey.

A. M. Tahir (✉) · G. A. Naselli · M. Zoppi
DIME PMAR, University of Genova, Via all'Opera Pia 15A, 16145 Genoa, Italy
e-mail: tahir@dimec.unige.it

G. A. Naselli
e-mail: giovanna_naselli@hotmail.com

M. Zoppi
e-mail: zoppi@dimec.unige.it

J.-H. Kim et al. (eds.), *Robot Intelligence Technology and Applications 5*,
Advances in Intelligent Systems and Computing 751,
https://doi.org/10.1007/978-3-319-78452-6_27

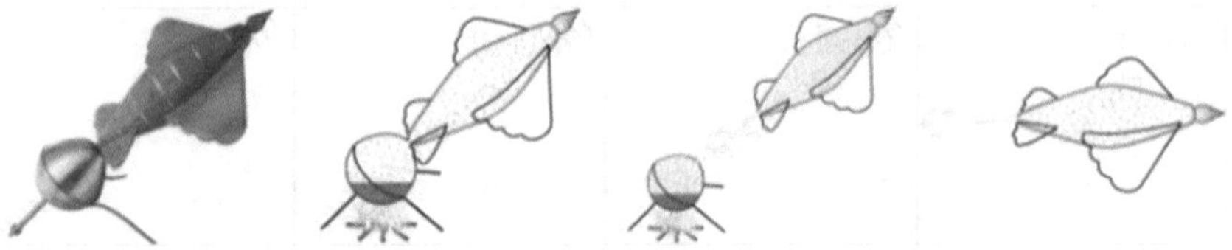

Fig. 1 Steam propelled design of "The Pigeon"

Leonardo Da Vinci (late 15th Century) can be considered as the modernizer of the field of Robotics. His human and animal inspired robot sketches and models were used by other innovators in the medieval times to build machines.

In contemporary past, the Electric Man (1865), the Steam Man (1883), the Automotive Man (1880s), and the Boiler Plate (1893), can be considered as some distinguished imaginations. Such efforts are evident that natural systems were considered imitative to come up with machines that can replace or support humans in various tasks.

1940s can be considered as the sliding partition between the imaginations and real time working robots while Isaac Asimov predicted the rise of a powerful robotized industry, that the world has witnessed today. Robotics law, drafted by Asimov in 1942 [4], narrated basic guidelines for a complex upcoming man-machine era in simple words, centered at human safety. Since then, robotics industry is expanding and currently with reducing prices and improved capabilities, the robotic industry revenue is being foresighted to become triple by 2025 [5, 6]. With the maturity of available technologies, research directions have become diverse in the field of robotics.

But the desire to produce systems like those existing naturally remains acquisitive. Researchers and scientists are now much more eager to pursue primeval human desire to imitate natural living beings as to develop flexible structures eliminating underlying rigid structures. This fresh episode of tremendous efforts has been recognized as Soft Robotics. Potentially, such robots are more flexible, and appeared to be more conforming to human safety than conventional robotic technology, owing to the softness and compliance. For instance, soft grippers similar to human finger tips offer stabilized grasping as compared to the rigid grippers and need reduced control system demand [7]. Furthermore, soft robots present a strong proposal to be more appropriate for socialization of robotic technology, where a harmless and compliant human-machine interaction is vital.

1.1 The Maiden Companionship

The recent state-of-the-art of bio-inspired soft robotics has been revolutionized from a state which was presented by way of fiction just a few decades ago. Although they are still abstract, however, now they are being transformed from fictional to factual

Fig. 2 Three channel FMA structure [8]

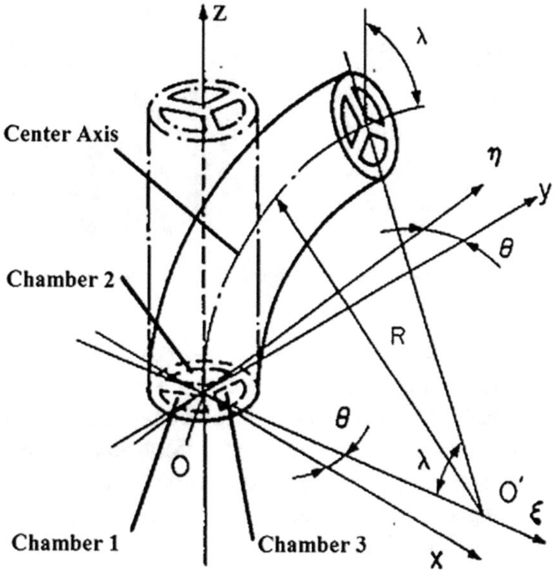

entities. The development of robotic actuators employing soft materials is tipped off in late 1980s.

Toshiba Corporation's flexible actuator can be considered as the foremost development in the field of soft robotics. Toshiba realized a flexible microactuator (FMA) for maneuvering micromanipulators and robotic mechanisms in 1989 [8–11], as shown in Fig. 2.

Made of fiber-reinforced rubber, the FMA was a three Degrees of Freedom (DoF) actuator capable of producing miniaturized movements pretending fingers, hand, arm and leg actuation (Fig. 3). The three degrees of freedom were designed through three channels which were discretely actuated by electro-pneumatic or electro-hydraulic pressure thus causing axial stretch or bending effect [12, 13]. Series of connected FMAs can produce increased DoFs for various applications [14–16].

Later on in late 90s, based on nature of application, material and FMA actuation characteristics; different fabrication techniques like molding [14], extrusion molding [17], and stereo lithography [18] were implemented to craft conforming FMAs [19].

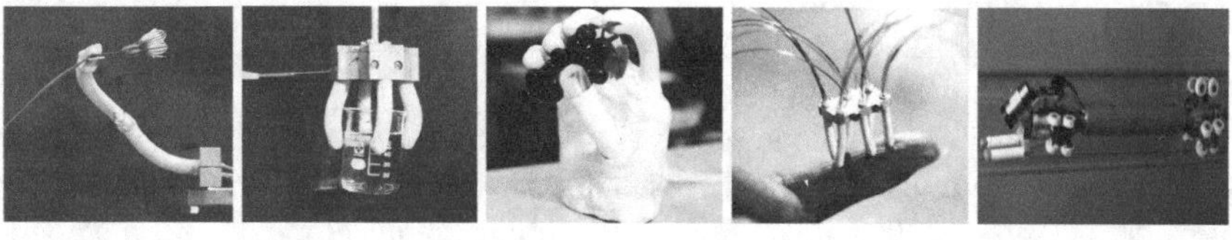

Fig. 3 Arm, finger, hand, legged, and mobile (pipeline inspection) robots based on FMA

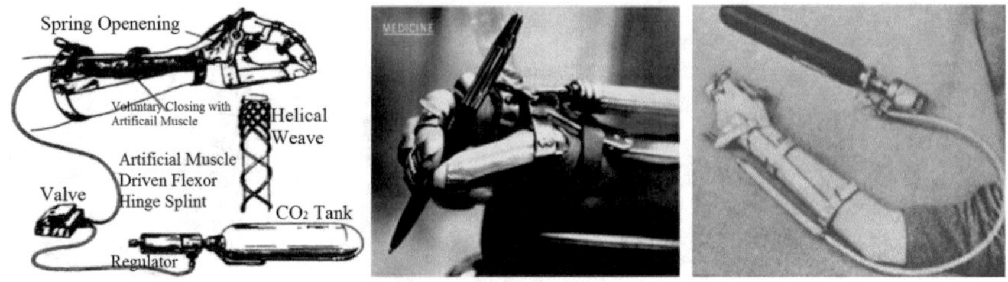

Fig. 4 McKibben muscle design and applications

While demonstrating FMA as the foremost soft robotics innovation, it is appropriate to recall, in the context of soft robotics, the invention of artificial muscle for prosthetics introduced as McKibben Muscle in 1957. Although some other fluidic actuators are even older, McKibben Muscle can be considered as the prominent invention in robotics. It was implemented in 1957–58 to actuate hands and fingers (Fig. 4) of polio affected Karen McKibben, the daughter of Dr. Joseph Laws McKibben [20, 21].

That innovation was based on a scheme developed by German scientists. The innovative design incorporated bellows cylinder inflated by carbon dioxide, thus acting as a muscle generated pinching movement of attached paralyzed fingers.

1.2 The Contemporary Apprehension

Pneumatic artificial muscles or McKibben muscle also known as rubbertuator may not be considered as the ground breaking for soft robotics, even if relevant, as that invention was purely developed for prosthetics. Furthermore, the muscle was a part of a system which was mechanically rigid at large. Rubbertuators were persistently implemented in the robotics systems for prosthetics and rehabilitation purposes [22–28]. Therefore, it is perceptible that the ground breaking work for soft robotics was accomplished in late 1980s in the form of FMA development in collaboration of Toshiba Corporation and Yokohama National University Japan.

During 1990s, major work was accomplished on the development of soft grippers like human finger tips [7], and on the soft skin development which was utilized both as skin and for touch sensing to improve the human-robot interaction [29, 30]. It is also worthy to mention that efforts were made to introduce soft joints making it possible for robotic structures to be more compliant and safer especially while interacting with humans [31]. Most of those primary undertakings were originated in Japan, USA and UK.

Research and development in the field of soft robotics can be witnessed today in various renowned universities and research institutes across the world. Today, the research in soft robotics is multifaceted, aiming at development of pragmatic materials, control techniques, design methodologies, and bio-inspired shaping and actuation. International training schools and competitions are being arranged at various levels and locations. Conferences, special conference issues, and dedicated journals have also been recognized acknowledging this vividly emergent field of technology.

1.3 Outfit of a Soft Robot

A soft robotic system generally consists of same elements like a conventional robot including body structure, actuation mechanism, sensors and transducers architecture, electronic and control interface, and power sources [32]. Some of the soft robotic systems may include Human Machine Interface (HMI) as well.

It is necessary to distinguish between soft bodied robots and hard bodied robots with soft-like actuation. Both categories may have impersonation of bio-inspiration and may represent flexible and compliant actuation, however, the objective of soft robots is to embody all elements of a robotic system into soft material avoiding rigid links and mechanisms. This ambition indeed set apart soft bodied robots, composed of soft materials of various grades, as soft-robots [32, 33]. More specifically, both of these categories may have similar morphologies, but they will be differentiated substantially based on of what materials their bodies are built from.

The soft robots are usually fabricated from elastomeric materials which are organic in nature [34]. The advantage of relying on soft materials for body construction of soft robots is the variability of conceivable motions that such materials can offer with a simple structural mechanism [34, 35]. Soft actuators which have been employed and are being investigated for soft robots are mainly of four types namely tendons, Pneumatic Artificial Muscles (PAMs), Fluidic Elastomer Actuators (FEAs), and Electroactive Polymers (EAPs) [32, 33]. Actuators are generally organized to generate bidirectional motion which offers an added advantage of better compliance. Conventional sensors and electronics are still in use though recently researches have been started to investigate and modernize soft, flexibles, stretchable and bendable sensors [32].

1.4 Paper Organization

This paper provides an insight into general, as well as applied soft robotics. This paper is organized in five sections including first section on Introduction: Sect. 2 presents overview of design techniques and materials used for the development of various soft robotic systems. Section 3 presents various actuators and sensors implemented in different soft robots. Some findings have been discussed in Sect. 4. Finally, Sect. 5 comments on the conclusion describing key challenges and the future work.

2 Soft Robotics Design Techniques and Materials

The core of soft robotics technology is to implant all required components of a robotic system in the body material making it dexterous enough to concert as brain and body simultaneously [32]. Furthermore, bio-inspiration does not necessarily copycatting natural system. The intention is to imitate capabilities of biological systems, conforming to their abilities to generate actions through body deformation ensuing their in-built capacities and constraints [36] to produce intrinsic mechanical compliance [37].

What do we need in a soft robot generally—a flexible body composed of soft material like polymers capable of producing continuum movements; soft actuators like tendons, PAMs or FEAs; sensors and driving electronics which are still conventional mostly; control techniques and power source—the specific components obviously depend upon a specific design and application of a robotic system. Together with material selection, it is indispensable to configure the shape and locomotion of the system. In soft robotics, it is crucial to address material selection, profile design and fabrication, and design of locomotion scheme in parallel to effectively utilize material properties to implement a conforming locomotion gait [38]. In this section, various design and modeling techniques have been discussed based on their respective robotic models as demonstrated in literature.

As already stated, FMA development and implementation in the late 1980s and 90s was the earliest effort presenting different design methodologies and fabrication techniques for different applications using the same actuator [8–15, 17–19]. In 1997, Toshiba Corporation and University of Tokyo, Japan implemented the Finite Element Method (FEM) to analyze characteristics of three different kinds of FMAs—optimized fiberless FMA; non-optimized fibreless FMA; and conventional fiber-reinforced FMA [39]. The optimized fiberless FMA was found to be practically more appropriate for low-cost FMA development through extrusion molding process and for the production of disposable micromanipulators for medical applications.

PAMs have also been widely implemented in designing different robotics systems in 1990s [24, 40, 41]. To realize improved bio-robotic muscle actuation using PAM, in 1999, researchers at University of Washington proposed PAM application with a hydraulic damper to achieve the required force-velocity level along with force-length relationship [37].

During that early phase, a group of scientists from Stanford University modeled soft fingertips consisted of a soft layer of viscoelastic fluid or powder covered by instrumented skin [42]. The design was based on the consideration that human fingertips have better compliance than hard or elastic fingertips and to develop task oriented stiff or soft fingertips. They were able to develop anthropomorphic fingers with 6 DOFs with soft manipulation for various kinematic finger-object couplings like rolling, compression and bending. The fingertip model developed and simulated exhibited a better grasp for sharp corner object owing to energy dissipation technique implemented in the soft fingertip.

Rubber structures follow simple operational principles and mimic natural motion capabilities, however, due to the material and geometric nonlinearities, the design process of pneumatic rubber actuators is tricky. The inherited complexity of these structures needs efficient analysis and simulation techniques to evaluate their deformations. One effective analysis method to solve these non-linear problems is FEM. In 2007, one of the developers of FMA along with other researchers at Okayama University and Osaka University implemented static analysis using non-linear FEM technique to consider correctly nonlinearities in the design and fabrication process of pneumatic rubber actuators [43]. This method provided optimal design solution for prototyping by a CAD/CAM based rubber molding process. A pneumatic rubber actuator with 2 DOF was developed and tested showing off that the characteristics corresponded to the achieved analytical results.

The soft robotic manipulators are being aimed to offer dexterousness maximizing the load capacity for available power or force to the manipulator. In 2000s, a consortium of various Universities and research agencies in USA, developed a series of octopus/elephant-trunk inspired pneumatic manipulator, OctArm I–VI [44–49]. Initially developed in 2004 [44] scientists at The Pennsylvania State University, USA further described an optimal design methodology for OctArm-VI prototype in [47]. The design of this manipulator incorporated the ability of the arm to manipulate objects. On the basis of its ability, a dexterity template was suggested defining wrap angles for different sections of the arm. The tube thickness and length of each section were calculated for base diameter and total arm length for desired configuration of the manipulator. It was realized that the optimal design depends on the ratio of maximum pressure applied to the manipulator to the Young's modulus of the manipulator's material. That result led to the conclusion that an optimal design would be valid for a family of manipulators with constant pressure to the young's modulus ratio. Certain limitations were also found including unstable optimal performance as load capacity maximized only for certain test pressures; and some design conditions were not addressed like elastic stability, out-of-plane deformations, and torsion. These researchers further investigated and proposed geometrically exact models for soft robotic manipulators [48] highlighting certain inaccuracies in the previous models of OctArm. Various shape estimation methods were further demonstrated in [49] employing the latest model of OctArm VI.

Single channel technique was implemented to develop an FMA based soft gripper (Fig. 5) accommodating asymmetric wall thickness of the tube in 2010 [50]. Symmetric neoprene tubes were cut and joined together to form asymmetric actuator. The strength of the tube was reinforced using a flat spring on the thicker side of the tube.

A novel quadruped soft robot (Fig. 6) was designed and fabricated implementing soft lithography technique in 2011. The mold was created by three-dimensional (3D) printing to fabricate the robot body. Pneu-net (PN) architecture [33] was employed to design pneumatic channels as it is highly compatible with soft lithography technique.

Considering the high dexterity, variable stiffness and complex behavior of octopus arm, Scuola Superiore Sant'Anna, Italy presented a novel design concept to

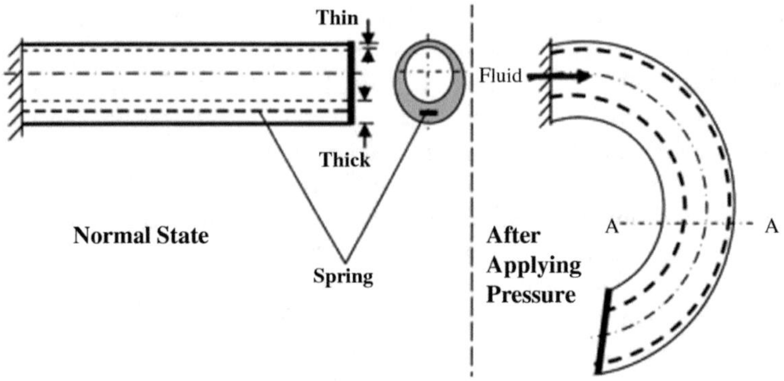

Fig. 5 Single channel FMA gripper

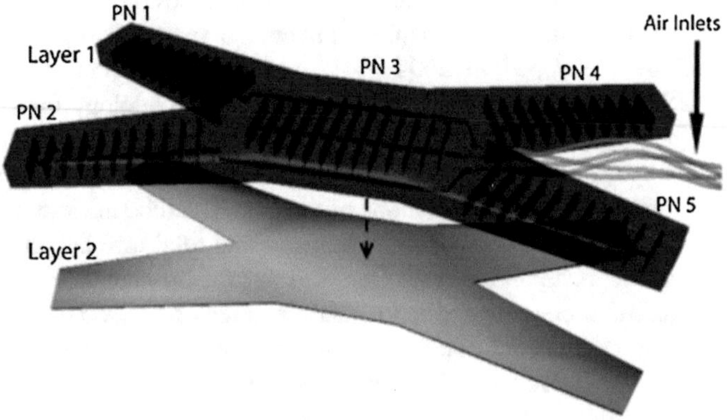

Fig. 6 Schematic of quadruped soft robot

imitate it in 2011 [51]. The design incorporated some of the important features of octopus arm like elongation ability, bending capability in all directions, and stiffness control (Fig. 7). The design represented an arrangement of longitudinal and transverse silicone based muscles forming an artificial muscular hydrostat. Different molding techniques implemented to fabricate and test different kinds of muscles for this design. The design and experimentation exhibited that the muscle placement is the key along with the actuation capabilities to achieve octopus like performance.

With the same inspiration, another group of the same institution deliberated another approach to mimic octopus arm in 2012. A cable driven conical shaped manipulator, made of silicone, was designed that was tested in water to take the advantage of the mechanical properties of the material [52]. To control and assess the design characteristics, direct kinetics model was evaluated by geometrically exact approach and jacobian was implemented for inverse kinetics model. The developed models were tested and found suitable with a good degree of accuracy

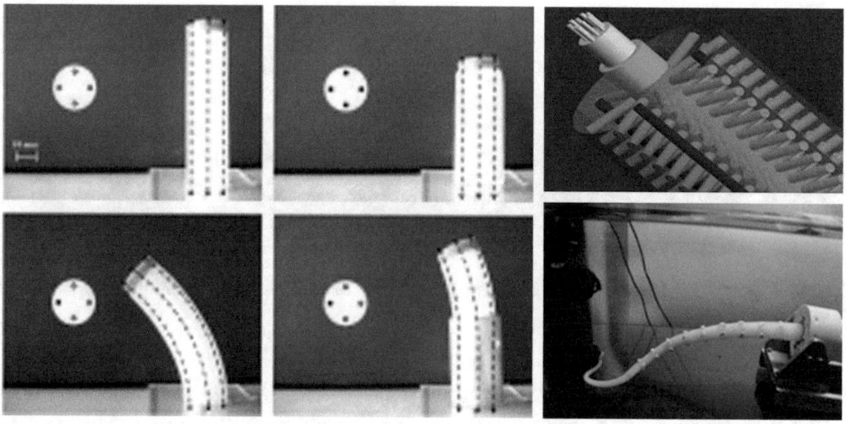

Fig. 7 Shortening, bending, CAD design and prototype of Bio-mimicking Octopus arm

for the fabricated manipulator. Some cable configuration based inaccuracies were also highlighted including friction, out of plan cable adjustment, and tip error.

A three-finger soft gripper (Fig. 8) with two DOF at each finger, mimicking human gripping, was realized by embedding three shape memory alloy (SMA) springs in silicone elastomeric tubes at University of Science and Technology, China in 2013 [53].

The design was developed to investigate the force distribution mechanism of the SMA actuators in the soft tubes and to testify that such actuators can substitute conventional solid state actuators. Two springs, placed on one side of the tube, bends into U-shape with applied current and the third spring acts as an auxiliary mechanism establishing a stable grip profile.

An adhesion based gripper employing microfibers, GeckoGripper, was designed at Carnegie Mellon University, USA in 2014. A 3D printed syringe pump was used to apply air pressure on a soft inflatable membrane to change the stretch

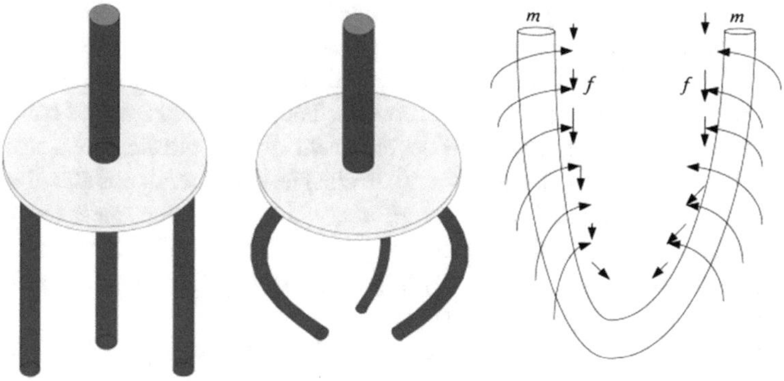

Fig. 8 Three-finger soft gripper model and its bending moment profile upon SMA actuation

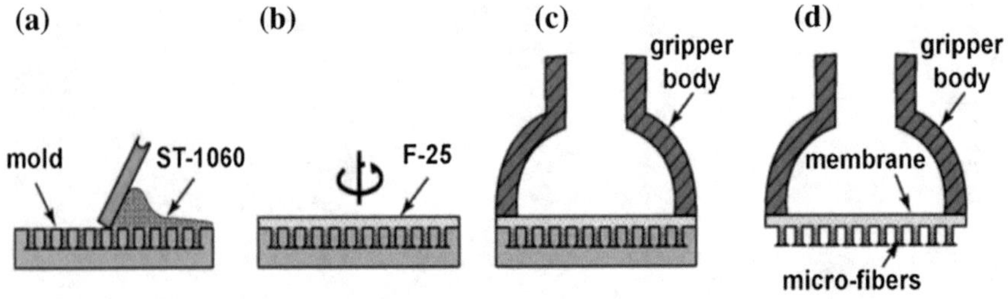

Fig. 9 **a** Molding micro-fibers, **b** spin-coating the membrane, **c** mounting gripper body, **d** GeckoGripper

configuration for gripping and un-gripping. The microfiber and the membrane were molded and spin coated respectively using polyurethane elastomers (Fig. 9). The 3D printed body was attached to the membrane during the curing time.

The above section has described a certain set of various designs and techniques reported in literature that were developed based on inspiration from different biological systems or species. The set of research reported in literature indicates an extensive shift towards flexible and soft continuum robotics. However, many systems are implementing similar design techniques exhibit similar kind of constraints and complexity in the control and manipulation. These techniques and methodologies should be further explored and experimented to realize smarter mechanisms. Furthermore, along with improvement in design methodologies, the control architecture approaches may require emphasis in order to accelerate the research in this novel area of robotics technology. The core of the soft robotic design is the embodied intelligence, manipulation and control, which is on its way with the extensive and dedicated efforts being made.

3 Sensors and Actuators in Soft Robotics

There are different kinds of commercially available sensors, being implemented in soft robotics, which are originally developed and used for conventional robotics. Actuators for soft robotics have a better recitation as more efforts can be witnessed towards actuators development in soft robotics. The nature and spirit of soft robotic mechanisms demand realization and development of soft and flexible sensors and actuators to make more dexterous and exclusively soft and compliant robotic systems. This section compliments the efforts, which have been made towards the development and experimentation of soft sensors and actuators, as reported in literature. Sensors are still in early development consideration in soft robotics, and therefore, actuators got more prominence.

We have already described FMA and PAM actuators. The FMA has been implemented widely for different kind of soft robotic systems [8–15, 17–19]. It is a rubber actuator incorporating stretching actuation mechanism and offers multiple

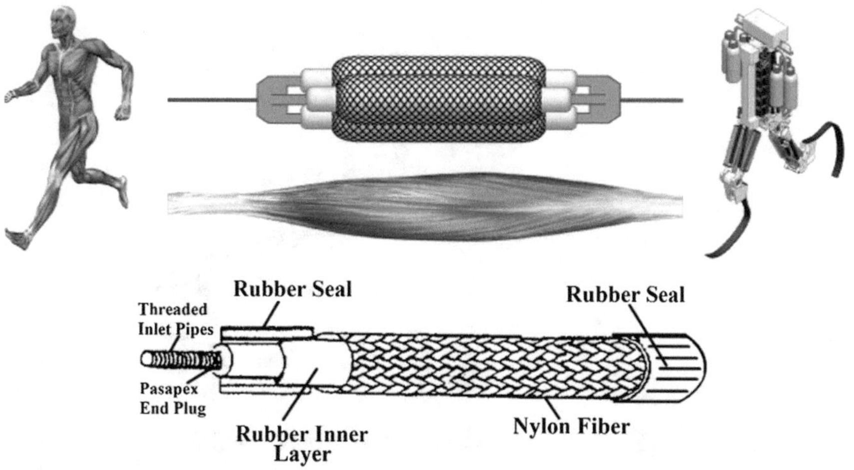

Fig. 10 Pneumatic muscle actuator [41, 54]

DOFs like pitch and yaw along with stretch; however, these are inter-dependent DOFs. It has many advantageous characteristics especially its capacity to act as the actuator as well as the body of the robot. Due to this property, miniaturized systems can easily be designed and constructed. Its actuation is smooth owing to its frictionless mechanism. FMA based systems can be developed using different fabrication/molding techniques for various system designs and applications.

The McKibben actuator or PAM is also very effective making a robotic system light, cheap, compliant and offers easy maintenance. It has wide range of application in anthropomorphic designs, and assistive robotic mechanisms (Fig. 10).

A plastic bellow tube actuator was presented at Bristol University for a gripper design in 1995. It can be considered a fine effort towards evolution of soft robotics as the idea was novel; however, the actuator was not completely soft. The airtight bellowed tube was attached to a steel strip, which upon air pressurization exhibits curling movement which could be utilized for grasping and gripping [40]. A customized tactile sensor was also designed for the gripper for strain measurement by arranging strain gages on the steel strip (Fig. 11).

A rotary-type soft actuator, ASSIST, was developed at Okayama University Japan in 2004–05 [55]. The actuator was composed of rubber tubes and fiber reinforced polyester bellows. The mechanism was controlled by pressure control system. The actuator was implemented in two different configurations; one employing McKibben muscles and the other without that. This actuator was designed for assistive motion of wrist. The McKibben muscles were used in the initial design and were attached to the palm mechanism for releasing the appliance control while it was nonoperational. The alternate design [56] was further implemented for wrist and elbow motion assistance.

Tactile soft sensor was implemented to calculate and detect the movement of the forearm. Bending angle of wrist and elbow, and human muscle power measured by

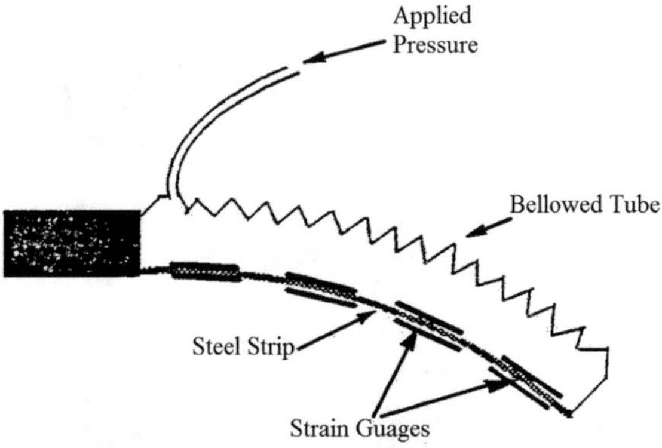

Fig. 11 Pneumatic actuator construction

electromyography (EMG) were incorporated to evaluate and determine the effectiveness of the actuator. An ASSIST equipped person can move the arm and hand with lesser muscular power as compared to an unequipped one.

OctArm design has already been described in Sect. 2. Its pneumatic actuator was designed considering five parameters: length based on target wrap angle; outer radius based on available mesh size and required mesh angle; structural stiffness and to meet required load capacity, appropriate tube thickness was selected; tube material was identified for desired extensibility and pneumatic permeability; and finally mesh angle was critically and experimentally selected to achieve desired actuation [44–49].

Another novel actuator, to drive an annelid robot, was developed at Sungkyunkwan University, Korea utilizing dielectric elastomer actuator (DEA) which has the maximum deformation amongst the available electroactive polymers (EAP) [57]. A thin layered sheet of the elastomer is connected to a rigid frame (Fig. 12). The elastomer expands in convex or concave profile with compressive force. This expansion generates actuation avoiding time-dependent and pre-strain actuation. These actuators were connected serially to develop the earthworm robot.

Another pneumatic rubber actuator, already mentioned in Sect. 2, was developed incorporating geometric, material and contact non-linearity characteristics for desired actuation by using non-linear FEM [43]. Nylon reinforced rubber was used to design a 2 DOF actuator with 2 independent tubes. Simple pressurized bending and stretching actuation has been generated. Where actuation technique is simple, design of such actuators is quite complex which is based on shape of tube cross-section, length and rubber elasticity (Fig. 13). The actuator was employed to design a Manta swimming robot.

Keeping in view the compliance required between soft actuators and sensors, and to avoid high stiffness in conventional potentiometers and optical encoders, scientists, including the developer of the FMA, at Okayama University, Japan developed a soft displacement sensor composed of piezoelectric polymer as deformation sensing element, conductive paste as electrodes, and base rubber (Fig. 14). The soft sensor was used to produce and intelligent FMA actuator, which

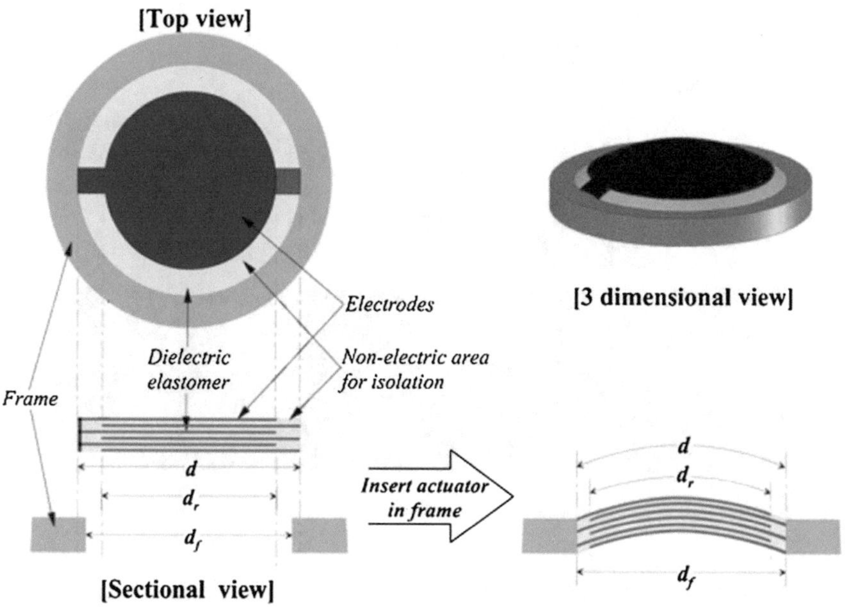

Fig. 12 Construction of DEA actuator

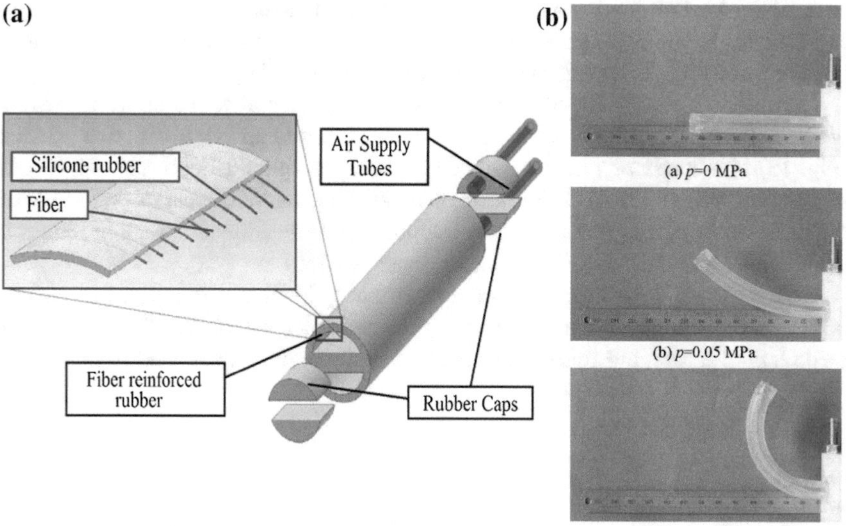

Fig. 13 **a** Actuator, **b** experimentation results

was capable of implementation in the development of flexible hand due to its improved grasping mechanism [58]. Experimental results varied positioning accuracy; however, hysteresis and noise were detected as well. The technique used

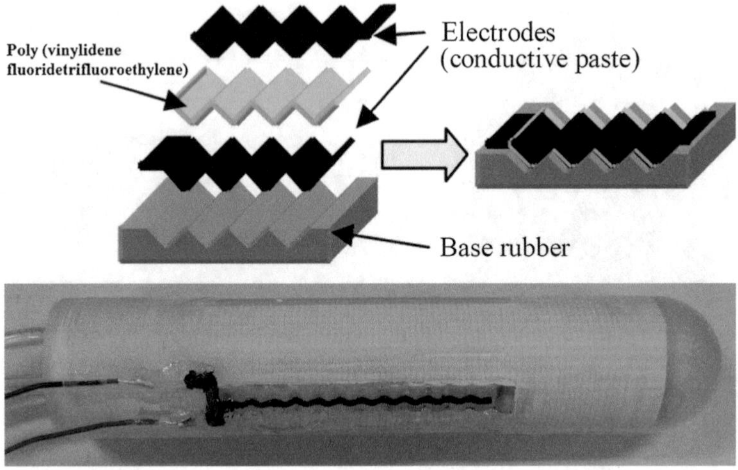

Fig. 14 Construction of piezoelectrci polymer sensor and its attachment on FMA

for the fabrication of the sensor and experimentation results exhibited that tactile or force sensors can also be developed for flexible mechanisms for higher accuracy and control.

The above sensor has a wave type pattern. These scientists developed and tested both straight pattern and waved sensors and found that straight sensor has stronger signal strength but with high signal to noise ratio. The wave patterned sensor output has high linearity, however, the signal strength is low [59].

Considering the need of softness and back-drivability required to realize bio-mimicking soft robots, scientists at SSSA proposed an electrostrictive actuator (Fig. 15) composed of platinum silicone rubber compound Ecoflex™ 01-10 and 00-30 (Smooth-on, USA) as dielectric material. The material selection was made on the basis of properties and compliance found during tensile-compression tests that were carried out on various elastomeric materials [60].

A thin film of the silicone, covered with thin gold film as compliant electrodes on both sides, was folded in a number of turns using a customized device to achieve well aligned layers and folds. Electrode connections were facilitated through little

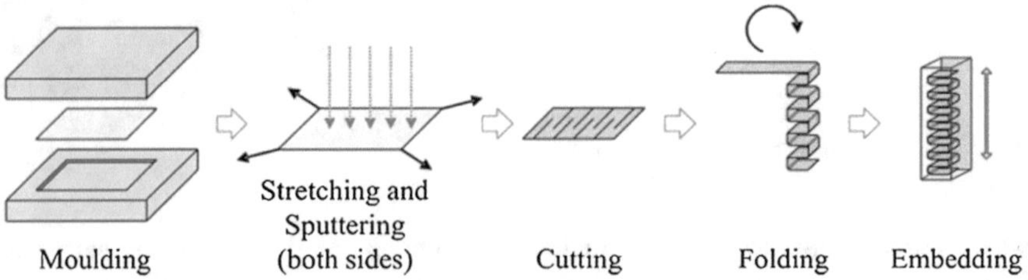

Fig. 15 Electrostrictive actuator fabrication and the prototype

strips of Kapton sheet with silver conducted glue CircuitWorks 7100. The whole package was finally wrapped in a thin layer of silicone to form the actuator. The actuator was employed to develop an Octopus arm model [61]. These scientists have developed octopus arm employing SMA based spring actuator as well [62, 63].

A flexible pneumatic robotic actuator (FPA) was proposed by Chinese scientists in 2013 composed of elastic silicone rubber with embedded spring for stiffness and radial deformation prevention. It was used to design different joints and grippers to manipulate different things and a climbing robot mechanism keeping in view the compliance, adaptability and safety that conventional rigid robots lack [64]. Although this actuator has been applied for rigid mechanisms, it is appropriate to report here keeping in view its potential of application towards soft mechanisms development based on properties it has exhibited.

A novel variable stiffness dielectric elastomer actuator (VSDEA) was developed by École Polytechnique Fédérale De Lausanne (EPFL), Switzerland in 2015 (Fig. 16). The actuator was composed of pre-stretched DEA attached on low-melting point-alloy (LMPA) substrate, capable of bending due to change in LMPA stiffness upon heating [65]. The actuator was employed as fingers for a gripper prototype. The self-sensing ability of the actuator, its stiffness changing capacity, and potential dynamic behaviors make it a potential actuator for flying, swimming, and surgical robotic applications.

To manipulate micro-soft objects like cells, scientists at University of Chung Cheng, Taiwan, realized an ioconic polymer metal composite actuator (IPMC) with arched bending actuation [66]. The actuator was employed for scissor mechanism for gripping force enhancement. An elastic pin, attached to the actuator joints, converted arched motion of the actuator into linear motion (Fig. 17). The actuator was fabricated using Nafion through a micro-mold which was developed using a die

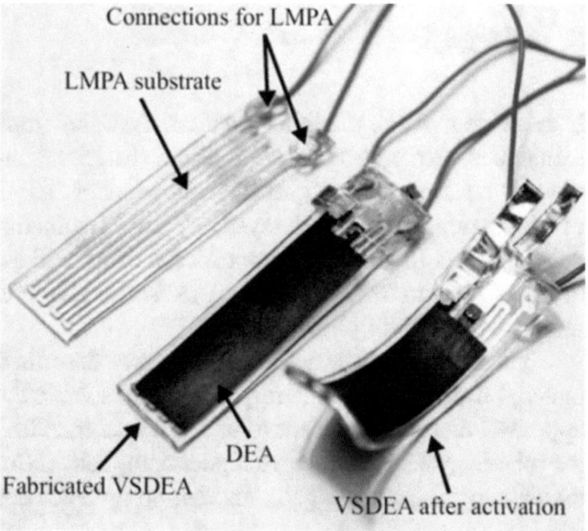

Fig. 16 VSDEA actuator construction

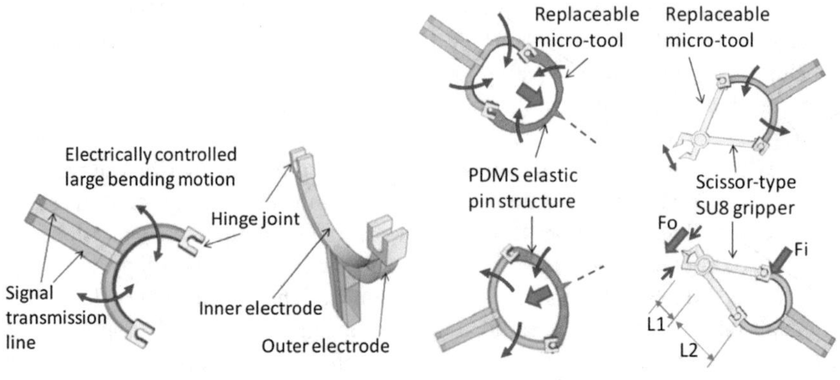

Fig. 17 IPMC actuator construction and applications

developed by photolithography. The model was selectively inactivated through photoresist to deploy platinum electrodes which were deployed through a chemical process. Ion exchange method applied to remove photoresist effect. The actuator was capable of producing 10 nM force and 100 μM displacement.

Bio-inspiration has driven researchers to comprehend actuators for desired locomotion and mechanisms development. Tendon driven, SMA based, pneumatic and fluidic actuators have been realized and employed for various bio-mimic systems. In contrast to actuators, flexible sensors development and realization has to go long way, as the conventional sensors are still in wide practice in soft robotics. For soft robotics, it seems ultimate to exploit new materials and techniques to move further towards the development of flexible and compliant sensors and more efficient actuators.

4 Discussion

It is evident from previous sections that the implementation of various novel techniques, exploration of appropriate materials, and efforts that are being made towards designing of soft sensors and actuators describe the functional importance of bio-inspired soft robotic systems. This fascination has steered some high-tech developments in this field. From McKibben based systems to FMA structures, and tendon driven to SMA based designs, substantial contributions have been made to imitate natural living beings.

Some of the earliest developments have described the relationship amongst the material properties and the respective conceivable compliance. These initial efforts appraised that it is pertinent to use soft materials for the structural development of the robotic systems to achieve higher compliance. Compliance can be controlled by regulating the feedback gain; and the compliance control enhances the stiffness of soft mechanisms for desired manipulation [14]. Predicting robotic behavior based on material and corresponding actuation pressure can be used for soft robots' gait and position control [67]. Non-linear actuation behavior suggests the development of predictive modeling techniques to effectively develop nonlinear motion control system for soft robots [68].

Another major orientation of soft robotics is to develop robots capable of generating unstructured motions as service tasks in daily life are quite unstructured in nature. Dexterity along with compliance makes the system more suitable for unstructured environments. Dexterity in soft robots, for instance in human fingers that has always been an inspiration and challenge for scientists, serves to handle thin and fragile items [69, 70]. Again, material and actuation technique plays very important role in achieving such characteristics.

Rubber actuators, the embodiment of most of the soft robots, are quite suitable not only for normal environment but under water as well owing to the water resistance, smooth deformation in compliance to water, high power density and lighter weight. Good performance underwater is one of the main catch imitating different aquatic species. The most inspirational living mechanism for soft roboticists that has multidirectional and multipoint bending, soft and deformable body that can stiffen itself, and has good strength of gripping as well, is the octopus [71]. Different soft mechanisms have been developed based on octopus arm morphology, using different techniques in recent years [71–81]. Mechanical skills have been evaluated that led to selection of suitable materials and actuation strategies to develop replicated robotic systems. Direct and inverse kinetic models for the control of arm helped to achieve good degree of accuracy demonstrating strength of manipulating algorithm. Successful grasping and manipulation of different shapes, sizes and weights have been demonstrated in these efforts. These systems include tendon driven, muscular hydrostat, PAM based and granular jamming based actuators in combination with especially designed fluidic actuation mechanisms. These efforts clearly indicate the importance of careful selection of the material, actuation scheme and control strategy to successfully robotizing a biologically inspired system.

5 Conclusion, Key Challenges and the Future Work

Soft robotics is a perfect prospect towards the realization and development of safe, flexible and compliant robotic systems. Together with bio-inspiration, it has inaugurated the era of biological duos. Underlying actuation principles of biological systems are being evaluated and respective robotics implementations are underway. Silicone is serving as the basic material for most developed soft mechanisms. EAPs, PAMs, SMAs and other applied structural materials have been successfully exploited to demonstrate the potential of these materials and the invented mechanisms. Actuation techniques are being fused to achieve optimal manipulation.

Evident efforts have been started towards the development of soft sensors for soft robotic systems.

Soft robotics is in the early phase of its exploration and investigations, and still facing fundamental challenges that any new technology would face. Different materials, actuation techniques, actuators and sensors upgrading, and suitable power sources are being evaluated for designing targeted projects and explicitly project based developments. Perhaps, this would be a key challenge to generalize actuation principles, sensing techniques and material selection that would fit for the development of new soft robots. Another contest in this innovative paradigm is to optimize flexibility and compliance of a developed system. For this purpose, diverse control strategies may require to be explored and examined. Nonetheless, materials need enhanced multifunctional properties that can be implemented to reproduce fully functional biological replicas.

Far future research may take a turn towards the successful development of smart materials capable of acting as body and brain of robotic systems. Such materials, with variety of constituent sensing and actuating properties may be serving as links, joints, skin and other organs of robots. Although the ongoing research is considering imitating underlying principles of biological systems, the future soft robots should be capable of replicating the physical systems in order to take optimal advantage of the natural flexibility and compliance. This would be possible through development of muscles based on smart materials. Furthermore, smart and embedded power sources are to be explored to make the soft robots fully autonomous. Smart materials, in combination with muscles and embedded power sources would make it possible to develop efficient ground, water and flying robots. That would further make it possible to realize proficient service, rehabilitation, prosthetics and industrial robotic manipulators and grippers.

References

1. Gellius, A., Beloe, W.: The Attic Nights of Aulus Gellius (Translated into English by the Rev. W. BELOE F.S.A., 1795, London), Book X, Chapter XII, vol. II, pp. 220–223. Printed for J. Johnson, London (1795)
2. Kostas, K.: The First Robot, Created in 400 BCE, Was A Steam-Powered Pigeon [Internet] (2014). Kostasvakouftsis.blogspot.it. https://kostasvakouftsis.blogspot.it/2014/04/first-robot-created-in-400-bce-was.html. Accessed 10 August 2016
3. Kotsanas, K.: The flying pigeon of archytas [Internet]. Museum of the ancient Greek technology. http://kotsanas.com/gb/exh.php?exhibit=2001001. Accessed 10 August 2016
4. Clarke, R.: Asimov's laws of robotics: implications for information technology-Part I. Computer **26**(12), 53–61 (1993)
5. Industrial Robot Revenue Will Nearly Triple by 2025, Fueled by Chinese Demand [Internet]. http://www.prnewswire.com/news-releases/industrial-robot-revenue-will-nearly-triple-by-2025-fueled-by-chinese-demand-300389443.html. Accessed 22 June 2017
6. Alex, G.: Sales of industrial robots are surging. So what does this mean for human workers? [Internet]. https://www.weforum.org/agenda/2017/05/sales-of-industrial-robots-are-surging-so-what-does-this-mean-for-human-workers/. Accessed 22 June 2017

7. Akella, P., Cutkosky, M.: Manipulating with soft fingers: modeling contacts and dynamics. In: International Conference on Robotics and Automation, pp. 764–769. IEEE (1989)
8. Suzumori, K., Iikura, S., Tanaka, H.: Development of flexible microactuator and its applications to robotic mechanisms. In: International Conference on Robotics and Automation, pp. 1622–1627. IEEE (1991)
9. Suzumori, K., Iikura, S., Tanaka, H.: Flexible microactuator for miniature robots. In: Micro Electro Mechanical Systems, MEMS'91, An Investigation of Micro Structures, Sensors, Actuators, Machines and Robots, pp. 204–209. IEEE (1991)
10. Suzumori, K.: Flexible microactuator: 1st report, static characteristics of 3 DOF actuator. Trans. Jpn Soc. Mech. Eng. Ser. C (in Japanese) **55**, 2547–2552 (1989)
11. Suzumori, K.: Flexible microactuator: 2nd report, dynamic characteristics of 3 DOF actuator. Trans. Jpn. Soc. Mech. Eng. Ser. C (in Japanese) **56,** 1887–1893 1990
12. Suzumori, K., Iikura, S., Tanaka, H.: Applying a flexible microactuator to robotic mechanisms. IEEE Control Syst. **12**(1), 21–27 (1992)
13. Suzumori, K., Kondo, F., Tanaka, H.: Micro-walking robot driven by flexible microactuator. J. Robot. Mechatron. **5**(6), 537–541 (1993)
14. Suzumori, K.: Elastic materials producing compliant robots. Robot. Auton. Syst. **18**(1–2), 135–140 (1996)
15. Toshiba Corporation: Press Releases 21 February, 1997 [Internet]. Toshiba.co.jp. (1997). https://www.toshiba.co.jp/about/press/1997_02/pr2101.htm. Accessed 13 September 2016
16. Suzumori, K., Miyagawa, T., Kimura, M., Hasegawa, Y.: Micro inspection robot for 1-in pipes. IEEE/ASME Trans. Mechatron. **4**(3), 286–292 (1999)
17. Suzumori, K., Maeda, T., Wantabe, H., Hisada, T.: Fiberless flexible microactuator designed by finite-element method. IEEE/ASME Trans. Mechatron. **2**(4), 281–286 (1997)
18. Suzumori, K., Koga, A., Haneda, R.: Microfabrication of integrated FMAs using stereo lithography. In: IEEE Workshop on Micro Electro Mechanical Systems, MEMS'94, pp. 136–141. IEEE (1994)
19. Suzumori, K., Koga, A., Kondo, F., Haneda, R.: Integrated flexible microactuator systems. Robotica **14**(05), 493 (1996)
20. Chou, C.-P., Hannaford, B.: Measurement and modeling of McKibben pneumatic artificial muscles. IEEE Trans. Robot. Autom. **12**(1), 90–102 (1996)
21. Hoggett, R.: 1957—"Artificial Muscle"—Joseph Laws McKibben (American) [Internet]. cyberneticzoo.com. (2012). http://cyberneticzoo.com/bionics/1957artificial-muscle-joseph-laws-mckibben-american/. Accessed 22 August 2016
22. Pack, R.T., Iskarous, M.: The use of the soft arm for rehabilitation and prosthetic. In: Proceedings of the Annual Conference RESNA 1994, pp. 472–475. RESNA Press (1994)
23. Google Patents: Robotic fluid-actuated muscle analogue. US Patent no. 5,021,064, 1991
24. Hamerlain, M.: An anthropomorphic robot arm driven by artificial muscles using a variable structure control. In: International Conference on Intelligent Robots and Systems 95, Human Robot Interaction and Cooperative Robots, pp. 550–555. IEEE (1995)
25. Groen, F., van der Smagt, P., Schulten, K.: Analysis and control of a rubbertuator arm. Biol. Cybern. **75**(5), 433–440 (1996)
26. Alford, W., Wilkes, D., Kawamura, K., Pack, R.: Flexible human integration for holonic manufacturing systems. In: Proceedings of the World Manufacturing Congress, pp. 53–62 (1997)
27. Wilkes, D., Pack, R., Alford, A., Kawamura, K.: HuDL, A design philosophy for socially intelligent service robots. American Association for Artificial Intelligence, AAAI Press, Technical Report, FS-97–02, pp. 140–145 (1997)
28. Cambron, M., Peters, II R., Wilkes, D., Christopher, J., Kawamura, K.: Human-centered robot design and the problem of grasping. In: The 3rd International Conference on Advanced Mechatronics ICAM'98-Innovative Mechatronics for the 21st Century, JSME, pp. 191–196 (1998)
29. Yamaha, Y., Iwanaga, Y., Fukunaga, M., Fujimoto, N., Ohta, E., Morizono, T., et al.: Soft viscoelastic robot skin capable of accurately sensing contact location of objects. In: IEEE/

SICE/RSJ International Conference on Multisensor Fusion and Integration for Intelligent Systems, MFI'99, pp. 105–110. IEEE (1999)
30. Hakozaki, M., Nakamura, K., Shinoda, H.: Telemetric artificial skin for soft robot. In: Transducers'99, pp. 844–847 (1999)
31. Bubic, F.: Flexible robotic links and manipulator trunks made thereform. US Patent no. 5,080,000 (1992)
32. Rus, D., Tolley, M.: Design, fabrication and control of soft robots. Nature **521**(7553), 467–475 (2015)
33. Iida, F., Laschi, C.: Soft robotics: challenges and perspectives. Procedia Comput. Sci. **7**, 99–102 (2011)
34. Ilievski, F., Mazzeo, A., Shepherd, R., Chen, X., Whitesides, G.: Soft robotics for chemists. Angew. Chem. **123**(8), 1930–1935 (2011)
35. Kim, S., Laschi, C., Trimmer, B.: Soft robotics: a bioinspired evolution in robotics. Trends Biotechnol. **31**(5), 287–294 (2013)
36. Trimmer, B.A., Ti Lin, H., Baryshyan, A., Leisk, G.G., Kaplan, D.L.: Towards a biomorphic soft robot: design constraints and solutions. In: 4th IEEE RAS and EMBS International Conference on Biomedical Robotics and Biomechatronics (BioRob), pp. 599–605. IEEE (2012)
37. Klute, G.K., Czerniecki, J.M., Hannaford, B.: McKibben artificial muscles: pneumatic actuators with biomechanical intelligence. In: Proceedings of the 1999 IEEE/ASME International Conference on Advanced Intelligent Mechatronics, pp. 221–226. IEEE (1999)
38. Rieffel, J., Knox, D., Smith, S., Trimmer, B.: Growing and evolving soft robots. Artif. Life **20**(1), 143–162 (2014)
39. Suzumori, K., Maeda, T., Wantabe, H., Hisada, T.: Fiberless flexible microactuator designed by finite-element method. IEEE/ASME Trans. Mechatron. **2**(4), 281–286 (1997)
40. Stone, R.S.W., Brett, P.N.: A flexible pneumatic actuator for gripping soft irregular shaped objects. In: IEE Colloquium on Innovative Actuators for Mechatronic Systems, pp. 13/1–13/3. IEE (1995)
41. Bowler, C.J., Caldwell, D.G., Medrano-Cerda, G.A.: Pneumatic muscle actuators: musculature for an anthropomorphic robot arm. In: IEE Colloquium on Actuator Technology: Current Practice and New Developments, (Digest No: 1996/110), pp. 8/1–8/6. IEE (1996)
42. Akella, P., Cutkosky, M.: Manipulating with soft fingers: modeling contacts and dynamics. In: International Conference on Robotics and Automation, pp. 764–769. IEEE (1989)
43. Suzumori, K., Endo, S., Kanda, T., Kato, N., Suzuki, H.: A bending pneumatic rubber actuator realizing soft-bodied manta swimming robot. In: International Conference on Robotics and Automation, pp. 4975–4980. IEEE (2007)
44. Pritts, M.B., Rahn, C.D.: Design of an artificial muscle continuum robot. In: International Conference on Robotics and Automation, ICRA'04, pp. 4742–4746. IEEE (2004)
45. McMahan, W., Chitrakaran, V., Csencsits, M., Dawson, D., Walker, I.D., Jones, B.A., et al.: Field trials and testing of the OctArm continuum manipulator. In: International Conference on Robotics and Automation, ICRA 2006, pp. 2336–2341. IEEE (2006)
46. Grissom, M.D., Chitrakaran, V., Dienno, D., Csencits, M., Pritts, M., Jones, B., et al.: Design and experimental testing of the OctArm soft robot manipulator. In: Defense and Security Symposium, Proceedings of the SPIE 6230, Unmanned Systems Technology VIII, pp. 62301F-62301F-10. International Society for Optics and Photonics (2006)
47. Trivedi, D., Dienno, D., Rahn, C.: Optimal, model-based design of soft robotic manipulators. J. Mech. Des. **130**(9), 091402 (2008)
48. Trivedi, D., Lotfi, A., Rahn, C.: Geometrically exact models for soft robotic manipulators. IEEE Trans. Rob. **24**(4), 773–780 (2008)
49. Trivedi, D., Rahn, C.: Model-based shape estimation for soft robotic manipulators: the planar case. J. Mech. Robot. **6**(2), 021005 (2014)
50. Udupa, G., Sreedharan, P., Aditya, K.: Robotic gripper driven by flexible microactuator based on an innovative technique. In: Workshop on Advanced Robotics and its Social Impacts (ARSO), pp. 111–116. IEEE (2010)

51. Cianchetti, M., Arienti, A., Follador, M., Mazzolai, B., Dario, P., Laschi, C.: Design concept and validation of a robotic arm inspired by the octopus. Mater. Sci. Eng. C **31**(6), 1230–1239 (2011)
52. Giorelli, M., Renda, F., Calisti, M., Arienti, A., Ferri, G., Laschi, C.: A two dimensional inverse kinetics model of a cable driven manipulator inspired by the octopus arm. In: 2012 International Conference on Robotics and Automation (ICRA), pp. 3819–3824. IEEE (2012)
53. Obaji, M.O., Zhang, S.: Investigation into the force distribution mechanism of a soft robot gripper modeled for picking complex objects using embedded shape memory alloy actuators. In: 6th IEEE Conference on Robotics, Automation and Mechatronics (RAM), pp. 84–90. IEEE (2013)
54. Robot and the Elastic Mind: Projects—Athlete Robot [Internet]. Isi.imi.i.u-tokyo.ac.jp. http://www.isi.imi.i.utokyo.ac.jp/~niiyama/projects/proj_athlete_en.html. Accessed 6 September 2016
55. Sasaki, D., Noritsugu, T., Takaiwa, M.: Development of active support splint driven by pneumatic soft actuator (ASSIST). In: International Conference on Robotics and Automation, pp. 520–525. IEEE (2005)
56. Sasaki, D., Noritsugu, T., Takaiwa, M., Kataoka, Y.: Development of pneumatic wearable power assist device for human arm "ASSIST". In: 2005 Proceedings of the JFPS International Symposium on Fluid Power, vol. 2005, no. 6, pp. 202–207 (2005)
57. Jung, K., Koo, J., Nam, J., Lee, Y., Choi, H.: Artificial annelid robot driven by soft actuators. Bioinspir. Biomim. **2**(2), S42–S49 (2007)
58. Yamamoto, Y., Kure, K., Iwai, T., Kanda, T., Suzumori, K.: Flexible displacement sensor using piezoelectric polymer for intelligent FMA. In: International Conference on Intelligent Robots and Systems IEEE/RSJ, pp. 765–770. IEEE (2007)
59. Kure, K., Kanda, T., Suzumori, K., Wakimoto, S.: Flexible displacement sensor using injected conductive paste. Sens. Actuators, A **143**(2), 272–278 (2008)
60. Cianchetti, M., Mattoli, V., Mazzolai, B., Laschi, C., Dario, P.: A new design methodology of electrostrictive actuators for bio-inspired robotics. Sens. Actuators B: Chem. **142**(1), 288–297 (2009)
61. Laschi, C., Mazzolai, B., Mattoli, V., Cianchetti, M., Dario, P.: Design of a biomimetic robotic octopus arm. Bioinspir. Biomim. **4**(1), 015006 (2009)
62. Follador, M., Cianchetti, M., Arienti, A., Laschi, C.: A general method for the design and fabrication of shape memory alloy active spring actuators. Smart Mater. Struct. **21**(11), 115029 (2012)
63. Cianchetti, M., Licofonte, A., Follador, M., Rogai, F., Laschi, C.: Bioinspired soft actuation system using shape memory alloys. Actuators **3**(3), 226–244 (2014)
64. Bao, G., Cai, S., Wang, Z., Xu, S., Huang, P., Yang, Q., et al.: Flexible pneumatic robotic actuator FPA and its applications. In: International Conference on Robotics and Biomimetics (ROBIO), pp. 867–872. IEEE (2013)
65. Shintake, J., Schubert, B., Rosset, S., Shea, H., Floreano, D.: Variable stiffness actuator for soft robotics using dielectric elastomer and low-melting-point alloy. In: 2015 IEEE/RSJ International Conference on Intelligent Robots and Systems (IROS), pp. 1097–1102. IEEE (2015)
66. Feng, G.H., Yen, S.C.: Micromanipulation tool replaceable soft actuator with gripping force enhancing and output motion converting mechanisms. In: 18th International Conference on Solid-State Sensors, Actuators and Microsystems (TRANSDUCERS), pp. 1877–1880. IEEE (2015)
67. Bertetto, A.M., Ruggiu, M.: In-pipe inch-worm pneumatic flexible robot. In: 2001 IEEE/ASME International Conference on Advanced Intelligent Mechatronics, pp. 1226–1231. IEEE (2001)
68. Shepherd, R., Ilievski, F., Choi, W., Morin, S., Stokes, A., Mazzeo, A., et al.: Multigait soft robot. Proc. Natl. Acad. Sci. **108**(51), 20400–20403 (2011)
69. Bogue, R.: Flexible and soft robotic grippers: the key to new markets? Ind. Robot: Int. J. **43** (3), 258–263 (2016)

70. Takashi, Y., Naoyuki, I., Makoto, M., Yoshinobu, A.: Picking up operation of thin objects by robot arm with two-fingered parallel soft gripper. In: Workshop on Advanced Robotics and its Social Impacts (ARSO), pp. 7–12. IEEE (2012)
71. Cianchetti, M.: The octopus as paradigm for soft robotics. In: 10th International Conference on Ubiquitous Robots and Ambient Intelligence (URAI), pp. 515–516. IEEE (2013)
72. Calisti, M., Arienti, A., Renda, F., Levy, G., Hochner, B., Mazzolai, B., et al.: Design and development of a soft robot with crawling and grasping capabilities. In: International Conference on Robotics and Automation (ICRA), pp. 4950–4955. IEEE (2012)
73. Margheri, L., Laschi, C., Mazzolai, B.: Soft robotic arm inspired by the octopus: I. From biological functions to artificial requirements. Bioinspir. Biomim. **7**(2), 025004 (2012)
74. Mazzolai, B., Margheri, L., Cianchetti, M., Dario, P., Laschi, C.: Soft-robotic arm inspired by the octopus: II. From artificial requirements to innovative technological solutions. Bioinspir. Biomim. **7**(2), 025005 (2012)
75. Renda, F., Cianchetti, M., Giorelli, M., Arienti, A., Laschi, C.: A 3D steady-state model of a tendon-driven continuum soft manipulator inspired by the octopus arm. Bioinspir. Biomim. **7** (2), 025006 (2012)
76. Giorelli, M., Renda, F., Calisti, M., Arienti, A., Ferri, G., Laschi, C.: A two dimensional inverse kinetics model of a cable driven manipulator inspired by the octopus arm. In: International Conference on Robotics and Automation (ICRA), pp. 3819–3824. IEEE (2012)
77. Kang, R., Branson, D., Zheng, T., Guglielmino, E., Caldwell, D.: Design, modeling and control of a pneumatically actuated manipulator inspired by biological continuum structures. Bioinspir. Biomim. **8**(3), 036008 (2013)
78. Kang, R., Guglielmino, E., Zullo, L., Branson, D., Godage, I., Caldwell, D.: Embodiment design of soft continuum robots. Adv. Mech. Eng. **8**(4), 1687814016643302 (2016)
79. Cianchetti, M., Ranzani, T., Gerboni, G., Nanayakkara, T., Althoefer, K., Dasgupta, P., et al.: Soft robotics technologies to address shortcomings in Today's minimally invasive surgery: the STIFF-FLOP approach. Soft Robot. **1**(2), 122–131 (2014)
80. Noh, Y., Sareh, S., Back, J., Würdemann, H.A., Ranzani, T., Secco, E.L., et al.: A three-axial body force sensor for flexible manipulators. In: International Conference on Robotics and Automation (ICRA), pp. 6388–6393. IEEE (2014)
81. Ranzani, T., Cianchetti, M., Gerboni, G., Falco, I., Menciassi, A.: A soft modular manipulator for minimally invasive surgery: design and characterization of a single module. IEEE Trans. Rob. **32**(1), 187–200 (2016)

Part IV
Intelligent Sensing and Control

An Observer Design Technique for Improving Velocity Estimation of a Gimbal System

Sangdeok Lee and Seul Jung

Abstract This paper presents filter design techniques for the improved estimation of the velocity of the gimbal axis in a control moment gyro (CMG) actuator. In the motion control applications, the velocity sensor is not available but the estimation filters are often used due to the cost of the sensor. The first derivative filter technique is used most, but the filtering performance is not reliable for the accuracy requirement. The velocity observer is designed to estimate the velocity based on the modelling technique by the recursive least squares (RLS) method. Experimental results of estimating the velocity of the gimbal system in a CMG are conducted. The estimation performances are compared through the experimental data.

1 Introduction

Gyroscopic actuation of a control moment gyroscope (CMG) is induced by the cross product of two velocity related physical quantities, angular momentum and the gimbal rate (velocity). In general, the angular momentum is set to a constant from the constantly rotating flywheel and the resultant gyro torque is dependent on the velocity of the gimbal axis [1]. Based on the concept of the gyroscopic force, it is obvious that the measurement of the gimbal velocity becomes very effective on obtaining the good performance. In reality, however, the high rotating flywheel of CMG can produce unexpected vibrations making it difficult to measure the accurate velocity of the gimbal axis.

An observer design is one of the promising techniques for the accurate estimation of the velocity of the gimbal axis since it is simple and practical [2, 3].

S. Lee · S. Jung (✉)
Intelligent Systems and Emotional Engineering (ISEE) Laboratory,
Department of Mechatronics Engineering, Chungnam National University,
99 Daehak-ro, Yuseong-gu, Daejeon 34134, Korea
e-mail: jungs@cnu.ac.kr

S. Lee
e-mail: sdcon.lee@cnu.ac.kr

J.-H. Kim et al. (eds.), *Robot Intelligence Technology and Applications 5*,
Advances in Intelligent Systems and Computing 751,
https://doi.org/10.1007/978-3-319-78452-6_28

The observer design technique only requires a model of the given system. When a numerically exact model of the system is given, the velocity observer can provide an exact state of the system. However, there are always modeling errors in dealing with an unknown system. Moreover, disturbances such as severe vibrations are present so that the accurate estimation cannot be guaranteed although a highly accurate model of the given system is obtained through the system identification procedure [4–6].

Therefore, to improve the velocity estimation performance of the disturbed gimbal, in this paper, we propose a complementary filter which consists of both a filtered encoder and a velocity observer. In the low frequency-domain, the filtered encoder data are dominantly used for the velocity estimation of the gimbal axis. On the other hand, in the high frequency-domain, the observed data are dominantly used for the velocity estimation of the gimbal axis. By doing so, the velocity estimation performance for the disturbed gimbal could be improved.

Firstly, the hardware configuration of the CMG is introduced. Secondly, the frequency characteristics of both the filtered encoder and the velocity observer are reviewed. Lastly, the proposed method is verified through the experimental data.

2 Configuration of a CMG

We developed a control moment gyro (CMG) system based on two brushless motor systems: a gimbal motor system and a flywheel motor system. The overall hardware configuration of the CMG is shown in Fig. 1.

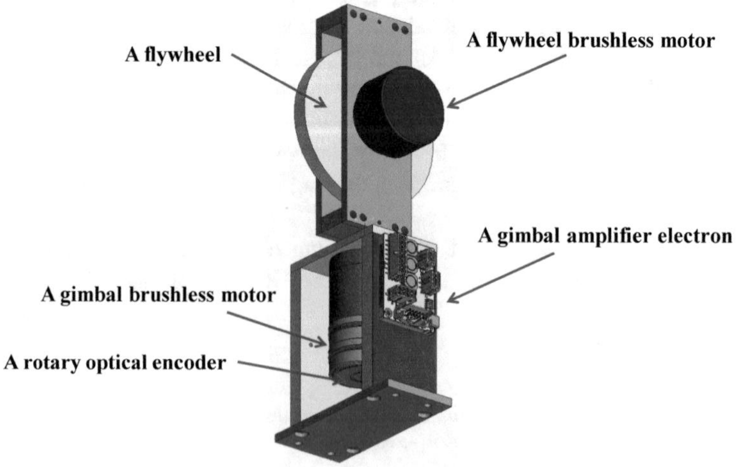

Fig. 1 Configuration of the CMG

In Fig. 1, the mass of the flywheel is 0.4 kg, the radius of the flywheel is 0.04 m, the inertia of the flywheel is 0.00032 kgm^2, the maximum velocity of the flywheel is 628.32 rad/s, and the maximum velocity of the gimbal axis is 21.74 rad/s.

The general mechanism of CMG is

$$\boldsymbol{\tau} = \mathbf{H} \times \mathbf{W} \tag{1}$$

where $\boldsymbol{\tau}$(Nm) is the gyroscopic torque, $\mathbf{H}$(Nms/rad) is the angular momentum of the flywheel system, and $\mathbf{W}$(rad/s) is the gimbal velocity. When the magnitude of $\mathbf{H}$ is constant, the magnitude of (1) can be rewritten as follows.

$$\tau = \mathrm{HW}, \tau \propto \mathrm{W}, \mathbf{H} \perp \mathbf{W} \tag{2}$$

The magnitude of the gyroscopic torque can be proportional to the velocity of the gimbal axis in (2). Consequently, the overall control performance of the CMG can be dependent on the control of the velocity of the gimbal axis. Therefore, the accurate estimation of the velocity of the gimbal system is important.

3 Review of a Velocity Observer

We consider the gimbal axis as a second-order system and the nominal model of the given system has two states such as angular velocity and angle. A recursive least squares (RLS) method is used for the identification of the given system [7]. Here we assume that the system model is available after the identification by RLS.

The nominal model for the gimbal axis can be shown as an adaptive infinite impulse response filter structure and it becomes

$$\frac{\hat{y}(z)}{u(z)} = \frac{c_1 + c_2 z^{-1} + c_3 z^{-2}}{1 + c_4 z^{-1} + c_5 z^{-2}} \tag{3}$$

where the parameter vector $\mathbf{c}[n] = [c_1\ c_2\ c_3\ c_4\ c_5]$ can be found adaptively and recursively through the minimization of the sum of the squared errors for the output estimation. Through the well-known RLS procedure, the filter coefficients of (3) can be updated in the time-domain [7].

From (3), a velocity observer can be easily derived. As a first step, (3) is transformed into the state-space representation. The state-space model for (3) after some modification of c_1 becomes

$$A = \begin{bmatrix} -c_4 & -c_5 \\ 1 & 0 \end{bmatrix} \tag{4}$$

$$B = [1 \quad 0]^T \tag{5}$$

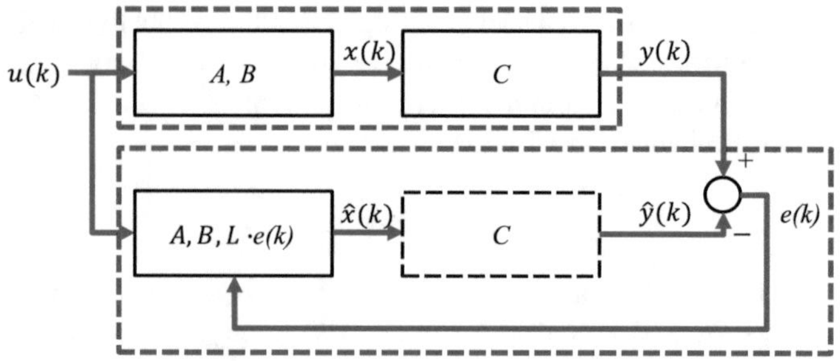

Fig. 2 State observer mechanism

$$C = \left[c_2 - c_1 c_4 \quad c_3 - c_1 c_5 \right] \tag{6}$$

$$D = c_1 \tag{7}$$

The state of the given system can be estimated through the following procedure shown in Fig. 2.

In Fig. 2, x(k) is the system state, $\hat{x}(k)$ is the estimated system state, e(k) is the output error, and $\hat{e}(k)$ is the estimated output error. The state dynamics can be described as

$$(x(k+1) - \hat{x}(k+1)) = (A - LC)(x(k) - \hat{x}(k)) \tag{8}$$

Error dynamics then can be

$$e_x(k+1) = (A - LC)e_x(k) \tag{9}$$

If $(A - LC)$ is stable and $[A, C]$ is observable, $e_x(k+1) \to 0$ when $k+1 \to \infty$. The observer gain L can be arbitrarily chosen.

For simplicity, the system model (3) is approximated as follows.

$$\frac{\hat{Y}(z)}{U(z)} = \frac{c_1 + c_2 z^{-1} + c_3 z^{-2}}{1 + c_4 c^{-1} + c_5 z^{-2}} \approx \frac{c_1}{1 + c_4 z^{-1} + c_5 z^{-2}} \tag{10}$$

In the time-domain, (10) becomes

$$u(k) = \frac{1}{c_1}\hat{y}(k) + \frac{c_4}{c_1}\hat{y}(k-1) + \frac{c_5}{c_1}\hat{y}(k-2) \tag{11}$$

When the system input is assumed as a sum of its states, the states can be defined as follows.

$$x_1(k) = K_1 \frac{1}{b_1} \hat{y}(k), x_2(k-1) = K_2 \frac{b_4}{b_1} \hat{y}(k-1), x_3(k-2) = K_3 \frac{b_5}{b_1} \hat{y}(k-2) \quad (12)$$

For the torque input, each state of (12) indicates its acceleration, velocity, and angle. To investigate the velocity state, the observer gain matrix L is chosen as follows.

$$L = \begin{bmatrix} \frac{1}{c_1} & \frac{c_4}{c_1} \end{bmatrix}^T \quad (13)$$

where $K_1 = K_2 = 1$ is empirically chosen in (12).

4 Experiment

4.1 Experimental Setup

In this section, we verify the proposed method. An experimental setup is shown in Fig. 3. Three velocity data are compared: Maxon corporation observed velocity data, time differentiated encoder data, and the proposed observer data. In the experiment, a high-end encoder sensor has 0.004185 (degrees/count) resolution.

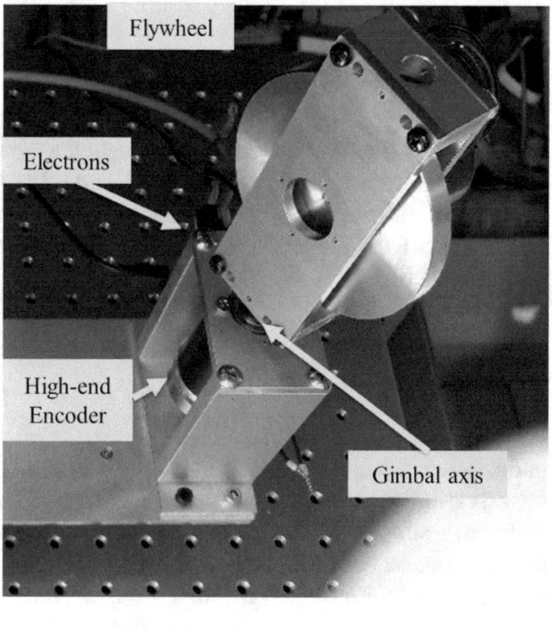

Fig. 3 Experimental setup for the verification

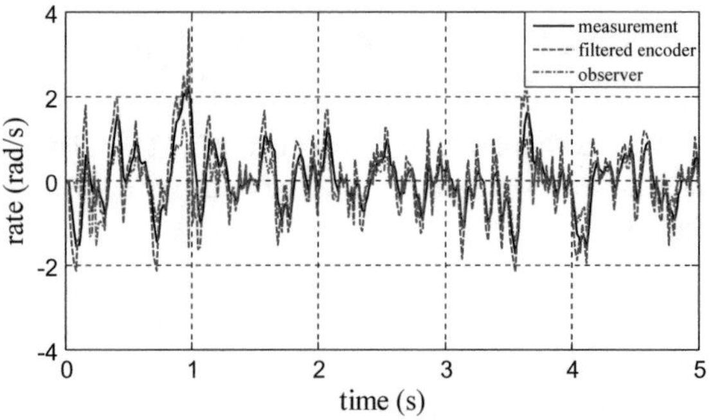

Fig. 4 Performance verification in the time-domain

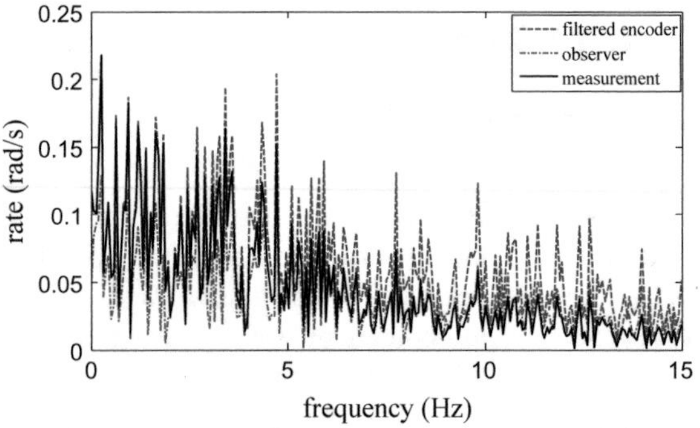

Fig. 5 Performance verification in the frequency-domain

4.2 Experimental Results

The designed observer is verified through the experiment. Figure 4 shows the experimental result in the time-domain and Fig. 5 shows the results in the frequency-domain. We see that the observer shows the better performance. The RMSE of the filtered encoder is 0.3842 and the RMSE of the observer estimation is 0.2635.

5 Conclusion

In this paper, we proposed a simple observer design technique to obtain the gimbal velocity estimation. A recursive least square (RLS) method could be directly used to design a simple velocity estimator by deriving approximated observer gains and the performance of the velocity observer was compared to that of the filtered encoder signals. Through the comparison, we found out that the observer has an advantage in the high frequency-domain and disadvantage in the low frequency-domain.

Acknowledgements This research has been supported by 2017 National Research Foundation of Korea (grant 2016R1A2B2012031).

References

1. Gollomp, B.: Gyroscopes. IEEE Instrum. Meas. Mag. **4**(3), 49–52 (2001)
2. Luenberger, D.: Observer for multivariable systems. IEEE Trans. Autom. Control **11**(2), 190–197 (1966)
3. Afri, C., Andrieu, V., Bako, V., Dufour, P.: State and parameter estimation: a nonlinear Luenberger observer approach. IEEE Trans. Autom. Control **62**(2), 973–980 (2017)
4. Reichbach, N., Kuperman, A.: Recursive-squares-based real-time estimation of supercapacitor parameters. IEEE Trans. Energy Convers. **31**(2), 810–812 (2016)
5. Tanelli, M., Piroddi, L., Savaresi, S.M.: Real-time identification of tire-road friction conditions. IET Control Theory Appl. **3**(7), 891–906 (2009)
6. Ljung, L.: Perspectives on system identification. Ann. Rev. Control **34**(1), 1–12 (2010)
7. Lee, S., Jung, S.: RLS model identification-based robust control for gimbal axis of control moment gyroscope. In: IEEE, pp. 584–589. AIM (2017)

Voice Controlled Multi-robot System for Collaborative Task Achievement

Chahid Ouali, Mahmoud M. Nasr, Mahmoud A. M. AbdelGalil and Fakhri Karray

Abstract This paper describes a framework for machine cooperation and interaction designed to promote collaborative work with humans in a natural and flexible manner. More specifically, we propose a system where all the interactions with the robots are achieved through voice using Alexa Voice Service (AVS). Besides the implementation of a voice interface channel, we develop a navigation module that enables a robot to autonomously navigate through a map to reach a specific destination while avoiding obstacles. We provide the system with the ability to choose and operate the appropriate robot in context of multi-robot cooperation for task achievement. In addition, a user can intervene during the process of executing a relatively complex task by interacting with the robots via voice commands. To validate the proposed framework, we conduct experiments to quantify the performance during cooperative action involving the human and the system.

1 Introduction

The Human-robot interaction has been an attractive subject in the field of robotics in the last few decades due to the positive effects that it can bring to human lives. There has been a great progress in the development of automated robotic systems that do not require human interventions. For example, robot manipulators have been

C. Ouali (✉) · F. Karray
Pattern Analysis and Machine Intelligence (PAMI) Research Group, Electrical and Computer Engineering Department, University of Waterloo, Waterloo, ON, Canada
e-mail: couali@uwaterloo.ca

F. Karray
e-mail: karray@uwaterloo.ca

M. M. Nasr · M. A. M. AbdelGalil
University of Science and Technology, Zewail City, 6th of October 12588, Egypt
e-mail: s-mahmoud.nasr@zewailcity.edu.eg

M. A. M. AbdelGalil
e-mail: s-mahmoud.abdelgalil@zewailcity.edu.eg

J.-H. Kim et al. (eds.), *Robot Intelligence Technology and Applications 5*,
Advances in Intelligent Systems and Computing 751,
https://doi.org/10.1007/978-3-319-78452-6_29

successfully used in large number of industrial applications. In contrast, the subject of robots operating in human-centric environments [1] is still relatively in its early days, despite the many technological advances witnessed in the past few years. In fact, this kind of robotic systems is more complex to design and implement because of the complexity and uncertainty encountered when operating in environments involving humans. In addition, implementing a system where a human can efficiently and safely interact is very challenging, especially when the interaction is in the form of spoken dialogue (through voice medium).

The resolution of complex real-world problems brings us to an important area in this domain: multi-robot cooperation. The main idea behind multi-robot cooperation suggests dividing the complex problem into smaller tasks allowing multiple robots with different capabilities to perform a complex task, otherwise impossible to be achieved by a single robot [2, 3]. This research topic has been very active since the early 90s [4], and many works were published on cooperative robots for task achievement like localization [5, 6], exploration [7–9], search [10, 11], and transportation [12–14].

On the other hand, robots operating in human-centric environments are required to continuously perceive humans and their intents, and must therefore be equipped with skills that allow them to naturally and efficiently communicate with them. Spoken dialogue is the most natural and privileged way of communication in this context (especially in hazardous or cluttered environments). Early research in this direction has not led to the development of many real word applications. However, recent advances in the field of automatic speech recognition and natural language processing made the implementation of such systems possible. For example, the work in [15] presents a spoken dialogue system that allows hospital staff to interact with a mobile robotic assistant that provides information and executes simple tasks. In a similar work [16], a spoken dialogue system is presented with the ability to handle multi-modal dialogs. Several other systems have been proposed in this context for tour-guide robot [17], robot lifting arm control [18] and learning robot for children [19].

In this paper, we propose a framework for robot cooperation and interaction operating in human-centric environments. This work is the first phase of a larger ongoing project intended to build customizable framework that provides augmented capabilities allowing natural human-robot interactions and heterogeneous multi-robot cooperation for task achievement.

The rest of the paper is organized as follows. Section 2 describes the architecture and the main modules of the proposed framework. In Sect. 3, we present the experimental results for robot navigation and task achievement. We conclude and summarize this paper in Sect. 4.

2 Framework for Machine Cooperation and Interaction

This section describes the overall architecture of the proposed framework as shown in Fig. 1. This framework is composed of different modules communicating with one another in real time. After explaining the role of each module in the first subsection, we will describe in detail the functionality of the main parts in the remaining subsections.

2.1 System Overview

An important aspect that should be considered when designing a multi-robot system is whether the system control is centralized or decentralized. Compared to the decentralized architecture, where any robot can be involved in the decision-making process, a centralized architecture possesses a central control agent (robot or computer) that has a global view of the environment and is responsible for supervising the team robots to achieve the planned task [2]. Each of these architectures has several advantages over the other.

Centralized architecture is a simple and efficient solution when it comes to controlling a limited number of robots that operate in known and unchanging environment [3], which is the case of our project. This led us to adopt the

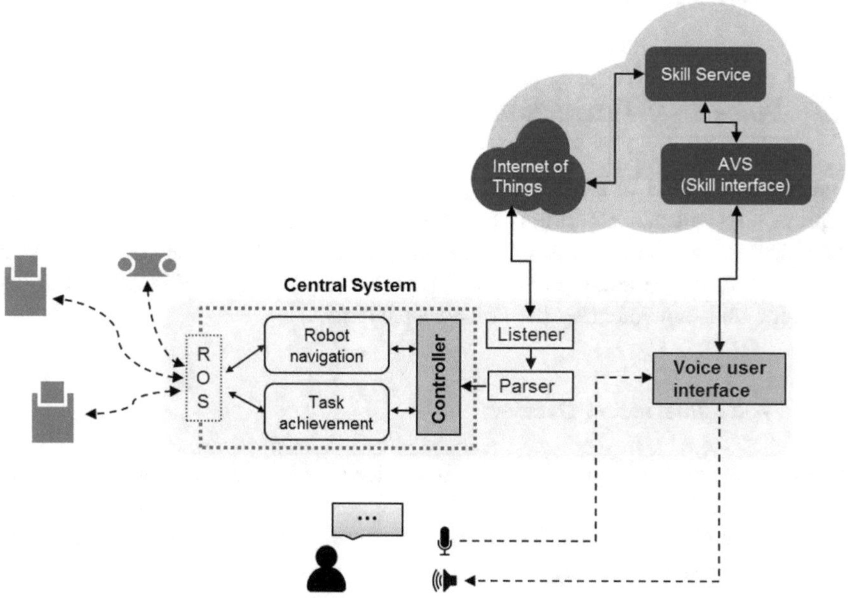

Fig. 1 System architecture

centralized structure as group architecture for the proposed system. Thus, the *controller* module, shown in Fig. 1, has the exclusive role of master who handles commands from the outside, organizes the team and decides which robot(s) would be enabled for the given task.

Another central part of the proposed framework is the human-computer interaction module. In this work, the main communication medium between the user and the system is achieved using voice. Our objective in this regard is to build a system that holds augmented intelligence that guarantees interactions with the user in a spontaneous and natural way.

Before going further in describing in more detail the different components of the framework, let's see how the system works. First, the user starts the process by vocally asking the system to execute a specific task. The voice query is handled in the cloud to determine the user's intent. The intent is then transformed into a message (command and parameters) and sent back to the *central system*. The role of the *listener* is to inform the *parser* of the arrival of a new message. This module parses the received message, extracts the command and the parameters, and then triggers the *controller*. Depending on the nature of the task, the controller executes the *task achievement* module to determine which robot(s) would accomplish the task. The *robot navigation* module allows the robot to navigate through the map to reach a particular location specified by the controller.

Another interesting part of the proposed framework is the use of the well-known Robot Operating System (ROS). ROS [20] is a generic software framework allowing the development of robot applications. It provides a huge number of libraries and tools that can be used in a variety of platforms. We will include in the next sections details about ROS packages we have used in this project.

2.2 Voice User Interaction

The implementation of a natural language user interface is a challenging task that requires sophisticated algorithms for speech recognition, natural language understanding and speech synthesis. Implementing these algorithms is outside this project scope. Thus, our strategy is to use Alexa Voice Service (AVS) and focus on designing and implementing the voice-user interface.

2.2.1 Voice Interaction Overview

Alexa Voice Service is an Amazon Web Service (AWS) that provides the ability to build voice-based applications (or skills) using Alexa Skills Kit (ASK), which is a collection of APIs and tools created for this purpose. More details about AVS and ASK can be found in [21–23].

A skill is a voice driven application composed of two main components: the *skill service* and the *skill interface*. The skill interface is responsible for processing the user spoken utterances, whereas the skill service determines how the skill behaves in response to the user speech.

When a user starts interacting with the system by stating a command, the voice query is sent to AVS, which resides in the cloud (see Fig. 1). Based on the skill interface model, the user's intent is identified after speech recognition and natural language understanding processing. The intent and the associated parameters are then sent to the skill service, which determines the appropriate actions to take. In our case, the skill interface implements different functions that handle the conversation with the user, and then send an MQTT messages using AWS IoT to the central system.

2.2.2 Voice Interaction Design

Table 1 summarizes the main user intents that can be handled by the voice interface we have designed. These intents, or operations, are divided into two groups according to their complexity. The first group represents simple operations that can be directly processed without the need of additional modules. In this case, the user needs only to specify the name of the robot, the type of operation and the associated parameters (e.g. distance, degree, direction, etc.).

On the other hand, complex group intents do not necessary imply a more complicated user interaction, but a relatively more difficult operation achievement. This requires a combination of simple operations (i.e. turn, move, etc.) or additional components (e.g. map, task achievement module, etc.) to accomplish the operation.

We should note here that for complex operation, and especially for task achievement operation, the task can be accomplished cooperatively by multiple robots. In addition, the user can intervene at any time during this process to assist the robots. The human-robot cooperation for task achievement is discussed in more detail in Sect. 3.

In order to make the human-computer voice interaction more natural and similar to a human-like way, we have adopted a multi-turn dialog model. Compared to a single-turn dialog where the user interaction is complete after a single utterance, a multi-turn dialog implies several interactions with the intention of gathering and

Table 1 Description of the operations handled by our voice interface

Complexity	Operation	Description
Simple	Move	Move the robot *forward* or *backward* a certain *distance*
	Turn	Turn the robot *right* or *left* with certain *degree*
	Gripper	*Open*, *close*, *up* or *down* the gripper
	Camera	*Open*, *close* or *rotate* the camera in specific *direction*
Complex	Go to	Move the robot to a predefined *location*
	Task achievement	Achieve a specific *task*

Table 2 Examples of multi-turn and single turn dialog

Model	Example
Single-turn	*user*: move Bobby forward 1 m *system*: moving Bobby forward 1 m
Multi-turn	*user*: move 1 m *system*: which robot would you like to move *user*: Roby *system*: robot name is invalid; please choose between Bobby and Alice. *user*: Bobby *system*: in which direction you want to move Bobby? *user*: move it forward *system*: moving Bobby forward 1 m

confirming values. Table 2 shows the difference between these two models using two conversation examples.

2.3 Robot Navigation

Getting a robot from point *A* to point *B* in a map is not an intuitive task. In fact, the robot should be able to first create its own map of the surrounding. However, creating a map for every operating environment is not always easy. In addition, this project presents some difficulties since working in an indoor environment means that no use of GPS is possible, and the availability of an indoor positioning system requires an infrastructure that is quite expensive.

Given the map, the robot should then be able to localize itself using the provided map and its available sensors by comparing the sensor readings to the map. From that, the robot knows its location within the map with an associated probability distribution. To be able to get it to another point in the map, a path planning algorithm has to come into play. Given the initial and final positions, the occupied, free, and unknown regions of the map, the algorithm works to find a path which links the two points though the free space.

So far, the robot believes it is operating in a perfect environment where the map it was given was not altered and no extra obstacles exist in the path. That, however, is not the actual case since we are operating in a dynamic environment. To be able to account for unknown obstacles coming in the way and to avoid unwanted collisions, the robot must be equipped with obstacle avoidance ability. Using the robot's sensors, it should be able to distinguish unknown obstacles in its way and plan its path around them.

2.3.1 Map Construction

The system in hand consists of two *Mobilerobots Peoplebot* which are based on *Pioneer 3* platforms equipped with a Linux operating system and a software responsible for controlling the robot; *ARIA*. These robots are equipped with a 180° laser scanner, bumpers, and ultrasonic sensors. The robot also has a short gripper with two degrees of freedom (up-down and open-close) and a *Pin-Tilt-Zoom* camera. With the available hardware, the system should be able to navigate through the map without any external sensors or GPS or other pre-installed infrastructure.

Since our navigation strategy relies on using a map, Simultaneous Localization and Mapping (SLAM) approach is employed in this study. As it can be understood from the name, SLAM is a combination between localization and mapping. In other words, the robot keeps updating its position (localization) with respect to the starting point, while simultaneously moving and constructing the map (mapping).

The SLAM problem is the estimation of the state trajectory of the robot along with the map, denoted as m, given the observations of the robot along its course and the given control inputs. It is computed as follows:

$$p(x_{1:t}, m|z_{1:t}, u_{1:t-1}) \tag{1}$$

The SLAM problem is currently being solved by two popular approaches, Smoothing SLAM and Filtering SLAM. Smoothing SLAM does not have much literature tackling it as it computationally expensive and is not yet optimized and can only be used for offline SLAM [24], whereas Filtering SLAM has all the research attention. Filtering SLAM can be further divided into two groups depending on the type of filter it uses; that is, Extended Kalman Filter (EKF) and Particle Filter (PF).

EKF [25] is a powerful solution as it is easy to use and has low computational cost. It, however, is merely an approximation of the real world and works only with Gaussian noise models and fails to converge in the presence of non-Gaussian noise. On the other hand, Particle Filter is based on Bayes' filter which uses random sampling for representing the estimated belief. Particle filter first spreads the particles, each used to present a probability of solution, all over the available space. It then works on reducing the number of particles in positions that are not supported by sensor readings and increasing them where the probability is expected to be higher. Although, this algorithm is computationally expensive since it needs a large number of particles (samples), it still better than EKF when it comes to dealing with non-gaussian noise.

Since all the other tasks (e.g. localization and navigation) depend on the map, it is of great importance to have a good representative map given any noise, which led us to use Particle Filter technique. More specifically, we used an implementation of the PF found in the *gmapping* [26] package available for ROS. This implementation uses Rao-Blackwellized particle filter (RBPF) whilst incorporating an adaptive technique to reduce the number of particles required by the filter to generate an Occupancy Grid Map (OGM); therefore, reducing the computational burden [27].

The grid map is constituted of either an occupied cell, an empty cell, or an unknown cell, each marked by a certain threshold.

To start mapping, the algorithm registers the initial laser scan received to the grid map. Then, after a threshold distance is traveled by the robot, it uses the laser scans to confirm the readings obtained by the odometry data and thus obtaining a posterior probability which corrects for the prior estimation from odometry. After this process, the weights of the particles are updated according to which samples gave more elective results. The formula used to find how well each particle represents the posterior probability is as follows:

$$N_{eff} = \frac{1}{\sum_{i=1}^{N} (\breve{\omega}^i)^2} \quad (2)$$

where $\breve{\omega}^i$ is the normalized weight of the particle i.

Through using the mentioned method and utilizing RBPF, it is possible to obtain results of good accuracy using low number of particles. This directly means that the computational requirements are lowered without sacrificing performance.

2.3.2 Localization

The localization problem is a reduced form of the SLAM problem. In localization, only the current state of the robot is estimated. Stated in mathematical terms, the localization problem is the estimation of the state belief trajectory of the system from time 1 to time t, denoted as $x_{1:t}$. When represented as a probability density function, given all of the observations of the surroundings of the environment $z_{1:t}$ and the control inputs issued to the robot $u_{1:t-1}$ it is found to be:

$$p(x_{1:t}, m|z_{1:t}, u_{1:t-1}, m) \quad (3)$$

This requires the availability of a so called "measurement model" and a "process model". The above probability distribution can be obtained via the Bayesian filter framework, exactly like the SLAM problem, only easier now because the map is a no longer a requirement, but an input to the system. The required inputs are the process model and measurement model. These are probabilistic density functions that give the probability density distribution for both location and observation. In mathematical terms, the process model is:

$$p(x_t|u_{t-1}, x_{t-1}) \quad (4)$$

and the measurement model is:

$$p(z_t|x_t, m) \quad (5)$$

where x_t is the current state, x_{t-1} is the previous state, u_{t-1} is the last control input and z_t is the current measurement and m is the map.

An important step that was done before that was to make sure the odometry is accurate and has no drift. Several methods can be found to test it in both translation and rotation. If the odometer was found to include a drift, it must be calibrated before going into the process of SLAM.

Once the odometer is calibrated, the robot can be set into localization mode. This can be done using several algorithms; the chosen algorithm in this study is the Monte Carlo Localization (MCL) [28, 29] which utilizes a particle filter to find the location with the highest probability. It starts off with uniform distribution and attempts to shift its probabilities towards the locations by comparing the sensors readings to the provided map. A ROS implementation of this algorithm is found in AMCL [30, 31] package. This package requires the previously generated OGM, the laser scans and the odometry readings in order to compute the most probable pose (position and orientation) for the robot as it moves.

2.3.3 Path Planning

As previously discussed, the next step after the robot is able to localize itself in a given map, is for it to be able to plan its path from its current position to a certain required point in the map (i.e. path planning). For this task, we used the *Navigation Stack* package [32] in ROS. This package depends on a concept of utilizing the *costmap*. The costmap is composed of three layers, namely the static layer, the inflation layer, and the obstacle layer.

The process of path planning is divided into global planning and local planning. The path planning problem can be first isolated from obstacle avoidance in the global planning. Without going into too much detail, here is how the path planning algorithm works. First, the global planner attempts to generate a path between the start and end point that runs through the cells with the least cost using the saved static map, referred to as the static layer. The inflation layer defines the radius for which the occupied cells extend and the rate at which the cost of the cells decreases within that radius. Tuning these parameters makes a great difference as it can cause the robot to either take sharp turns with jerky movement, or smoothly pass right between occupied cells. It can also make the robot believe it cannot path through spaces that are obviously free, due to a mistake in the inflation layer set.

In order to obtain the optimum global plan, we tested the algorithm with several different parameters [33]. The obtained global plan is then divided into small segments that the local planner attempts to execute. The local planner, attempts to follow the points on the global path while taking into consideration the obstacle layer. The obstacle layer gets real-time readings from the sensors to the local planner to plan the path around them. Just like the static layer, the obstacle layer has inflation radius to give a safe distance between the robot and the obstacles in its way.

2.4 Task Achievement

The proposed system includes the autonomy of robot task achievement and the interaction with humans. Certain tasks, which are out of the robot's capabilities, can be done in cooperation with a human. In this situation, the human takes the lead by commanding the robot in simple voice commands to do simple actions such as move forward for a certain distance or move the gripper to a certain height and close the gripper or open it. The global task can therefore be achieved using this cooperation between the human and robot through voice medium.

As mentioned in Sect. 2.2, the user can either interact with a specific robot individually (by specifying the robot name) or lets the system decide which robot(s) would be involved (task achievement). In other words, the user can (1) command a certain robot to move to a certain location for example; (2) command the system as a whole to achieve a certain task (e.g. bring an object). In the latter case, the system evaluates the task and determines which robot should perform it, or if the task needs more than one robot to coordinate the interaction between them.

The selection function is based on several parameters such as the distance between each robot and the location where the task should be performed, the battery status of each robot, the localization certainty, etc. The cost function used in the process is flexible and can be improved and designed to achieve the desired system behavior. This cost function is defined as follows:

$$F(COV, dist) = a * COV + b * dist \tag{6}$$

where, *dist* is the distance between the robot and the target and *COV* is the covariance, resembling the uncertainty in localization. The constants *a* and *b* are chosen to give relative weights between the parameters.

The robots are interfaced to ROS through the package *RosAria*. This package provides support to Pioneer-based platforms. It does so by providing ROS wrappers around Aria API to publish robot sensor readings and control the robot actuators through standard ROS communication methods (topics and services).

Peoplebot is a modified version of Pioneer robot. More specifically, the Peoplebot has a gripper that can be controlled via Aria API. However, the gripper is not supported by RosAria package. Thus, we modified RosAria source code to support controlling the gripper through the voice commands directly from the voice interaction system. Also, we modified RosAria to provide direct motion commands through voice interaction with the system, such as move forward or set heading.

As illustrated in Fig. 1, the block responsible for interfacing the listener node to the ROS network is the parser node. This node is responsible of parsing the output string from the listener node and sending commands to the targeted node in the ROS system. This design makes the system independent from the user input part and does not limit the human-robot interaction to voice. It could be through mobile apps or other front-ends that can construct the supported commands.

Once the command has been interpreted by the voice interfacing node and sent as a string message on the ROS network, the command parser node receives the contents of the message, parses the information and directs the command to the responsible ROS topics to execute the required actions. Through this, string command messages are executed by the robot.

3 Experiments

In this section, we present a few experiments to test and show the functionality of the proposed framework. These experiments are conducted using two Peoplebot robots named *Alice* and *Bobby*. They are both equipped with the same set of sensors mentioned before: an 180° SICK LMS 200, a set of 24 ultrasonic sensors (8 at the bottom on each side and 8 at the top front), and a PTZ camera. We ran three experiments on our system, and we describe them in the following order: map generation, robot navigation and task achievement.

3.1 Map Generation

The first experiment consists of generating a map of the environment where the robots would operate as described in Sect. 2.3.1. We need to build this map only one time using a single robot, and it can be used thereafter by any robots connected to the system. We first ran the SLAM package that allows us to manipulate the laser scanner and to move the robot in parallel with teleoperation. We slowly guided the robot through the environment to construct a map of the desired area. The generated map is then saved on an external desktop computer running *Debian Jessie* with ROS *Kinetic* that is connected to the same ROS network as the mapping robot. The result of this operation is shown in Fig. 2. This figure represents the occupancy grid map for three rooms surrounded by corridors. The small dark and shaded areas within the rooms represent chairs and tables.

3.2 Navigation

To be able to navigate through the map, we provided the map to Bobby and Alice and tested their localization using AMCL. We then configured the navigation stack which is responsible for the path planning process. After tuning the parameters as mentioned in the previous section, the result obtained for the localization and path planning are shown in Fig. 3.

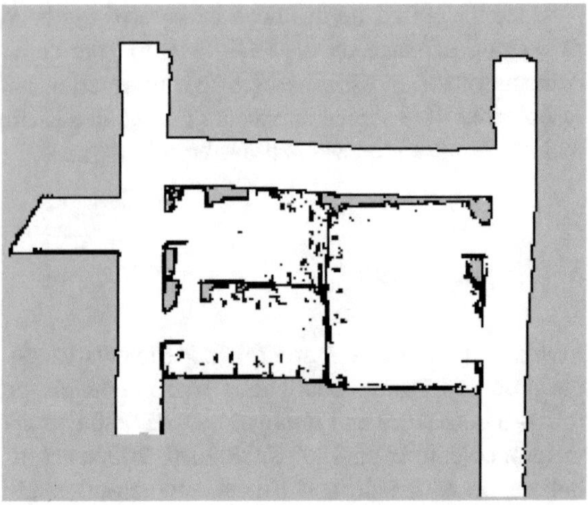

Fig. 2 Generated occupancy grid map

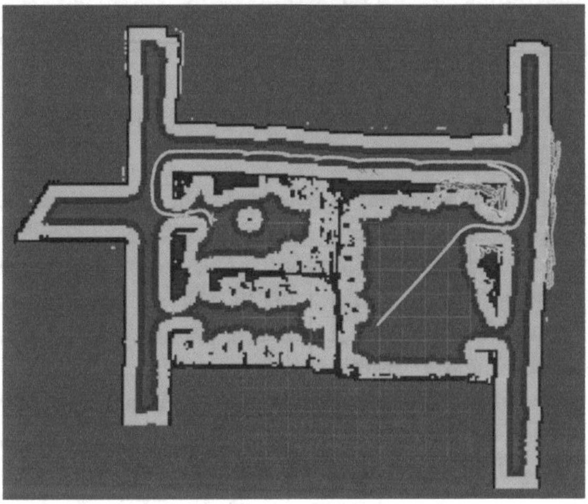

Fig. 3 Navigation sample visualization using RVIZ

In Fig. 3, the red arrows indicate the position estimate as calculated by AMCL, with the highest probability being in the center. The cost map is represented in the range of colors around the occupied cells. The sky-blue color indicates the occupied cells inflation layer and it can be seen to fade into red, then purple and then dark blue as the cost decrease. The gray area indicates the least cost and therefore represents the preferred areas where the robot should navigate. Regarding the motion planning, the green line shows the global path the robot is taking, and the yellow line on top shows the local path currently being followed.

3.3 Task Achievement

The last experiment is to test the entire system functionality for task achievement, where the task is to bring a specific object. This part involves a voice interaction with the system which decides about how the task should be tackled.

We started the experiment by asking the system to bring a *bottle of water* without specifying the location of this object. For the purpose of this experiment, we defined three objects: a *bottle of water*, a *red box* and a *book*. The system knows the exact location of each object in the map, which were manually defined for the purpose of the experiment. A later stage of the project would be to incorporate an *object detection* module that will allow the system to search and find the object of interest on its own. According to the selection function defined previously (that considers the distance to the object, the robot localization uncertainty, etc.) the system could determine which one of the two robots is best suited to bring the object.

We run two experiments where the task is to bring an object. The results of these experiments are shown in Table 3. The values we assigned for the constants were $a = 0.3$ and $b = 0.7$ since more importance was given to the distance than the uncertainty (covariance) of the localization. In the first experiment, the system was commanded to bring a bottle of water (position predefined). Alice was selected to achieve this task since it had a lower value for the cost function. In the second experiment, the system was asked to get a book and Bobby was found to be more suitable to retrieve it. Although Bobby was closer than Alice, the localization uncertainty made cost function scores very close. However, since we gave a higher weight to the distance, Bobby's cost function still returned a lower value than Alice and therefore Bobby was commanded to do the task.

The voice command directed at the system was interpreted as a navigation goal. During the robots' (in both experiments) traversal through the map, a live camera feed was streamed from the camera to the central system so the user can monitor the robot and intervene when necessary. When the robot arrived at the destination, we tested the system ability to respond to voice commands to perform simple actions

Table 3 Experimental results for task achievement

	Experiment 1	Experiment 2
Parameters	Values	Values
Voice command	*Bring me a bottle of water*	*Bring me a book*
Alice distance	0.6130	2.366
Alice covariance	0.0844	0.04836
Bobby distance	3.280	1.758
Bobby covariance	0.0635	0.13
Alice cost function	0.45442	1.6707
Bobby cost function	2.31505	1.2696
System decision	*Send Alice to bring bottle*	*Send Bobby to bring book*

such as move/turn and gripper actions. We used these voice commands and the live camera feed to guide the robot through the process of picking the object. At this level, the intervention of the user is necessary since the robot has not yet the ability to manipulate the gripper and move the object without any assistance. Finally, we commanded the robot to go back to our position (with the bottle of water and book) through voice commands.

4 Conclusion

In this paper, we described the basic architecture of a multi-robot framework for task achievement. This framework is suitable for an environment where humans and machines closely interact. To make the interactions with the system in a human-like way, we built a voice interface using Alexa Voice Service (AVS) allowing the user to have dynamic interactions with the system in form of multi-turn dialogue. Besides, we described our design and implementation of a robot navigation module that allows the robot to autonomously move through an Occupancy Grid Map (OGM) while avoiding obstacles. The purpose of the proposed framework is to provide a solution for the problem of task achievement involving multiple robots. In this scenario, the system decides how the robots can cooperate to achieve the required task based on some criteria, like the battery status, localization accuracy and the location of the robots. We experiment our system for a task that involves using only one robot selected automatically by the system based on the criteria mentioned above. This framework is the first step for the development of a more sophisticated platform where heterogonous robots would cooperate to achieve a complex task. Thus, our future work will be devoted to extend this work by enhancing the voice interaction and including other kind of machines (e.g. drone) with the goal of setting up a collaborative team for complex task achievement.

References

1. Pyo, Y., Nakashima, K., Kuwahata, S., Kurazume, R., Tsuji, T., Morooka, K.I., Hasegawa, T.: Service robot system with an informationally structured environment. Robot. Auton. Syst. **74**, 148–165 (2015)
2. Yan, Z., Jouandeau, N., Cherif, A.A.: A survey and analysis of multi-robot coordination. Int. J. Adv. Robot. Syst. **10**(12), 399 (2013)
3. Gautam, A., Mohan, S.: A review of research in multi-robot systems. In: 2012 7th IEEE International Conference on Industrial and Information Systems (ICIIS), pp. 1–5. IEEE (2012)
4. Laengle, T., Lueth, T.C.: Decentralized control of distributed intelligent robots and subsystems. Annu. Rev. Autom. Program. **19**, 281–286 (1994)
5. Roumeliotis, S.I., Bekey, G.A.: Distributed multirobot localization. IEEE Trans. Robot. Autom. **18**(5), 781–795 (2002)

6. Martinelli, A., Pont, F., Siegwart, R.: Multi-robot localization using relative observations. In: Proceedings of the 2005 IEEE International Conference on Robotics and Automation, ICRA 2005, pp. 2797–2802. IEEE (2005)
7. Rekleitis, I., Dudek, G., Milios, E.: Multi-robot collaboration for robust exploration. Ann. Math. Artif. Intell. **31**(1–4), 7–40 (2001)
8. Yan, Z., Jouandeau, N., Cherif, A.A.: Multi-robot decentralized exploration using a trade-based approach. In: Proceedings of ICINCO (2), pp. 99–105 (2011)
9. Andre, T., Bettstetter, C.: Collaboration in multi-robot exploration: to meet or not to meet? J. Intell. Robot. Syst. **82**(2), 325 (2016)
10. Jennings, J.S., Whelan, G., Evans, W.F.: Cooperative search and rescue with a team of mobile robots. In: Proceedings of 8th International Conference on Advanced Robotics 1997, ICAR'97, pp. 193–200. IEEE (1997)
11. Dadgar, M., Jafari, S., Hamzeh, A.: A PSO-based multi-robot cooperation method for target searching in unknown environments. Neurocomputing **177**, 62–74 (2016)
12. Alami, R., Fleury, S., Herrb, M., Ingrand, F., Robert, F.: Multi-robot cooperation in the MARTHA project. IEEE Robot. Autom. Mag. **5**(1), 36–47 (1998)
13. Kube, C.R., Bonabeau, E.: Cooperative transport by ants and robots. Robot. Auton. Syst. **30** (1), 85–101 (2000)
14. Farinelli, A., Zanotto, E., Pagello, E.: Advanced approaches for multi-robot coordination in logistic scenarios. Robot. Auton. Syst. **90**, 34–44 (2017)
15. Spiliotopoulos, D., Androutsopoulos, I., Spyropoulos, C.D.: Human-robot interaction based on spoken natural language dialogue. In: Proceedings of the European Workshop on Service and Humanoid Robots, pp. 25–27 (2001)
16. Toptsis, I., Li, S., Wrede, B., Fink, G.A.: A multi-modal dialog system for a mobile robot. In: Eighth International Conference on Spoken Language Processing (2004)
17. Drygajlo, A., Prodanov, P.J., Ramel, G., Meisser, M., Siegwart, R.: On developing a voice-enabled interface for interactive tour-guide robots. Adv. Robot. **17**(7), 599–616 (2003)
18. Imam, A.T., Al-Rousan, T., Odeh, A.: Developing of natural language interface to robot-an Arabic language case study. Comput. Softw. 1256 (2014)
19. Han, J., Jo, M., Park, S., Kim, S.: The educational use of home robots for children. In: IEEE International Workshop on Robot and Human Interactive Communication, 2005, ROMAN 2005, pp. 378–383. IEEE (2005)
20. Quigley, M., Conley, K., Gerkey, B., Faust, J., Foote, T., Leibs, J., Ng, A.Y.: ROS: an open-source robot operating system. In: ICRA Workshop on Open Source Software, vol. 3, no. 3.2, p. 5 (2009)
21. Developer.amazon.com.: Alexa. https://developer.amazon.com/alexa (2017). Accessed 2 Sept 2017
22. Developer.amazon.com.: Alexa voice service. https://developer.amazon.com/alexa-voice-service (2017). Accessed 2 Sept 2017
23. Developer.amazon.com.: Alexa skills kit—build for voice with amazon. https://developer.amazon.com/alexa-skills-kit (2017). Accessed 2 Sept 2017
24. Fernández-Madrigal, J.A. (ed.): Simultaneous Localization and Mapping for Mobile Robots: Introduction and Methods: Introduction and Methods. IGI Global
25. Huang, S., Dissanayake, G.: Convergence and consistency analysis for extended Kalman filter based SLAM. IEEE Trans. Robot. **23**(5), 1036–1049 (2007)
26. Grisetti, G., Stachniss, C., Burgard, W.: Improved techniques for grid mapping with rao-blackwellized particle filters. IEEE Trans. Robot. **23**(1), 34–46 (2007)
27. Grisettiyz, G., Stachniss, C., Burgard, W.: Improving grid-based slam with Rao-Blackwellized particle filters by adaptive proposals and selective resampling. In: Proceedings of the 2005 IEEE International Conference on Robotics and Automation, 2005, ICRA 2005, pp. 2432–2437. IEEE (2005)
28. Thrun, S., Burgard, W., Fox, D.: Probabilistic Robotics. Curriculum Vitae (2005)

29. Fox, D., Burgard, W., Dellaert, F., Thrun, S.: Monte carlo localization: efficient position estimation for mobile robots. In: Proceedings of AAAI/IAAI, vol. 1999, no. 343–349, pp. 2–2 (1999)
30. Wiki.ros.org.: amcl—ROS Wiki. http://wiki.ros.org/amcl (2017). Accessed 9 Sept 2017
31. Li, T., Sun, S., Duan, J.: Monte Carlo localization for mobile robot using adaptive particle merging and splitting technique. In: 2010 IEEE International Conference on Information and Automation (ICIA), pp. 1913–1918. IEEE (2010)
32. Marder-Eppstein, E., Berger, E., Foote, T., Gerkey, B., Konolige, K.: The office marathon: robust navigation in an indoor office environment. In: 2010 IEEE International Conference on Robotics and Automation (ICRA), pp. 300–307. IEEE (2010)
33. Zheng, K.: ROS navigation tuning guide (2017). arXiv:1706.09068

The Control of an Upper Extremity Exoskeleton for Stroke Rehabilitation by Means of a Hybrid Active Force Control

Zahari Taha, Anwar P. P. Abdul Majeed, Muhammad Amirul Abdullah, Kamil Zakwan Mohd Azmi, Muhammad Aizzat Bin Zakaria, Ahmad Shahrizan Abd Ghani, Mohd Hasnun Arif Hassan and Mohd Azraai Mohd Razman

Abstract This paper evaluates the efficacy of a hybrid active force control in performing a joint based trajectory tracking of an upper limb exoskeleton in rehabilitating the elbow joint. The plant of the exoskeleton system is obtained via system identification method whilst the PD gains were tuned heuristically. The estimated inertial parameter that enables the AFC disturbance rejection effect is attained by means of a non-nature based metaheuristic optimisation technique known as simulated Kalman filter (SKF). It was demonstrated that the proposed PDAFC scheme outperformed the classical PD algorithm in tracking the prescribed trajectory in the presence of disturbance attributed by the limb weight.

1 Introduction

It has been recorded that almost 21,000 stroke cases are reported in Malaysia each year, in which 9.52% accounts death whilst the remaining survivors are left with significant disabilities [1]. The survivors would consequently require assistance with their activities of daily living (ADL). A considerable amount of literature has suggested that continuous and repetitive rehabilitation activities allow them to relearn and regain their complete or partial loss of motor control [2–4].

Z. Taha · A. P. P. Abdul Majeed · M. A. Abdullah · K. Z. M. Azmi · M. A. B. Zakaria (✉)
A. S. A. Ghani · M. H. A. Hassan · M. A. M. Razman
Innovative Manufacturing, Mechatronics and Sports Laboratory, Faculty of Manufacturing Engineering, Universiti Malaysia Pahang, 26600 Pekan, Pahang, Malaysia
e-mail: maizzat@ump.edu.my

Z. Taha
e-mail: zaharitaha@imamslab.com; ztrmotion@gmail.com

A. P. P. Abdul Majeed
e-mail: anwarmajeed@imamslab.com

J.-H. Kim et al. (eds.), *Robot Intelligence Technology and Applications 5*,
Advances in Intelligent Systems and Computing 751,
https://doi.org/10.1007/978-3-319-78452-6_30

However, conventional rehabilitation therapy is believed to be costly and labour demanding, hence limiting the patients' rehabilitation activities [4]. The adoption of robotics, conversely, has the ability to mitigate the drawbacks of conventional rehabilitation therapy [3–5]. The utilisation of exoskeletons may progressively phase out the long hours taken for rehabilitation and consultation sessions, subsequently permitting the therapist to cater a larger pool of patients [6].

Joint-based trajectory tracking control is of particular importance, specifically in the early stage of rehabilitation whereby passive mode is non-trivial in order to improve the patient's mobility via continuous and repetitive exercise on the affected limb. Different types of control techniques have been employed for such passive rehabilitation device for the upper limb, namely proportional-derivative (PD) [7], proportional-derivative (PID) [7, 8], nonlinear sliding mode control (SMC) [9], computed torque control (CTC) [7, 10], amongst others. However, the aforementioned control schemes have its inherent problems.

Active Force Control (AFC) has been demonstrated both numerically as well experimentally to be robust in disturbance rejection on different applications [11–19]. It is worth noting that such desirable trait is an essential requirement for an exoskeleton that has to accommodate different patients anthropometric parameters, in particular, the weight of the affected limb. This paper examines the effectiveness of a hybrid PD simulated Kalman filter (SKF) optimised active force control (PDAFC) in performing joint tracking objectives as well as its ability in rejecting constant disturbance (in the form of weight) of an exoskeleton that is aimed at rehabilitating the flexion and extension of the elbow joint.

2 Control Architecture

The notion of AFC was initially conceived by Hewit and Burdess in the early eighties established on the principle of invariance and Newton's second law of motion [20]. The efficacy of this method relies on the appropriate approximation of the inertial parameters of the dynamic system. Various intelligent methods have been applied in estimating the inertial matrix of the dynamic system, namely iterative learning [15, 21], neural networks [22, 23], genetic algorithm [24], fuzzy logic [25] and particle swarm optimisation [2, 19] to name a few.

In the present study, SKF is utilised to acquire the estimated inertial value of the exoskeleton system. SKF is a relatively new non-nature based metaheuristic optimisation algorithm developed by Ibrahim et al. in 2015 based on the estimation capability of the Kalman Filter [26]. The individual agents in SKF are considered to be individual Kalman Filter and based on the mechanism of Kalman filtering, i.e. prediction, measurement and estimation, the global maximum or minimum may be estimated based on an objective function [27]. The actual measurement process that is required in Kalman filtering is mathematically modelled and simulated instead. The agents then communicate amongst themselves to update and improve the solution during the search process. The readers are encouraged to refer to [26] for

Fig. 1 The SKF algorithm

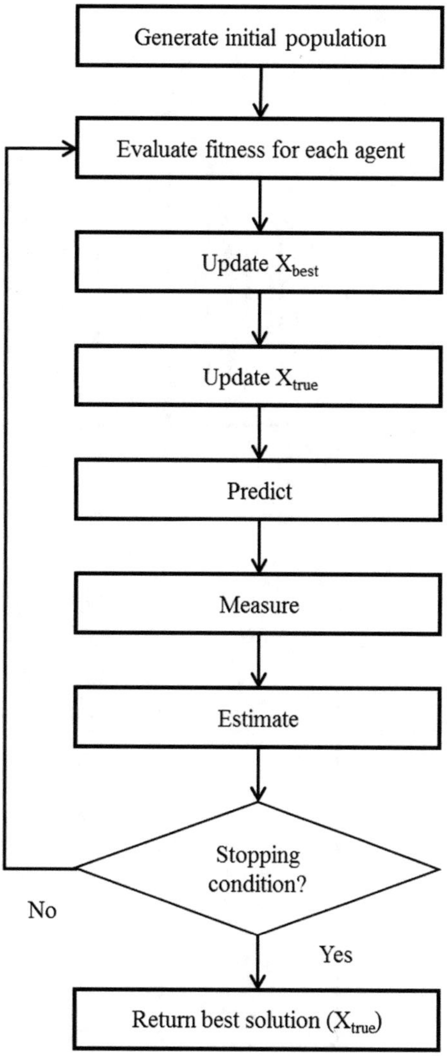

the mathematical treatment of the subject matter. Figure 1 illustrates the flowchart of the aforementioned algorithm.

A MATLAB Simulink diagram of the AFC architecture with the PD controller applied to the exoskeleton is shown in Fig. 2. The PDAFC control scheme is activated upon the initiation of the AFC loop, without its commencement the system is controlled by the classical PD control law. The actual system is depicted in Fig. 3. The estimated disturbance torque τ_{d}^{*} derived from Newton's second law is computed to reject the effect of the actual disturbance, may be written as

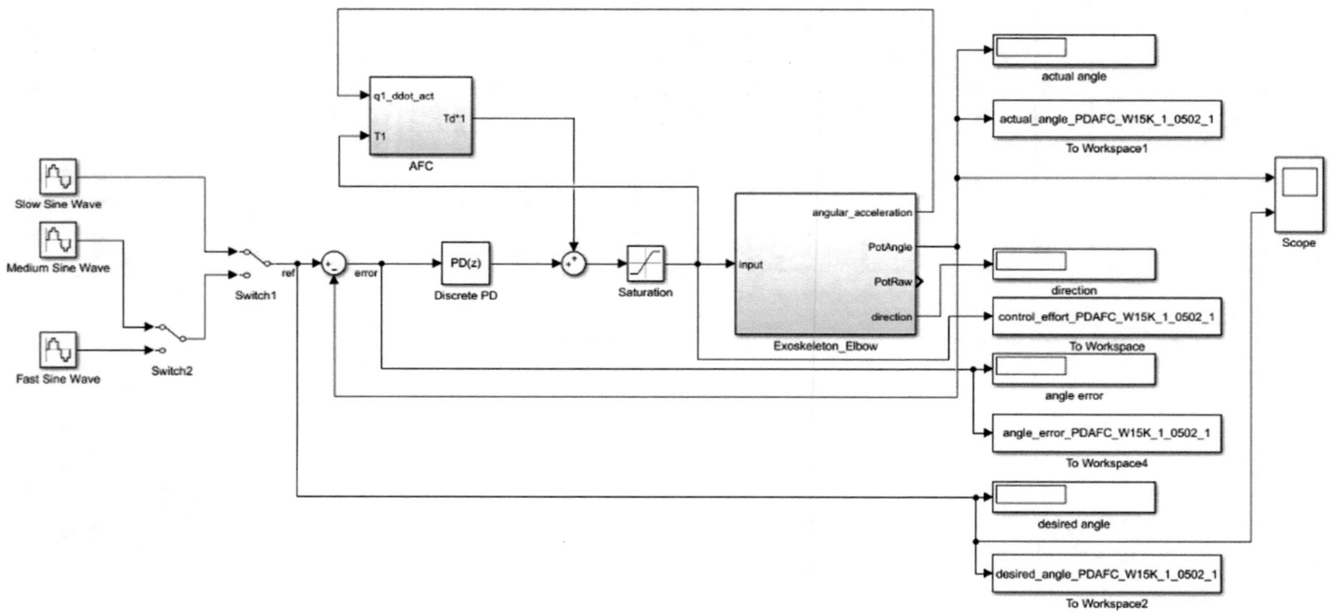

Fig. 2 PDAFC scheme for the control of the upper limb exoskeleton

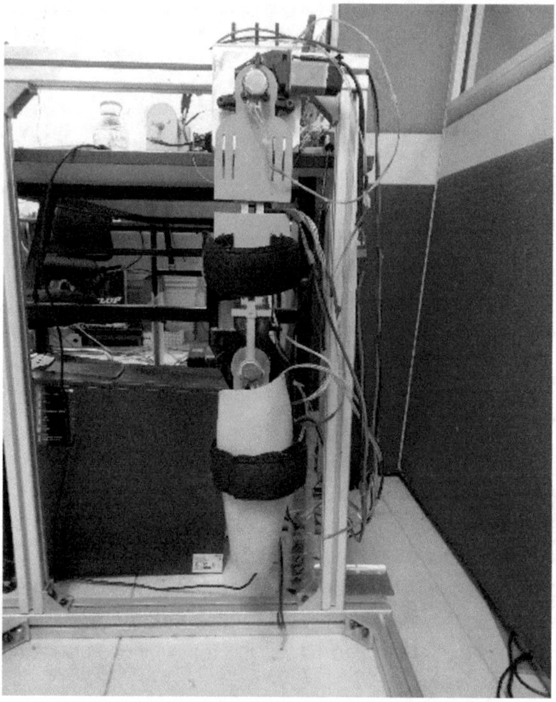

Fig. 3 The upper limb exoskeleton system

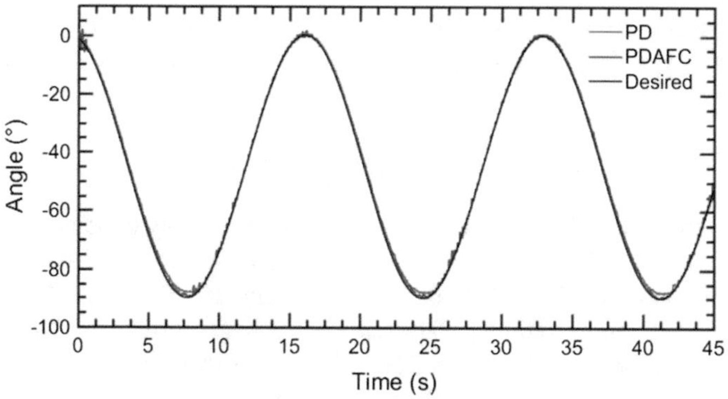

Fig. 4 Elbow trajectory tracking with no disturbance

$$\tau_{d}^{*} = \tau - \mathbf{IN}\dot{\theta} \quad (1)$$

where **IN** is the estimated inertial matrix term, whilst τ and $\dot{\theta}$ are the measured actuation torque and the measured acceleration signal, respectively. In the present

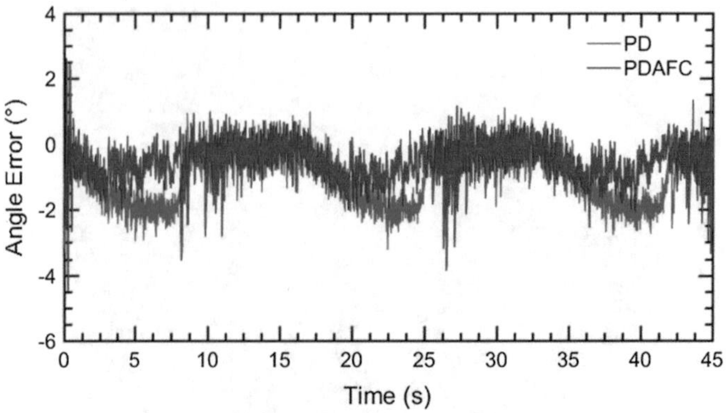

Fig. 5 Elbow tracking error with no disturbance

study, the torque is measured indirectly by measuring the current supplied to the motors via a current sensor (BB-ACS756) and by multiplying it with the motor torque constant, whilst the angular acceleration signal is obtained by numerically computing the data measured from a potentiometer (Honeywell 53C3 20 k). Arduino Mega 2560 was used as the DAQ for the study, and MATLAB Simulink Arduino Support Package was used to realise the proposed controller.

3 Results and Discussion

The plant of the system was obtained via system identification. The details on the attainment of the plant are deliberated in [28]. The plant identified is as follows

$$G(s) = \frac{386.5}{s^3 + 14.06s^2 + 65.01s + 9.182} \tag{2}$$

Furthermore, the PD parameters were tuned appropriately to achieve reasonable trajectory tracking. The PD gains are heuristically tuned from the identified model are 53.4771 and 0.71227 for K_p and K_d, respectively. The **IN** for the AFC loop was obtained via the minimisation of the root mean squared error (RMSE) for a population size of 20 and an iteration of 100. The initial value of the error covariance estimate, P(0), the process noise, Q and the measurement noise, R parameters for the SKF algorithm are taken as 1000, 0.5 and 0.5, respectively [26]. Through the proposed optimisation technique, the appropriate IN value attained is 0.000913289 A s^2.

A sinusoidal signal with an amplitude of 90° and a frequency of 0.375 rad/s for a period of 45 s is supplied to the system to mimic a common rehabilitation exercise for the elbow joint [9]. A forearm mannequin with a mass of 1.5 kg was attached to the exoskeleton to investigate the efficacy of the controllers, i.e. PD and PDAFC in

Table 1 Trajectory tracking performance

Error metrics	No disturbance		Disturbance, 1.5 kg	
	PD	PDAFC	PD	PDAFC
ME	0.9625	0.5753	6.9533	0.8126
RMSE	1.2398	0.8492	11.0446	1.1262
MAE	0.9989	0.6622	6.9633	0.8542

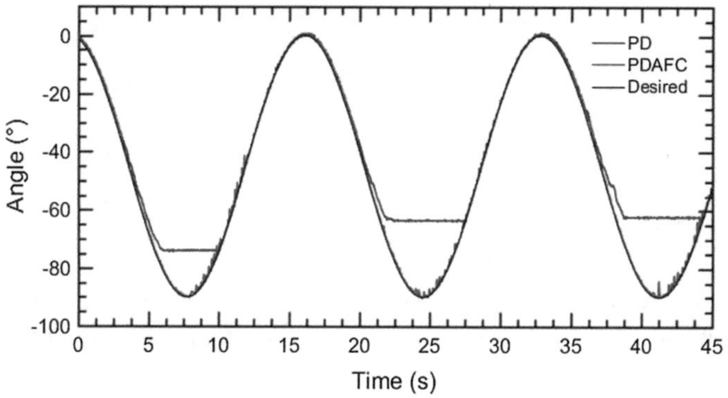

Fig. 6 Elbow trajectory tracking with disturbance

performing the desired trajectory in the presence of constant disturbance (the weight of the forearm). Table 1 lists the error metrics, i.e. mean error (ME), RMSE and mean absolute error (MAE) that are used to evaluate the performance of the aforementioned controllers.

It is evident from the error metrics that the PDAFC scheme is far more superior in comparison to the classical PD control architecture. The effectiveness of the proposed scheme in catering the effect of disturbance, i.e. the weight of the forearm is more pronounced if the weight is 1.5 kg. Furthermore, it could be seen from Fig. 6 that the classical PD scheme is unable to track the desired trajectory, nonetheless, the PDAFC scheme is able to mitigate the gravitational effect of the forearm. It is worth to note that the erratic spikes are primarily due to the limitations of the physical structure of the exoskeleton prototype. A similar observation was also made by Jahanabadi et al. concerning this issue [25]. Nevertheless, this shortcoming could not negate the fact on the ability of the proposed control architecture in catering different weights whilst providing reasonably accurate trajectory tracking as demonstrated in both Figs. 4 and 6, respectively (Figs. 5 and 7).

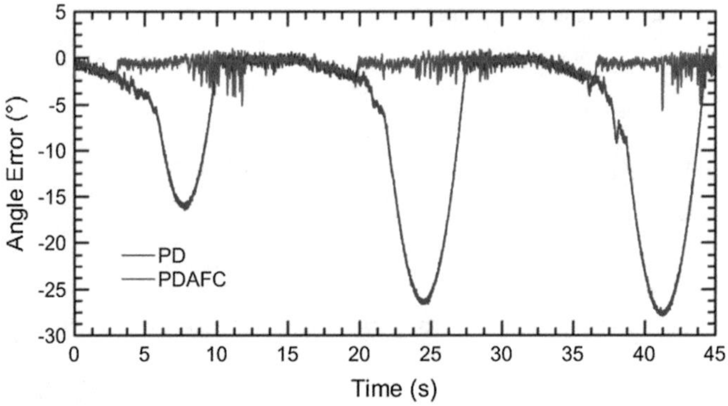

Fig. 7 Elbow tracking error with disturbance

4 Conclusion

It could be concluded from the present investigation that the proposed AFC-based control scheme quite robust in the wake of disturbance arising from the limb weight. The simplicity of the proposed control algorithm further for its ease of implementation. Future studies would include the different types of rehabilitation regimes to improve the motor skills of the impaired joint as well as the ability of the proposed controller in accomplishing it.

Acknowledgements The authors would like to thank Universiti Malaysia Pahang for providing the grant (RDU160106) in supporting this research.

References

1. Nazifah, S.N., Azmi, I.K., Hamidon, B.B., Looi, I., Zariah, A.A., Hanip, M.R.: National stroke registry (NSR): Terengganu and Seberang Jaya experience. Med. J. Malays. **67**, 302–304 (2012)
2. Abdul Majeed, A.P.P., Taha, Z., Mohd Khairuddin, I., Wong, M.Y., Abdullah, M.A., Mohd Razman, M.A.: The control of an upper-limb exoskeleton by means of a particle swarm optimized active force control for motor recovery. In: Ibrahim, F., Cheong, J.P.G., Usman, J., Ahmad, M.Y., Razman, R., Selvanayagam, V.S. (eds.) In: Proceedings of IFMBE, pp. 56–62. Springer Singapore, Singapore (2017)
3. Loureiro, R.C.V., Harwin, W.S., Nagai, K., Johnson, M.: Advances in upper limb stroke rehabilitation: a technology push. Med. Biol. Eng. Comput. **49**, 1103–1118 (2011)
4. Volpe, B.T., Ferraro, M., Krebs, H.I., Hogan, N.: Robotics in the rehabilitation treatment of patients with stroke. Curr. Atherosclerosis Rep. **4**, 270–276 (2002)
5. Lo, H.S., Xie, S.Q.: Exoskeleton robots for upper-limb rehabilitation: state of the art and future prospects. Med. Eng. Phys. **34**, 261–268 (2012)

6. Majeed, A.P.P.A., Taha, Z., Abidin, A.F.Z., Zakaria, M.A., Khairuddina, I.M., Razman, M.A. M., Mohamed, Z.: The control of a lower limb exoskeleton for gait rehabilitation: a hybrid active force control approach. Procedia Comput. Sci. (2017)
7. Nef, T., Mihelj, M., Riener, R.: ARMin: a robot for patient-cooperative arm therapy. Med. Biol. Eng. Comput. **45**, 887–900 (2007)
8. Garrido, J., Yu, W., Li, X.: Modular design and control of an upper limb exoskeleton. J. Mech. Sci. Technol. **30**, 2265–2271 (2016)
9. Rahman, M.H., Saad, M., Kenné, J.P., Archambault, P.S.: Nonlinear sliding mode control implementation of an upper limb exoskeleton robot to provide passive rehabilitation therapy. In: Proceedings of Lecture Notes in Computer Science (including subseries Lecture Notes in Artificial Intelligence and Lecture Notes in Bioinformatics), pp. 52–62. Springer, Berlin, Heidelberg (2012)
10. Rahman, M.H., Rahman, M.J., Cristobal, O.L., Saad, M., Kenné, J.P., Archambault, P.S.: Development of a whole arm wearable robotic exoskeleton for rehabilitation and to assist upper limb movements. Robotica **33**, 19–39 (2015)
11. Tavakolpour Saleh, A.R., Mailah, M.: Control of resonance phenomenon in flexible structures via active support. J. Sound Vib. **331**, 3451–3465 (2012)
12. Hashemi-Dehkordi, S.M., Abu-Bakar, A.R., Mailah, M.: Stability analysis of a linear friction-induced vibration model and its prevention using active force control. Adv. Mech. Eng. **6**, 251594 (2014)
13. Ismail, Z., Varatharajoo, R.: Satellite cascade attitude control via fuzzy PD controller with active force control under momentum dumping. In: IOP Conference Series: Materials Science and Engineering, vol. 152, p. 12030 (2016)
14. Mailah, M., Jahanabadi, H., Zain, M.Z.M., Priyandoko, G.: Modelling and control of a human-like arm incorporating muscle models. In: Proceedings of the Institution of Mechanical Engineers, Part C: Journal of Mechanical Engineering Science, vol. 223, pp. 1569–1577 (2009)
15. Kwek, L.C., Wong, E.K., Loo, C.K., Rao, M.V.C.: Application of active force control and iterative learning in a 5-link biped robot. J. Intell. Robot. Syst. **37**, 143–162 (2003)
16. Varatharajoo, R., Wooi, C.T., Mailah, M.: Two degree-of-freedom spacecraft attitude controller. Adv. Space Res. **47**, 685–689 (2011)
17. Tahmasebi, M., Rahman, R., Mailah, M., Gohari, M.: Roll movement control of a spray boom structure using active force control with artificial neural network strategy. J. Low Freq. Noise Vib. Control. Act. **32**, 189–202 (2013)
18. Noshadi, A., Mailah, M., Zolfagharian, A.: Intelligent active force control of a 3-RRR parallel manipulator incorporating fuzzy resolved acceleration control. Appl. Math. Model. **36**, 2370–2383 (2012)
19. Taha, Z., P.P. Abdul Majeed, A., Zainal Abidin, A.F., Hashem Ali, M.A., Mohd Khairuddin, I., Deboucha, A., Wong Paul Tze, M.Y.: A hybrid active force control of a lower limb exoskeleton for gait rehabilitation (2017). https://www.degruyter.com/view/j/bmte.ahead-of-print/bmt-2016-0039/bmt-2016-0039.xml
20. Hewit, J.R., Burdess, J.S.: Fast dynamic decoupled control for robotics, using active force control. Mech. Mach. Theory **16**, 535–542 (1981)
21. Mailah, M., Hooi, H.M., Kazi, S., Jahanabadi, H.: Practical active force control with iterative learning scheme applied to a pneumatic artificial muscle actuated robotic arm. Int. J. Mech. **6**, 88–96 (2012)
22. Tahmasebi, M., Rahman, R.A., Mailah, M., Gohari, M.: Roll movement control of a spray boom structure using active force control with artificial neural network strategy. J. Low Freq. Noise Vib. Act Control. **32**, 189–201 (2013)
23. Tahmasebi, M., Rahman, R.A., Mailah, M., Gohari, M.: Sprayer boom active suspension using intelligent active force control. Int. J. Mech. Aerosp. Ind. Mechatron Manuf. Eng. **6**, 1277–1281 (2012)
24. Mailah, M., Yee, W.M., Jamaluddin, H.: Intelligent active force control of a robotic arm using genetic algorithm. J. Mek. (2002)

25. Jahanabadi, H., Mailah, M., Zain, M.Z.M., Hooi, H.M.: Active force with fuzzy logic control of a two-link arm driven by pneumatic artificial muscles. J. Bionic Eng. **8**, 474–484 (2011)
26. Ibrahim, Z., Aziz, N.H.A., Aziz, N.A.A., Razali, S., Shapiai, M.I., Nawawi, S.W., Mohamad, M.S.: A Kalman filter approach for solving unimodal optimization problems. ICIC Express Lett. **9**, 3415–3422 (2015)
27. Muhammad, B., Ibrahim, Z., Ghazali, K.H., Mohd Azmi, K.Z., Ab Aziz, N.A., Abd Aziz, N. H., Mohamad, M.S.: A new hybrid simulated Kalman filter and particle swarm optimization for continuous numerical optimization problems. ARPN J. Eng. Appl. Sci. **10**, 17171–17176 (2015)
28. Taha, Z., Majeed, A.P.P.A., Abdullah, M.A., Khairuddin, I.M., Zakaria, M.A., Hassan, M.H. A.: The identification and control of an upper extremity exoskeleton for motor recovery. In: BinAbdollah, M.F., Tuan, T.B., Salim, M.A., Akop, M.Z., Ismail, R., Musa, H. (eds.) Proceedings of Mechanical Engineering Research Day 2017 (MERD), pp. 483–484. Centre Advanced Research Energy-CARE, Fac Mechanical Engineering, Univ Teknikal Malayasia Melaka, Hang Tuah Jaya, Durian Tunggal, Mekala 76100, Malaysia (2017)

Intellectual Multi Agent Control of Tripod

Anatoliy Gaiduk, Sergey Kapustyan, Alexander Dyachenko, Ryhor Prakapovich and Ivan Podmazov

Abstract The analytical design method of the intellectual multi agent control for the mechanism of parallel structure of type the manipulator-tripod is considered in this paper. Tripod is intended for maintenance of the spatial technological movements of the a gripper with a working tool. A tripod's construction is simple, but if the tripod's control is intellectual then the big variety of technological operations can be realized. The tripod's control system includes three agents. The algorithms of these agents are designed here. Application of the multi agent system simplifies the algorithm of the tripod's control, in particular, with using of the intellectual approach.

Keywords Multi agent control systems · Manipulator-tripod · Spatial movements · Intellectual algorithms of agents

---

A. Gaiduk
Department of Automatic Control Systems, Southern Federal University, Taganrog, Russia
e-mail: gaiduk_2003@mail.ru

S. Kapustyan (✉)
Southern Scientific Center of the Russian Academy of Sciences, Rostov-on-Don, Russia
e-mail: kap56@mail.ru

A. Dyachenko
Scientific Research Institute of Multiprocessor Computer Systems, Southern Federal University, Taganrog, Russia
e-mail: aleksandernet@yandex.ru

R. Prakapovich · I. Podmazov
United Institute of Informatics Problems NAS of Belarus, Minsk, Belarus
e-mail: rprakapovich@robotics.by

I. Podmazov
e-mail: podmazov@gmail.com

J.-H. Kim et al. (eds.), *Robot Intelligence Technology and Applications 5*,
Advances in Intelligent Systems and Computing 751,
https://doi.org/10.1007/978-3-319-78452-6_31

1 Introduction

Mechanisms of parallel structure find more and more wide application in various units and devices, in particular, in mechatronic devices. They have a line of useful properties (rather small weight, increased carrying capacity, rigidity and the high accuracy of positioning) which provide wide application of similar mechanisms in technical equipment. As a rule, the mechanisms of parallel structure are intended for realization of the spatial forward or rotary movements of the robot grippers, in which a working tool is placed. The movements of such types are necessary at assembly, welding, painting, measurements and many other technological operations which are carried out on robotized manufactures [1–6].

The distinctive feature of the mechanisms of parallel structures (MPS) is a complex mathematic model. This complexity is caused by presence in MPS of several parallel kinematic chains and essentially complicates the solution of the control problem of the MPS movements. On the other hand, the coordinated control of MPS provides technical realization of complex spatial gripper movements [4–7].

The design problem of MPS control systems was considered in [7–10] and in others. More often, the control systems of MPS are created on a basis of the optimal control theory. The optimal control on speed [6] and the control in the meaning of a minimal square-law function are used [9–11]. Mathematical models of MPS are dynamic and described by the systems of nonlinear differential equations of rather high orders [12–14].

The development of control systems of complex dynamic objects, working in the uncertainty conditions and the intensive external disturbances, demands the using of no conventional approaches. New methods of the knowledge processing, new types of feedback, modern intellectual, information and telecommunication technologies and neural networks are no conventional. These and a set of other methods are united under the common name “artificial intellect”. One of the most perspective directions is use multi agent systems. These systems allow decoupling a control task on several subtasks, and intellectual agents carry out their solution in parallel [15, 16].

In this paper the analytical design problem of the intellectual, multi agent control systems for the mechanism of parallel structure like a tripod is considered. The control system is realized as a set of the several agents, each of which operates the corresponding executive device of the tripod [15]. The primary goal of the paper is development of intellectual agent algorithms. Realization of these algorithms is supposed on a basis of self-organizing control systems or neural networks [6, 16].

2 Statement of the Problem

The construction of considered tripod is shown in Fig. 1. Its image is taken from [10]. The manipulator-tripod consists of the basis 5, representing a rigid isosceles triangle. Three links 1, 2, 3 have variable lengths and are connected with the basis 5 by the pivot-type coupling. The ends of these links are fastened in the point 6 by the pivot-type coupling which has five degrees of freedom. The gripper with the working tool is fastened to this pivot-type coupling. If these links change the length, gripper will makes necessary technological movements at the working zone, which is determined by the minimal and maximal lengths of these links. The working zone can be increased by the changing of the inclination corner of the basis 5 at the expense of change of the link 4 length. All the links of variable length have the electric direct current motors. The motors voltages of the links 1, 2 and 3 are controls of the considered manipulator. The length of the link 4 can be change if it is necessary, but in this case it is fixed.

The kinematics circuit of the tripod and the accepted system of coordinates $Oxyz$ are shown in Fig. 2 [10, 14]. The beginning of the coordinate system is the point in the middle of the section BC of the triangular basis ABC. The axis Oz is directed vertically upwards, the axis Ox—to the right, and the axis Oy is parallel to longitudinal axis of the manipulator and the robot. The basis ABC turns around of the section BC by change of φ, that is the inclination corner, at the expense of change l_4, which is length of the link 4.

The point M makes movement in the working space of the tripod at the expense of change of the lengths l_1, l_2, l_3 of the controlled links 1, 2, 3. Lengths l_1, l_2, l_3 are related to the coordinates of this point $x_M = x_1(t)$, $y_M = x_3(t)$, и $z_M = x_5(t)$, by the following expressions:

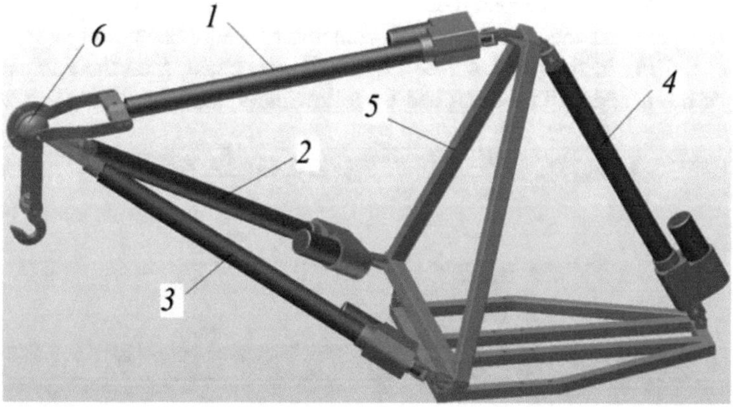

Fig. 1 Manipulator-tripod

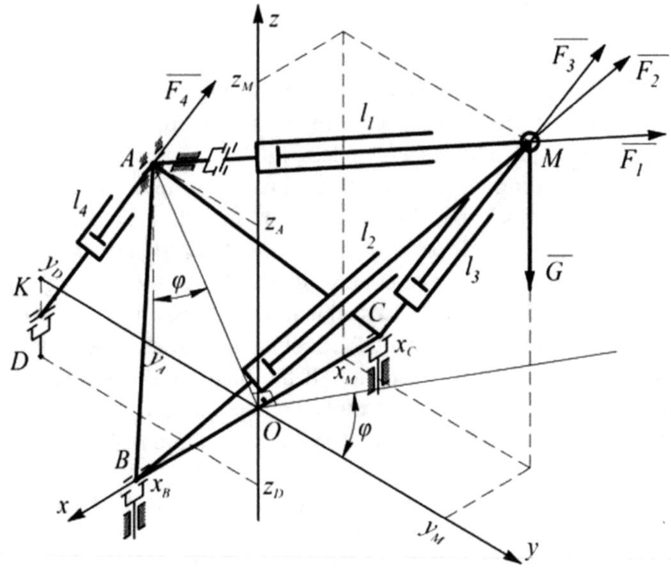

Fig. 2 Kinematic circuit of the manipulator-tripod

$$l_1 = \sqrt{x_1^2 + (x_3 + a)^2 + (x_5 - c)^2},\ l_2 = \sqrt{(x_1 - b)^2 + x_3^2 + x_5^2},$$
$$l_3 = \sqrt{(x_1 + b)^2 + x_3^2 + x_5^2}, \tag{1}$$

where $a = OA \sin \varphi$, $b = OB$, $c = OA \cos \varphi$; OA, OB are geometrical parameters of the manipulator basis 5 (see Fig. 1).

The tripod's mathematical model, which is used for the solution of the problem, has been received in [10, 14] on the basis of the Lagrange equations with uncertain multipliers. To obtain a simpler model, the real mechanism has been replaced by the dynamic equivalent mechanism, which contains two equivalent weights m in the point M and m_A in the point A (see Fig. 2). If φ = const, then the mathematical model of the tripod will be described by following equations:

$$\dot{x}_1 = x_2,\ \dot{x}_2 = \frac{F_1}{m\,l_1} x_1 + \frac{F_2}{m\,l_2}(x_1 - x_B) + \frac{F_3}{m\,l_3}(x_1 + x_B), \tag{2}$$

$$\dot{x}_3 = x_4,\ \dot{x}_4 = \frac{F_1}{m\,l_1}(x_3 + d_3) + \frac{F_2}{m\,l_2} x_3 + \frac{F_3}{m\,l_3} x_3, \tag{3}$$

$$\dot{x}_5 = x_6,\ \dot{x}_6 = \frac{F_1}{m\,l_1}(x_5 - d_5) + \frac{F_2}{m\,l_2} x_5 + \frac{F_3}{m\,l_3} x_5 - g, \tag{4}$$

$$y = [x_1 \quad x_3 \quad x_5]^T = [x_M \quad y_M \quad z_M]^T. \tag{5}$$

Here $x = [x_1 \quad x_2 \quad x_3 \quad x_4 \quad x_5 \quad x_6]^T$ is a state vector; $F = F(t) = [F_1(t)\ F_2(t)\ F_3(t)]^T$ is the vector of the control actions which are created by the corresponding electro motors of the tripod; g is the free fall acceleration.

Let's emphasize that in (1)–(5) the variables $x_1 = x_M$, $x_3 = y_M$, $x_5 = z_M$ are the coordinates of the point M, in which the gripper is fastened, and the variables $x_2,\ x_4,\ x_6$ are the projections of the speed of this point on the coordinate axes Ox, Oy, Oz [10]. The problem consists in defining the control laws of the actions $F_i = F_i(t)$, $i = 1, 2, 3$ in view of the restrictions on the allowable changes of the controlled links lengths:

$$l_{\min,i} \le l_i \le l_{\max,i},\ i = \overline{1,\ 3}. \tag{6}$$

3 Decupling of the Model Tripod

To construct the multi agent control system, first of all, the tripod model (2)–(6) is decupled on the three separate problems of the control systems design. With this purpose, the state variables are designated as follows:

$$x_1 = \tilde{x}_1,\ \ x_2 = \tilde{x}_4,\ \ x_3 = \tilde{x}_2,\ \ x_4 = \tilde{x}_5,\ \ x_5 = \tilde{x}_3,\ \ x_6 = \tilde{x}_6. \tag{7}$$

In the vector-matrix form and in the new designations (2)–(5) become

$$\dot{\tilde{x}}_I = E\tilde{x}_{II},\quad \dot{\tilde{x}}_{II} = B(\tilde{x})\,F + e_3\,g, \tag{8}$$

where $\tilde{x}_I = [\tilde{x}_1 \quad \tilde{x}_2 \quad \tilde{x}_3]^T = [x_1 \quad x_3 \quad x_5]^T$; $\tilde{x}_{II} = [\tilde{x}_4 \quad \tilde{x}_5 \quad \tilde{x}_6]^T = [x_2 \quad x_4 \quad x_6]^T$; E is an 3×3 identity matrix, e_3 is the 3-rd column of matrix E; the vector $\tilde{x}_I$ is the vector of the point M positions (see Fig. 2), and $\tilde{x}_{II}$ is the vector of its speed projections on the coordinate axis. The matrix

$$B(\tilde{x}) = \begin{bmatrix} \frac{\tilde{x}_1}{m\,l_1} & \frac{\tilde{x}_1 - x_B}{m\,l_2} & \frac{\tilde{x}_1 + x_B}{m\,l_3} \\ \frac{\tilde{x}_2 + d_3}{m\,l_1} & \frac{\tilde{x}_2}{m\,l_2} & \frac{\tilde{x}_2}{m\,l_3} \\ \frac{\tilde{x}_3 - d_5}{m\,l_1} & \frac{\tilde{x}_3}{m\,l_2} & \frac{\tilde{x}_3}{m\,l_3} \end{bmatrix}. \tag{9}$$

The system (8) can be decupled only when the determinant of the matrix $B(\tilde{x})$ (9) is not equal to zero. Calculating this determinant, we shall receive

$$\det B(\tilde{x}) = \frac{2x_B(d_3\tilde{x}_3 + d_5\tilde{x}_2)}{m^3 l_1\, l_2\, l_3} = D_B(x). \tag{10}$$

Let's suppose further, that a condition

$$d_3\tilde{x}_3 + d_5\tilde{x}_2 = d_3x_5 + d_5x_3 = OA(z_M \sin\varphi + y_M \cos\varphi) \neq 0 \tag{11}$$

is carried out in all points of the tripod working space. Therefore, the matrix $B^{-1}(\tilde{x})$ exists, and the equations system (8) is decupled by the control

$$F = B^{-1}(\tilde{x})\, v, \tag{12}$$

where $F = [F_1, F_2, F_3,]^T$; $v = [v_1 \quad v_2 \quad \tilde{v}_3]^T$ is the new control vector. We should note that the matrix $B^{-1}(\tilde{x})$ exists by virtue of the condition (11). Taking into account the control (12) the equations system (8) becomes

$$\dot{\tilde{x}}_I = E\tilde{x}_{II}, \quad \dot{\tilde{x}}_{II} = B(\tilde{x})\,B^{-1}(\tilde{x})\,v + e_3 g. \tag{13}$$

It is well-known that, $B \times B^{-1} = E$ [17, 18], therefore from (13) new system follows

$$\dot{\tilde{x}}_I = E\tilde{x}_{II}, \quad \dot{\tilde{x}}_{II} = Ev + e_3 g. \tag{14}$$

Taking into account the control vector $v = [v_1 \quad v_2 \quad \tilde{v}_3]^T$ the equations system (14) concerning variables $\tilde{x}_i, \quad i = \overrightarrow{1,\ 6}$ looks like:

$$\dot{\tilde{x}}_1 = \tilde{x}_4, \quad \dot{\tilde{x}}_4 = v_1, \quad \dot{\tilde{x}}_2 = \tilde{x}_5, \quad \dot{\tilde{x}}_5 = v_2 \quad \dot{\tilde{x}}_3 = \tilde{x}_6, \quad \dot{\tilde{x}}_6 = \tilde{v}_3 - g. \tag{15}$$

Taking into account the designations (7) the equations system (15) concerning a vector $x = x(t)$ is equivalent to the next system

$$\dot{x}_1 = x_2, \quad \dot{x}_2 = v_1, \quad \dot{x}_3 = x_4, \quad \dot{x}_4 = v_2, \quad \dot{x}_5 = x_6, \quad \dot{x}_6 = \tilde{v}_3 - g. \tag{16}$$

The state variables in (16) are replaced as follows

$$w_1 = [x_1 \quad x_2]^T, \quad w_2 = [x_3 \quad x_4]^T \quad w_3 = [x_5 \quad x_6]^T. \tag{17}$$

The equations system (16) takes vector-matrix forms at the designations (17)

$$\begin{gathered}\dot{w}_1 = \begin{bmatrix} 0 & 1 \\ 0 & 0 \end{bmatrix} w_1 + \begin{bmatrix} 0 \\ 1 \end{bmatrix} v_1, \quad \dot{w}_2 = \begin{bmatrix} 0 & 1 \\ 0 & 0 \end{bmatrix} w_2 + \begin{bmatrix} 0 \\ 1 \end{bmatrix} v_2, \\ \dot{w}_3 = \begin{bmatrix} 0 & 1 \\ 0 & 0 \end{bmatrix} w_3 + \begin{bmatrix} 0 \\ 1 \end{bmatrix} \tilde{v}_3 - \begin{bmatrix} 0 \\ 1 \end{bmatrix} g. \end{gathered} \tag{18}$$

Obviously, the third system in (18) differs from the first two. The structure of the third system in (18) will be similar to the structure of the first two if the control v_3 look like

$$\tilde{v}_3 = v_3 + g. \tag{19}$$

Taking into account the expression (19) all three Eqs. (18) can be written down in such a way:

$$\dot{w}_i = \begin{bmatrix} 0 & 1 \\ 0 & 0 \end{bmatrix} w_i + \begin{bmatrix} 0 \\ 1 \end{bmatrix} v_i, \quad i = 1, \ 2, \ 3. \tag{20}$$

According to the Eqs. (18) and (20) the problem of tripod control (see Fig. 1) can be solved by use of three almost similar agents. Each of these agents solves the control task of one system from (20). Formally, these three systems and their solution, do not depend on each other, however their realization can be carried out only taking into account the expressions (8)–(10) and the restrictions (6).

4 Solution of the Design Task

This task consists in definition of the control laws $v_i = v_i(t)$, $i = 1, 2, 3$, i.e. in definition of the algorithms determining actions of the agents for control of the tripod. Let this task consists in moving the point M (see Fig. 2) from an initial position $\tilde{x}_{I0} = [x_{M0} \quad y_{M0} \quad z_{M0}]$, $\tilde{x}_{II0} = 0$ to a final position $\tilde{x}_{I1} = [x_{M1} \quad y_{M1} \quad z_{M1}]$, $\tilde{x}_{II1} = 0$ so that some square-law criterion of quality has a minimal value, and (6) is carried out.

For the task solution, we shall proceed to equations in deviations. Let $z_i(t) = w_i(t) - w_{i0}$ are the deviations of the state vectors, and $u_i(t) = v_i(t) - v_{i0}$ are the control deviations, $i = 1, 2, 3$. The systems (20) are linear, therefore the systems of equations in deviations coincide at form with (20), i.e.

$$\dot{z}_i = \begin{bmatrix} 0 & 1 \\ 0 & 0 \end{bmatrix} z_i + \begin{bmatrix} 0 \\ 1 \end{bmatrix} u_i, \quad i = 1, \ 2, \ 3, \tag{21}$$

where $z_{i0} = w_i(0) - w_{i0} = 0$, $u_{i0} = 0$.

The control task of (21) consist in moving these systems from the initial positions $z_{i0} = [w_{i10} \quad 0]^T \neq 0$ to the final positions $z_{i1} = [z_{i11} \quad 0]^T = 0$ so that the next conditions

$$J_i = \int_0^\infty [z_i^T Q_i z_i + r_i u_i^2] dt \to \min_{u_i}, \quad Q_i = \begin{bmatrix} q_{i1} & 0 \\ 0 & q_{i2} \end{bmatrix}, \quad i = 1, \ 2, \ 3 \tag{22}$$

and the restriction (6) are satisfied.

It is known [11, 19], that optimal control u_i, at which (22) is satisfied, determined by the expression

$$u_i^\circ = \gamma_i^\circ g_i - k_{i1}^\circ z_{i1} - k_{i2}^\circ z_{i2} = \gamma_i^\circ g_i - k_i^{\circ T} z_i, \tag{23}$$

where g_i are the desired values of the variables w_{i1}, $i = 1, 2, 3$. The values of the factors k_i° of the vector k° from (23) are determined by the solution of the corresponding equation Riccati [18, 19], and the factor $\gamma_i = k_{i1}$. The values $\gamma_i = k_{i1}$ are chosen from the condition: $C_{i0} = 0$, where C_{i0} is the factor of mistake, caused by the input signal g_i of the system (21), (23) [11].

Using the results of [18], it is easy to show that in case of (21) the optimal vectors k_i° are determined by the expressions

$$k_i^{\circ T} = \left[\sqrt{q_{i1}} \quad \sqrt{q_{i2} + 2\sqrt{r_i q_{i1}}}\right] / \sqrt{r_i}, \quad i = 1,\ 2,\ 3. \tag{24}$$

The received expressions (24) with the given criterion (22) uniquely determine values of the factors k_{i1}°, k_{i2}° of the optimal control (23). However, these controls can be realized, only if the values $g_i = w_{i1}$ are known. These values are equal to the coordinates of the final position of the point M. The restriction (6) also should be taken into account at the definition of the vectors k_i°, $i = 1, 2, 3$.

5 Control Algorithm of Tripod

The algorithm of the optimal control of the tripod (see Fig. 1) ensue from the expressions resulted above.

For moving the point M of tripod from position M_0 to final position M_1 the initial dates include the following parameters: OA, OB, $l_{i\min}$, $l_{i\max}$, m, x_B, φ, g, q_{11}, q_{12}, r_1, q_{21}, q_{22}, q_{23}, r_2, q_{31}, q_{32}, q_{33}, r_3, x_{M0}, y_{M0}, z_{M0}, x_{M1}, y_{M1}, z_{M1}, v_{i0}, $i = 1, 2, 3$. Here v_{i0} are the values of the controls ν_i, by which the tripod point M has been moved to the point M_0. The considered algorithm includes the following steps.

Step 1. Let's accept: $g_1 = x_{M1} - x_{M0}$, $g_2 = y_{M1} - y_{M0}$, $g_3 = z_{M1} - z_{M0}$, $w_{110} = x_{M0}$, $w_{210} = y_{M0}$; $w_{310} = z_{M0}$, and $z_{i0} = [z_{i10} \quad z_{i20}] = [0 \quad 0]$. The optimal controls $u_i^\circ(t)$ are calculated by (24) and (23). The variables $z_i(t)$ are determined as the results of integrating the systems (21) with the controls $u_i^\circ(t)$ $i = 1, 2, 3$ on intervals from $z_{i0} = 0$ up to $g_{i,ycm}$(actually, from the point M_0 up to the point M_1).

Step 2. The vectors $w_i(t)$ and the controls $v_i(t)$ are calculated under the formulas: $w_i(t) = z_i(t) + w_{i0}$, and $v_i(t) = u_i(t) + v_{i0}$, $i = 1, 2, 3$. Then the variables $x_i(t)$ are found under the formulas: $[x_1 \quad x_2]^T = [w_{11} \quad w_{12}]^T$, $[x_3 \quad x_4]^T = [w_{21} \quad x_{22}]^T$, $[x_5 \quad x_6]^T = [w_{31} \quad w_{32}]^T$, and the variables $\tilde{x}_i(t)$ are found by (7).

The control $\tilde{v}_3(t)$ is calculated by (19) and the vector $v(x)$ is made as $v(x) = [v_1(x) \quad v_2(x) \quad \tilde{v}_3(x)]$.

Step 3. The satisfaction of the condition (6) is checked. If this condition is not carried out the point M_0 or the point M_1 (or both) is outside of the tripod working space. In this case, the coordinates of the point being outside of the working space, change, and go to the step 1. If the condition (6) is satisfied, the transition to the step 4 is carried out.

Step 4. The matrix $B(\tilde{x})$ and the vector of the controlling forces $F = F(x)$ are calculated according to (9), (10) and (12), and the lengths $l_1^\circ(x)$, $l_2^\circ(x)$, $l_1^\circ(x)$ are calculated by (1).

Step 5. The variables $x_i = x_i(t)$, $i = \overline{1,\ 6}$ are the functions of time, therefore the found functions $l_1^\circ(x)$, $l_2^\circ(x)$, $l_1^\circ(x)$ are the optimal control laws of the lengths of the links 1, 2, and 3. The agents should realize these laws with the purpose of moving the point M from the initial position M_0 to the final position M_1 of the tripod. Therefore, the agents send these functions, and also their initial values, on the inputs of the corresponding control systems links on the given step.

In the steady-state mode, the condition: $x_{1\,ycm} = x_{M\,1}$, $x_{3\,ycm} = y_{M\,1}$, $x_{5\,ycm} = z_{M\,1}$ are carried out, i.e. the tripod point M is in the necessary position M_1.

The control systems of tripod links can be as usual control systems or as the self-organizing systems, or the intellectual systems realized on the basis of neural networks.

Simulation. All points of the gripper's trajectory should satisfy to conditions (11) and (6). Therefore the algorithm of the tripod's control is carried out by steps, and all calculations are carried out by a computer.

Efficiency of the suggested multi agent control system of tripod was estimated by computer simulation. The gripper has been moved from the point $\boldsymbol{O}_0$ with coordinates x_0 = 5 cm, y_0 = 10 cm, z_0 = 5 cm to the point $\boldsymbol{O}_1$ with coordinates x_1 = 20 cm, y_1 = 22 cm, z_1 = 25 cm on the linear trajectory, and from the point $\boldsymbol{O}_0$ with coordinates x_0 = 5 cm, y_0 = 10 cm, z_0 = 0 cm to the point $\boldsymbol{O}_1$ with coordinates x_1 = 20 cm, y_1 = 22 cm, z_1 = 30 cm—on the curvilinear trajectory. Only the schedules of the variables $x(t)$ and $z(t)$ are shown on Fig. 3 for brevity.

The schedules of the variables $x(t)$ and $y(t)$ have a similar kind. Corresponding trajectories of the gripper's movement are shown on Fig. 4 in the first and in the second cases.

The resulted schedules testify that the gripper makes the desired spatial movements precisely enough with use of the suggested multi agent control system of the tripod.

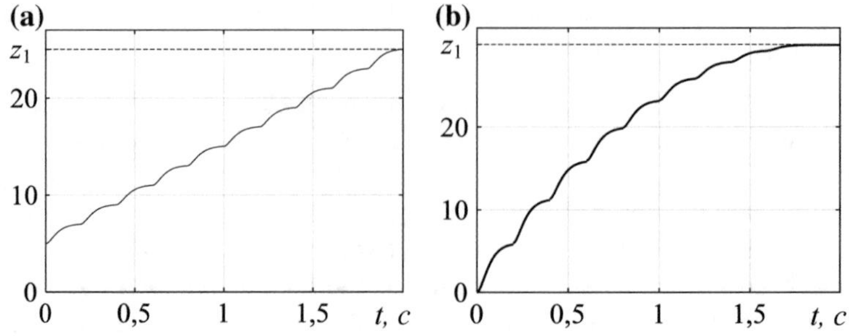

Fig. 3 Changes of the variable z (t) of the tripod's gripper: **a** linear trajectory; **b** curvilinear trajectory

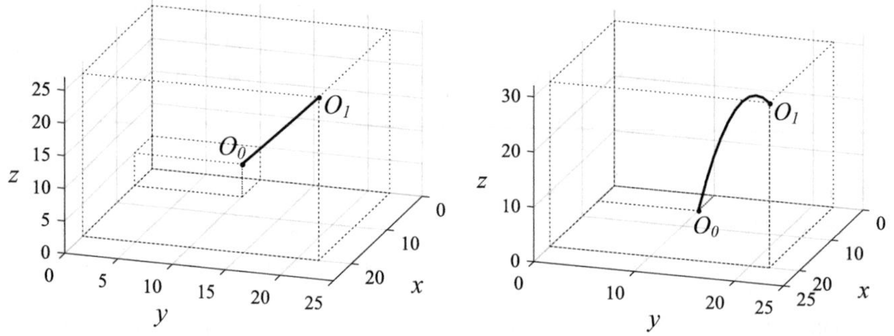

Fig. 4 Trajectories of the tripod's gripper

6 Conclusion

The method suggested in this paper, allows carrying out the design of the multi agent control systems for the tripod, which is a mechanism of parallel structure with three controlled links of variable length. The intellectual, coordinated control of the tripod links provides any spatial movements of the gripper with the working tool. These movements are necessary to realize various technological operations: assembly, weldings, paintings, measurements and many others.

The multi agent control system of tripod includes three independent agents. These agents provide the movement of the tripod gripper along the required trajectories. In particular, the agents provide moving the tripod gripper from some point to the required point of the tripod working space. The multi agent systems with the intellectual control can be applied for automation of many other technical objects.

Acknowledgements This work is supported by RFBR, project number 16-58-00226.

References

1. Glazunov, V.A., Koliskor, A.Sh., Krayinev, A.F.: Spatial Mechanisms of Parallel Structure. Moscow, Nauka, 96 p. (1991) (in Russian)
2. Bushuev, V.V., Holshev, I.G.: Mechanism of parallel structure in mechanical engineering. STIN (1), 3–8 (2001)
3. Gregorio, R.D.: Kinematics of the 3-UPU wrist. Mech. Mach. Theory **38**, 253–263 (2003)
4. Kong, X., Gosselin, C.: Type synthesis of 3-DOF translational parallel manipulators based on screw theory. J. Mech. Des. **126**, 82–126 (2004)
5. Kong, X., Gosselin, C.: Type Synthesis of Parallel Mechanisms, 275 p. Springer (2007)
6. Gaiduk, A.R., Kapustyan, S.G., Shapovalov, I.O.: Self-organization in groups of intelligent robots. Adv. Intell. Syst. Comput. **345**, 171–181 (2015)
7. Markovich, S.V., Tyves, L.I.: Optimal on speed control of the robot movement on own trajectory. Probl. Mech. Eng. Mach. Reliab. (5), 76–82 (1993) (in Russian)
8. Glazunov, V.A., Dugin, E.B., Kistanov, V.A., Vu Ngok Bik: Optimization of the mechanisms of parallel structure parameters on the basis of the simulation working range. Prob. Mach. Build. Mach. Reliab. (6), 12–16 (2005) (in Russian)
9. Gerasun, V.M., Nesmiyanov, I.A.: Control system for manipulators on the basis of spatial executive mechanisms. Mechatron. Autom. Control (2), 24–28 (2010) (in Russian)
10. Dyashkin-Titov, V.V., Pavlovsky, V.E.: Problem of optimal control of moving the gripper of manipulator-tripod. Izvestiya Nizhne-Volzhskogo Agro-University Complex **3**(36), 231–236 (2014)
11. Gaiduk, A.R.: Theory of Automatic Control: Tutorial, p. 415. Moscow, Vysshaya shkola (2010). ISBN 978-5-06-006055-3 (in Russian)
12. Berthomieu, T., Reboulet, C.: Dynamic model of parallel manipulator with six degree of freedom. In: Proceedings of the International CAR, Pise, pp. 1153–1157, 19–22 June 1991
13. Volkomorov, S.V., Kaganov, J.T., Karpenko, A.P.: Simulation and optimization of some parallel mechanisms. In: Information Technologies, Vyp. 5, pp. 1–32 (2010) (in Russian)
14. Zhoga, V.V., Gerasun, V.M., Nesmiyanov, I.A., Vorobjeva, N.S., Dyashkin-Titov, V.V.: Dynamic synthesis of optimum program movements of the manipulator-tripod. Probl. Mach. Build. Mach. Reliab. (2), 85–92 (2015) (in Russian)
15. Skobelev, P.O.: Multi-agent systems for real time adaptive resource management, Chapter 12. In: Industrial Agents: Emerging Applications of Software Agents in Industry, 24 p. Elsevier (2015)
16. Krot, A.M., Prakapovich, R.A.: Nonlinear analysis of the Hopfield network dynamical states using matrix decomposition theory. Chaotic Model Simul **1**, 133–146 (2013)
17. Gaiduk, A.R.: Continuous and Discrete Dynamic Systems. Moscow, Uchlitvuz (2004). ISBN-5-8367-0025-X (in Russian)
18. Gaiduk, A.R.: Mathematical Methods of the Analysis and Design of Dynamic Systems. Lap Lambert Academic Publishing, Saarbrücken, Deutschland (2015) (in Russian)
19. Kaliaev, I.A., Gaiduk, A.R.: Gregarious principles of control in group of objects. Mechatron. Autom. Control (12), 29–33 (2004) (in Russian)

Estimation and Recognition of Motion Segmentation and Pose IMU-Based Human Motion Capture

Phan Gia Luan, Nguyen Thanh Tan and Nguyen Truong Thinh

Abstract Talking about motion capture systems, we will think of a system that has a lot of white markers distributed on the suit worn on the human body, can record and simulate the motion of human or any other object in the software. However, these systems are worth a lot of money, can only operate in a wide and fixed space with many cameras attached around. Therefore, only large animation filmmakers or graphic designers are capable of purchasing this type of system. In this paper, the wireless IMU-based motion capture system is researched and developed with low cost, moderate accuracy, high speed, portable as well as easy-to-use, that is our main contribution. The full-featured hardware, the very simple operation program, and controlling software are focused and built on using the low cost components. The designed system is based on a network of small inertial measurement units (IMU) called "node" distributed on the human body. In essence, the MCU is the core of the board. It collects measured data from the sensors, perform orientation filter based on a quaternion-based Madgwick orientation filter and transfer data to host via Wi-Fi or store it in the memory for later use. After that, the node's processed data was simulated by the program called "SHURIKEN launcher". The nodes' behavior also is controlled by this program. All of these activities are incorporated into the operation of the system in this project. The result of experiments on accuracy demonstrated the feasibility and advantages also a few shortcomings of the system. The advantages and limitations of the system, hardware and software architecture in more detail will be discussed.

P. G. Luan · N. T. Tan (✉) · N. T. Thinh
Department of Mechatronics, Ho Chi Minh City University of Technology and Education, Ho Chi Minh City 70000, Vietnam
e-mail: tannt@hcmute.edu.vn

P. G. Luan
e-mail: luan13146298@gmail.com

N. T. Thinh
e-mail: thinhnt@hcmute.edu.vn

J.-H. Kim et al. (eds.), *Robot Intelligence Technology and Applications 5*,
Advances in Intelligent Systems and Computing 751,
https://doi.org/10.1007/978-3-319-78452-6_32

1 Introduction

Motion capture (Mocap for short) is the process of capturing the movements of the object and apply the captured movements to three-dimensional model so as to simulate real object's movements. From the early 19th to the late 20th Century, traditional animation was the most popular in the movie industry. During this time, drawing frame by frame picture which was either projected or recorded was the way people form animation. However, this technique spent too much time and effort of producers. To solve this problem, the motion capture technology was invented. The motion capture technology first appeared in the late 1970's [1]. Then, the motion capture tech was gradually developed through various methods, expanded phenomenally since 1980's [1]. Nowadays, the Mocap has played an essential part in our life contributing in various fields. It has been used in tracking the performance of athletes, creating more realistic characters in cartoons, games or movies for entertainment purposes. Besides, it has also been used in tracking head movement of pilots in military, supporting the training of anatomy etc. Generally, motion capture has two main methods. The first method is to use either cameras, the image sensor or depth sensor for motion tracking called optical method. The second method is to measure inertia and motion called non-optical method [2, 3]. The overview of motion capture methods is shown in Fig. 1.

Unlike optical methods, this method does not require a lot of space and time to set up a large number of cameras. These devices are unfixed and portable, so that they are very convenient and comfortable to the users. To be more specific, the users can wear and take off these devices in a second, at the same time they also capture the data quickly and put it into the computer right away. Unlike mechanical systems, these devices are light-weight and not bulky. It takes for ages, the costs a lot of money as well as makes all effort to build a complete system. This paper mainly focuses on the key sections involving a moderate system with qualified technical parameters. The research objectives include three main tasks. The first one is the design and implement the completed hardware. Having more profound basic knowledge of inertial

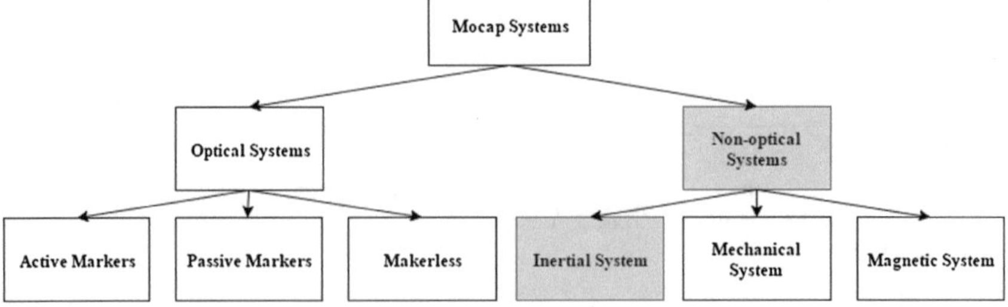

Fig. 1 Motion capture methods

sensors, magnetometer, quaternion, kinetics, etc. is next task. Finally, based on transmitting the data from IMUs and computer via router using the wireless network in order to apply 3D modeling tool so as to build a simple appliance.

2 Node Architecture

In this section, “node” represents a device as shown in Fig. 2 with the baseboard housed inside the node. In the Fig. 3, it described a hardware block diagram of the node integrated devices. The node is a wireless tracking motion device that has three separated parts housed in 3D printed box: SHURIKEN, battery and motor. Battery provides power for baseboard. Baseboard provides power and control the motor. Figure 3 illustrates the SHURIKEN baseboard interconnections. SHURIKEN baseboard consists of an integrated WiFi chip 32-bit MCU, a sensor that house 3-axis gyroscope, 3-axis accelerometer and 3-axis magnetometer, external memory (up to 128 GB microSD card), vibration motor and the on-board power circuit. The MCU completely controls the operation of sensor, external memory and motor. The angular rate, acceleration and magnetic field are measured

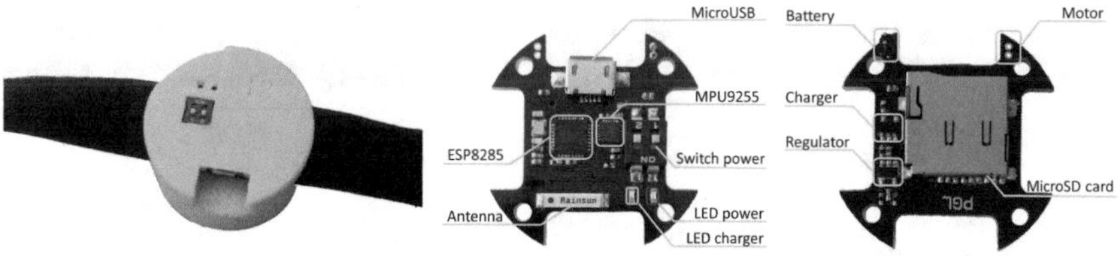

Fig. 2 The structure of SHURIKEN

Fig. 3 SHURIKEN interconnections block diagram

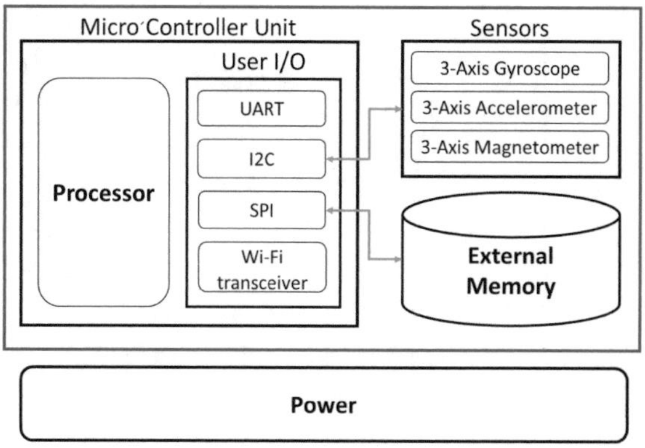

by the sensor and transferred to MCU via I2C protocol. SHURIKEN also has one microUSB port to communicate with computer via UART, upgrade firmware or charge the battery. Because SHURIKEN is the wearable and wireless device, integrating power on board is essential. In order to have a compact device, chargeable and a long battery life, proper battery selection is also a problem.

3 Kinematics

Based on the Madgwick algorithm, the orientation of the device in the global frame can be obtained. However, in order to be able to use these data to simulate in modeling software, it should be processed the data through several steps. Each node measures the orientation of its frame relatives to global frame. To measure the orientation of other objects or object's parts, we need tighten nodes to them. However, these objects have its non-defined frame called body frame, this problem can be solved by attaching these nodes so that device frame coincide to body frame but it is sometimes impossible. Like the real world, we also have two main frames in modeling system. And model frame is equivalent to body frame, it represents the body frame of model relatives to unity frame. These frames were visualized in the Fig. 4. For convenient computing, we will denote u as Unity frame, c as model frame and b as body frame. Expected results of this project is the orientation of model frame relatives to unity frame, q_c^u, based on the orientation of device frame

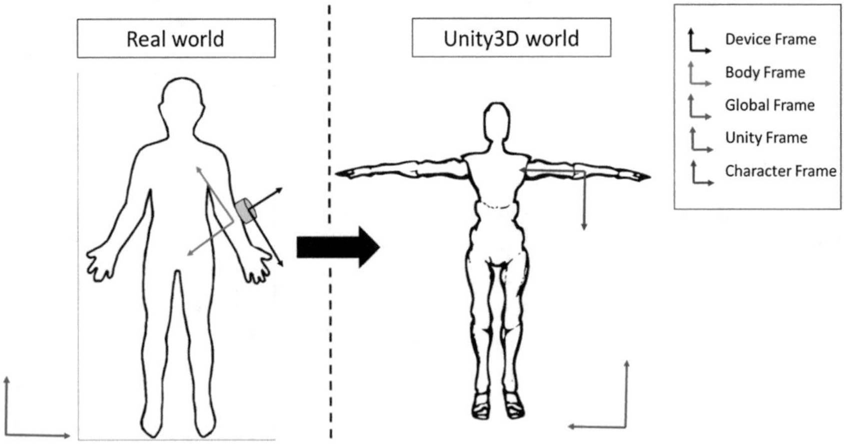

Fig. 4 Five frames need to perform simulation in virtual environment

relatives to global frame, q_d^g while body frame coincides to model frame. After processing the node's data, we obtain its orientation in the global frame, q_d^g. The coordination of virtual environment has a Y-up orientation. The X and Z axes perpendicular to each other on the "floor", and the Y run straight up to the sky. To resolve these issues while mapping the player's motions to the model, the quaternion was converted from the global frame to the virtual frame in order for the rotation directions to match correctly [4]. We assume that the player stands in the matching initial pose and direction of the model—this makes the body frame almost identical to the model frame—until q_d^u is stable (k times which is recovered time from any sensor over-ranging). After k times, $q_{u_k}^d$ represents the orientation of device frame relatives to unity frame, using it as the initial quaternion. Implement below equations, orientation of model frame relatives to unity frame based on orientation of device frame relatives to unity frame can be archived:

$$\begin{cases} q_{c_t}^u = q_{d_t}^u \otimes q_{d_k}^{u*} \otimes q_{c0}^u \\ q_{c_t}^u \otimes q_{b_t}^{u*} = 1 \end{cases}$$

4 Communications

The MCU initializes a Wi-Fi transceiver and all sensors if the power is active. Then system try to connect to the router, which SSID and password has been given, until they are connected. Next, node wait the flag (signal form in byte) given by the host. If flag equals 100, node will send flag back and the program will start the loop until it lost connection or power off. In the loop, MCU will get measured data from sensor, process and send it to the user via router. If flag equals 105, node will instantly calibrate the gyroscope and accelerometer. Similarly, if it equals 112, it will calibrate the magnetometer. After calibration, node will send flag back to user and assign the flag to zero. The SHURIKEN launcher is the application that connects user and nodes via Wi-Fi, receives data from nodes, reprocesses data and simulates data. With UDP protocol, User can control the operation of the nodes. As mentioned above, user sends a specific flag to a node (unicast) or to all nodes (broadcast) by using the buttons in the interface so that the nodes exactly perform a specific function. After done the task, by returning the same flag from the nodes to user, software can perform a closed loop control—Detail described in Fig. 5. To perform the calibration, user have to select the "node simulate scene" in main scene. For once process, software just can request only one node perform calibrate that is slightly inconvenient but it has specific reason based on the way performing calibration. The processes for the performing gyroscope and the accelerometer calibration can show like as: To archive successful calibration, before calibration user

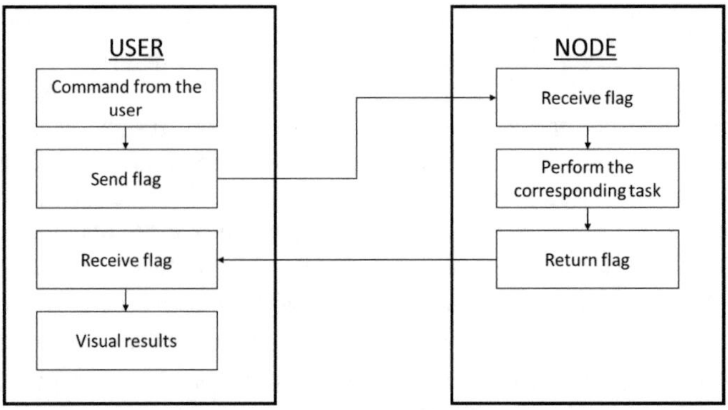

Fig. 5 Block diagram represents closed loop control of user and node

need lie node on a flat surface. Then user click the button "Calibrate MPU" and the process is started. The calibration process will have done after a few seconds. While performing the calibration process, external movements greatly affect the result of the calibration process. Performing magnetometer calibration: It completely contrast to the performance of gyro and accelerometer calibration, the magnetometer calibration requires user rotates the node continuously in different directions over 10 s for the good results. Because of difference from local magnetic flux, calibrate magnetometer at each working area is required.

5 Experiments and Discussions

Due to the limitations of techniques, the measurement of the orientation errors including static error and dynamic error obtained from the nodes will be calculated by comparing the data from step motor. The term static error or static accuracy refers to measurements made when the inertial unit is not moving and the on-board orientation algorithm has recovered from any sensor over-ranging. This error is obtained by leaving node on the surface after k times which is recovered time from any sensor over-ranging. Then measure the data over several seconds. Because of Data measured in quaternion form, to make it easier to visualize the error, this data will be passed into the Euler form. The below table shows parameters used in the experiment. With Δt, the cycle takes data from the sensor, equals 1 ms and μ equals 1, we get the result in Fig. 6.

The term dynamic error or accuracy refers to measurements made while the inertial unit is moving and not exceeding the measurement range of the individual on-board sensors. This error is obtained by leaving node on the surface after k

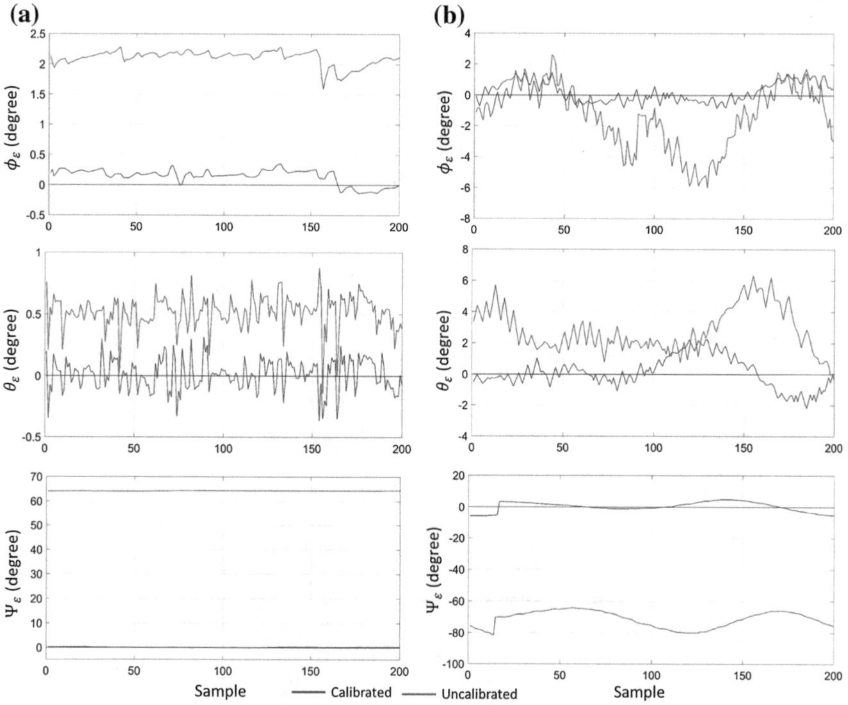

Fig. 6 Static error (**a**) and dynamic errors (**b**) obtained by leaving node on the surface by rotating around Zenith with constant speed

times. Then measure the data over several seconds while the step motor rotate around specific axis with constant speed. This experiment is completed with Δt equals 1 ms, μ equals 1, and the stepper rotates around the Zenith axis at 60RPM presented in Fig. 6. The results of an investigation into the effect of the filter adjustable parameter, μ, on the time response of output data represented by the fourth element of quaternion and steady-state error represented by the third element are summarized in Fig. 7. The experiment is based on the time that the algorithm achieves the node's current orientation when the node is stationary. The results were based on three experiments with three different μ parameters. The larger convergence coefficient, the smaller the settling time and the larger the steady-state error. Based on empirical tests, μ equals 1 gives an acceptable error and settling time is reasonable to ensure that able to capture motion with angular movement is less than the speed limit of the gyroscope sensor.

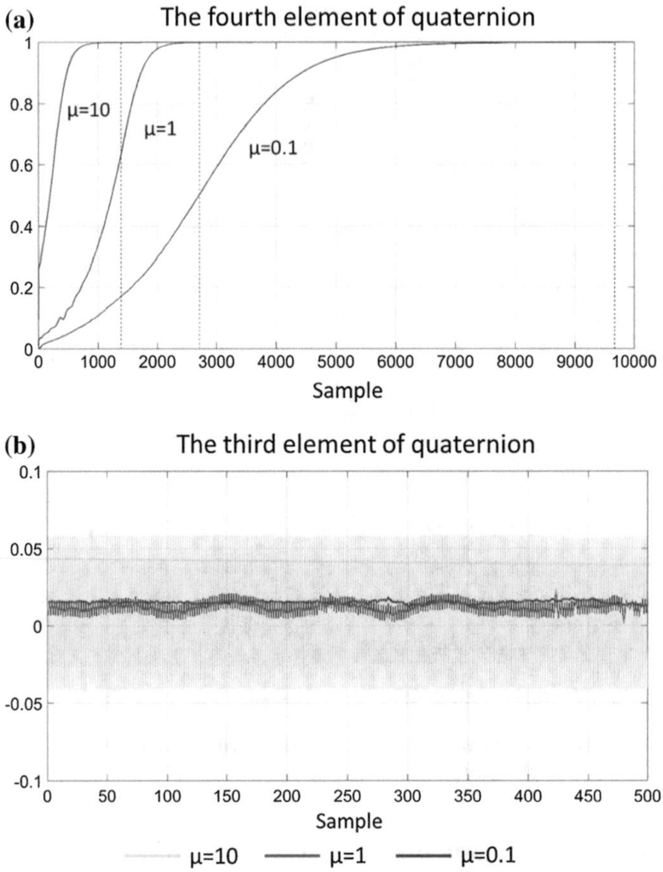

Fig. 7 Time response (**a**) and steady-state error (**b**) with different convergence coefficient

6 Conclusion

In this work, the complete moderate system that can capture motion of the human or other objects was built. This system does not required any cable for communication. It transfer data from measuring device to the end device via Wi-Fi and simulate this data on software made by Unity in real-time. Based on various experiments, mocap system is stable. It can perform continuous simulation about an hour and 30 min. The result of the error test only have moderate accurately because testing and evaluation methods are not accurate enough. However, specific simulation experiments on human body give us result which is usable. Currently, this motion capture system has many disadvantages and limitations. The biggest problem is the effect of linear motion and change of local magnetic flux. They greatly influence on the accuracy of the system. The second problem concern me which is the battery charger IC. This problem makes one third of the nodes can't charge. There are many ways to extend system development. If good enough results was obtained from nodes, gesture recognition via some mathematical algorithms could be performed. In practical applications, users can use simple gestures to control or interact with devices without physically touching them.

Acknowledgements This study was financially supported by Ho Chi Minh city University of Technology and Education, Vietnam.

References

1. Sturman, D.J.: A brief history of motion capture for computer character animation. SIGGRAPH94, Course9 (1994)
2. Sharma, A., et al.: Motion capture process, techniques and applications. Int. J. Recent Innov. Trends Comput. Commun **1**, 251–257 (2013)
3. Brigante, C.M.N. et al.: Towards miniaturization of a MEMS-based wearable motion capture system. IEEE Trans. Ind. Electron. **58.8**, 3234–3241 (2011)
4. Alavi, S., Arsenault, D., Whitehead, A.: Quaternion-based gesture recognition using wireless wearable motion capture sensors. Sensors **16**(5), 605, 6–7 (2016)

Fault Diagnosis of a Robot Manipulator Based on an ARX-Laguerre Fuzzy PID Observer

Farzin Piltan, Muhammad Sohaib and Jong-Myon Kim

Abstract This paper presents a stable ARX-Laguerre fuzzy proportional-integral-derivative observation (FPIDO) system for fault detection and identification (FDI) of actuator and sensor faults in a multi-degrees of freedom robot manipulator. An ARX-Laguerre technique is used in this paper to improve the system modeling in the presence of uncertainty and disturbance in a robot manipulator. The proposed FPIDO is applied to the ARX-Laguerre procedure to modify fault detection, estimation and identification to reduce the system's order. Fuzzy coefficient scheduling is utilized to modify the convergence with respect to the minimum error.

Keywords Fault diagnosis · ARX method · ARX-Laguerre technique
Observation fault diagnosis · PID observation technique · Fuzzy coefficient scheduling

1 Introduction

The main contribution of this work is the design of a robust observation technique to detect, estimate and identify faults for multi-input, multi-output (MIMO) nonlinear uncertain dynamical robot manipulators. Robot manipulators have been extensively used in industrial manufacturing to replace humans or for many other complex specialized applications. These complex specialized applications require robots with nonlinear mechanical architectures, which create multiple challenges for diagnosis, prognosis and fault tolerance in various systems. Several types of

F. Piltan · M. Sohaib · J.-M. Kim (✉)
School of Electrical Engineering, University of Ulsan, Ulsan 680-749, South Korea
e-mail: jongmyon.kim@gmail.com

F. Piltan
e-mail: piltan_f@iranssp.org

M. Sohaib
e-mail: krengr.msohaib@gmail.com

J.-H. Kim et al. (eds.), *Robot Intelligence Technology and Applications 5*,
Advances in Intelligent Systems and Computing 751,
https://doi.org/10.1007/978-3-319-78452-6_33

faults have been defined in robot manipulators, which are divided into three main categories: actuator faults, plant (robot manipulator) faults, and sensor faults [1, 2].

Fault diagnosis is one of the main challenges in different electrical and mechanical tools. According to the literature [3], fault diagnosis techniques have been divided into two main categories: (1) hardware-based fault diagnosis and (2) functional-based fault diagnosis. Hardware-based fault diagnosis is reliable and stable, but it is also very expensive for different types of industries. Functional-based fault diagnosis has been recommended to solve this challenge. The main challenge in functional-based fault diagnosis is reliability and stability. In the last few decades, different techniques have been introduced that focus on functional-based fault diagnosis, but they are divided into three different categories: (1) model-free fault diagnosis, (2) knowledge-based fault diagnosis, and (3) model reference fault diagnosis. The main idea of model-free fault diagnosis is data analysis, which has been improved by researchers over the years [4–6]. Apart from several advantages, this method has challenges associated with system reliability. In knowledge-based fault diagnosis, knowledge of the system is needed to perform reliable fault diagnosis. Artificial intelligence and machine learning are the main tools in this area [6, 7]. Knowledge-based fault diagnosis also has challenges associated with system reliability. Model reference fault diagnosis is recommended to improve stability and reliability. System identification and observation techniques are two famous methods in model reference fault diagnosis [8–10].

System identification techniques have been introduced to model and estimate the systems based on input and output data. Numerous techniques have been introduced by researchers, and these are divided into two main categories; linear system identification and nonlinear system estimation. The concept of linear and nonlinear system identification is the regression. The key to this method is the definition of a function to estimate the output based on input/output data and minimize the cost function. Linear estimation techniques have simple structures and well-understood dynamic responses. Therefore, they are widely used in many applications, but they fail to model noisy and highly nonlinear systems. To solve this challenge, nonlinear intelligent techniques (e.g., fuzzy logic and neural networks) or mathematical techniques (e.g., the Laguerre method) have been used in different applications [11–15]. Based on the literature [15, 16], the Laguerre method has been used to reduce the model order. To improve the input and output performance of ARX, the Laguerre method is applied to ARX to filter the input and output.

The main goal of model reference fault diagnosis is model estimation, fault detection and fault identification with minimum error in the presence of disturbance and uncertainty [3, 8–10, 17]. To address this challenge, several researchers have designed nonlinear observer techniques such as the high order sliding mode observer [10]. The sliding mode observer is a powerful observation method, especially in the presence of noise and uncertainty, but it has a high computation cost, especially for a highly nonlinear system such as a robot manipulator. In this research, the proposed fuzzy PID observation method is used to modify the traditional PID observer by applying an ARX-Laguerre method to estimate the output without a state estimation. Fuzzy coefficient scheduling is recommended to optimize coefficients. This research has two main objectives: (1) modeling the robot manipulator and reducing the system's order, filtering the input and output and fault detection based on the ARX-Laguerre method, and (2) designing a fuzzy PIDO and applied to the ARX-Laguerre method to improve the performance of ARX-Laguerre and input/output fault diagnosis in robot manipulator. This paper has the following sections. The second section outlines the problem statements. The proposed ARX-Laguerre fuzzy PID observation method for system modeling, fault detection, and fault diagnosis is presented in the third section. In the fourth section, we analyzed the fault detection, estimation, and identification of the proposed method. In the final section, we provide conclusions.

2 Problem Statements

One of the main challenges in a nonlinear and time variant system (robot manipulator) is fault detection and identification (FDI). The main objective of this research is to detect, estimate and identify the input (actuator) and output (sensor) fault in a multi degrees of freedom robot manipulator. The dynamic formulation of a robot manipulator in the presence of uncertainty is written as [1]:

$$q = \iint [A^{-1}{}_{(q)} \times (\tau - \{B_{(q)}[\dot{q} \quad \dot{q}] + C_{(q)}[\dot{q}]^2 + G_{(q)} + \tau_d\}) + \eta(t - t_f)]. \tag{1}$$

The nonlinearity and uncertainty of a robot are defined by:

$$\begin{aligned} I_{(q,\dot{q})} &= B_{(q)}[\dot{q} \quad \dot{q}] + C_{(q)}[\dot{q}]^2 + G_{(q)} \\ \Delta_{(q,\dot{q},t)} &= A_{(q)}^{-1} \times (-\tau_d), \end{aligned} \tag{2}$$

where, $q, A_{(q)}, \tau, B_{(q)}, C_{(q)}, G_{(q)}, \tau_d, \eta(t - t_f)$, and t_f are the joint variable, inertia symmetric matrix, Coriolis matrix, centrifugal matrix, gravity matrix, disturbance and uncertainty vector, fault (actuator, plant, sensor) vector, and time of fault occurrence, respectively. According to (1) and (2), the dynamic equation of the robot manipulator in the presence of faults and uncertainty is:

$$q = \iint [A^{-1}{}_{(q)} \times (\tau - I_{(q,\dot{q})}] + \Delta_{(q,\dot{q},t)} + \eta(t - t_f). \tag{3}$$

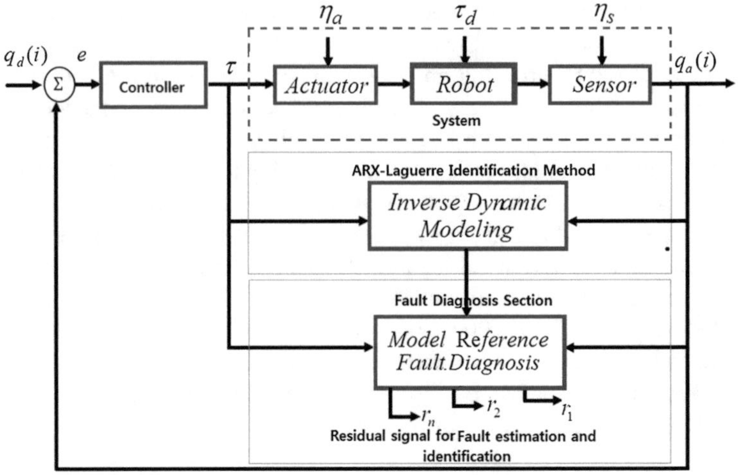

Fig. 1 Proposed technique for the model reference FDI method

Based on (3), the main challenge in this paper is actuator/sensor FDI using the proposed ARX-Laguerre fuzzy PID observation method. To detect and identify the faults based on observation techniques, the state space definition from ARX-Laguerre is defined by:

$$\begin{cases} X(k+1) = [(A+\Delta A)X(k) + B_y(y(k)+\eta_s(k)) + B_u(u(k)+\eta_a(k)) + B_d d(k)] \\ y(k) = (K)^T X(k) + B_s \eta_s(k), \end{cases} \quad (4)$$

where, $X(k), y(k), u(k), \eta_a(k), \eta_s(k)$, and $d(k)$ are the system state, measured output, control input, actuator fault, sensor fault, and process disturbance, respectively. A, B_y, B_u, B_d, B_s, and K are known coefficient matrices, and ΔA is an unknown modeling parameter. Based on (3) and (4), we have two main objectives:(1) inverse dynamic identification of a robot manipulator based on the ARX-Laguerre technique and extracting the state space equation, and (2) fault detection, estimation and identification for an actuator/sensor robot manipulator based on the ARX-Laguerre fuzzy PID observation technique. Figure 1 shows the proposed method for FDI in a multi degrees of freedom robot manipulator.

3 ARX-Laguerre Fuzzy PID Observation Method

According to the model-reference proposed FDI principle, a fuzzy PID observer (FPIDO) is formulated using the ARX-Laguerre method to estimate the system uncertainty, improve the stability and convergence by modifying the coefficients.

3.1 Dynamic Modeling Based on ARX-Laguerre

The linearized ARX state space model in the presence of faults and disturbances is described by the following equation:

$$\begin{cases} X(k+1) = [(A+\Delta A)X(k) + B_u(u(k)+\eta_a(k)) + B_d d(k)] \\ y(k) = (K)^T X(k) + B_s \eta_s(k) \end{cases} \tag{5}$$

The ARX-Laguerre model for robot manipulators has been introduced to improve the modeling and FDI as follows [15, 16]:

$$\begin{array}{l} y(k) = \sum\limits_{0}^{N_a-1} K_{(n,a)} x_{(n,y)}(k) + \sum\limits_{0}^{N_b-1} K_{(n,b)} x_{(n,u)}(k) \\ X(k) = \begin{bmatrix} x_{(n,u)}(k) & x_{(n,y)}(k) \end{bmatrix} \\ x_{(n,y)}(k) = L_n^a(k,\xi_a)^* y(k) \\ x_{(n,u)}(k) = L_n^b(k,\xi_b)^* u(k) \end{array}, \tag{6}$$

where $y(k), u(k), (K_{n,a}\,\&\,K_{n,b}), (N_a, N_b),\ x_{n,y}(k), x_{n,u}(k), x_{n,y}(k), (L_n^a(k,\xi_a),$ and $L_n^b(k,\xi_b))$ are the system output, system input, Fourier coefficients, output signal filter, input signal filter, and Laguerre-based orthonormal functions, respectively. Based on (6), the state space ARX-Laguerre in the presence of the actuator and sensor faults and disturbances are obtained by:

$$\begin{cases} X_\alpha(k+1) = [(A+\Delta A)X_\alpha(k) + B_y(y_\alpha(k)+\eta_s(k)) + B_u(u(k)+\eta_a(k)) + B_d d(k)] \\ y(k) = (K)^T X_\alpha(k) + B_s \eta_s(k) \end{cases}. \tag{7}$$

The fault can be detected using the following equations if a robot manipulator has an actuator fault.

$$\begin{array}{l} e_y(k) = y_\alpha(k) - y(k) \\ e_X(k) = \begin{bmatrix} x_{\alpha(n,u+\eta_a)}(k) - x_{(n,u)}(k) \\ x_{\alpha(n,y_a)}(k) - x_{(n,y)}(k) \end{bmatrix} \end{array} \tag{8}$$

In this condition,

$$x_{\alpha(n,u+\eta_a)}(k) \neq x_{(n,u)}(k) \to X_\alpha(k) \neq X(k) \to y_\alpha(k) \neq y(k) \to e_y(k) \neq 0. \tag{9}$$

The output (sensor) fault for a robot manipulator can be detected as follows:

$$\begin{array}{l} e_y(k) = y_\alpha(k) - y(k) \\ e_X(k) = \begin{bmatrix} x_{\alpha(n,u)}(k) - x_{(n,u)}(k) \\ x_{\alpha(n,y_a+\eta_s)}(k) - x_{(n,y)}(k) \end{bmatrix} \end{array}. \tag{10}$$

Consequently, to analyze the fault detection,

$$x_{\alpha(n,y_a+\eta_s)}(k) \neq x_{(n,y)}(k) \rightarrow X_\alpha(k) \neq X(k) \rightarrow y_\alpha(k) \neq y(k) \rightarrow e_y(k) \neq 0. \tag{11}$$

Normal and abnormal robot manipulators can be identified based on (8) to (11), and the proposed fuzzy PID observation technique is applied to the ARX-Laguerre method to diagnose the fault.

3.2 Fault Diagnosis

The proposed ARX Laguerre fuzzy PID observations method is recommended for estimating and identifying the actuator and sensor faults in the robot manipulator. This method is defined by the following equations in the presence of input/output faults for a robot manipulator:

$$\begin{cases} \widehat{X}(k+1) = A\hat{X}(k) + B_y(\hat{y}(k) + \hat{\eta}_s(k)) + B_u(u(k) + \hat{\eta}_a(k)) + K_p e(k) \\ e_s(k) = (q_s(k) - \hat{q}_s(k)) \\ e_a(k) = (\tau_a(k) - \hat{\tau}_a(k)) \\ \hat{\eta}_a(k+1) = \hat{\eta}_a(k) + K_{ia}e_a(k) + K_{d_a}(e_a(k+1) + e_a(k) + e_a(k-1)) \\ \hat{\eta}_s(k+1) = \hat{\eta}_s(k) + K_{i_s}e_s(k) + K_{d_s}(e_s(k+1) + e_s(k) + e_s(k-1)) \\ \hat{y}(k+1) = (K)^T \widehat{X}(k+1) + \beta_s \hat{\eta}_s(k) \end{cases}. \tag{12}$$

Based on (12), there are three main parts: process states $(\widehat{X}(k))$, sensor/actuator faults $(\eta(k))$, and system output $(\hat{y}(k))$. The advantage of the proposed fuzzy PID observation technique is that it performs fault estimation based on two important factors. The first factor is the integral part used to estimate the integral of error estimation, and the second factor is the derivative part to estimate and control the rate of error. This paper focuses on three cases based on (12) and the type of fault:

Case A: if $\eta_a \neq 0, \eta_s = 0$ and $\hat{\eta}_a(k) \neq \eta_a(k)$, the fault diagnosis algorithm is:

$$\begin{aligned} &(y(k+1) - \hat{y}(k+1) \neq 0) \;\&\; (X(k+1) - \hat{X}(k+1)) \neq 0 \Rightarrow \left[X_1^T(k+1) \quad X_2^T(k+1)\right]^T \\ &\qquad - \left[\hat{X}_1^T(k+1) \quad \hat{X}_{2,\eta_a}^T(k+1)\right]^T \neq 0 \Rightarrow \begin{cases} x_{(n,u)}(k) - \hat{x}_{(n,u+\eta_a)}(k) \neq 0 \\ x_{(n,y)}(k) - \hat{x}_{(n,y)}(k) \neq 0 \end{cases} \end{aligned} \tag{13}$$

and if $\eta_a \neq 0, \eta_s = 0$ and $\hat{\eta}_a(k) = \eta_a(k)$,

$$\begin{aligned} &(y(k+1) - \hat{y}(k+1) = 0) \;\&\; (X(k+1) - \widehat{X}(k+1)) \neq 0 \Rightarrow \left[X_1^T(k+1) \quad X_2^T(k+1)\right]^T \\ &\qquad - \left[\widehat{X}_1^T(k+1) \quad \widehat{X}_{2,\eta_a}^T(k+1)\right]^T \neq 0 \Rightarrow \begin{cases} x_{(n,u)}(k) - \hat{x}_{(n,u+\eta_a)}(k) \neq 0 \\ x_{(n,y)}(k) - \hat{x}_{(n,y)}(k) = 0 \end{cases} \end{aligned} \tag{14}$$

Based on (13) and (14), if the error of ARX-Laguerre fuzzy PIDO is near zero, the rate of fault diagnosis is extremely high. Based on the above equations, the residual signal for actuator faults is:

$$\hat{\eta}_a = \eta_a \rightarrow q_a - \hat{q}_a \cong 0 \,\&\, \tau - \hat{\tau} \neq 0 \rightarrow r = \tau - \hat{\tau}. \tag{15}$$

The rate of residual signal for actuator faults is much larger than normal conditions.

Case B: if $\eta_s \neq 0, \eta_a = 0$ and $\hat{\eta}_s(k) \neq \eta_s(k)$, the fault diagnosis algorithm is obtained by:

$$\begin{aligned}(y(k+1) - \hat{y}(k+1) \neq 0) \,\&\, (X(k+1) - \widehat{X}(k+1)) \neq 0 \Rightarrow \left[X_1^T(k+1) \quad X_2^T(k+1)\right]^T \\ - \left[\widehat{X}_{1,\eta_s}^T(k+1) \quad \hat{X}_2^T(k+1)\right]^T \neq 0 \Rightarrow \begin{cases} x_{(n,u)}(k) - \hat{x}_{(n,u)}(k) \neq 0 \\ x_{(n,y)}(k) - \hat{x}_{(n,y+\eta_s)}(k) \neq 0 \end{cases}\end{aligned} \tag{16}$$

and if $\eta_s \neq 0, \eta_a = 0$ and $\hat{\eta}_s(k) = \eta_s(k)$, then:

$$\begin{aligned}(y(k+1) - \hat{y}(k+1) \neq 0) \,\&\, (X(k+1) - \hat{X}(k+1)) \neq 0 \Rightarrow \left[X_1^T(k+1) \quad X_2^T(k+1)\right]^T \\ - \left[\hat{X}_1^T(k+1) \quad \hat{X}_{2,\eta_a}^T(k+1)\right]^T \neq 0 \Rightarrow \begin{cases} x_{(n,u)}(k) - \hat{x}_{(n,u)}(k) = 0 \\ x_{(n,y)}(k) - \hat{x}_{(n,y+\eta_s)}(k) \neq 0 \end{cases}\end{aligned} \tag{17}$$

According to (16) and (17), the ARX-Laguerre fuzzy PIDO error performance has an important role in improving the performance of sensor fault diagnosis in a robot manipulator. The sensor residual signal for a sensor fault is obtained by:

$$\hat{\eta}_s = \eta_s \rightarrow \tau - \hat{\tau} \cong 0 \,\&\, q_a - \hat{q}_a \neq 0 \rightarrow r = q_a - \hat{q}_a \tag{18}$$

Case C: In the final steps, if $\eta_s \neq 0, \eta_a \neq 0$ and $\hat{\eta}_s(k) \neq \eta_s(k) \,\&\, \hat{\eta}_a(k) \neq \eta_a(k)$, then:

$$\begin{aligned}(y(k+1) - \hat{y}(k+1) \neq 0) \,\&\, (X(k+1) - \widehat{X}(k+1)) \neq 0 \Rightarrow \left[X_1^T(k+1) \quad X_2^T(k+1)\right]^T \\ - \left[\widehat{X}_{1,\eta_s}^T(k+1) \quad \hat{X}_2^T(k+1)\right]^T \neq 0 \Rightarrow \begin{cases} x_{(n,u)}(k) - \hat{x}_{(n,u+\eta_a)}(k) \neq 0 \\ x_{(n,y)}(k) - \hat{x}_{(n,y+\eta_s)}(k) \neq 0 \end{cases}\end{aligned} \tag{19}$$

if, $\eta_s \neq 0, \eta_a \neq 0$ and $\hat{\eta}_s(k) = \eta_s(k) \,\&\, \hat{\eta}_a(k) = \eta_a(k)$, then:

$$\begin{aligned}(y(k+1) - \hat{y}(k+1) \neq 0) \,\&\, (X(k+1) - \widehat{X}(k+1)) \neq 0 \Rightarrow \left[X_1^T(k+1) \quad X_2^T(k+1)\right]^T \\ - \left[\widehat{X}_1^T(k+1) \quad \hat{X}_{2,\eta_a}^T(k+1)\right]^T \neq 0 \Rightarrow \begin{cases} x_{(n,u)}(k) - \hat{x}_{(n,u+\eta_a)}(k) \neq 0 \\ x_{(n,y)}(k) - \hat{x}_{(n,y+\eta_s)}(k) \neq 0 \end{cases}\end{aligned} \tag{20}$$

Based on (20), when the robot manipulator has sensor and actuator faults, the torque residual signal and joint variable residual signal are used to identified the faults. The sensor and actuator residual signals for actuator and sensor faults are obtained by:

$$\hat{\eta}_a = \eta_a \,\&\, \hat{\eta}_s = \eta_s \to r_1 = \tau - \hat{\tau} \gg 0 \,\&\, r_2 = q_a - \hat{q}_a \gg 0. \tag{21}$$

To modify the residual signal and fault estimation based on ARX-Laguerre fuzzy PIDO, fuzzy coefficient scheduling is recommended. The main coefficients for an optimal observer are $K_{p_a}, K_{i_a}, K_{d_a}, K_{p_s}, K_{i_s},$ and K_{d_s}. The coefficients defined by the following formulae provide the best convergence.

$$\begin{aligned} K_{i_a} &= \frac{K_{p_a}}{T_{i_a}}, K_{d_a} = K_{p_a}.T_{d_a} \\ K_{i_s} &= \frac{K_{p_s}}{T_{i_s}}, K_{d_s} = K_{p_s}.T_{d_s} \end{aligned} \tag{22}$$

Here, $T_{i_a}, T_{i_s}, T_{d_a},$ and T_{d_s} are the integral risetimes for actuator fault, integral risetime for sensor fault, derivative risetime for actuator fault, and derivative risetime for sensor fault, respectively. To design an ARX-Laguerre fuzzy PID observation (FPIDO) for robot manipulator fault diagnosis, $K_{p_a}, K_{p_s}, K_{d_a}, K_{d_s}, \alpha_a, \alpha_s$ are defined based on the following equations.

$$\begin{aligned} \alpha_a &= \frac{T_{i_a}}{T_{i_a}}, K_{i_a} = \frac{(K_{p_a})^2}{\alpha_a K_{d_a}} \\ \alpha_s &= \frac{T_{i_s}}{T_{i_s}}, K_{i_s} = \frac{(K_{p_s})^2}{\alpha_s K_{d_s}} \end{aligned} \tag{23}$$

To normalize the above parameters:

$$\begin{aligned} K'_{p_a} &= \frac{K_{p_a} - K_{p_a(\min)}}{K_{p_a(\max)} - K_{p_a(\min)}} \in [0,1], K'_{d_a} = \frac{K_{d_a} - K_{d_a(\min)}}{K_{d_a(\max)} - K_{d_a(\min)}} \in [0,1], 2 \le \alpha_a \le 5 \\ K'_{p_s} &= \frac{K_{p_s} - K_{p_s(\min)}}{K_{p_s(\max)} - K_{p_s(\min)}} \in [0,1], K'_{d_s} = \frac{K_{d_s} - K_{d_s(\min)}}{K_{d_s(\max)} - K_{d_s(\min)}} \in [0,1], 2 \le \alpha_s \le 5. \end{aligned} \tag{24}$$

To design ARX-Laguerre FPIDO, two-input fuzzy logic is recommended based on the following equation:

$$\begin{aligned} K'_{p_a} = (\textstyle\sum \theta_a^T \xi_{p_a}(x)), K'_{i_a} = (\sum \theta_a^T \xi_{i_a}(x)), \alpha_a = (\sum \theta_a^T \xi_{\alpha_a}(x)) \\ K'_{p_s} = (\textstyle\sum \theta_s^T \xi_{p_s}(x)), K'_{i_s} = (\sum \theta_s^T \xi_{i_s}(x)), \alpha_s = (\sum \theta_s^T \xi_{\alpha_s}(x)) \end{aligned}, \tag{25}$$

where θ^T is the updating factor, $\eta = \frac{\sum_i \mu(x_i) \cdot x_i}{\sum_i \mu(x_i)}$ and $\mu(x_i)$ is a membership function.

4 Results and Discussion

Case Study: To test and analyze ARX-Laguerre fuzzy PID observation method for actuator and sensor faults diagnosis in a robot manipulator, a PUMA robot manipulator was selected. This robot manipulator is one of the well-known industrial robot arms, which has serial links, six degrees of freedom, nonlinear dynamics, time variant properties, MIMO, uncertainty, and strong coupling effects between the joints. Consequently, diagnosis of the input/output faults for this system is important. To examine the power of fault diagnosis based on the ARX-Laguerre fuzzy PID observation method in a robot manipulator, four types of cases are presented for the first three joints.

4.1 Case 1 (Normal Condition)

In this case, the state robot manipulator works under ideal conditions. The input and output residual signals for a healthy system are calculated using the following formulae:

$$\begin{aligned} r(\tau) &= \tau - \hat{\tau} \rightarrow r(\tau) = \tau - (\tau_{Observer} + \eta_a) \rightarrow r(\tau) = \tau - (\tau_{Observer} + 0) \cong 0 \\ r(\theta) &= \theta - \theta \rightarrow r(\theta) = \theta - (\theta_{Observer} + \eta_s) \rightarrow r(\theta) = \theta - (\theta_{Observer} + 0) \cong 0 \end{aligned} \quad (26)$$

Figures 2 and 3 demonstrate the input and output residual signals for a robot manipulator in a healthy condition. According to Fig. 2, the threshold of joint variable residual signals for the normal state is $\pm 0.02^{Rad}$.

Figure 3 illustrates the torque residual signal under normal conditions for the first three links of a robot manipulator. The actuator and sensor faults are defined by the following equation:

$$\begin{aligned} Fault_{actuator} &= \eta_a(t - t_a) = \begin{cases} 0, t < t_a \\ \eta_{a,} t > t_a \end{cases} \\ Fault_{sensor} &= \eta_s(t - t_s) = \begin{cases} 0, t < t_s \\ \eta_{s,} t > t_s \end{cases} \end{aligned} \quad (27)$$

According to (27), Fig. 3 shows the residual torque for the first three links, when $t < t_a$. The threshold torque for the normal state is $\pm 15^{Nm}$.

4.2 Case 2 (Actuator Fault Condition)

In this case, the robot manipulator has an actuator fault in the first three links as follows:

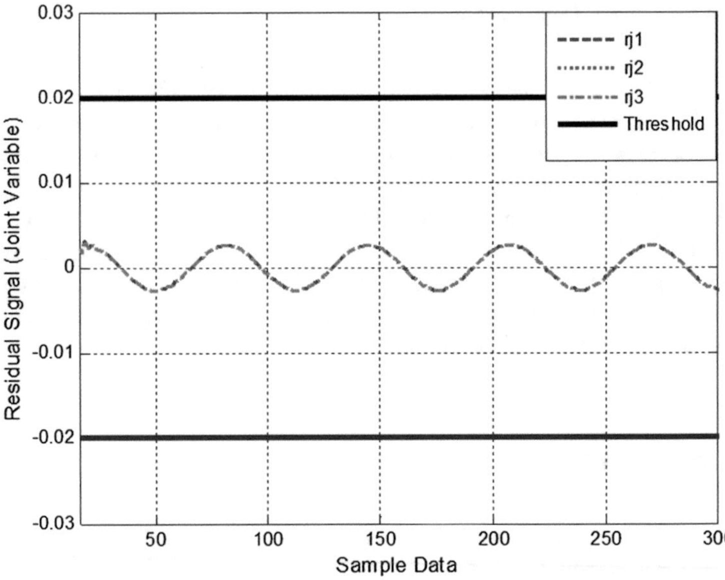

Fig. 2 Joint variable residual signals and threshold for the first three joints (Case 1)

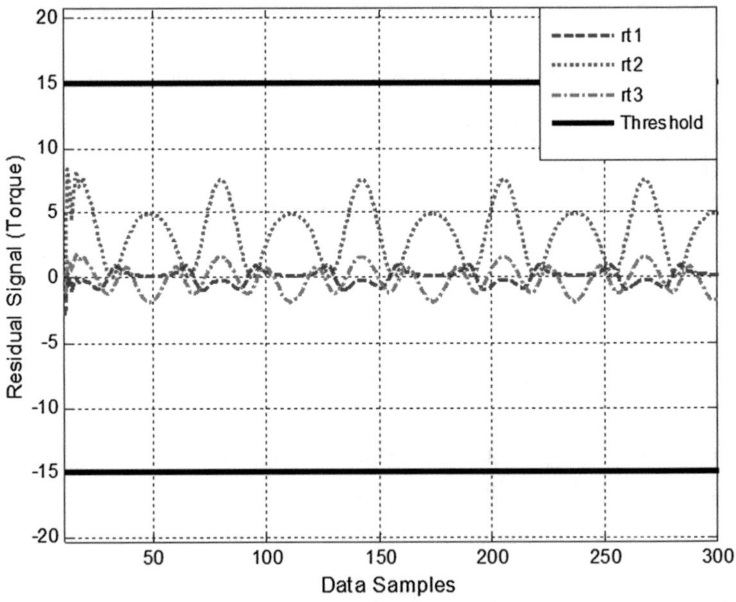

Fig. 3 Torque residual signals and threshold for the first three joints (Case 1)

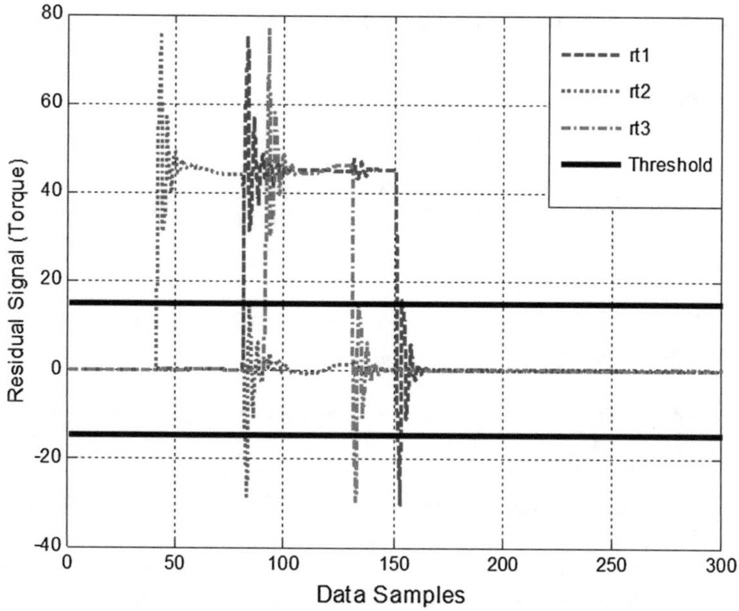

Fig. 4 Torque residual signals and threshold for the first three links in the actuator fault condition (Case 2)

$$\tau 1_{\eta_a}(\mathrm{Nm}) = \begin{cases} 45, & 8 \le t \le 15 \\ 0, & otherwise \end{cases}, \tau 2_{\eta_a}(\mathrm{Nm}) = \begin{cases} 45, & 4 \le t \le 8 \\ 0, & otherwise \end{cases}, \tau 3_{\eta_a}(\mathrm{Nm}) = \begin{cases} 45, & 9 \le t \le 13 \\ 0, & otherwise \end{cases}. \tag{28}$$

According to the above, the residual signal for the joint variable and torque in ARX-Laguerre fuzzy PID observation is:

$$\begin{array}{l} r(\tau) = \tau - \hat{\tau} \rightarrow r(\tau) = \tau - (\tau_{Observer} + \eta_a) \gg 0 \\ r(\theta) = \theta - \theta \rightarrow r(\theta) = \theta - (\theta_{Observer} + \eta_s) \rightarrow r(\theta) = \theta - (\theta_{Observer} + 0) \cong 0 \end{array}. \tag{29}$$

Figure 4 illustrates the effect of actuator fault on the torque residual signal. Based on Fig. 4 and (28) (for $8 < t < 15$, $4 < t < 8$ and $9 < t < 13$), the ARX-Laguerre fuzzy PID observation detected and estimated the faults in the first, second, and third link, respectively. Based on (27), $t < 8 \,\&\, t > 15$ for the first link, $t < 4 \,\&\, t > 8$ for the second link, and $t < 9 \,\&\, t > 13$ for the third link, while the fault value was equal to zero.

4.3 Case 3 (Sensor Fault Condition)

In this case, the first three joints of the robot manipulator have a sensor fault as follows:

$$\theta_{1_{\eta_s}}(\text{Rad})=\begin{cases}0.8, 8\leq t\leq 15\\0, \ \ otherwise\end{cases}, \theta_{2_{\eta_s}}(\text{Rad})=\begin{cases}0.8, 4\leq t\leq 8\\0, \ \ otherwise\end{cases}, \theta_{3_{\eta_s}}(\text{Rad})=\begin{cases}0.8, 9\leq t\leq 13\\0, \ \ otherwise\end{cases} \tag{30}$$

According to the relations above, the residual signal for joint variable and torque in ARX-Laguerre fuzzy PID observation is:

$$\begin{matrix} r(\tau)=\tau-\hat{\tau}\rightarrow r(\tau)=\tau-(\tau_{Observer}+\eta_a)\rightarrow r(\tau)=\tau-(\tau_{observer}+0)\cong 0 \\ r(\theta)=\theta-\theta\rightarrow r(\theta)=\theta-(\theta_{Observer}+\eta_s)\gg 0 \end{matrix}. \tag{31}$$

Figure 5 shows the joint variable residual signal and threshold joint variable to detect, estimate and identify the faults in the first three links of a robot manipulator. According to Fig. 5 and (30), the first three links of a robot manipulator have the same fault at different times. For instance, at $8<t<15$, $4<t<8$ and $9<t<13$, the ARX-Laguerre fuzzy PID observation detects and estimates the faults in the first, second, and third link, respectively. According to Fig. 5, ARX-Laguerre fuzzy PID observation detects and identifies the sensor fault and the time of fault as well. According to (27) $t<8\,\&\,t>15$ for the first link, $t<4\,\&\,t>8$ for the second link, and $t<9\,\&\,t>13$ for the third link, the fault value was equal to zero.

4.4 Case 4 (Actuator and Sensor Fault Condition)

In this case, a robot manipulator has two types of faults at the same time, a sensor fault and an actuator fault.

$$\begin{matrix} \tau_{1_{\eta_a}}(\text{Nm})=\begin{cases}45, 8\leq t\leq 15\\0, \ otherwise\end{cases}, \tau_{2_{\eta_a}}(\text{Nm})=\begin{cases}45, 4\leq t\leq 8\\0, \ otherwise\end{cases}, \tau_{3_{\eta_a}}(\text{Nm})=\begin{cases}45, 9\leq t\leq 13\\0, \ otherwise\end{cases} \\ \theta_{1_{\eta_s}}(\text{Rad})=\begin{cases}0.8, 8\leq t\leq 15\\0, \ \ otherwise\end{cases}, \theta_{2_{\eta_s}}(\text{Rad})=\begin{cases}0.8, 4\leq t\leq 8\\0, \ \ otherwise\end{cases}, \theta_{3_{\eta_s}}(\text{Rad})=\begin{cases}0.8, 9\leq t\leq 13\\0, \ \ otherwise\end{cases} \end{matrix} \tag{32}$$

To detect, estimate and identify of this fault, the following equations are introduced.

$$\begin{matrix} r(\tau)=\tau-\hat{\tau}\rightarrow r(\tau)=\tau-(\tau_{Observer}+\eta_a)\gg 0 \\ r(\theta)=\theta-\theta\rightarrow r(\theta)=\theta-(\theta_{Observer}+\eta_s)\gg 0 \end{matrix} \tag{33}$$

Figures 6 and 7 illustrate the power of ARX-Laguerre fuzzy PID observation to detect, estimate and identify the sensor and actuator faults.

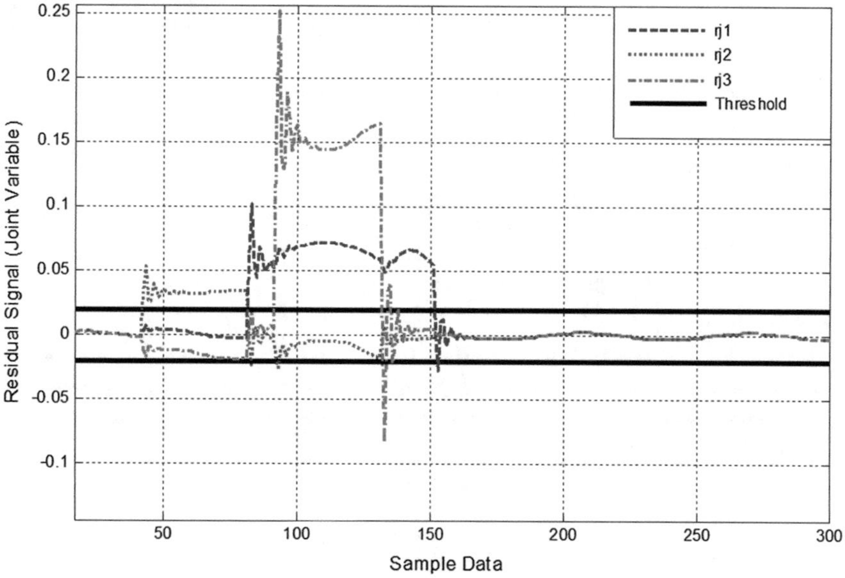

Fig. 5 Joint variable residual signals and threshold for the first three links in a sensor fault condition (Case 3)

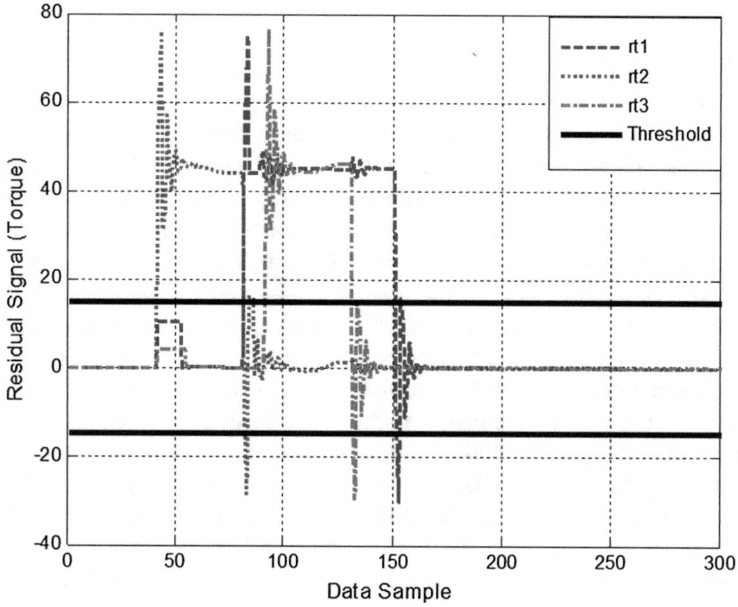

Fig. 6 Torque residual signals and threshold for the first three links in a sensor +actuator fault condition (Case 4)

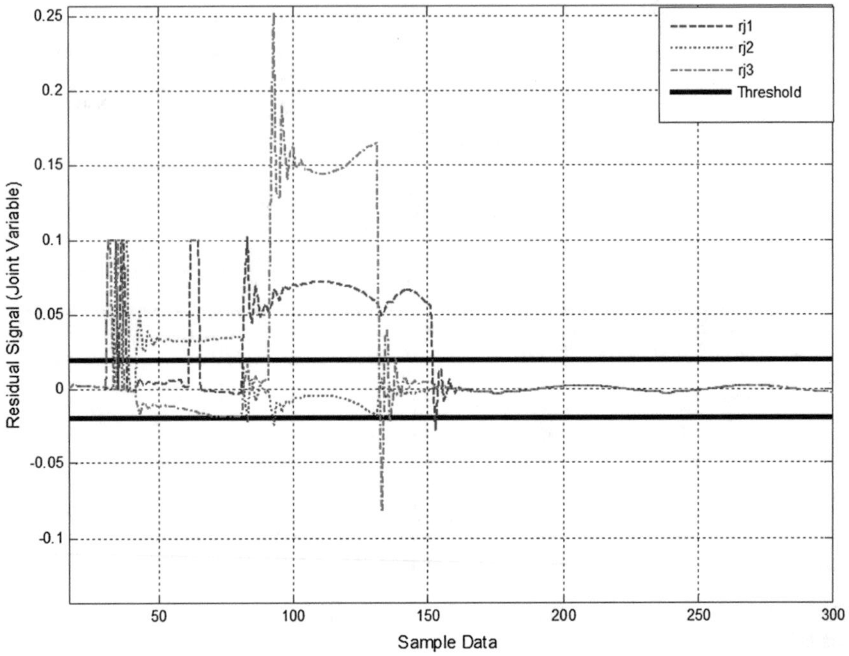

Fig. 7 Joint variable residual signals and threshold for the first three links in a sensor +actuator fault condition (Case 4)

5 Conclusions

In this paper, we analyzed a technique proposed for fault diagnosis in robot manipulators. ARX-Laguerre dynamic modeling was applied to a fuzzy PID observation technique to detect, estimate and identify sensor and actuator faults in a robot manipulator. In the first step, an ARX-Laguerre method was applied to improve the performance of the linear ARX method for modeling the robot manipulator. In the second step, the fuzzy PID observation method was designed and applied to the ARX-Laguerre algorithm. The proposed observation technique has two main advantages: (1) the integral part reduces the estimated steady state error and improves the observation stability, and (2) the derivative of the fault was used to tune the rate of estimation fault, which is used to reduce the transient estimation error. Fuzzy coefficient scheduling was proposed to design an optimal ARX-Laguerre PID observer. Based on our results, the sensitivity of fault diagnosis for actuator faults was about ± 2 Nm and for sensor faults it was about ± 0.02 Rad. In the future, we will improve the performance of FDI technique and practical application using a field programmable gate array (FPGA) based ARX-Laguerre fuzzy PID observer.

Acknowledgements This work was supported by the Korea Institute of Energy Technology Evaluation and Planning (KETEP) and the Ministry of Trade, Industry and Energy (MOTIE) of the Republic of Korea (No. 20162220100050, No. 20161120100350, and No. 20172510102130). It was also funded in part by the Leading Human Resource Training Program of Regional Neo Industry through the National Research Foundation of Korea (NRF) funded by the Ministry of Science, ICT and future Planning (NRF-2016H1D5A1910564), and in part by the Basic Science Research Program through the National Research Foundation of Korea (NRF) funded by the Ministry of Education (2016R1D1A3B03931927).

References

1. Siciliano, B., Khatib, O.: Springer Handbook of Robotics. Springer (2016)
2. Ngoc Son, N., Anh, H.P.H., Thanh Nam, N.: Robot manipulator identification based on adaptive multiple-input and multiple-output neural model optimized by advanced differential evolution algorithm. Int. J. Adv. Robot. Syst. **14**, 1729881416677695 (2016)
3. del Titolo, T.P.I.C., di Dottore, D.R.: Model-Based Fault Diagnosis in Dynamic Systems Using Identification Techniques (2002)
4. Lafont, F., Balmat, J.-F., Pessel, N., Fliess, M.: A model-free control strategy for an experimental greenhouse with an application to fault accommodation. Comput. Electron. Agric. **110**, 139–149 (2015)
5. Badihi, H., Zhang, Y., Hong, H.: Fault-tolerant cooperative control in an offshore wind farm using model-free and model-based fault detection and diagnosis approaches. Appl. Energy (2017)
6. Wang, X., Li, X., Wang, J., Fang, X., Zhu, X.: Data-driven model-free adaptive sliding mode control for the multi degree-of-freedom robotic exoskeleton. Inf. Sci. **327**, 246–257 (2016)
7. Khalastchi, E., Kalech, M., Rokach, L.: A hybrid approach for improving unsupervised fault detection for robotic systems. Expert Syst. Appl. **81**, 372–383 (2017)
8. Stavrou, D., Eliades, D.G., Panayiotou, C.G., Polycarpou, M.M.: Fault detection for service mobile robots using model-based method. Auton. Robot. **40**, 383–394 (2016)
9. López-Estrada, F.R., Ponsart, J.-C., Theilliol, D., Zhang, Y., Astorga-Zaragoza, C.-M.: LPV model-based tracking control and robust sensor fault diagnosis for a quadrotor UAV. J. Intell. Rob. Syst. **84**, 163–177 (2016)
10. Van, M., Franciosa, P., Ceglarek, D.: Fault diagnosis and fault-tolerant control of uncertain robot manipulators using high-order sliding mode. Math. Probl. Eng. (2016)
11. Anh, H.P.H., Nam, N.T.: Novel adaptive forward neural MIMO NARX model for the identification of industrial 3-DOF robot arm kinematics. Int. J. Adv. Rob. Syst. **9**, 104 (2012)
12. Alavandar, S., Nigam, M.: Neuro-fuzzy based approach for inverse kinematics solution of industrial robot manipulators. Int. J. Comput. Commun. Control **3**, 224–234 (2008)
13. Jami'in, M.A., Hu, J., Marhaban, M.H., Sutrisno, I., Mariun, N.B.: Quasi-ARX neural network based adaptive predictive control for nonlinear systems. IEEJ Trans. Electr. Electron. Eng. **11**, 83–90 (2016)
14. Hartmann, A., Lemos, J.M., Costa, R.S., Xavier, J., Vinga, S.: Identification of switched ARX models via convex optimization and expectation maximization. J. Process Control **28**, 9–16 (2015)
15. Bouzrara, K., Garna, T., Ragot, J., Messaoud, H.: Online identification of the ARX model expansion on Laguerre orthonormal bases with filters on model input and output. Int. J. Control **86**, 369–385 (2013)
16. Bouzrara, K., Garna, T., Ragot, J., Messaoud, H.: Decomposition of an ARX model on Laguerre orthonormal bases. ISA Trans. **51**, 848–860 (2012)
17. Busawon, K.K., Kabore, P.: Disturbance attenuation using proportional integral observers. Int. J. Control **74**, 618–627 (2001)

A Two-Level Approach to Motion Planning of Soft Variable-Length Manipulators

Dengyuan Wang, Xiaotong Chen, Gefei Zuo, Xinghua Liu, Zhanchi Wang, Hao Jiang and Xiaoping Chen

Abstract Soft variable-length continuum manipulators have emerged as ideal agents in common human interaction scenarios owing to their flexible movements, large workspace and safety assurance. Besides, it's also their variable-length property that leads to a much larger configuration space, which makes it more difficult to solve the motion planning problem using state-of-the-art sampling-based methods. In this paper, we propose an algorithm that fully exploits the variable-length property of these manipulators. The algorithm directly generates and selects feasible configuration nodes in task space. Before that, a path of the manipulator's end effector is pre-generated in task space according to the positions of the goal and obstacles, which provides approximately accurate guiding direction of the whole manipulator. During the simulation experiments, the two-level algorithm is validated with efficiency in comparison of Jacobian-based methods and other extensional tasks.

Keywords Soft robotics · Motion planning

D. Wang · X. Chen (✉) · G. Zuo · H. Jiang · X. Chen
Department of Computer Science, University of Science and Technology of China, Hefei, China
e-mail: chenxt11@mail.ustc.edu.cn

D. Wang
e-mail: dy960220@mail.ustc.edu.cn

G. Zuo
e-mail: alkaid@mail.ustc.edu.cn

H. Jiang
e-mail: jhjh@mail.ustc.edu.cn

X. Chen
e-mail: xpchen@ustc.edu.cn

X. Liu · Z. Wang
Department of Nuclear Science, University of Science and Technology of China, Hefei, China
e-mail: lxh94@mail.ustc.edu.cn

Z. Wang
e-mail: zkdwzc@mail.ustc.edu.cn

J.-H. Kim et al. (eds.), *Robot Intelligence Technology and Applications 5*, Advances in Intelligent Systems and Computing 751,
https://doi.org/10.1007/978-3-319-78452-6_34

1 Introduction

Nowadays, soft continuum manipulators have various application environments, especially for human interaction scenes where safe and friendly motions are primary requirements [1, 2], and those with variable lengths are able to undertake more complicated tasks taking advantage of a larger workspace. Nevertheless, their highly redundant DoFs and uncertain motions make it difficult to figure out an appropriate motion plan in complex environments within an applicable time. Besides, obstacles in the working environment as well as the manipulators' inherent configuration constraints (also called kinematics singularity) further complicate the problem. Therefore, more effective motion planning algorithms suitable for soft robots are in highly demand.

We focus on the single query motion planning problem with static environment. The problem can be described briefly as to find a configuration sequence of the manipulator that connects the given initial state and goal, within the configuration space free from obstacles and configuration constraints. This can be seen as a search problem, whose basic strategy is to balance and combine the two procedures: explore unknown parts and obtain information of the working environment, and exploit information to reach the goal.

Considering the algorithm's cost, it's an intuitive inclination to exploit more and explore less, but incomplete exploration surely causes defect. For instance, direct pseudoinverse of Jacobian [3] always selects the fastest way towards the goal, yet it cannot avoid obstacles. Artificial potential field [4] based on gradient descent take the obstacles into account, however, it is still susceptible to local minima and saddle points in the potential. Besides, these methods also lack consideration of the configuration constraints of manipulators. At the other extreme, some algorithms spend much to explore the working environment. Navigation functions [5] avoid local minima by considering the global information, whose computational cost is intolerable in realistic settings. Original sampling-based algorithms, PRM [6] and RRT [7], etc. are also inefficient in high dimensional configuration space.

In recent years, researchers have developed improvements based on sampling-based methods, which exploit randomly and ensure completeness. Bertram et al. [8] implement the goal-biased strategy on RRT by selecting the configuration node closest to the goal according to the workspace distance metric, where the extension direction is still randomized. Other works integrate direction-biased heuristic methods by exploiting workspace information. Similar to artificial potential field, the pseudoinverse of Jacobian is utilized by Vahrenkamp et al. [3] to guide the manipulator steps towards the direction of goal, but its computational cost is high. Weghe et al. [9] use the transpose of Jacobian instead, but the convergence rate is much slower. Besides, Diankov et al. [10] and Yao et al. [11] grow an additional end effector space tree to guide the growth of the joint space tree. Both approaches are troubled with local minima. Besides, there are many other implementations based on the improvement of RRT*, PRM, RRM, etc. [12–15]. These methods combine the advantages of searching maps and trees, or grow two trees separately from goal and start point. A novel approach outperforming the above methods called EET [16] balances the exploration and exploitation well by dynamically adjusting the leverage according to the available workspace information.

In this work, we aim to build a motion planner for soft multi-segment extensible continuum manipulators, which contain more than five segments, thus, most of aforementioned approaches will be too computational costly to be used here. Therefore, instead of pursuing an optimal solution, we need a method with much more well-directed exploration as well as thorough exploitation that can provide acceptable solutions in most common situations.

Note that the variable-length property of most soft manipulators provides an essential basis for realizing direct exploitation in task space rather than high-dimensional configuration space. Specifically, a rigid manipulator's shape in task space can be represented by a combination of all the segment tips' positions, so it's almost impossible to directly select that combination to reach a goal without forward kinematics, let alone predicting the movements in task space. On the other side, the variable-length property of segments in a soft manipulator ensures a much larger workspace and makes it possible to realize similar things. Substantially, it's feasible to directly select a projection of configuration in task space for soft manipulators, which expedites the searching process. Moreover, if a certain manipulator's shape is known, we can also easily find a feasible new shape in neighbor area within a few attempts. Thus, a configuration can be directly generated as a projection in task space, which leads to a way for motion planning completely different from those based on Jacobian matrix or complete random sampling methods.

Apart from direct motion generation and selection in task space, we also need a guiding direction towards the goal with consideration of obstacles. To solve this problem, we develop a method similar to the prior ideas that create a 'safe space' in task space (represented by adjoining spheres in most works, [16] etc.), but our design is much more time-saving and practicable in common scenarios. Since soft manipulators are highly flexible and we have discovered that feasible motions can be easily generated in task space, we only need to provide a much simpler and intuitional guideline to ensure a roughly correct direction for extension.

From the prior ideas, our algorithm is proposed and it is composed of two parts. A guiding path of the manipulator's end effector is first generated using a mature path planning algorithm, which ensures an approximately accurate searching direction. Then, we propose a specific method employing the extensibility of the manipulator, which transforms the problem of searching a feasible node in configuration space into a problem of selecting an explicit pose/curve (corresponding projection of configuration) in workspace, where it is simple and intuitive to implement heuristic strategy for searching available poses. Benefited from the deeper exploitation of workspace information, the algorithm's time cost is acceptable for soft manipulators with high DoFs.

2 Motion Planning Algorithm

2.1 Problem Formulation

In this paper, we focus on the motion planning problem of variable-length continuum manipulators in static workspace. Assume the manipulator has n segments and each segment can be actuated individually in 3D space. The positions of obstacles, and the configuration constraints of the manipulator are also assumed to be known.

Let $\mathbf{C} \subseteq \mathbb{R}^{3n}$ be the configuration space of the manipulator and $\mathbf{C}_{free} \subset \mathbf{C}$ denote the subspace of the configuration space where the manipulator is not in collision with an obstacle, or breaks configuration constraints. Let $\boldsymbol{q} \in \mathbf{C}$ denote a certain configuration, and $\mathbf{T} \subseteq \mathbb{R}^3$ denote the task space of the manipulator, and $x \in \mathbf{T}$ denote a certain position in task space. Let $\mathbf{T}_{ob} \subseteq \mathbf{T}$ denote directly the obstacles' position in task space. The motion planner requires $\boldsymbol{q}_{init}$ as input of the start configuration, and x_{goal} as the goal position, and it returns a configuration sequence $\boldsymbol{Q}_{aim} = (\boldsymbol{q}_1, \boldsymbol{q}_2, \dots, \boldsymbol{q}_m)$, where the distance between two neighbor configurations is within a preset limit, denoted as $\delta(\boldsymbol{q}_i, \boldsymbol{q}_{i+1}) < \delta_c$. $\boldsymbol{q}_1$ is exactly $\boldsymbol{q}_{init}$, while the position of the manipulator's end effector corresponding to $\boldsymbol{q}_m$ in task space is at x_{goal}.

To simplify our description in the following subsections, we introduce another concept: the projection of a certain configuration $\boldsymbol{q}_i$, represented as a set of points in task space according to forward kinematics, $\boldsymbol{s}_i = forward(\boldsymbol{q}_i) = (x_{i1}, x_{i2}, \dots, x_{in}) \in \mathbb{R}^{3n}$, where $x_{ij} \in \mathbb{R}^3$ denotes the tip position in task space of segment j of the ith configuration state. Thus the initial end effector position and x_{goal} can be denoted as x_{1n} and x_{mn} respectively. Note that $\boldsymbol{s}$ and $\boldsymbol{q}$ can be uniformly mapped using the invertible forward kinematics.

2.2 Algorithm Overview

In our algorithm, the motion planning problem is separated into two parts (shown in Algorithm 1). The first part is designed based on the following fact. If an available configuration sequence exists, there must be also a path of the manipulator's end effector in task space that connects the initial position and goal (shown in Fig. 2). Therefore, an interval space $\boldsymbol{\Phi}_{inter}$ can be selected in task space according to $\boldsymbol{T}_{ob}$, $\boldsymbol{C}_{free}$, $\boldsymbol{q}_{init}$ and x_{goal}, where a feasible end effector path can be found inside with large possibility.

Then, the second part generates the configuration sequence $\boldsymbol{Q}_{aim}$ with guidance that interval space, which is implemented in the extension of an RRT in configuration space. Besides, heuristic strategy is employed during the randomized searching process.

Algorithm 1 Motion Planner

function MOTIONPLANNER($\mathbf{T}$, $\mathbf{T}_{ob}$, $\mathbf{C}_{free}$, $\boldsymbol{q}_{init}$ and x_{goal})
 $\boldsymbol{\Phi}_{inter} \leftarrow$ InterimTargetGenerator($\mathbf{T}$, $\mathbf{T}_{ob}$, $\boldsymbol{q}_{init}$, x_{goal}, $\mathbf{C}_{free}$)
 $\boldsymbol{Q}_{aim} \leftarrow$ ConfigurationGenerator($\boldsymbol{\Phi}_{inter}$, $\mathbf{C}_{free}$, $\boldsymbol{q}_{init}$, x_{goal})
 return $\boldsymbol{Q}_{aim}$
end function

2.3 *Interim Target Generator*

Function InterimTargetGenerator (Algorithm 2) returns the interval space $\boldsymbol{\Phi}_{inter}$, which can be denoted as a set of the end effector's position in task space as interim goals, $\boldsymbol{X}_{inter}$ with corresponding safe radii, $\boldsymbol{D}_{inter}$, regardless of manipulator's configuration. As mentioned in the last section, the algorithm requires a fundamentally safe area, consistent with $\mathbf{T}_{ob}$, $\mathbf{C}_{free}$, as well as the start and end point, $\boldsymbol{q}_{init}$, x_{goal}.

Algorithm 2 Interim Target Generator

function INTERIMTARGETGENERATOR($\mathbf{T}$, $\mathbf{T}_{ob}$, $\boldsymbol{q}_{init}$, x_{goal}, $\mathbf{C}_{free}$)
 $\mathbf{T}'_{ob} \leftarrow$ ExpandObstacles($\mathbf{T}_{ob}$, d) ▷ d is half width.
 $\mathcal{P} \leftarrow$ SinglePointPathPlanning($\boldsymbol{q}_{init}$, $\mathbf{T}$, $\mathbf{T}'_{ob}$, x_{goal})
 $cost_C \leftarrow$ ReachableArea($\mathbf{C}_{free}$)
 $cost_T \leftarrow$ ShortestLength($\mathcal{P}$)
 $\boldsymbol{p} \leftarrow$ SelectBestPath($\mathcal{P}$, $cost_C$, $cost_T$)
 $\boldsymbol{X}_{inter} = (x_1^*, x_2^*, \ldots, x_k^*) \leftarrow$ Discretization($\boldsymbol{p}$, λ)
 ▷ n is segment number while k is interim goal number.
 $\boldsymbol{D}_{inter} \leftarrow min$(MinDistance($\boldsymbol{X}_{inter}$, $\mathbf{T}_{ob}$), R) ▷ R is safe radius.
 $\boldsymbol{\Phi}_{inter} \leftarrow \boldsymbol{X}_{inter}$ and $\boldsymbol{D}_{inter}$
 return $\boldsymbol{\Phi}_{inter}$
end function

In practice, the manipulator's self shape should also be considered, so $\mathbf{T}_{ob}$ is pre-expanded to $\mathbf{T}'_{ob}$ with a certain radius d. For the manipulator shown in Fig. 1, d can be half of its width. Then a certain single point path planning algorithm can be utilized to generate randomly a set $\mathcal{P}$ of end effector paths.

Next, we use two cost functions to select a best path $\boldsymbol{p} \in \mathcal{P}$. As shown in Algorithm 3, they consider the reachable space of end effector as well as the path length respectively. $\boldsymbol{p}$ is then discretized to a set of points, $\boldsymbol{X}_{inter} = (x_1^*, x_2^*, \ldots, x_k^*)$ as interim

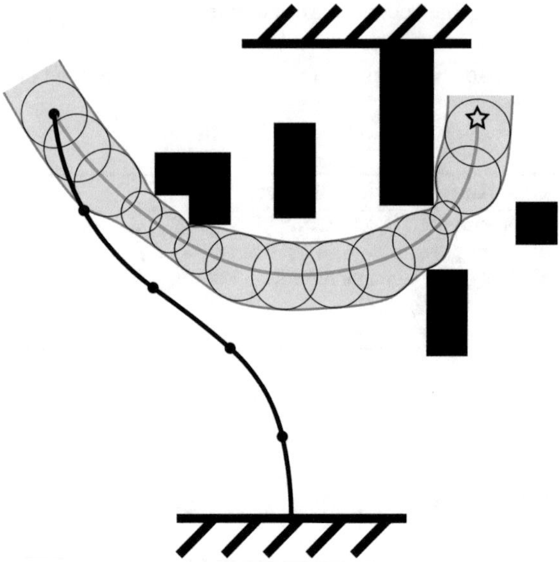

Fig. 1 This figure illustrates the interval space $\boldsymbol{\Phi}_{inter}$ in shaded area, denoted as the interim goals $\boldsymbol{X}_{inter}$ and their corresponding $\boldsymbol{D}_{inter}$. A selected end effector path is shown as a solid blue line, and $\boldsymbol{X}_{inter}$ are found on the line. Then, $\boldsymbol{D}_{inter}$ are figured out from the smaller one of safe radius R and the minimum distance to obstacle

goals (shown in Fig. 1), specially, x_k^* is exactly x_{goal}. We calculate the minimum distance of each x_i^* to the obstacles, denoted as $d_i^* \in \boldsymbol{D}_{inter}$ using a Voronoi diagram as in [17], similar to the artificial potential field algorithm. Note that d_i^* cannot be more than a preset relatively large safe radius R (related to the manipulator's shape). Thus, we can get a pair of sets, $\boldsymbol{X}_{inter}$ and $\boldsymbol{D}_{inter}$.

It should be mentioned that although we have generated many path samples and selected a best one, it's still possible to figure out a 'bad' path, as shown in Fig. 2a. This path will lead to a configuration boundary where it's already at its maximum length yet haven't reached the goal. This condition will be dealt in configuration sampling process in next subsection.

2.4 Configuration Generator

Function Configuration generator (Algorithm 3) calculates the desired $\boldsymbol{Q}_{aim}$ according to $\boldsymbol{X}_{inter}$, $\boldsymbol{D}_{inter}$ and $\mathbf{C}_{free}$ based on RRT algorithm. An RRT $\mathcal{T}$ in configuration space is built from $\boldsymbol{q}_{init}$, and the algorithm starts from searching for the first interim goal x_1^*.

Algorithm 3 Configuration Generator

function CONFIGURATIONGENERATOR($\mathbf{T}_{ob}$, $\boldsymbol{q}_{init}$, x_{goal}, $\boldsymbol{\Phi}_{inter}$)
  RRT $\mathcal{T} \leftarrow \boldsymbol{q}_{init}$
  for $i = 1, i \leq k, i = i + 1$ **do** ▷ k is goal number in $\boldsymbol{X}_{inter}$.
    times $\leftarrow$ 0
    $x_{near} \leftarrow \boldsymbol{q}_{init}$
    while $\delta(x_{near}, x_i^*) > d_i^*$ *and* times < threshold **do**
      $p \leftarrow$ Random(0, 1)
      $\boldsymbol{q}_{near} \leftarrow$ FindNearestNode($\mathcal{T}$)
      ($\boldsymbol{s}_{near}$, x_{near}) $\leftarrow$ ForwardKinematics($\boldsymbol{q}_{near}$)
      if $p < p_g$ **then**
        ($\boldsymbol{q}_{new}$,tag) $\leftarrow$ ExtendToGoal($\boldsymbol{s}_{near}$,x_{near},x_i^*,$\mathbf{T}_{ob}$,$\mathbf{C}_{free}$)
        if tag == *True* **then**
          AddNode($\mathcal{T}$, $\boldsymbol{q}_{new}$)
        end if
      else
        $\boldsymbol{q}_{new} \leftarrow$ ExtendRandomly($\boldsymbol{q}_{near}$, $\mathbf{T}_{ob}$, $\mathbf{C}_{free}$)
        AddNode($\mathcal{T}$, $\boldsymbol{q}_{new}$)
      end if
      times $\leftarrow$ times + 1
    end while
  end for
  $\boldsymbol{Q}_{aim} \leftarrow$ GetPath($\mathcal{T}$)
  Return $\boldsymbol{Q}_{aim}$
end function

During the RRT extension towards x_i^*, when the distance between the end effector's position x_{near} of the nearest node $\boldsymbol{q}_{near}$ and x_i^* is within a preset limit, or the circulation times breaks a preset threshold, the interim goal will update to x_{i+1}^*. Regarding the extension mode, our algorithm conforms to the classical goal-biased method. Specifically, RRT $\mathcal{T}$ has a constant probability p_g to extend towards x_i^* detailed in function ExtendToGoal in next subsection, otherwise, it extends randomly. At last, the goal will progress to x_{goal}, and a feasible node path $\boldsymbol{Q}_{aim}$ can be found in $\mathcal{T}$.

As mentioned before, the pre-generated path could be at relatively bad positions (see Fig. 2). This will lead to the abandon of some of the interim goals (shown in Fig. 2b) expressed as *tag* == *False* in Algorithm 3 and retraction in RRT $\mathcal{T}$. In an extreme case, the algorithm degrades to the classical goal-biased RRT algorithms whose extension is biased only with x_{goal}.

2.5 Heuristic Motion Selection in Task Space

Function ExtendToGoal (Algorithm 4) is the highlight of our algorithm, where heuristic strategy is directly utilized in task space during the searching process.

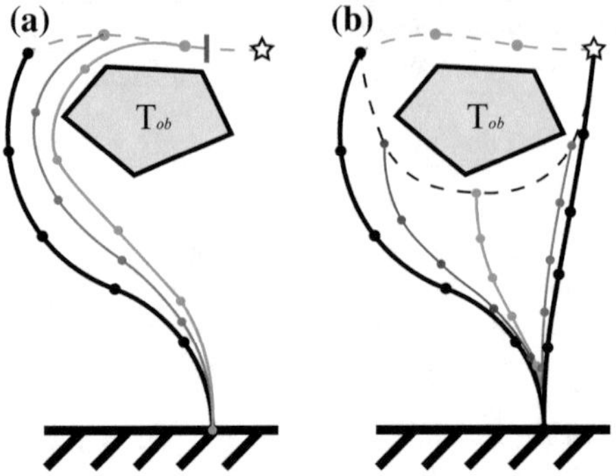

Fig. 2 Blue dotted line in (**a**) illustrates a badly generated end effector path considering two costs, where the manipulator will exceed its maximum length (the red mark) when extending along it, although the path is shorter ($cost_T$) and the goal is constantly within the reachable area ($cost_C$). The manipulator will exceed its maximum length if it goes along the generated path above the obstacle, in fact, the path below is a feasible path. The impact of these two paths can be avoided during the path selection

The algorithm provides rational moving guidance for each segment, which composes the next feasible configuration, in the following steps.

A virtual translation of the manipulator is firstly conducted, as shown in Fig. 3a, by adding scaled vector of the end effector's displacement x_{disp} onto each segment's tip $\boldsymbol{s}_{near}$, where the vector is oriented to next interim goal x_i^*. Then, an interpolation curve $\boldsymbol{s}_{test}$ similar to the current shape $\boldsymbol{s}_{near}$ is depicted as an estimation of the manipulator's position. We can see that there is a tip and part of neighbor segments lying in $\mathbf{T}_{ob}$, which indicates a collision.

From this, the manipulator's segments can be divided into two parts: free segments and constrained segments (separately colored with blue and red in (b)). The free segments are those from the end effector to the first segment, denoted as t in Algorithm 4, that collides in the prior virtual translation, and the rest are regarded as constrained ones. If there is no collisions, all the segments are considered free.

Then, the two parts of segments are imparted with different modes of movements. As shown in Fig. 3b, intuitionally, we can assume that the free segments can move as safely as the translation illustrates, so these segments are imparted with the same displacements as in (a) for next step. For those constrained segments, they are expected to provide necessary extension or retraction for those free segments while almost stay at their current positions in order to avoid latent collisions. So we focus on the tip at the connection segment t of free and constrained parts (see Fig. 3c), where the scaled displacement generates a projection vector on the tangent direction, denoted as $\mathbf{x}_{proj}$. Intuitionally, the constrained parts are expected to provide exactly the displacement of $\mathbf{x}_{proj}$ while remain almost static, so we uniformly separate the length of $\mathbf{x}_{proj}$ for each constrained segment, and convert their directions to each tangent's direction respectively at corresponding segment tip, denoted as $\mathbf{x}_{tan}$. In this way, we can predict a natural and appropriate configuration in task space, as shown in Fig. 3b.

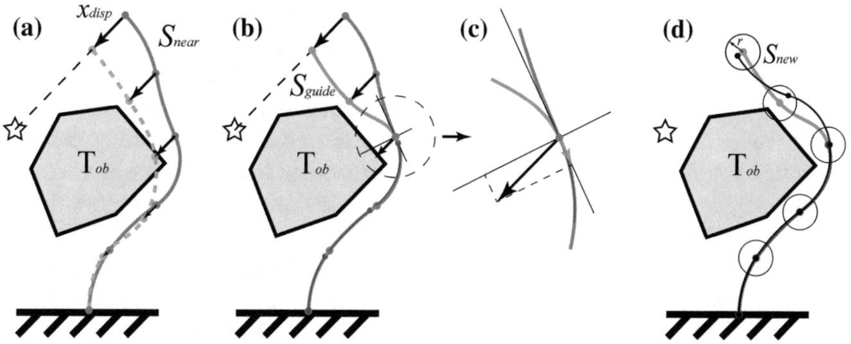

Fig. 3 This figure illustrates the critical parts of finding a new feasible configuration directly in task space. The virtual translation is illustrated in (**a**). A goal-directed displacement x_{disp} is scaled and added onto each segment tip of current shape $\boldsymbol{s}_{near}$, where we can see that the third tip is predicted to collide with obstacles. Thus, the manipulator is divided into free segments and constrained segments, as shown in (**b**), and the predicted shape $\boldsymbol{s}_{guide}$ is also depicted. **c** details the projection at the connection point, and **d** shows the sampling area as well as the found configuration for the new node

Algorithm 4 Extend To Goal

function EXTEND TO GOAL($\boldsymbol{s}_{near}$, x_{near}, x_i^*, $\mathbf{T}_{ob}$, $\mathbf{C}_{free}$)
  $x_{disp} \leftarrow \alpha(x_i^* - x_{near})$
  $\boldsymbol{s}_{test} \leftarrow \boldsymbol{s}_{near} + (1/n, 2/n, \ldots, 1)x_{disp}$ ▷ n is segment number.
  $(t, \mathbf{x}_{proj}) \leftarrow$ FirstConstrainedSegment($\boldsymbol{s}_{test}$, $\mathbf{T}_{ob}$)
  $\mathbf{x}_{tan} \leftarrow$ ConvertAndSeparate($\mathbf{x}_{proj}$, t)
  $\boldsymbol{s}_{guide} \leftarrow$ ReallocateDisplacement(x_{disp}, $\mathbf{x}_{tan}$, $\boldsymbol{s}_{near}$)
  for i = 1, i ≤ k, i = i + 1 **do**
    $S \leftarrow$ SelectNeighborSamples($\boldsymbol{s}_{guide}$, r)
    $cost_1 \leftarrow$ DistanceToObstacles(S, $\mathbf{T}_{ob}$)
    $cost_2 \leftarrow$ DistanceToConfigurationConstraints(S, $\mathbf{C}_{free}$)
    $cost_3 \leftarrow$ UniformShape(S)
    ($\boldsymbol{s}_{new}$, tag) $\leftarrow$ FindBestSample(S, $cost_1$, $cost_2$, $cost_3$)
    ▷ The best sample is found according to three costs.
  end for
  if tag == *True* **then**
    $\boldsymbol{q}_{new} \leftarrow$ ForwardKinematics'($\boldsymbol{s}_{new}$) ▷ Calculation is reversed.
    return ($\boldsymbol{q}_{new}$, tag)
  else
    return (***NIL***, tag)
  end if
end function

Owing to the flexibility of soft structures, each segment of the manipulator has a variable length in a relatively large range, which guarantees an area of reachable space rather than an arc, as shown in Fig. 3d. Thus, we implement sampling around the tips s_{guide} derived in last step within a radius of r. Three evaluation metrics are employed for assessing the samplings, realized in three cost functions. $cost_1$ and $cost_2$ are compulsory demands: they filter the points that lead to collisions with obstacles or breaking configuration constraints, and they also select relatively more distant from the two limits.

Comparatively, $cost_3$ presents a requirement on a higher level: a manipulator's motion composed of all the segments with similar lengths and curvatures are regarded most natural. In addition, this is a state with most space to adjust itself and adapt other situations, because no segment has touched the boundary of configuration constraints unless the whole manipulator has reached its workspace boundary.

From the three cost functions, we can find a best in the samples and calculate the corresponding configuration state by using forward kinematics reversely in large possibility. If there is no qualified samples, we do not sample more and consider that the current step is mislead into an unpleasant region in the working map.

3 Simulation Results

3.1 Setup

The multi-segment manipulator's curve model built in simulation is conformed to the piecewise constant curvature assumption, thus, the forward kinematics calculation used in later experiments can be derived from iterative arc equation and coordinate transformation [18]. The position and orientation of the manipulator's base are supposed to be fixed. Specifically, the model parameters conform to the manipulator designed by authors in [19]. Each segment is of 55 mm original length and 50 mm width, 0.013 m^{-1} maximum curvature, and 85 mm maximum length.

For simplification, in this paper, our simulation experiments are conducted in 2D plane. The benchmarks were run on a 2.80 GHz Core2 processor with 8 GB RAM in MATLAB R2016a. Considering the algorithm's applicability, we set the threshold number of configuration nodes in RRT as 5,000, and the maximum tolerable time is 60 s. The path planning algorithm of end effector is implemented based on the original PRM, whose average running cost is 1–2 s in later scenarios.

In the first part, a scenario with an obstacle placed upwards in steps, is set as the testing benchmark (shown in Fig. 4), where our algorithm competes with the Jacobian-based pseudoinverse of Jacobian (IK-RRT) and the transpose of Jacobian method (JT-RRT). Then, more experiments are conducted using our algorithm with different segment numbers with three levels of difficulty (see Fig. 5), and the algorithm's potential in high DoFs is shown in the result. At last, the algorithm is tested in extensional scenarios where it exhibits decent adaptability.

3.2 Comparison Experiments

In this experiment, the task space is set with an obstacle whose position is adjusted upwards in 8 steps (from (a) to (h) in Table 1). At start the obstacle lies below the connection of the end effector's initial position and the goal, and at last it is placed above the connections of the manipulator's base and the goal. Figure 4 shows three typical conditions (b), (d) and (h), whose solutions or failure points are illustrated as the manipulator's shape sequence in task space.

The performance of three algorithms are shown in Table 1, where each task is tried 20 times and 50,000 nodes. From the variations of the node numbers and time cost, we can find a dramatical change between (a) and (b) for the Jacobian-based methods, while the obstacle just moves a small step, as shown in Fig. 4b. The main reason is that Jacobian matrix always heads for the fastest direction towards the goal, regardless of obstacles and configuration constraints, so when the obstacle disturbs the path, the matrix was trapped, and it relies completely on the sampling process to

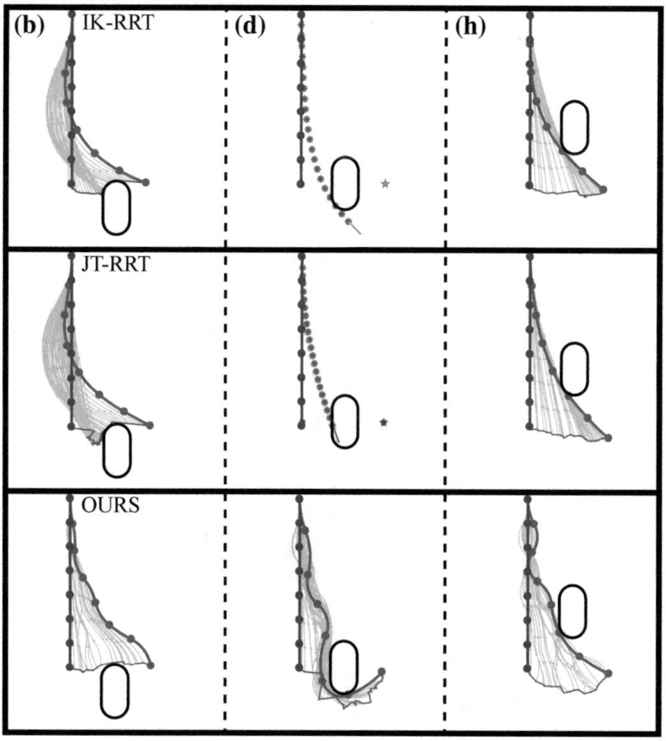

Fig. 4 Three typical conditions, (**b**), (**d**) and (**h**) are illustrated with three algorithms' solution or failure. The red solid lines are end effector paths. In (**b**), the straight movement towards the goal generated by Jacobian-based methods is disturbed. In (**d**), the Jacobian methods are trapped in local minima, whose shapes are depicted as red dotted lines. In (**h**), all the three algorithms perform well

Table 1 Performance comparison

Task		(a)	(b)	(c)	(d)	(e)	(f)	(g)	(h)
SuccessRate	IK-RRT	100%	100%	95%	0%	0%	100%	100%	100%
	JT-RRT	100%	100%	100%	0%	0%	95%	100%	100%
	OURS	100%	100%	100%	100%	100%	100%	100%	100%
NodeNumber	IK-RRT	49.95	9785.3	23725.8	N_A^a	N_A	14055.6	3452.4	828.6
	JT-RRT	47.58	12172.6	22846.6	N_A	N_A	10792.2	3712.0	627.0
	OURS	38.6	328.26	1357.3	2785.68	409.98	109.12	90.48	81.6
TimeCost(s)	IK-RRT	0.0343	31.1812	93.2126	N_A	N_A	44.0190	8.0162	1.5530
	JT-RRT	0.0250	34.2780	86.5216	N_A	N_A	31.5810	8.8722	1.1124
	OURS	2.181	6.253	16.408	24.718	4.529	1.993	2.291	2.003

[a] N_A means exceeding time cost or node numbers searched in RRT

escape (see Fig. 4b). As the trap gets deeper, it would be impossible to escape only by sampling (see Fig. 4d). On the other side, (h) corresponds the performance between (a) and (b) in Table 1, while in Fig. 4h we can find that the configuration distance between the trapped state and escaped state is smaller than that in (b), shown as the state's distribution. Therefore, we can conclude that Jacobian-based methods are entirely restrained by the trap's distribution in configuration space corresponding to the obstacles and configuration constraints. Comparatively, our algorithm properly avoids the traps during the searching process, thanks to the guidance of the interim targets.

3.3 Performance Evaluation

In order to evaluate the performance of our algorithm under different segment numbers of the manipulator, we set three levels of standard scenarios according to the segment number: Let n be the segment number, l be the segment length, thus the total static length of the manipulator is $L = nl$; let d be the segment width. In easy level, there is no obstacles; in medium level, the obstacles are $\frac{1}{2}L$ away from the manipulator and leave a hole with a width of $2d$ at the vertical position of $\frac{3}{4}L$; in hard level, the obstacles still lie $\frac{1}{2}L$ away from the manipulator yet the hole is at the vertical position of $\frac{1}{2}L$. The performance statistics are listed in Table 2.

From the statistics, we can conclude that:

- The time cost increases approximately linearly as the segment number increases, with a small constant factor. Specifically in the comparison of easy and medium levels, when the segment number is relatively low, the node numbers in easy level are much smaller than those in medium level, yet the time cost in easy level is a little longer. It can be explained by the interim targets' generation process using PRM algorithm, where a more open area will lead to more feasible paths and elongate the time of selection.
- When the segment number is low, as the working environment grows more complex, the time cost and success rate varies a lot, because the manipulator with less segments has a smaller reachable space and obstacles greatly reduce the reachable space. On the other side, the variation of time cost as well as success rate is not notable when the segment number is high. The main reason is that the manipulator now has a much greater reachable space, whose reduction ratio for obstacles is little.
- For the more complicated working environments, the increase number of segments dramatically promotes the success rate, while the computational cost is not large and acceptable.

Table 2 Performance evaluation

SegmentNumber		5	6	7	8	9	10	11
HardLevel	SuccessRate	84%	98%	92%	96%	100%	96%	100%
	NodeNumber	2260.18	1997.26	1833.50	1644.92	1385.04	1840.58	1361.66
	TimeCost(s)	32.18	30.25	31.13	31.42	31.72	44.43	38.33
MediumLevel	SuccessRate	100%	100%	100%	100%	100%	100%	100%
	NodeNumber	68.76	74.98	85.22	86.86	133.92	130.02	242.36
	TimeCost(s)	3.18	3.70	4.54	5.11	7.54	7.92	9.47
EasyLevel	SuccessRate	100%	100%	100%	100%	100%	100%	100%
	NodeNumber	36.16	38.96	37.06	38.30	57.46	50.84	119.42
	TimeCost(s)	3.90	5.03	5.52	6.25	7.00	5.49	6.29

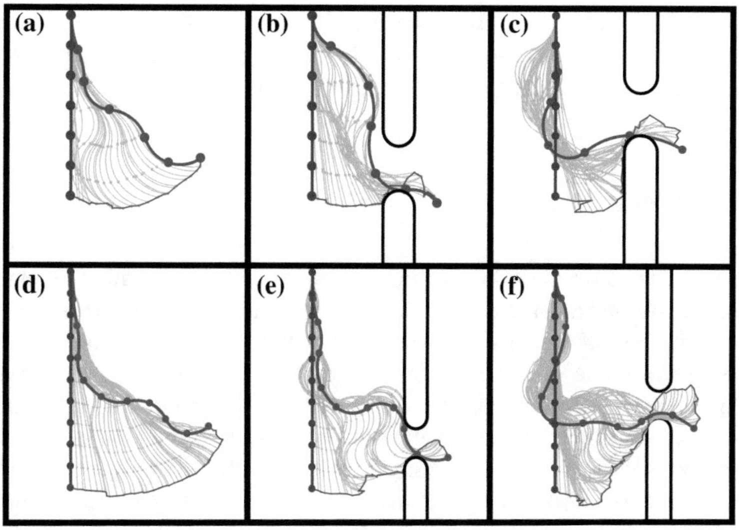

Fig. 5 Some typical conditions, including three levels of difficulty separately for 6 and 10 segments. Maps for easy, medium and hard level are on the left, middle and right column; 6 and 10 segments are on the first and second row

3.4 Extensional Experiments

To evaluate the algorithm in more complex environments with obstacles, we conduct extensional experiments. As shown in Fig. 6, the algorithm performs decent reaching rate in various situations. The tasks can be separated into three types: task (d) illus-

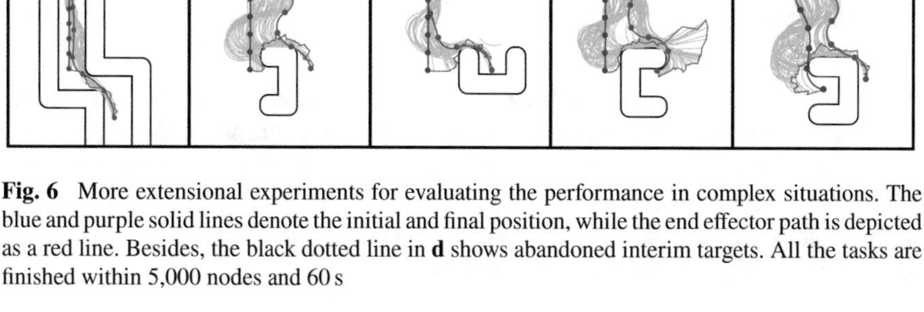

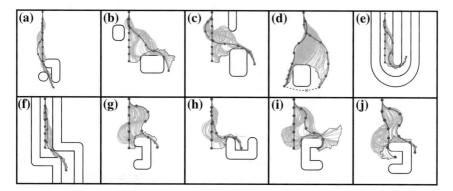

Fig. 6 More extensional experiments for evaluating the performance in complex situations. The blue and purple solid lines denote the initial and final position, while the end effector path is depicted as a red line. Besides, the black dotted line in **d** shows abandoned interim targets. All the tasks are finished within 5,000 nodes and 60 s

trates a misled situation of interim targets; tasks (g) and (f) shows the unexpected cases; other complex tasks can be easily finished in high success rate within 20 s and 2000 nodes. The prior two types will be further discussed in next section.

4 Discussions

We have compared the algorithm's performance in experiments with different working environment and manipulators with different segment numbers, where it exhibits high adaptability as well as efficiency. From the comparison experiment with Jacobian-based methods, we can conclude that local minima can be reasonably avoided using the interim targets. On the other hand, as the interim targets are generated regardless of whole manipulator configuration, it doesn't ensure practicable configuration sequence along these targets. In this situation, our algorithm performs as the classical goal-biased sampling methods. As shown in Fig. 6d, the manipulator manages to reach the goal after the black dotted guideline is abandoned. Considering the low cost to generate the interim targets and their success in most cases, it is a rational strategy for the motion planning of soft variable-length manipulators.

Besides, in the performance evaluation experiment, we notice there is an unusual change of the performance when the segment number is relatively low: the node number has a negative correlation with the segment number in hard level, different from easy and medium levels. As shown in the state's distribution in Fig. 5c, f, when the manipulator has less segments, the configuration distance between the trapped state and escaped state is greater than that with more segments. There are two main reasons. The first one is that when the segment number is low, the manipulator's reachable space is relatively small, which makes it hard to avoid the configuration gap of obstacles between the initial state and goal. And the other is the algorithm's limitation during the heuristic configuration generation in task space (Algorithm 4); we restrain the movements of constrained segments, which aggravates the lack of reachable space, makes it more dependent on the sampling process to escape from the trap and degrades the performance. Even though, our configuration selection strategy works efficiently in most practical scenes.

5 Conclusion

We present a two-level approach to motion planning problem for soft manipulators, which fully exploits workspace information as well as their property. The algorithm's performance is satisfactory in various scenarios, where it also enables the manipulators to avoid local minima. Besides, it is capable of dealing high dimensional configuration space, which makes the classical methods (Jacobian-based, etc.) inefficient yet is common in practical soft manipulators.

In our algorithm, as we cannot consider the information of the whole manipulator during the generation of interim targets, the guiding information is incomplete and can lead to traps in the heuristic configuration selection process. In these conditions, its performance is impaired because it relies more on the sampling part.

In future, we need to balance better the prior two levels in the planning algorithm. Potential methods can be developed as to dynamically correct the guideline when falling into traps. Besides, the algorithm is to be extended to the scenarios with dynamic obstacles.

Acknowledgements This research is supported by the National Natural Science Foundation of China under grant 61573333.

References

1. Rus, D., Tolley, M.T.: Design, fabrication and control of soft robots. Nature **521**(7553), 467–475 (2015)
2. Kim, S., Laschi, C., Trimmer, B.: Soft robotics: a bioinspired evolution in robotics. Trends Biotechnol **31**(5), 287–294 (2013)
3. Vahrenkamp, N., Berenson, D., Asfour, T., Kuffner, J., Dillmann, R.: Humanoid motion planning for dual-arm manipulation and re-grasping tasks. In: IEEE/RSJ International Conference on Intelligent Robots and Systems, 2009. IROS 2009, pp. 2464–2470. IEEE (2009)
4. Khatib, O.: Real-time obstacle avoidance for manipulators and mobile robots. In: Autonomous Robot Vehicles, pp. 396–404. Springer (1986)
5. Koditschek, E.: Exact robot navigation using artificial potential functions. IEEE Trans. Robot. Autom. **8**(5), 501–518 (1992)
6. Lydia, E., Kavraki, Svestka, P., Latombe, J.-C., Overmars, M.H.: Probabilistic roadmaps for path planning in high-dimensional configuration spaces. IEEE trans. Robot. Autom. **12**(4), 566–580 (1996)
7. LaValle, S.M.: Rapidly-Exploring Random Trees: A New Tool for Path Planning (1998)
8. Bertram, D., Kuffner, J., Dillmann, R., Asfour, T.: An integrated approach to inverse kinematics and path planning for redundant manipulators. In: Proceedings 2006 IEEE International Conference on Robotics and Automation, 2006. ICRA 2006, pp. 1874–1879. IEEE (2006)
9. Weghe, M.V., Ferguson, D., Srinivasa, S.S.: Randomized path planning for redundant manipulators without inverse kinematics. In: 2007 7th IEEE-RAS International Conference on Humanoid Robots, pp. 477–482. IEEE (2007)
10. Diankov, R., Ratliff, N., Ferguson, D., Srinivasa, S., Kuffner, J: Bispace planning: concurrent multi-space exploration. In: Proceedings of Robotics: Science and Systems IV, vol. 63 (2008)
11. Yao, Z., Gupta, K.: Path planning with general end-effector constraints: using task space to guide configuration space search. In: 2005 IEEE/RSJ International Conference on Intelligent Robots and Systems, 2005. (IROS 2005), pp. 1875–1880. IEEE (2005)
12. Denny, J., Amato, N.M.: Toggle prm: simultaneous mapping of c-free and c-obstacle-a study in 2D. In: 2011 IEEE/RSJ International Conference on Intelligent Robots and Systems (IROS), pp. 2632–2639. IEEE (2011)
13. Leo Keselman, Verriest, E., Vela, P.A.: Forage rrtan efficient approach to task-space goal planning for high dimensional systems. In: 2014 IEEE International Conference on Robotics and Automation (ICRA), pp. 1572–1577. IEEE (2014)
14. Littlefield, Z., Li, Y., Bekris, K.E.: Efficient sampling-based motion planning with asymptotic near-optimality guarantees for systems with dynamics. In: 2013 IEEE/RSJ International Conference on Intelligent Robots and Systems (IROS), pp. 1779–1785. IEEE (2013)

15. Alterovitz, R., Patil, S., Derbakova, A.: Rapidly-exploring roadmaps: weighing exploration vs. refinement in optimal motion planning. In: 2011 IEEE International Conference on Robotics and Automation (ICRA), pp. 3706–3712. IEEE (2011)
16. Rickert, M., Sieverling, A., Brock, O.: Balancing exploration and exploitation in sampling-based motion planning. IEEE Trans. Robot. **30**(6), 1305–1317 (2014)
17. Bhattacharya, P., Gavrilova, M.L.: Roadmap-based path planning-using the voronoi diagram for a clearance-based shortest path. IEEE Robot. Autom. Mag. **15**(2) (2008)
18. Webster III, R.J., Jones, B.A.: Jones. Design and kinematic modeling of constant curvature continuum robots: a review. Int. J. Robot. Res. **29**(13), 1661–1683 (2010)
19. Jiang, H., Liu, X., Chen, X., Wang, Z., Jin, Y., Chen, X.: Design and simulation analysis of a soft manipulator based on honeycomb pneumatic networks. In: 2016 IEEE International Conference on Robotics and Biomimetics (ROBIO), pp. 350–356. IEEE (2016)

A Grasp Strategy for Polygonal Objects Using a Honeycomb Pneumatic Network Soft Gripper

Bing Tang, Rongyun Cao, Haoyu Zhao and Xiaoping Chen

Abstract In this paper, we propose a grasping strategy based on Honeycomb Pneumatic Network (*HPN*) soft gripper. Theoretically, the soft gripper has infinite degrees of freedom (*DOF*) so it can fit surface of objects efficiently. The strategy is used to judge whether a certain status meets the conditions of relative form closure theory. In order to verify the effectiveness of the strategy and obtain optimal grasping position of objects, our experiment is carried out in Bullet Physics Engine. We change the variables that affect the success rate of grasping objects and add stochastic noises to simulated environment so that we can get the success rate of grasping in different positions. Besides, we also put forward an evaluation function that can verify the effectiveness of our strategy. Combining the strategy with evaluation function, we can greatly improve the success rate of grasping on a variety of ordinary geometrical objects.

Keywords HPN · Simulated environment · Grasping strategy
Evaluation function

1 Introduction

Over the past decades, an increasing number of researchers have focused on robotic grasping. Diverse algorithms, which are based respectively on analytical and

B. Tang (✉) · R. Cao · H. Zhao · X. Chen
Department of Computer Science, University of Science and Technology of China, No. 96, JinZhai Road, Baohe District, Hefei 230026, Anhui, People's Republic of China
e-mail: tb9527@mail.ustc.edu.cn

R. Cao
e-mail: ryc@mail.ustc.edu.cn

H. Zhao
e-mail: zhy4567@mail.ustc.edu.cn

X. Chen
e-mail: xpchen@ustc.edu.cn

J.-H. Kim et al. (eds.), *Robot Intelligence Technology and Applications 5*,
Advances in Intelligent Systems and Computing 751,
https://doi.org/10.1007/978-3-319-78452-6_35

empirical approaches, have been developed for synthesizing robotic grasps so as to acquire stable grasping or meet the conditions of force-closure [1]. Analytical approaches take primarily kinematics and dynamics formulations into account in determining grasps. On the basis of the property that each point in 2D space can be parameterized with two parameters linearly, Ponce et al. develop necessary linear conditions for force-closure grasping of three and four-finger and implement them as a set of linear inequalities in the contact positions [2, 3]. The author in [4] presents an approach which is based on a linear matrix inequality formalism for computing a task-oriented quality measure, dealing with friction cone constraints without the pyramidal approximation. Empirical approaches refer to the technique based on classification and learning methods that avoids the computational complexity of analytical ones. Aleotti and Caselli also propose an approach for programming task-oriented grasping by feat of user-supplied demonstrations [5]. In article [6], the author develops some techniques which learn to identify grasping regions in an object image.

Although there are many demonstrations, analytical approaches suffer from computational complexity and empirical approaches require plenty of experimental data. In recent years, the emergence and development of soft robot technology [7–11] have exploited a new field for the research and application of grasping because of self-adaptibility brought by infinite *DOF*. In our previous work, Hao Sun et al. put forward a novel structure *HPN* which is composed of honeycomb-shaped pneumatic network and prove that it can accomplish elongation and bend in any direction by controlling each airbag to proper pressure [12]. Inspired by *HPN*, we propose a prototype of soft gripper which possesses theoretically infinite *DOF*, as shown in Fig. 1. Due to the complexity of modeling caused by nonlinearity and viscoelasticity of soft grippers, it is difficult to make use of analytical approaches for precise manipulation. There are two main disadvantages of empirical approaches: a high demand of a large amount of experimental data and poor portability. Both of two approaches are not completely suitable for our proposed soft gripper. Therefore, we present a strategy hybridizing analytical approaches and empirical approaches: 1. Obtain a feasible grasping position set based on relative form closure theory. 2. Acquire the optimal grasping position based on an evaluation function derived from experiments. In the above two steps, a simple mathematical model is used to obtain feasible grasping position by analytical approaches, afterwards, the optimal grasping position can be acquired with the evaluation function, which is adjusted with a little data. Consequently, it is not necessary for the strategy to obtain a precise kinematic model and numerous experimental data. Besides, we build a simulated experiment on the basis of [13] and verify the effectiveness of the strategy.

The main contribution of this paper is a demonstration of grasping strategy for polygonal objects considering the characteristic (infinite *DOF*) of soft gripper. Another contribution is that we present an effective evaluation function to reflect the success rate of grasping objects in different positions and provide evidence for optimizing evaluation function. We add stochastic noise as disturbance when grasping objects in the simulated environment so that we can evaluate grasping stability. The performance of our strategy and evaluation function is demonstrated with numerous experiments.

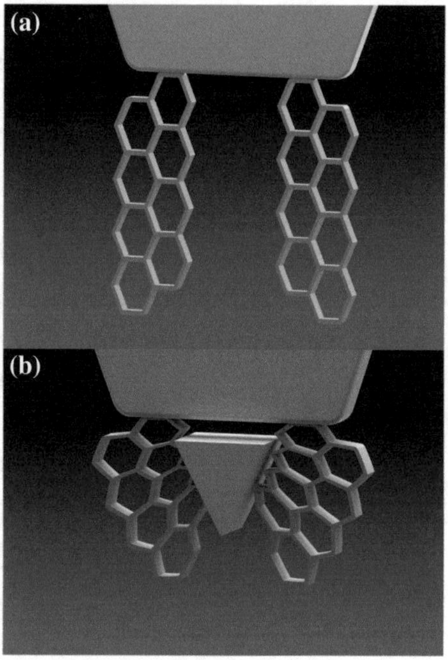

Fig. 1 Prototype of honeycomb pneumatic soft gripper

The rest of the paper is organized as follows. We present characteristics and physical parameters about the *HPN* soft gripper in Sect. 2. Section 3 describes the modeling process of grasping theory primarily and introduces concrete process of obtaining feasible grasping solution. Section 4 utilizes simulated experiment to verify the grasping theory mentioned above. Finally, conclusion is shown in Sect. 5.

2 Honeycomb Pneumatic Network Soft Gripper

HPN is inspired by natural honeycomb structure. The structure is composed of two parts: compressible honeycomb structure and pneumatic network system. The former has many merits, such as extensible and bendable properties and crush resistance, while pneumatic network system is lightweight and environmentally benign. Combining the properties, we place pneumatic units inside the honeycomb-shaped structure. Each pneumatic unit is controlled individually so that the whole *HPN* is able to elongate or bend when pressurized. The structure has many hexagonal chambers which can change their geometrical shape without soft material deforming dramatically, so the main body frame of *HPN* can be made of relatively stiffer material. These characteristics ensure its stability and a better structural strength, which greatly broaden the scope of its applications.

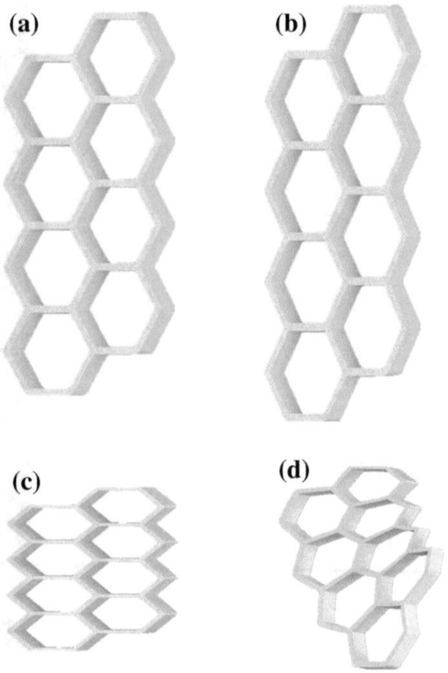

Fig. 2 Four states of honeycomb pneumatic network: **a** Relaxation **b** Extension **c** Contraction **d** Bend

In our work, the soft gripper is designed and fabricated based on *HPN*. As is shown in Fig. 2, each finger consists of parallel *HPN* units and each unit can be embedded with airbag. When the airbag is inflated or deflated, there is a certain degree of deformation so that the whole framework can achieve diverse elongation and bend by controlling the pressure of each unit.

3 Grasp Planning of Soft Gripper

3.1 Grasp Planning

Grasp planning refers to taking uncertainty in shape or position of objects into account to seek the optimal grasping position when manipulating objects. Soft gripper is under-actuated, so it is difficult to control the process of grasping object precisely. There are a series of constraints that increase the difficulty of accurate grasping, such as the geometry of target object and disturbances in manipulation environment. Consequently, grasping task is so complex that effective grasping strategy should be applied to it. Based on *HPN*, the optimal grasping position can be determined as follows:

Input: Geometric shape of object, Parameter of gripper(Length)
Output: Optimal grasping position

---

1. Obtain grasping position set according to geometric shape of object and parameter of gripper.
2. For each element of grasping position set.
   (a) Obtain final status of soft gripper by calculation.
   (b) Judge whether the status meets requirements of relative form closure theory.
   (c) Put the grasping position corresponding to eligible status into feasible grasping position set.
3. For each element of feasible grasping position set
   (a) Calculate its evaluation function value.
   (b) Put the value into evaluation function value set.
4. Obtain optimal value of evaluation function value set.
5. Acquire the grasping position corresponding to the optimal value.

3.2 Feasible Grasping Solution

In the modeling process of grasping theory, we simplify conditions which have little influence on experimental results but dramatically increase difficulty in controlling. We make following assumptions:

1. Objects are lightweight and rigid.
2. Grasp rigidly and take no account of contact surface's friction.
3. Soft gripper has infinite *DOF* in reachable workspace.
4. There must be the optimal grasping distance ($D^{center}_{optimal}$), which is the distance between optimal grasping position and centroid of object.

When objects are lightweight and rigid, soft gripper can execute rigid grasping on objects and we don't need to consider friction of contact surface. Then, we can acquire stable grasping based on relative form closure theory. Besides, soft gripper has infinite *DOF* so that it can fit surface of objects well.

Grasping manipulation can be divided into two categories according to grasping method: grasping manipulation in 2D space and grasping objects by cladding in 3D space. Soft gripper we design is composed of two parallel *HPN* fingers, possessing high *DOF* in 2D space. It is suitable to do some matching and planning when cross section of object is determined. In this article, we concentrate on planning problem after determining grasping cross section. According to geometrical knowledge, for an object with polyhedral structure, its grasping cross section must be polygonal structure. In particular cases, some objects contain curved surface, which results in grasping cross section of curved surface containing curved side. The most typical examples are sphere and cylinder. Their grasping cross sections are usually circles. Each vertex of inscribed hexagon of circle is on the circle so that we can approximate the circle by using its inscribed hexagon. For given object meeting the conditions of feasible grasping, we can assume that its grasping cross section is polygonal structure so that we can utilize vertex coordinate set V to represent shape of polygonal structure.

$$V = \{A_1, A_2, \ldots, A_N\} \quad (1)$$

$A_{1,2,\ldots,N}$ are respectively vertexes of polygonal structure and N is the number of vertexes. If we use T_l and T_r to represent contact point sets of two fingers individually, d_{lleft} and d_{rleft} represent remnant length of two fingers that don't contact with the object.

$$T_{l,r} = \{A_i, A_{i+1}, \ldots, A_{i+n-1}\}, \quad N - n + 1 \geq i \geq 1 \quad (2)$$

n is the number of contact points and i is subscript of initial point.

Solution model of feasible grasping solution is as follows:

1. Use formula $\mathbf{V}_{center} = \frac{1}{N}\sum_{i=1}^{N}\overrightarrow{PA_i}$ to represent geometric center vector. Under the premises that result is not affected, regard centroid of object as the origin O by shifting and rotating the coordinate system.
2. Acquire vertex set V of object and position P of gripper, lengths of two fingers are G_{len}. Regard the point that is nearest to position P as A_1, remanent points are labeled in clockwise direction. As Fig. 3 shows.
3. Initialize relevant parameters. Originally, soft gripper is not in contact with object, therefore, T_l, T_r, and $D_{l,r}$ which are distance sets between remanent points and contact points are initialized to empty and d_{lleft} and d_{rleft} are set to G_{len}.
4. Obtain unit normal vector $\boldsymbol{n}$ which is situated in the flat. There is a relationship between $\mathbf{V}_{center}$ and $\boldsymbol{n}$:

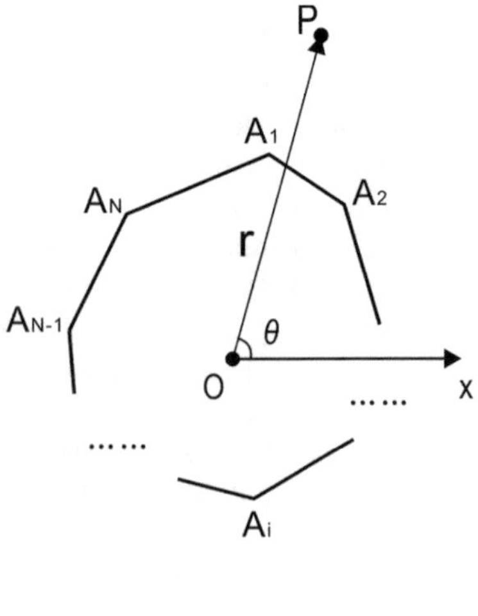

Fig. 3 P is the position of soft gripper, and vertexes are labeled in clockwise direction, the point nearest to P is regarded as A_1

$$\mathbf{V}_{center} \cdot \boldsymbol{n} = 0 \quad (\|\boldsymbol{n}\| = 1) \tag{3}$$

5. Divide vertex set V into two subsets according to (4): U_l and U_r, whose elements are located on either side of geometric center vector $\mathbf{V}_{center}$.

$$\begin{cases} U_l = \{A_i\}, & \overrightarrow{PA_i} \cdot \boldsymbol{n} > 0 \\ U_r = \{A_i\}, & \overrightarrow{PA_i} \cdot \boldsymbol{n} < 0 \end{cases} \tag{4}$$

6. For each element of set S, calculate angle sets ψ_l and ψ_r between vector $\overrightarrow{PA_i}$ and geometric center vector $\mathbf{V}_{center}$, and distance sets D_l and D_r.

$$\begin{cases} \psi_l = \left\{ \arccos \dfrac{\overrightarrow{PA_i} \cdot \mathbf{V}_{center}}{\left\|\overrightarrow{PA_i}\right\| \cdot \|\mathbf{V}_{center}\|}, \quad A_i \subset U_l \right\} \\ \psi_r = \left\{ \arccos \dfrac{\overrightarrow{PA_i} \cdot \mathbf{V}_{center}}{\left\|\overrightarrow{PA_i}\right\| \cdot \|\mathbf{V}_{center}\|}, \quad A_i \subset U_r \right\} \end{cases} \tag{5}$$

$$\begin{cases} D_l = \left\{ \left\|\overrightarrow{PA_i}\right\|, \quad A_i \subset U_l \right\} \\ D_r = \left\{ \left\|\overrightarrow{PA_i}\right\|, \quad A_i \subset U_r \right\} \end{cases} \tag{6}$$

7. Obtain subscript of next contact point according to angel sets $\psi_{l,r}$ and distance sets $D_{l,r}$.

$$\begin{cases} I_l = \{i, \theta_i = \max \psi_l, d_i < d_{lleft}, d_i \subset D_l\} \\ I_r = \{i, \theta_i = \max \psi_r, d_i < d_{rleft}, d_i \subset D_r\} \end{cases} \tag{7}$$

8. If I_l and I_r are not empty at the same time, the final status has not reached yet, then redirect to step 9. Otherwise, we have gotten T_l, T_r, d_{lleft} and d_{rleft} which are used to describe the final status. Based on T_l and T_r, we judge whether a group of contact points are satisfied with relative form closure theory. If the result meets the conditions of the theory, this group of contact points are considered as a feasible grasping solution. Otherwise, this grasping is not stable.
9. I_l and I_r are not empty at the same time, the farthest feasible vertex in intersection is next extensible one. The definition of subscript l and r is as follows:

$$\begin{cases} d_l = \max d_i, i \subset I_l \\ d_r = \max d_i, i \subset I_r \end{cases} \tag{8}$$

add the new extensible vertex into T_l or T_r.

$$\begin{cases} T_l = T_l \cup \{A_l\} \\ T_r = T_r \cup \{A_r\} \end{cases} \tag{9}$$

update d_{lleft} and d_{rleft}

$$\begin{cases} d_{lleft} = d_{lleft} - d_l \\ d_{rleft} = d_{rleft} - d_r \end{cases} \tag{10}$$

10. Based on former contact points, calculate ψ_l, ψ_r, D_l and D_r.

$$\begin{cases} \psi_l = \left\{ \arccos \dfrac{\overrightarrow{A_lA_{l+1}} \cdot \overrightarrow{A_lA_{l+i}}}{\left\|\overrightarrow{A_lA_{l+1}}\right\| \cdot \left\|\overrightarrow{A_lA_{l+i}}\right\|}, \quad r - l > i > 1 \right\} \\ \psi_r = \left\{ \arccos \dfrac{\overrightarrow{A_rA_{r-1}} \cdot \overrightarrow{A_rA_{r-i}}}{\left\|\overrightarrow{A_rA_{r-1}}\right\| \cdot \left\|\overrightarrow{A_rA_{r-i}}\right\|}, \quad r - l > i > 1 \right\} \end{cases} \tag{11}$$

$$\begin{cases} D_l = \left\{ \left\|\overrightarrow{A_lA_{l+i}}\right\|, \quad r - l > i > 1 \right\} \\ D_r = \left\{ \left\|\overrightarrow{A_rA_{r-i}}\right\|, \quad r - l > i > 1 \right\} \end{cases} \tag{12}$$

Redirect to step 6.

3.3 Evaluation Function

In order to obtain superior grasping solution from feasible grasping solution set, we need to judge which is optimal. Therefore, it is necessary to find an efficient grasp quality measure in terms of structural features. In our work, we take primarily the equation set (13) into our account. We assume that A_i and A_j are respectively the elements of T_l and T_r, A_{left} and A_{right} are vector sum of vectors which are respectively located on either side of the vector $\overrightarrow{PO}$ (*from position P to centroid of object O*), as shown is Fig. 4, so evaluation function is shown below:

$$Q_{score} = Q_{score1} - Q_{score2} + Q_{score3}$$
$$Q_{score1} = \frac{S}{r^2}$$

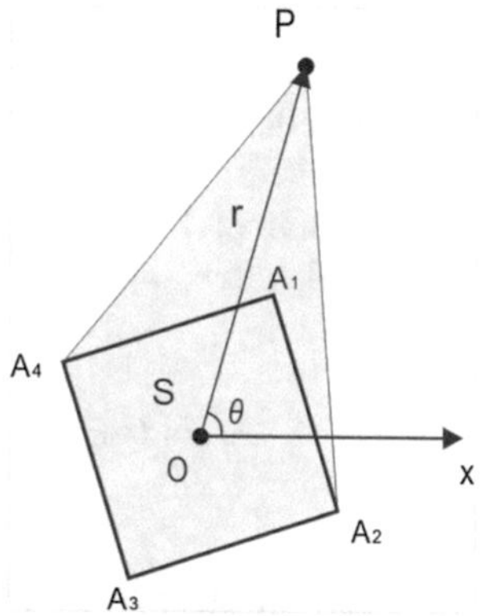

Fig. 4 Diagrammatic sketch of evaluation function

$$Q_{score2} = \frac{r^2}{c_1\left(r - R_{max}\right)^2} \tag{13}$$

$$Q_{score3} = c_2 \frac{S}{r^2} e^{-c_3\left(\frac{\left|\theta_{left}-\theta_{right}\right|}{\theta_{left}+\theta_{right}} + \frac{\left|l_{left}-l_{right}\right|}{l_{left}+l_{right}}\right)}$$

where Q_{score} is ultimate evaluation result, Q_{score1} represents impact exerted on result by S and L; Q_{score2} is to reduce the impact brought by the small value of $\left\|\overrightarrow{PO}\right\|$; Q_{score3} is used to characterize the effect of symmetrical degree about $\overrightarrow{PO}$ on result; For more details, S is the maximum area of polygons that are composed of A_i, A_j, and optional vertex set. r represents the distance between centroid of object and position P. R_{max} represents the maximum distance from centroid of object to vertexes of optional vertex sets. θ_{left} and θ_{right} represent individually the angles between vector sum and $\overrightarrow{PO}$. l_{left} and l_{right} are respectively the corresponding length of A_{left} and A_{right}. c_1, c_2, and c_3 are nonnegative constants which are determined by a small number of experimental verifications and are used to adjust the weight of corresponding items. What we need to emphasize is that c_3 is also used to adjust probability distribution of score on degree of symmetry.

We use the evaluation function to calculate corresponding scores of two different statuses for the same object. The status that gets higher score is regarded as better.

3.4 Solution Procedure of Grasping

According to above algorithm, specific solution is as follows:

1. Acquire vertex set V of object and length G_{len} of soft gripper, evaluation function value set G_{score} of feasible grasping position is initialized to be empty.
2. Determine grasping position set G_{Pos} according to vertex set V.

$$G_{Pos} = \{P_1, P_2, \dots P_{2N}\} \tag{14}$$

The vertex nearest to position P is A_1. Remanent vertexes are labeled in clockwise direction. Especially, vertex A_1 is adjacent to vertex A_N and we can make an assumption: $A_{N+1} = A_1, A_{1-1} = A_N$. The relationship between grasping position P_i and vertex A_i is shown in the following equation.

$$\overrightarrow{OP_i} = \overrightarrow{OA_i} + \lambda_i \left(\frac{\overrightarrow{A_{i+1}A_i}}{\left\| \overrightarrow{A_{i+1}A_i} \right\|} + \frac{\overrightarrow{A_{i-1}A_i}}{\left\| \overrightarrow{A_{i-1}A_i} \right\|} \right), \quad \lambda_i \geq 0, N \geq i \geq 1 \tag{15}$$

λ_i is determined by $\left\| \left\| \overrightarrow{OP_i} - \mathbf{V}_{center} \right\| - D_{optimal}^{center} \right\|$ when the minimum of the latter is obtained and P_i is not inside the polygon.

$$\overrightarrow{OP_{N+i}} = \overrightarrow{OA_i} + \frac{1}{2}\overrightarrow{A_iA_{i+1}} + \lambda_{N+i}\boldsymbol{n}, \quad \lambda_{N+i} \geq 0, N \geq i \geq 1 \tag{16}$$

$\boldsymbol{n}$ is unit normal vector of $\overrightarrow{A_iA_{i+1}}$, pointing to the outside of the polygon. Similarly, λ_{N+i} is determined by $\left\| \left\| \overrightarrow{OP_{N+i}} - \mathbf{V}_{center} \right\| - D_{optimal}^{center} \right\|$ when the latter gets the minimum value and P_{N+i} is not inside the polygon.

3. For each element in collection G_{Pos}, the process of calculating feasible grasping solution is as follows:

   (a) If collection G_{Pos} is traversed completely, then redirect to step 5.
   (b) If we obtain a group of feasible solution T_l and T_r, then redirect to step 4.
   (c) If there is no feasible solution, then recalculate.

4. For results that step 3 outputs, calculate corresponding evaluation function values, put them into collection G_{score}.

$$G_{score} = G_{score} \cup \{Q_{score}\} \tag{17}$$

redirect to step 3.

5. The position corresponding to maximum value of collection G_{score} is optimal grasping position.

3.5 Relative Form Closure Theory

In a given reference system, grasping **G** is depicted by a group of line vectors $\boldsymbol{p}$[14], which are represented by outward unit normal vector $\boldsymbol{n}$ and radius vector $\boldsymbol{r}$ of contact point on rigid body.

$$\boldsymbol{p} = \begin{bmatrix} \boldsymbol{n} \\ \boldsymbol{r} \times \boldsymbol{n} \end{bmatrix} \tag{18}$$

$$\mathbf{G} = \{\boldsymbol{p}_1, \boldsymbol{p}_2, \cdots, \boldsymbol{p}_m\} \tag{19}$$

The space composed of line vectors $\boldsymbol{p}$ is called spin space and expressed by R^6, constraint cone of grasp $\mathbf{G}$:

$$\boldsymbol{P}[\boldsymbol{G}] = \left\{ \boldsymbol{p} = \sum_{i=1}^{m} \lambda_i \boldsymbol{p}_i, \lambda_i \geq 0 \right\} \tag{20}$$

Rigorous definition of form closure: if spin space R^6 is fulfilled with constraint cone $\boldsymbol{P}[\boldsymbol{G}]$ of grasping $\mathbf{G}$, we affirm that $\mathbf{G}$ meets the conditions of form closure.

$$\boldsymbol{P}[\boldsymbol{G}] = \mathbf{R}^6 \tag{21}$$

It means that grasping $\mathbf{G}$ will keep stable even if it encounters a external force. Nevertheless, in practical application, it is difficult to meet the conditions completely. Meanwhile, it has been proved by Mishra that rotating surface, spiral surface and flat are impossible to achieve form closure [15], which results in emergence of relative form closure theory.

A method that is applicable to detect and judge relative form closure theory is to define judgement function $\boldsymbol{J}_0(\boldsymbol{G})$ of grasping $\mathbf{G}$:

$$\boldsymbol{J}_0(\boldsymbol{G}) = \min_{\omega_0} \sum_{i=1}^{n+1} y_i \tag{22}$$

Feasible set ω_0 is determined by solution set of system of inequalities:

$$\begin{cases} \sum_{i=1}^{m} \lambda_i \boldsymbol{p}_i + \boldsymbol{y} = \boldsymbol{0} \\ \sum_{i=1}^{m} \lambda_i + y_{n+1} = 1 \\ \lambda_i \geq 0, \quad i = 1, 2, \ldots, m \\ y_j \geq 0, \quad j = 1, 2, \ldots, n, n+1 \end{cases} \tag{23}$$

$\mathrm{y} = [y_1, y_2, \ldots, y_n]^{\mathrm{T}}$ and y_{n+1} are artificial variables. According to algebra equivalence theorem, we can draw a conclusion that when judgement function $\boldsymbol{J}_0(\boldsymbol{G})$ is equal to zero, this group of contact points satisfy relative form closure theory.

4 Experimental Verification

This section presents a few results of our approach and makes a detailed analysis of the results. In reality, thousands of experiments are needed to verify strategy so that it is not proper to verify our strategy by artificial approaches. On the contrary, the grasping strategy and theory can be not only verified quickly in the simulated environment, but it is also convenient for us to change some properties of experimental environment. In this paper, we perform theoretical verification and analysis of results, which are carried out on the basis of previous work [13]. In experimental verification, we perform open-loop grasping based on finger-closing soft mechanism and the result of each grasping is judged by artificial handling.

4.1 Simulated Environment

In our experiment, we use Bullet Physics Engine which is an open source software based on position constraint to accomplish experimental simulation. The simulator which can be used to simulate soft structure must contain following parts:

1. Integrator: It is used to calculate location data according to load conditions of object. Its performance has a direct impact on the accuracy of simulator.
2. Simulation of material properties: Whether performance is consistent with reality when objects collide each other is determined by the accuracy of simulation of material properties, such as, height of the bounce when object falls to ground and distance when objects glide freely on ground.
3. Constraint stability: Constraint refers to combining two or more objects in simulator into a whole in some way. This kind of constraint is generally performed in the form of a joint in simulator.
4. Collision detection system: It is the most important part in simulator and describes the interaction among objects in real world and has significant embodiment in classical physics.
5. Collision detection performance: A perfect collision detection system is not only capable of detecting collision situation between two objects but also is effectively applied to the situation where lots of objects collide simultaneously.

However, in computer graphics, it is impossible to simulate how objects move in a real environment just by replying on geometric change of position of graph. The best way is to calculate physical parameters of object or particle by utilizing Newton's second law. In physics simulation engine, soft material is usually represented by a group of mass point network structures which are combined with springs. When mass points are compressed or stretched, elasticity of the whole object is charactered by elastic coefficient of springs. In order to simulate effect in reality, discrete time is used to simulate continuous time. Consequently, we can calculate current acceleration according to internal forces and external forces imposed on object at any

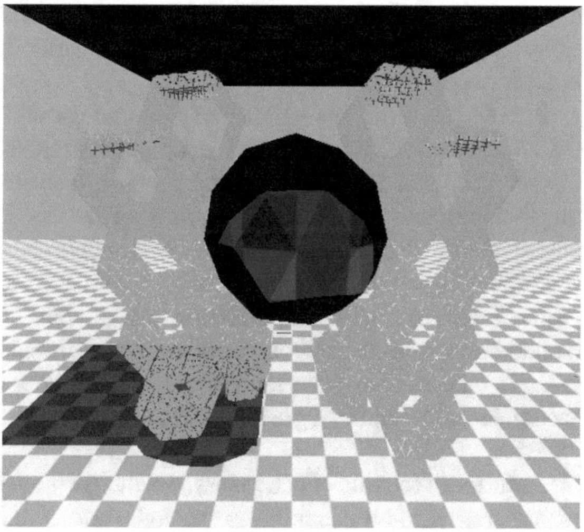

Fig. 5 Overview of grasping pose in simulated environment

time so that we can acquire velocity and displacement at next moment. In this way, continuous physical simulation effects are displayed. However, if the whole process takes place in a disturbed environment, it will result in an unstable and inaccurate result. Therefore, we use simulated method based on position [16], which increases interaction between users and simulated environment and the stability of simulation. Figure 5 is one of effect pictures of grasping in simulation.

4.2 Experimental Result and Discussion

In our simulated environment, both fingers consist of 8 honeycomb pneumatic units which are regular hexagon and are composed of a set of mass points and springs used to connect mass points. Expansion or contraction of airbags is essentially a change in the distance among mass points. The distances among mass points increase with pressures of airbags increasing. On the contrary, the distances decrease as the pressures get lower. Based on the theoretical basis, we can apply physical simulation to the whole *HPN* structure.

Under the circumstance of grasping position determined before grasping object in simulated environment, for each object, experimental input can be described from two aspects: 1. distance between centroid of object and position *P* of gripper, 2. rotation angle of object around its centroid. As important theme to notice, once variables which influence our experiment are fixed, we will get the same simulated results no matter how many experiments we implement. On this occasion, we have to introduce some stochastic noises: 1. distance noise: vertical distance between gripper's position *P* and centroid of object, 2. angle noise: rotation angle of object around its centroid, 3. deviation noise: deviation distance of object in horizontal direction,

4. mass noise: mass of object varies stochastically within a given range. Throughout experiment, we only need to control these two variables accompanied with stochastic noise for specified object.

In order to verify the validity of our strategy and the evaluation function mentioned above. We choose five objects: quadrangular prism whose cross section is isosceles trapezoid, cube, triangular prism whose cross section is equilateral triangle, cylinder, and polygonal prism whose cross section is dodecagon. In simulated environment, we conduct about 120 experiments on each of these five objects in the corresponding optimal grasping position obtained according to our grasping strate. Figure 6 depicts relationship between success rate and corresponding object. Horizontal coordinate is category of object and vertical coordinate represents success rate. As can be seen from the picture, success rate reach 100% when we operate each object at its optimal position. To some extent, It also shows that the strategy is effective.

To further verify its effectiveness, we choose three objects: cube, triangular prism whose cross section is equilateral triangle, and cylinder in simulated environment. For each object, we perform 100 experiments, calculate its success rate and evaluate corresponding score in different positions. The results are shown below:

Figures 7a, 8a, and 9a describe respectively success rates of grasping cube, triangular prism and cylinder when they are in different position. Especially, in Figs. 7 and 9, θ-axis represents polar angle, r-axis represents polar radius in polar coordinate system, vertical axis represents success rate when object is at corresponding position. In Figs. 7b and 9b, the meaning of θ-axis and r-axis is the same as that of

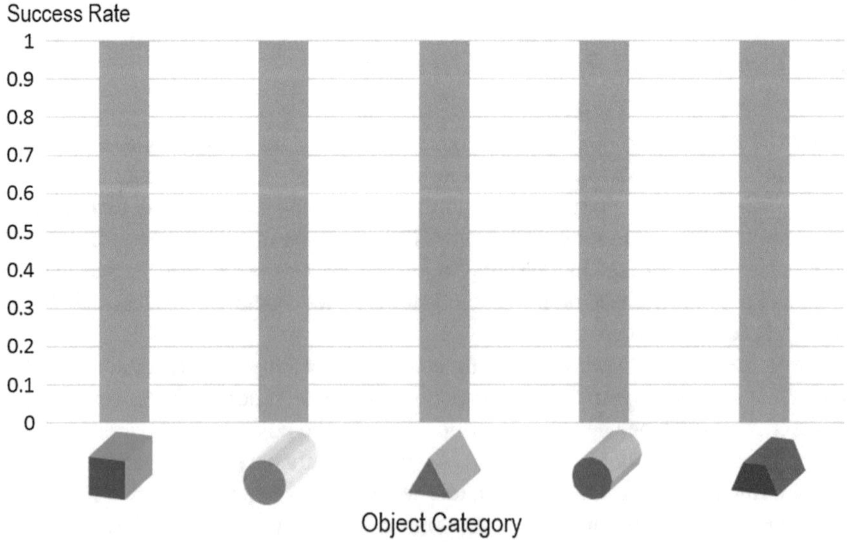

Fig. 6 Result of objects grasping in optimal grasping position obtained from our strategy. Horizontal coordinate represents different shapes: cube, cylinder, triangular prism, dodecagon, trapezoid

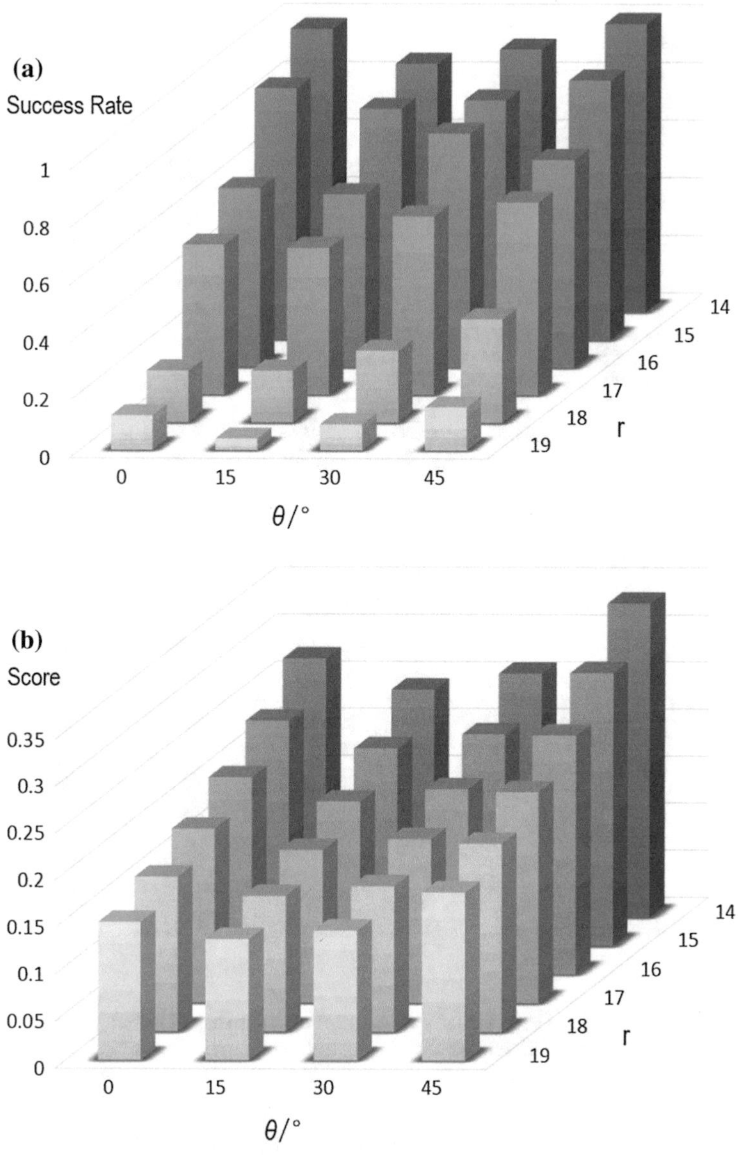

Fig. 7 Result of changing the grasping position of a rectangle-shaped object. **a** Success rate in different positions. **b** Score in different positions

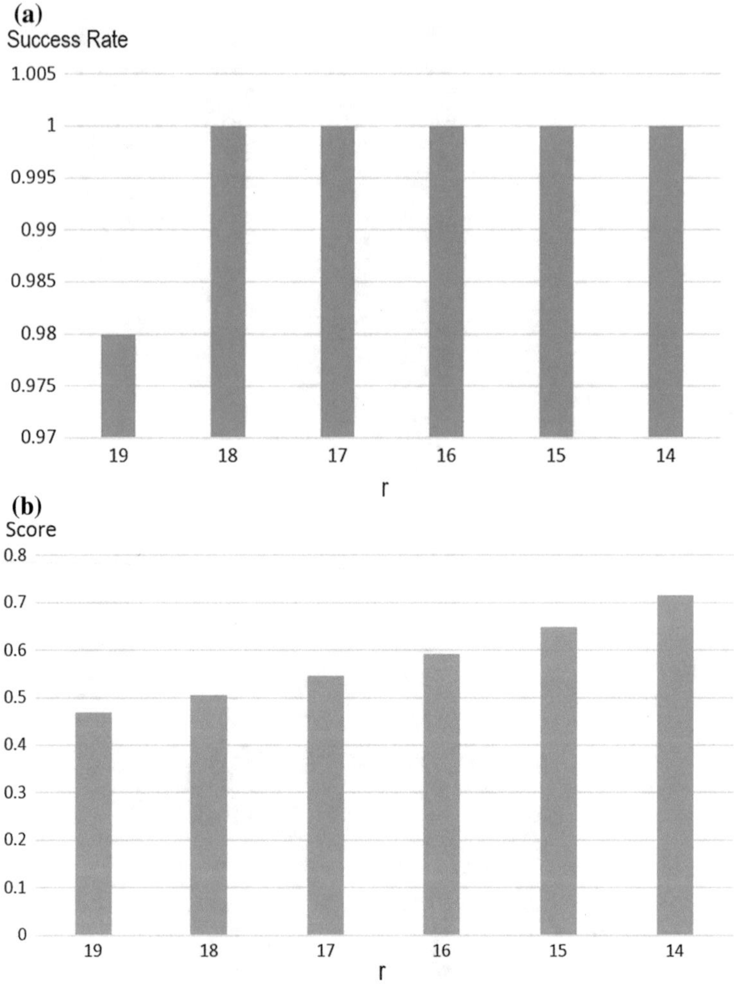

Fig. 8 Result of changing the grasping position of a cylinder-shaped object. **a** Success rate in different heights. **b** Score in different heights

Figs. 7a and 9a, vertical axis denotes scores obtained by calculation of evaluation function. What needs to be emphasized is that there is no θ-axis in Fig. 8 because of particularity of circle. Comparing two histograms in Figs. 7, 8, and 9, we can find out that success rate of grasping object is related to its score partially. For example, it is not difficult for us to find out that there is basically similar variation tendency in θ-axis or r-axis. It indicates that our grasping strategy and evaluation function are valid.

The reason why we introduce simulation rather than experiment on physical robots is that we can change experimental environment and the parameters more

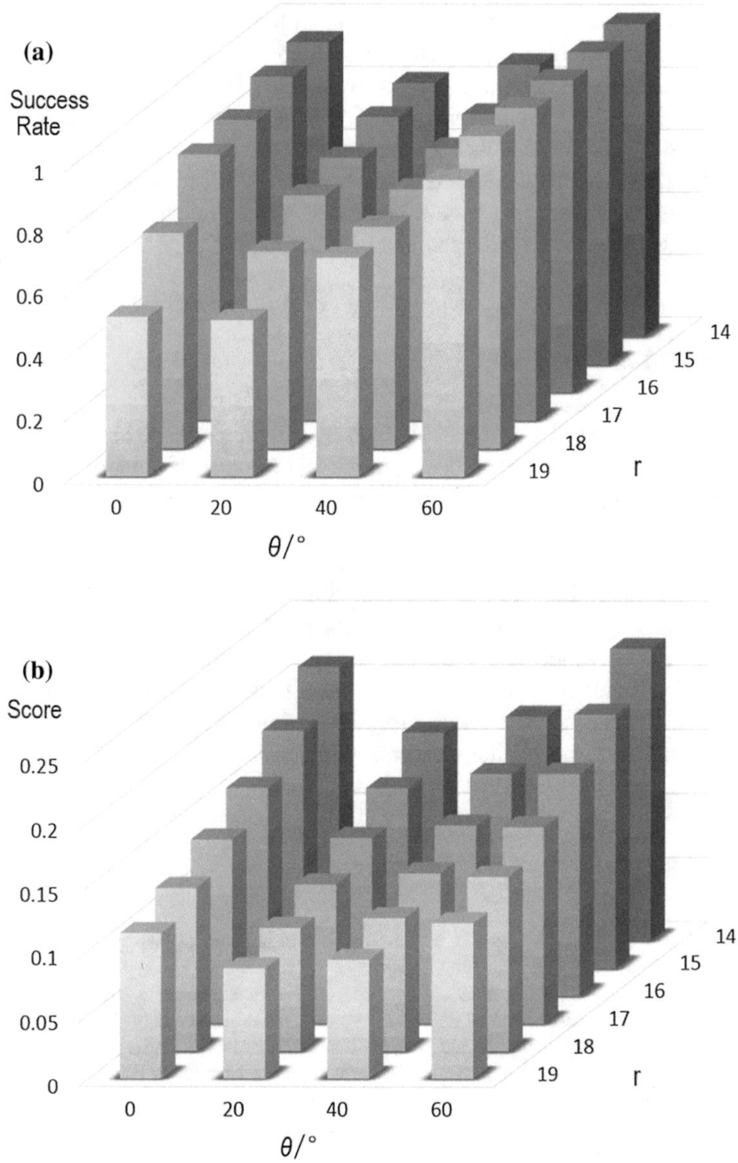

Fig. 9 Result of changing the grasping position of a triangle-shaped object. **a** Success rate in different positions. **b** Score in different positions

easily so that we can do plenty of experiments in limited time. In addition to introducing simulation, we have constructed an evaluation function based on the experimental situation, which gives us a perspective to verify the effectiveness of grasping strategy. Although the grasping strategy and evaluation function improve the success rate significantly and provide us with evaluation criterion for grasping effect, there is also deficiency. The whole process of our experiment is implemented in 2D space, which leads to the flaw that we have to know the cross section of objects. This makes our current grasping strategy limited in a certain range of application.

5 Conclusion

This paper puts forward a grasping strategy hybridizing analytical approaches and empirical approaches for soft gripper, presents an evaluation function according to characteristics of the structure and acquires the position that has maximum success rate. Our experiment performed in a simulated environment verifies the effectiveness of our strategy. Experimental results show that the grasping strategy can be applied to grasping manipulation and evaluation function can significantly improve success rate of grasping objects for a variety of objects. Meanwhile, the structure and grasping strategy can be widely expanded into grasping task of robots and have a highly promising development. In our future work, we will focus on promoting the grasping strategy into 3D space as well as enriching evaluation function so that we can obtain better evaluation criterion to increase grasp efficiency.

References

1. Sahbani, A., El-Khoury, S., Bidaud, P.: An overview of 3D object grasp synthesis algorithms. Robot. Auton. Syst. **60**(3), 326–336 (2012)
2. Ponce, J., Sullivan, S., Boissonnat, J.D., et al.: On characterizing and computing three-and four-finger force-closure grasps of polyhedral objects. In: Proceedings of the 1993 IEEE International Conference on Robotics and Automation, 1993, 821–827. IEEE (1993)
3. Ponce, J., Sullivan, S., Sudsang, A., et al.: On computing four-finger equilibrium and force-closure grasps of polyhedral objects. Int. J. Robot. Res. **16**(1), 11–35 (1997)
4. Haschke, R., Steil, J.J., Steuwer, I.: Task-oriented quality measures for dextrous grasping, Computational Intelligence in Robotics and Automation, et al.: In: Proceedings of the 2005 IEEE International Symposium on CIRA 2005, vol. 2005, pp. 689–694. IEEE (2005)
5. Aleotti, J., Caselli, S.: Programming task-oriented grasps by demonstration in virtual reality. In: Proceedings of IEEE/RSJ International Conference on Intelligent Robots and Systems, WS on Grasp and Task Learning by Imitation (2008)
6. Saxena, A., Driemeyer, J., Ng, A.Y.: Robotic grasping of novel objects using vision. Int. J. Robot. Res. **27**(2), 157–173 (2008)
7. Jiang, H., Liu, X., Chen, X., et al.: Design and simulation analysis of a soft manipulator based on honeycomb pneumatic networks. In: 2016 IEEE International Conference on Robotics and Biomimetics (ROBIO), pp. 350–356. IEEE (2016)

8. Deng, T., Wang, H., Chen, W., et al.: Development of a new cable-driven soft robot for cardiac ablation. In: 2013 IEEE International Conference on Robotics and Biomimetics (ROBIO), pp. 728–733. IEEE (2013)
9. Ilievski, F., Mazzeo, A.D., Shepherd, R.F., et al.: Soft robotics for chemists. Angew. Chem. **123**(8), 1930–1935 (2011)
10. Chou, C.P., Hannaford, B.: Measurement and modeling of McKibben pneumatic artificial muscles. IEEE Trans. Robot. Autom. **12**(1), 90–102 (1996)
11. Marchese, A.D., Rus, D.: Design, kinematics, and control of a soft spatial fluidic elastomer manipulator. Int. J. Robot. Res. **35**(7), 840–869 (2016)
12. Sun, H., Chen, X.P.: Towards honeycomb pneunets robots. Robot Intell. Technol. Appl. **2**, 331–340 (2014)
13. Cheng, B., Sun, H., Chen, X.P.: Evolving honeycomb pneumatic finger in bullet physics engine. Robot Intelligence Technology and Applications vol. 3, pp. 411–423. Springer, Cham (2015)
14. Xiong, Y.L.: Theory of point contact restraint and qualitative analysis of robot grasping. Sci. China Ser. A **5**, 629–640 (1994)
15. Mishra, B., Schwartz, J.T., Sharir, M.: On the existence and synthesis of multifinger positive grips. Algorithmica **2**(1), 541–558 (1987)
16. Mller, M., Heidelberger, B., Hennix, M., et al.: Position based dynamics. J. Vis. Commun. Image Represent. **18**(2), 109–118 (2007)

A Parameterized Description of Force Output of Soft Arms in Full Workspace

Chengkai Xia, Yiming Li, Xiaotong Chen, Zhanchi Wang, Yusong Jin, Hao Jiang and Xiaoping Chen

Abstract In this paper, aiming at fully taking advantage of soft manipulators working ability, we propose a parameterized approximating method to characterize their force output in full workspace. We define the Workspace-Load bearing capacity Cloud (WLC) of soft arms and present the method to calculate WLC in the three-dimensional space by linear fitting. At last, finite element analysis is used to validate its effectiveness in characterizing the force output of soft arms in full workspace.

Keywords Soft robot · Force output description · Workspace

C. Xia · Y. Li
School of Engineering Science, University of Science and Technology of China, Hefei, China
e-mail: xiack@mail.ustc.edu.cn

Y. Li
e-mail: ll100718@mail.ustc.edu.cn

X. Chen (✉) · Z. Wang · Y. Jin · H. Jiang · X. Chen
School of Computer Science, University of Science and Technology of China, Hefei, China
e-mail: chenxt11@mail.ustc.edu.cn

Z. Wang
e-mail: zkdwzc@mail.ustc.edu.cn

Y. Jin
e-mail: Roy08@mail.ustc.edu.cn

H. Jiang
e-mail: jhjh@mail.ustc.edu.cn

X. Chen
e-mail: xpchen@ustc.edu.cn

J.-H. Kim et al. (eds.), *Robot Intelligence Technology and Applications 5*, Advances in Intelligent Systems and Computing 751,
https://doi.org/10.1007/978-3-319-78452-6_36

1 Introduction

Compared with traditional rigid robots, soft robots can serve as better solutions in unstructured environments. Their continuous deformation and theoretically infinite degrees of freedom (DoFs) allow for adaptive movements in confined space and flexible interactions with unpredictable environments [1–4]. In this sense, soft robots have contributed to widening robots' application scenarios.

Generally, except for the fixed ones or those finished by trial-and-error methods, the robot manipulator requires awareness of its working ability of finishing tasks (such as a limit of carrying loads), which usually varies in different end-effector positions. Without proper awareness (or estimation), the manipulator may cause serious damage to itself and the environment during working, for instance, it will break down when its load bearing capacity is overestimated, or hurt humans during interaction by excessive forces. Accordingly, the awareness of working ability is substantial ensurance for robot manipulators to fully exploit their performance as well as work regularly, further, it is the precondition for robot manipulators to be widely applied.

As for the manipulation system, there are reasons influencing the working ability at a certain end-effector position, such as the manipulator's configuration [5, 6]. Due to their common highly-redundant DoFs, it is extremely difficult to get accurate solution in all configurations. In this sense, we focus on the force output at different positions, as a representative. For different manipulator prototypes, researchers have developed methods to explore its distribution among the workspace.

For traditional rigid manipulators, their force output relies on the coupled output torques. There exist two main definitions. The first is the maximum load under which the manipulator is able to freely manipulate in full workspace [7], and the other is the maximum load under which the manipulator is able to perform a predetermined trajectory [8]. The former is frequently adopted in tasks in unstructured environments, which has little effect on manipulator's performance, while the latter describes the maximum load requirement in repetitive motions. The definitions are derived based on an important property of traditional manipulators: their configurations vary little under different loads, which makes it possible to separately study the dexterity (workspace) and force output.

For soft manipulators, their working ability relies on the combination of actuation torques and structural characteristics [2], and the task influences the workspace. Therefore it is essential to learn their working ability in varying tasks in order to perform best in application scenes. Michael D. et al. designed OctArm by a combination of cabletendon and McKibben actuators. They defined vertical and horizontal load capacity, maximum rotation angle and maximum extension to characterize its performance in lab environment [9]. Daniel M. et al. designed an inflatable pneumatic manipulator and developed an evaluation method relying on sampling many randomly generated configurations and collecting custom metrics at each pose related to the dexterity and load bearing capacity [10]. H. Jiang et al. design and fabricate a soft manipulator based on honeycomb pneumatic networks. To measure its potential to application, they propose a combined evaluation criteria of load bearing capacity and flexibility [11]. In summary, these definitions regard the working ability as the manipulator's workspace or load bearing capacity in certain configurations, so they cannot represent real performance and ability in executing tasks.

In this paper, aimed at fully taking advantage of soft manipulators' working ability, we propose an approximating method to find their working ability in workspace, and we validate its effectiveness in simulation. The compliance of soft manipulators make them exhibit structural deformation and reach a completely new configuration, instead of breaking down, when executing a task beyond its ability. Therefore, for soft manipulators, we do not need a such accurate working ability model as their rigid counterparts, which ensures our approximation reliable.

2 Load Ability and Workspace

2.1 Workspace-Load Cloud

Soft robot arm has a natural advantage that they have common highly-redundant DoFs. However, this feature also makes its workspace and load ability (the load are the random direction cause what we most concerned is the mass that the arms can lift up so we use it represent the force output) difficult to measure by actuator and length of the arm. To deal with this issue, we propose a Workspace-Load Cloud (WLC), which can be used to judge the load ability and workspace of the arm. At first, we assume robots arms load ability in one location can be represented as the force can be output in this location along any direction. Therefore we define the aggregate of the location where the tip of arm can reach with the force enable to output and the direction as the WLC which is a six-dimensional space including a three-dimensional space of location where the arm can output a forces that are greater than zero in any three vertical directions and another three-dimensional of force, which the arm can output maximally at that location. Therefore, every location corresponds a three-dimensional force space. In other words, the space of that the tip can reach and the state of the arm are corresponding one-by-one. Therefore if we know the WLC of an arm, we can calculate the workspace with a given load at one direction or the load ability with a given the workspace. In the WLC, based on the assumption, we use $F(r_1, r_2)$ to stand for the maximum force the arm can output at location r_1 and direction r_2. The maximum force F the arm can output along the direction r_2.

$$F = \frac{F\left(r_1, \dfrac{r_2}{|r_2|}\right)}{|r_2|} \cdot r_2$$

And then when the arm carries a known payload G, we can know the new $F_{new}(r_1, r_2)$.

$$F_{new} = F + G$$

$$F_{new} = \frac{F\left(r_1, \frac{r_2}{|r_2|}\right)}{|r_2|} \cdot r_2 + G$$

Therefore $F_{new}(r_1,\ r_2)$ can be written into

$$F_{new}\left(r_1, \frac{r_2}{|r_2|}\right) = F\left(r_1, \frac{r_2}{|r_2|}\right) + \frac{G \cdot r_2}{|r_2|}$$

Hence, there is the problem of how to get an arm's WLC. Getting a real WLC of an arm is too hard of getting any force in any locations. Luckily, in view of the fact that the space and the force are consecutive in theory. Actually, the real WLC must be measured cause the materials and constructions are not standard, but we don't need accurate result so we use the fitting method to get the WLC. First, we build a three dimensional coordinate system in the center of the workspace. The three axis are three directions which are vertical to each other. Then we use an octahedron to characterize the workspace for example. After that we can get six locations that are the maximum along the six directions (including three directions of axis and their opposite directions) in the coordinate system. We know that in these locations, the force the arm can output is only along the opposite direction, because if it can output the force along another direction, that location would not be the maximum along the six directions. Still then we just need to measure only one force at one location. As we get these data of the arm, we can use the fitting method to get the WLC. For example, when there is an arm, we define the arm tip's original location as the origin of coordinates. The direction of the arm is x-axis. The direction of vertical is y-axis. And the horizontal direction is z-axis. We use $P_1(x_1, y_1, z_1)$ to $P_6(x_6, y_6, z_6)$ to stand for the location of the maximum deformation at x+, x-, y+, y- and z+, z- (shown in Fig. 1). And the $F_{x+}, F_{x-}, F_{y+}, F_{y-}, F_{z+}, F_{z-}$ stand for the force along the direction opposite the corresponding location the arm could bear. Therefore, we can calculate the maximum force the arm can output at any location. For example, the direction along the x-, the maximum force F_{x-} can be calculated with using the following way. We used the $P(a, b, c)$ to represent the location where we need to calculate the force arm can bear.

When the P is in the tetrahedron $P_1P_2P_3P_6$, the maximum force record as F'_{x-} (α, β, γ) and γ are indexes. They have no specific meaning.

$$\begin{pmatrix} x_2 - x_1 & x_3 - x_1 & x_6 - x_1 \\ y_2 - y_1 & y_3 - y_1 & y_6 - y_1 \\ z_2 - z_1 & z_3 - z_1 & z_6 - z_1 \end{pmatrix} \cdot \begin{pmatrix} \alpha \\ \beta \\ \gamma \end{pmatrix} = \begin{pmatrix} \alpha - x_1 \\ \beta - y_1 \\ \gamma - z_1 \end{pmatrix}$$

The force at the surface $P_2P_3P_6$ was considered as zero. Therefore,

$$F'_{x-} = (\alpha + \beta + \gamma) \cdot F_1$$

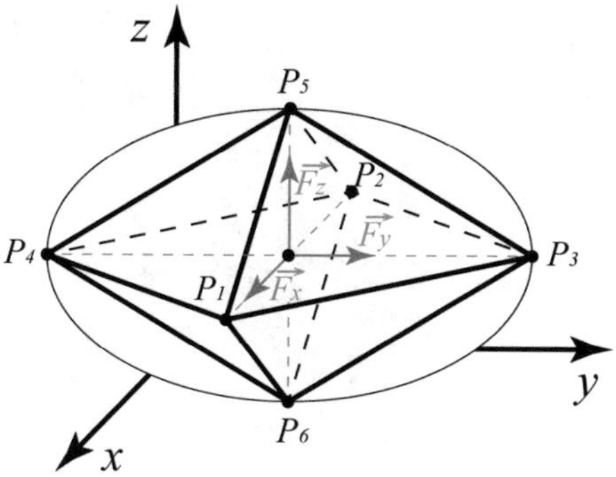

Fig. 1 This figure illustrates the 3D force output distribution, where P_1P_6 stand for the locations with maximum deformations on x, y, and z axis, and F_x, F_y, F_z stand for the force along these directions. Inside the octahedron $P_1P_2P_3P_4P_5P_6$, each 3D point has its maximum force output,. The force output on surfaces can be considered as zero along the opposite direction, from which the force output at points inside can be calculated using linear combinations in this part

In the same way, the output force along the x- direction in other tetrahedrons can be solved. All results (including F'_{x-}) are recorded as F_{x-}. As while as the force along other direction $F_{x+}, F_{y+}, F_{y-}, F_{z+}, F_{z-}$ can be solved. And then these six forces were considered as the force's max component along those directions. Therefore the maximum force along any direction is known. In addition, the P isn't a specific location so it is expressed as the $F(r_1, r_2)$ above.

What above is just a way to solve the WLC of an arm. The data which used to fit have no definite requirement. But the data on the boundary of feasible region was recommend, because the force has no component in the outward direction of the normal.

If high accuracy is needed, we could use high order fitting and measure more force at more locations along more directions.

3 Simulation Result

Test I, Test II and Test III are used to measure the data which is use to fit and test the error between estimation and fact. Test IV and Test V, we try to test our method's performance in the three-dimensional space.

3.1 Setup

To evidence that our method can reflect the fact to a certain extent, we use the finite element analysis (ANSYS 17.0). We use the Festo arm [12] model to run the static

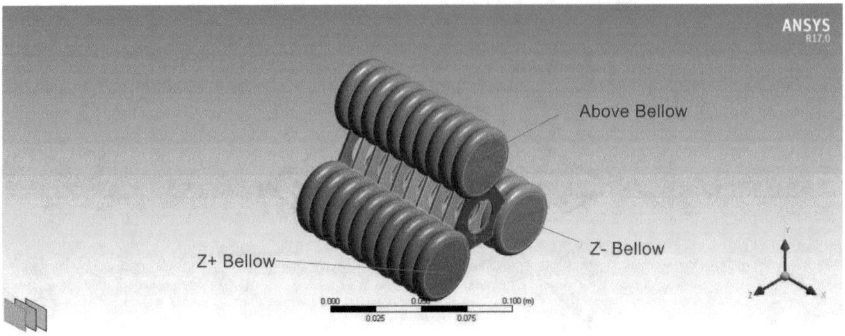

Fig. 2 This figure is the model we used in the simulator environment. The red surface is where the force applied at

Table 1 Force setting of each test

Test	Load direction	Load force	Bellow pressure
Test I	X-	0–100 N	All 100 Kpa
Test II	Y-	0–100 N	Above 100 Kpa other 0 pa
Test III	Z-	0–100 N	Above 50 Kpa Z − 100 Kpa
Test IV	Shown in Fig. 3	0–00 N	Z + 0 Other 100 Kpa
Test V	X-	50 N	Above 100Kpa Z − 0 to 100 Kpa Z + 0

structural simulator (Fig. 2). In the simulator we use the deformation to represent the location of tip. All results presented here assume the simulation is run on a single worker thread of an Intel(R) Core(TM) i7-5500U CPU at 2.40 GHz. At this simulator, the arm's rated load is 100 N along the X, Y, Z three directions. We set the material of bellows as nylon (Young modulus: 1G pa, Poisson's ratio: 0.35),and triangular connecting plates as structural steel (Young modulus: 200G pa, Poisson's ratio: 0.3). In these tests, we all use the fixed support at the farthest surface in the negative direction of the X axis. And the load setting shows in Table 1.

3.2 *Fitting and Prognosis Experiments*

In Test I, we use the payload forces that are 0 and 100 N with deformation to fit the whole payload and deformation at X axis. It shows that if only two endpoints are used, it still can reflect the relationship between the deformation and payload force (shown in Fig. 4). There is a phenomenon that the data points near the 45 N deviated from the fitting line, mostly because the WLC reflects the displacement of the tip of

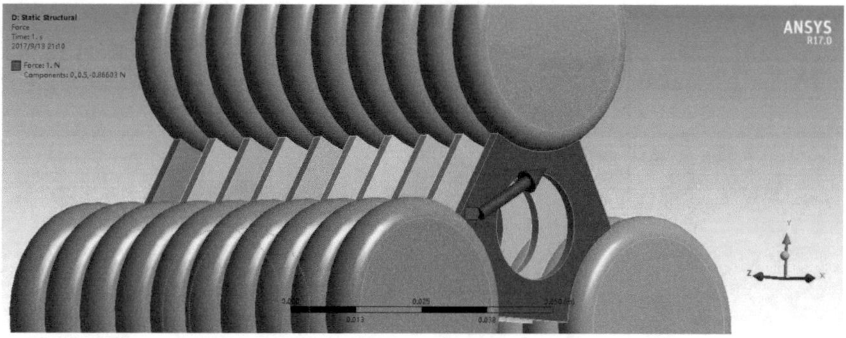

Fig. 3 This figure shows the direction of force in Test IV

the arm nor the deformation as the simulator calculated. When it is at the direction of Y, it also shows that the WLC can reflect the payload force and deformation (shown in Fig. 5). And when it is at the direction of Z, the result still shows the WLC can reflect the payload force and deformation (shown in Fig. 6). Above all, from the result (shown in Table 2), this method has high accuracy in the direction where the data is used to fit at.

Next, we tried to test this method in three-dimensional location space. With using the six data used to fit in what mentioned above, we calculated the force enable to output. We tested at a series of location (y represented the projection of location in the line at the y axis). These location were not at the axis we used to test, and the

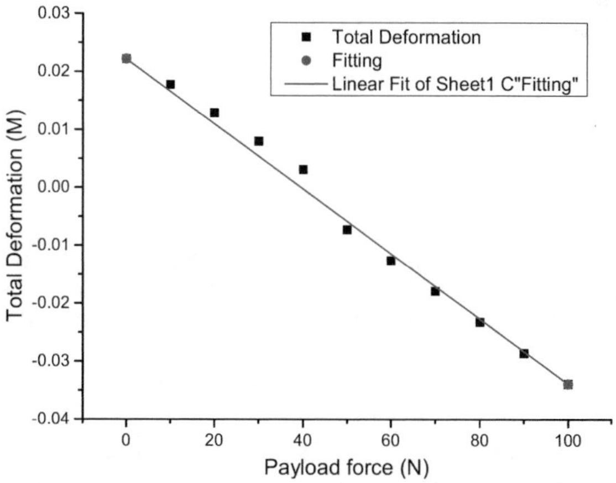

Fig. 4 This figure is the result of the situation when payload force was along the negative direction of x-axis. The black points are total deformation (include three directional deformations at X, Y and Z, but the directional deformations at Y and Z are too little to ignore) when the payload force on the arm in the simulate environment. The red points are the data we used to fitting. And the red line is the result of fitting

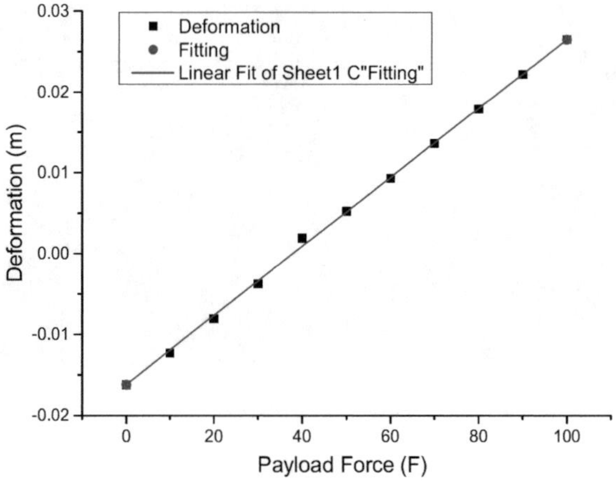

Fig. 5 This figure is the result of the situation when payload force was along the negative direction of y-axis. The black points are directional deformation at Y when the payload force on the arm in the simulate environment. The red points are the data we used to fitting. And the red line is the result of fitting

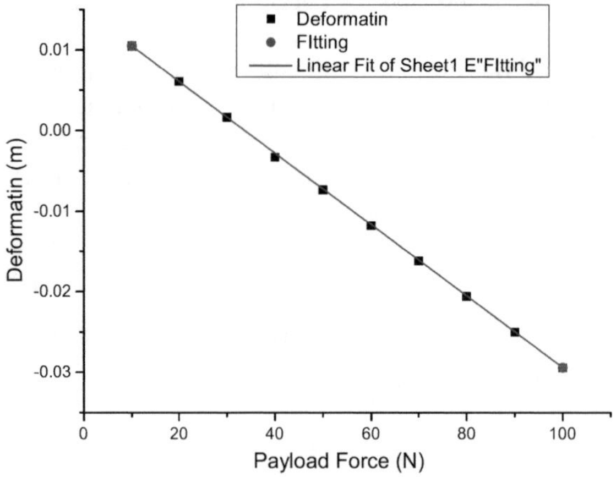

Fig. 6 This figure is the result of the situation when payload force was along the negative direction of z-axis. The black points are directional deformation at Z when the payload force on the arm in the simulate environment. The red points are the data we used to fitting. And the red line is the result of fitting

direction were not the direction we used to test (shown in Fig. 7). It shown the error appears at whole directions and especially at the location near 0 m, because in reality, the feasible space isn't an octahedrite. Therefore the large deviation appears near the angular point. Besides we simplified it as linear in whole space. But it still reflected that the method can use the location to calculate the force.

Table 2 Force setting of each test

Payload force direction	Standard deviation	Figure
Negative direction of X-axis	1.524E-03	Fig. 4
Negative direction of Y-axis	3.956E-04	Fig. 5b
Negative direction of Z-axis	1.600E-04	Fig. 6b

Fig. 7 This figure shows the deviation between reality force and theoretical force

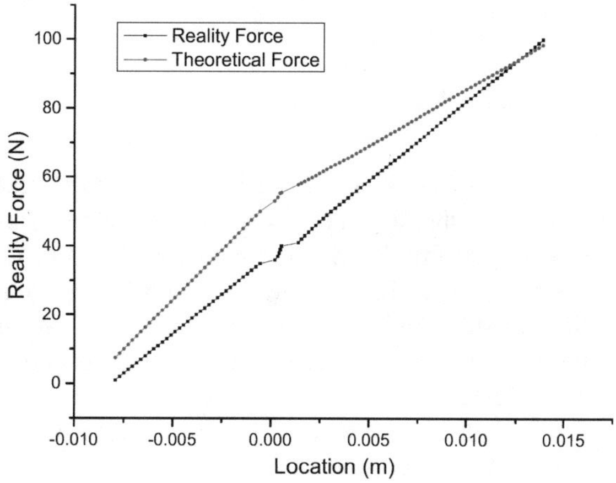

Fig. 8 The figure shows error distance and reality total deformation

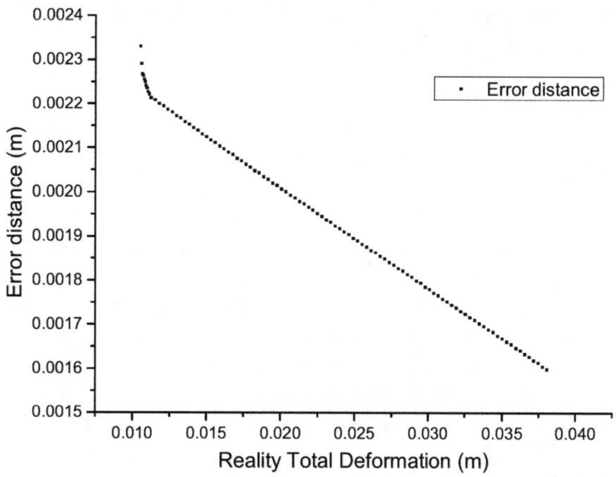

And then we also test the estimation of the location with a given force (50 N, vertically downward). Drawn the Fig. 8. It showed the error distance within the acceptable level and it is related to total deformation. This result reflected that this method can be used to calculate the feasible region when knowing the payload.

In general, the simulation could prove that this method can reflect the fact in the three-dimensional space. If only the linear fitting is used, according the result, it still represents the relationship between location and the force output. And we can conclude that:

When this method was used to calculate the force output, the accuracy was related to location. If the location was not at the line connected to the data point measured, the accuracy would be lower than at it. And this method was used to calculate the location with a given the force, there is negative correlation between error and deformation.

4 Conclusion

We present a method of description of soft arms which can fully describe the force (including payload) available with the location (workspace) under our assumption. This way of description was designed to the situation that output force and workspace are coupling. In addition, comparing with the former standard of workspace or load ability, this description is more complete and can reflect clearly the relationship between them. Meanwhile, it can also be applied to maximize the tradition arms' performance.

In the future, we need to add some dynamic indexes to enhance the descriptive power of movement performance.

Acknowledgements This research is supported by the National Natural Science Foundation of China under grant 61573333.

References

1. Rus, D., Tolley, M.T.: Design, fabrication and control of soft robots. Nature **521**(7553), 467–475 (2015)
2. Trivedi, D., Rahn, C.D., Kier, W.M., Walker, I.D.: Soft robotics: biological inspiration, state of the art, and future research. Appl. Bionics Biomech. **5**(3), 99–117 (2008)
3. Kim, S., Laschi, C., Trimmer, B.: Soft robotics: a bioinspired evolution in robotics. Trends Biotechnol. **31**(5), 287–294 (2013)
4. Majidi, C.: Soft robotics: a perspective current trends and prospects for the future. Soft Robot. **1**(1), 5–11 (2014)
5. Patel, S., Sobh, T.: Manipulator performance measures-a comprehensive literature survey. J. Intell. Robot. Syst. **77**(3–4), 547 (2015)
6. Korayem, M.H., Ghariblu, H.: Analysis of wheeled mobile flexible manipulator dynamic motions with maximum load carrying capacities. Robot. Auton. Syst. **48**(2), 63–76 (2004)
7. Wang, L.-T., Ravani, B.: Dynamic load carrying capacity of mechanical manipulators part i: problem formulation. J. Dyn. Syst. Meas. Control **110**(1), 46–52 (1988)
8. Yao, Y.L., Korayem, M.H., Basu, A.: Maximum allowable load of flexible manipulators for given dynamic trajectory. Robot. Comput-Integr. Manuf. **10**(4), 301–309 (1993)
9. Grissom, M.D., Chitrakaran, V., Dienno, D., Csencits, M., Pritts, M, Jones, B., McMahan, W., Dawson, D., Rahn, C., Walker, I.: Design and experimental testing of the octarm soft robot manipulator. In: Unmanned Systems Technology VIII, vol. 6230, p. 62301F. International Society for Optics and Photonics (2006)

10. Bodily, D.M., Allen, T.F., Killpack, M.D.: Multi-objective design optimization of a soft, pneumatic robot. In: 2017 IEEE International Conference on Robotics and Automation (ICRA), pp. 1864–1871. IEEE (2017)
11. Jiang, H., Liu, X., Chen, X., Wang, Z., Jin, Y., Chen, X.: Design and simulation analysis of a soft manipulator based on honeycomb pneumatic networks. In: 2016 IEEE International Conference on Robotics and Biomimetics (ROBIO), pp. 350–356. IEEE (2016)
12. Grzesiak, A., Becker, R., Verl, A.: The bionic handling assistant: a success story of additive manufacturing. Assem. Autom **31**(4), 329–333 (2011)

Intelligent Multi-fingered Dexterous Hand Using Virtual Reality (VR) and Robot Operating System (ROS)

Aswath Suresh, Dhruv Gaba, Siddhant Bhambri and Debrup Laha

Abstract The concept of using robots for reducing human effort is now a buzzword since the last decade. Now, with the advancement of the current technologies the robots could be remotely controlled with much greater ease and precision than before in domains, where we could have never even thought of their applications. The paper describes a new user friendly interface which allows the control of the robot manipulator (hand) in the most effective and simplest manner possible. By making use of 360° camera the vision is created which is displayed on the user's mobile phone (which is placed inside a VR-goggles) and viewed as a virtual reality environment by the user. The user and the robot manipulator are situated far apart and the user will feel like as if the robotic arms are his own arms. Moreover, the user can control the robotic arm placed remotely in any high risk area with or without any internet connectivity. This methodology is very useful for controlling the robotic arm for dealing with radioactive and hazardous materials. Therefore, the user doesn't have to be in the place of risk and he/she is still able to operate in the most convenient manner. In this paper, a detailed description of the control of the open-source 3D printed robotic hand/forearm from InMoov is discussed. Basically, the InMoov hand/forearm is modified into a useable all-purpose dexterous hand with the application of Virtual Reality to enhance user ease. There are two modes in which the system could be used. First is the hand mimicking mode, in which the user's hand movements are being replicated onto the robotic hand/forearm,

A. Suresh (✉) · D. Gaba · D. Laha
Department of Mechanical and Aerospace Engineering, New York University,
New York, NY, USA
e-mail: as10616@nyu.edu; aswathashh10@gmail.com

D. Gaba
e-mail: dg3035@nyu.edu

D. Laha
e-mail: dl3515@nyu.edu

S. Bhambri
Department of Electronics and Communication Engineering,
Bharati Vidyapeeth's College of Engineering, New Delhi, India
e-mail: siddhantbhambri@gmail.com

J.-H. Kim et al. (eds.), *Robot Intelligence Technology and Applications 5*,
Advances in Intelligent Systems and Computing 751,
https://doi.org/10.1007/978-3-319-78452-6_37

and the second mode is the voice recognition mode in which simple voice commands from the user are processed and are used to control the robotic arm. Voice recognition enables hands free usage of the robotic hand for people with disability. The system makes use of ROS (Robot Operating System) as its baseline program management system. The development is done by making use of ROS catkin workspace on Raspberry Pi 3 with Ubuntu Mate with the implementation of the necessary packages and nodes for every functionality.

Keywords Catkin · 360 vision · Gesture robot · Raspberry pi
ROS · Virtual reality · Voice recognition

1 Introduction

Millions of people across the world are working in environments that have hazardous materials and could prove to be deadly as well. The major concern of handling hazardous materials by humans that are working in a place filled with radiations or having considerable amounts of radioactive materials is life threatening for the workers. These preventive measures are never 100% safe. In 2012, 1133 people died as a result of hazardous waste accidents. In 2014, 4679 people died on the job in the United States of America; on an average, 13 deaths per day were reported. The statistics show a considerable increase in the annual deaths of people worldwide due to hazardous wastes. Thus, to decrease this radical increase in the number of deaths some serious steps should be taken. Recently there have been a lot of advancement in the preventive measures to avoid these accidents, but they are not capable of eradicating life threat to the safety of the individual. Thus, there is a need to create a system that is capable of handling hazardous materials remotely instead of human on-site dealing with the radioactive materials. With the advent of new technologies (like Virtual Reality), the ease of use of these robots has tremendously increased which has led to a boost of their use in a variety of domains. These advancements will eventually increase the quality and safety of the workers at such work places.

This paper introduces a complete integrated system that combines every necessary aspect of a human hand to act as an Intelligent Multi-fingered Dexterous Hand which is capable of doing tasks that required humans in the past. The system consists of a two-way communication between two Raspberry Pi B3 with internet and through Xbee pro. Both the communications together provide better network conditions and ensure interruption-free communication between the hand and the human. The robotic hand can be controlled from anywhere on earth and helps to perform the task without visiting the site. (Virtual Reality) VR technology is used to give a virtual video feedback of the robotic arm environment in which the robotic hand is working. This technology enables us to operate the prosthetic hand in the same manner in which human beings use their own hands, thus making the task easy to perform. The most important aspect of the system is that it doesn't require any human presence at the place of work thus preventing the direct exposure of human life to the hazardous environment. The master control of the robotic hand is made using a globe having multiple IMU's and multiple flex sensors attached to it for gesture recognition. The accelerometer is used to give the acceleration of the hand in the three axes and a gyroscope is used to give the angular rotation and angular velocity of the hand. Flex sensors are helpful in defining the pressure applied through hands and finger movements. The combination of both IMU and flex sensor help to mimic our hand movements perfectly. The system is also capable of receiving voice commands and act accordingly. This feature makes the whole unit work in a hands-free manner and thus makes it more user-friendly and convenient to use. A 360 view camera is used to give the real-time video feedback on the environment of the workplace.

These kind of systems are also very useful in day-to-day activities beside the industrial applications. Consider a problem when there is shortage of time and you want to cook food without wasting time or you are coming from office and you want to utilise the travelling time in preparing food in your kitchen. In these conditions, this kind of a prosthetic arm is beneficial. The system is capable of creating a virtual environment of the kitchen and helps the user to perform the cooking task while travelling or studying. The prosthetic arm in the kitchen conducts the cooking work without having the need of the human being to be physically present there. VR technology replicates the whole kitchen environment virtually in front of the user. Machine learning is applied to make the system capable of performing the tasks using the voice commands like defining the positions of different components of the kitchen, the procedure to operate the gas burner, and to pick up the utensils by its own using the voice commands. Thus, this system is a salutary contribution in helping the humans in saving time and preventing many lives.

With the development of modern science and technology, robots have been used in a variety of applications, while robotic teleoperation technology can span space, place people, machines, and task objects in a closed loop to achieve the human and the objective world of synchronous machine interaction, and to improve people's perception and behaviour in a large extent [1]. For robots in complex operating environments (such as home services), sometimes robotic vision may be difficult to provide enough information to successfully perform such tasks. Additionally, some occlusion of the situation in the real indoor environment often occurs, which makes relying solely on the robot itself is difficult to complete and correct perception of the surrounding environment, and to make decisions. In this case, it is an effective way to solve these problems by putting human, robot and task object in a closed loop and introducing human experience to control the robot [2, 3]. In order to deal with the complex operating tasks, the operator must always behave according to the actual situation of the transformation, the robot coordinated control, and constitute a human-robot working environment system. Since human beings are good at performing perceptual understanding, action planning, action resolution, and making decisions based on experience, people play a crucial role in the process. At this point, people are required to complete the intelligent analysis part, and then complete the underlying work through the body language control robot. By this process, this man-machine interaction can achieve better results. The manipulator is the main operating mechanism of the robot. Therefore, it is of great significance to carry out the research of the remote control system based on gesture control.

The Micro-Engineering Department (Institut de Systèmes Robotiques: ISR) of the Swiss Federal Institute of Technology (EPFL) is involved in robotic design and development, with a special focus on industrial applications. The classical methods for robotic systems programming (off-line as well as on-line) lack user friendless and performance. This is why, since 1990 people have been developing Virtual Reality (VR) interfaces to simplify robot task planning, supervision and control [4]. In addition, the Intelligent Mechanisms Group (IMG) of the NASA Ames Research Center (developers of the Virtual Environment Vehicle Interface [5, 6], a user interface to operate science exploration robots) has shown that a tool to generate rapidly VR interfaces for new robots arm manipulators would provide great benefits.

If a robotic manipulator is to be controlled, it is usually done using the RF (Radio Frequency) remote control or by gesture recognition and the recording of the camera is being displayed onto a 2D screen. But the problem with conventional systems is that they have a very difficult and cumbersome user interface, which inhibits its usability by a great deal. The interactive user interface (360° VR-Vision) and the voice control of the robotic hand using the voice recognition module are the two features of our system which sets it apart from the prior art. Our control algorithm integration with the voice recognition node (ROS node) makes it beneficial for handicapped people. This is because mostly paralytic or handicapped people require help from such tools and using a hand to mimic its motion is actually useless for disabled people. Adding voice recognition nodes into the ROS subsystem makes it much more user friendly such that disable people could simply control it with simple voice commands, which could be decoded by the controller into movement commands. This way we could control our robotic arm in households as well as in places of hazardous materials. Apart from this, our system is made in such a manner that it could also address the issue of cost effectiveness.

2 Design and Manufacturing of Hand/Forearm

The mechanical design of the seven degree of freedom (DOF) robotic arm is done using 3D printing and its properties are as shown in the Table 1.

The right hand and forearm was designed using SolidWorks as shown in Fig. 1a, b. The design was redesigned 3 times in SolidWorks to bring flexibility to the design and looks more like a human hand. Once the CAD files were created for every part, those were converted to a STL files and imported in the Mojo 3D Printer's file processing software—Print Wizard which scaled the parts as necessary in proper orientations. For setting up the printer, all the parts were given an infill of 30%. The five fingers were printed with 2 shell and without any support or raft. The

Table 1 Table defining the properties of the material used

Material—Acrylonitrile Butadiene Stryene (ABS)	
Density	1040 kg/m^3
Symmetry	Isotropic
Stress—strain response	Linear
Young's modulus	2390 MPa
Poisson's ratio	0.399
Coefficient of thermal expansion	9.54/C
Specific heat capacity	1720 J/Kg-K
Thermal conductivity	0.258 W/(m K)

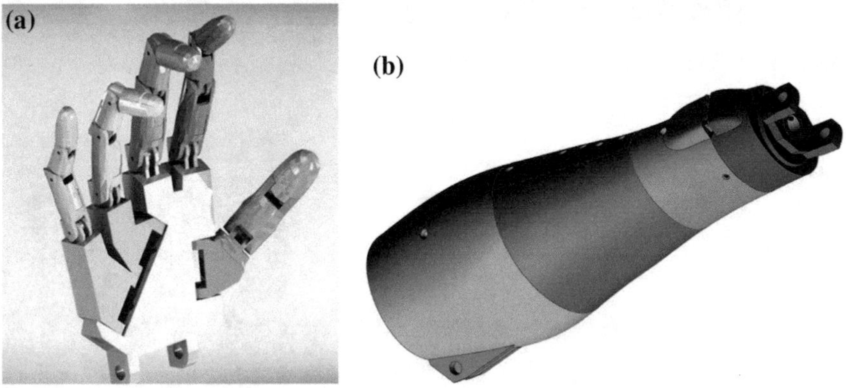

Fig. 1 **a** CAD model of right hand made in solidworks and **b** CAD design of forearms

portion of the hand that is connected to the wrist and the five fingers were printed with 3 shells without any support or raft. Once the parts were printed of ABS material using the Fused Deposition Modeling (FDM) process, the parts with the support were put in WaveWash 55 which is an automatic support removal system. Once it was plugged in, an EcoWorks tablet was dropped in there along with the parts with the support. After that it was filled with tap water and the machine was turned on. It took considerable amount of time to remove the supports and once the supports were gone, the solution was poured down the drain and the parts were rinsed off.

Once all the parts were printed, sandpaper was used to file the edges of the parts to make those smoother. Then the parts were assembled together using various types of joints and fasteners that includes, nuts, bolts, washers, and glue. The final 3D printed prototype of right hand and forearm is shown in Fig. 2a, b. Figure 3 shows the final assembly merging hand and forearm respectively.

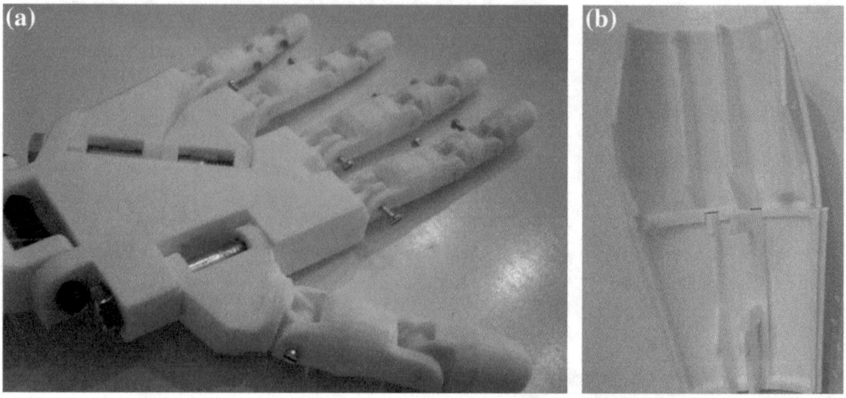

Fig. 2 **a** 3D printed prototype of right hand and **b** 3D printed prototype of forearms

Fig. 3 Assembled hand and forearm

3 Working Principle

The whole system mainly consists of two parts,—the first being the control part and the other is the robotic arm actuation part as shown in Fig. 4. The control part consists of a Samsung Gear VR which gives virtual video feedback of the robotic arm environment and its surrounding which gives the user the feeling as if he is

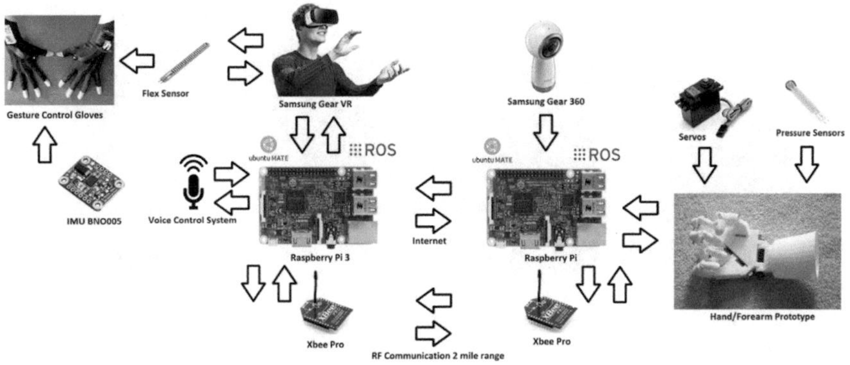

Fig. 4 Shows the working principle of the system

present in the hand environment itself. Then the gesture control gloves use IMU for 2 DOF of the hand manipulator which is rotational base and up-down motion which is great for pick and place robot and it also uses flex sensor for the finger movements. The combination of both IMU and flex sensor help to mimic our hand movements perfectly. We also use voice recognition shield to receive verbal commands and then make corresponding motions with the hand. The advantage of voice control is that it doesn't require muscles and is more flexible to different physical conditions (Hands free usage). The Raspberry Pi 3 running on ubuntu mate with Robot Operating System (ROS) is the brain of the system. The Raspberry Pi uses either Xbee RF communication channel or internet channel depending on the distance between the transmitter and receiver part. If the distance of separation is less than 2 miles, we use Xbee pro and for higher distances, we use internet which makes our device to be controlled from anywhere on the Earth. The receiver side consists of Samsung Gear 360 video camera which records live footage and sends to the user via raspberry pi interface. The robotic hand consists of five servos to achieve the normal five finger movements. It also uses additional two servos for the base rotation and up-down motion. It also consists of five pressure sensors with active feedback to the raspberry pi which monitors how much pressure to apply based on objects to be picked up. The overall experience of product is great because it makes the user feel as if he is in the hand actuation environment even though he is miles away.

4 Electrical System and Control of the Robotic Arm

The electrical system majorly consists of two control units:

1. Master Control Unit
2. Slave Control Unit

Fig. 5 **a** Samsung gear VR **b** samsung gear 360

Samsung Gear VR is a virtual reality platform as shown in Fig. 5a. It is used as a medium to see the real environment of the workplace virtually. VR technology helps in mimicking the working environment in which the robotic hand is working. It also helps in improving the user experience and makes the control of the robotic hand much more comfortable for the operator. The VR is interfaced with Samsung gear 360 camera (Fig. 5b) which is capable of recording 360° videos and hence gives real-time streaming of video of the working environment. This helps the operator to understand the real conditions of the job and make them capable of doing the task without being physically present.

4.1 Master Control Unit

Master control unit as shown in Fig. 6a is the transmitter part of the system which is responsible for controlling the slave control unit. The Master control unit consists of the following components:

1. Microprocessor (Raspberry Pi B3): Raspberry Pi B3 is used as the microprocessor in the master control unit because of its high processing speed and it contains inbuilt WiFi. Two Raspberry Pi B3 are connected together to establish a communication between the master control unit and the slave control unit. Raspberry Pi B3 is responsible for every single processing operation in the system. Raspberry Pi is connected to WiFi for establishing a network between the user and the robotic hand.
2. Xbee Pro: Xbee Pro is used to establish a fast and reliable network between the master and slave control unit. Xbee works on ZigBee protocol which enables it to establish an uninterrupted network for shorter distances. The combination of Xbee communication and WiFi communication prevents the system from network failure.

Baud Rate of Xbee: 115200 Baud

Frequency: 2.4 GHz

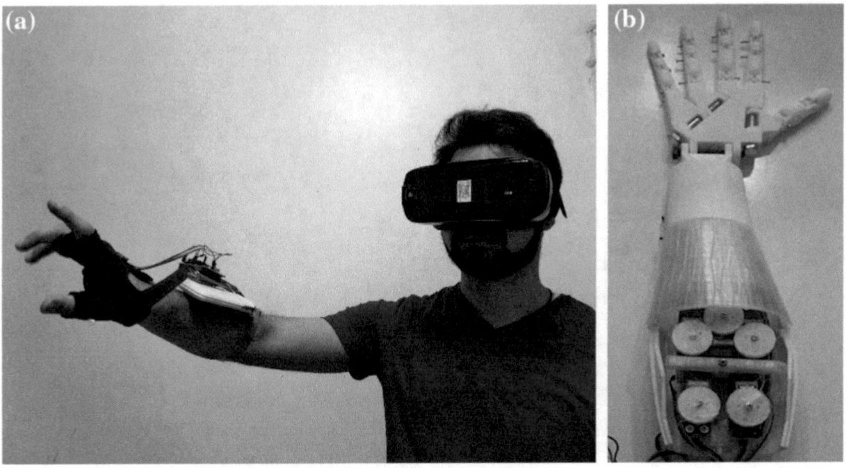

Fig. 6 **a** Master control unit prototype and **b** slave control unit robotic hand prototype

3. Sensors: Flex sensor: Five flex sensors are used in the hand for replicating the gesture and motion of the fingers. Every finger has one flex sensor connected with the microprocessor and the data of flex sensor is used by the microprocessor to analyse and give results.

BNO005 IMU: BNO005 is used to define the acceleration and angular motion of the robotic hand. It consists of an accelerometer which provides the acceleration of the robotic hand in the three axes, a gyroscope which is responsible for defining the angular velocity of the hand, and a magnetometer. The flow chart of Gesture control is shown in Fig. 7.

4. Voice Recognition: This part of the system is responsible for the recording of the voice commands and then sending the encapsulated data to the receiver end. Where it is decoded and then the command is performed. The flow chart of the voice recognition system is shown in the Fig. 8.

4.2 Slave Control Unit

Slave control unit as show in Fig. 6b is the most important part of the whole system as it is responsible for conducting the job on the behalf of human. The slave control unit consists of a robotic arm which is capable of doing every single task that a human hand can perform. The robotic hand consists of the following components:

1. Microprocessor (Raspberry Pi B3): In the slave control unit, Raspberry Pi is used as the controlling unit of the prosthetic arm. With the help of an inbuilt

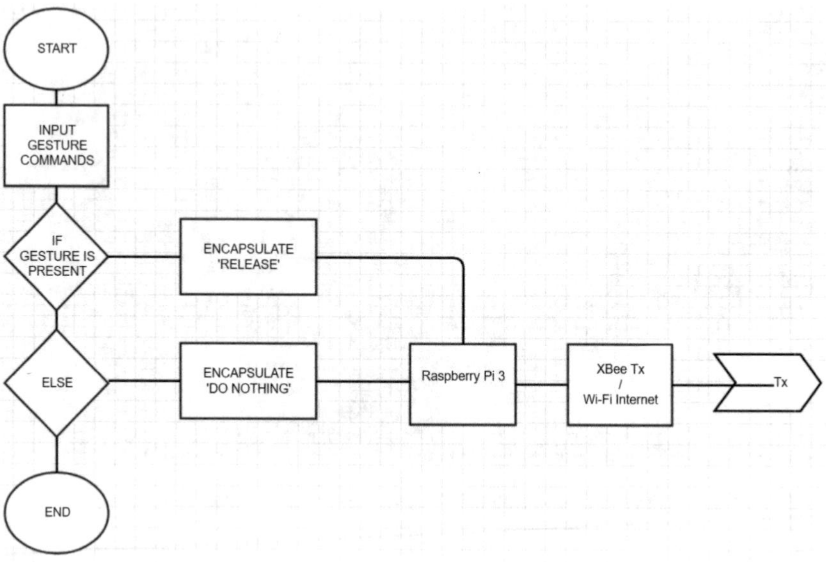

Fig. 7 Flow chart of gesture control

WiFi, a wireless communication is established between the slave and the master control unit to control the robotic hand from longer distances.

2. Xbee Pro: is used to establish a fast and reliable network between the master and slave control unit. In slave control unit coordinator mode of Xbee is used as receiver. Xbee works on ZigBee protocol which enables it to establish an uninterrupted network for shorter distances. The combination of Xbee communication and WiFi communication prevents the system from network failure.
3. Servo motors: To control four fingers and one thumb, five different servo motors are used as shown in the Fig. 9. The servos are connected to the fingers and the thumb with the help of strings. Every servo is connected with the help two strings with one finger to control the to and fro motion of the finger. One string is responsible for pulling the finger inwards and the other string is responsible for pulling the same finger outwards to make the hand work exactly the same way in which the human hand works.
4. 3D Printed Arm: The prosthetic arm consists of four fingers and one thumb just in the same manner the human hand has. It also has equal number of degrees of freedom the human hand has.
5. Pressure sensors: Five Pressure sensors are placed in the robotic hand fingers to provide the feedback of the pressure applied by the user. The system gives an alert to the user when it exceeds the desired prescribed pressure.

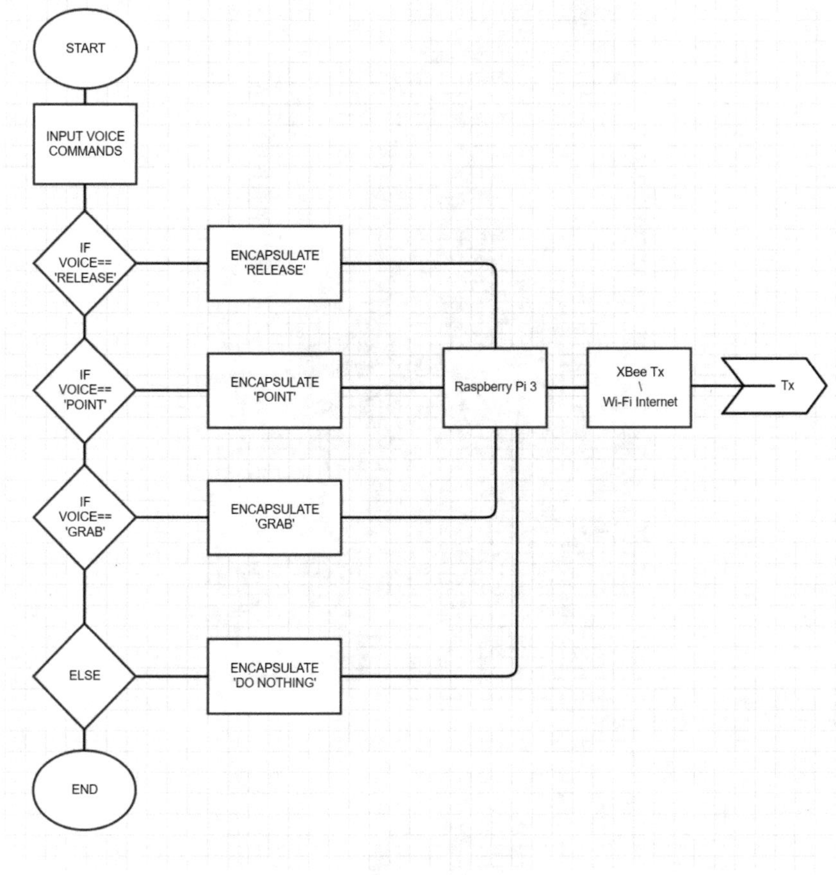

Fig. 8 Flow chart of voice control

IMU Control for sensing the position of the palm of the user:

Complementary filters are used in this case as shown in Fig. 10. Accelerometer values are augmented with the gyroscope values in all to get the filtered value from the IMU (Inertial Measurement Unit) which are transmitted to the receiver end where it is processed and utilized to replicate or mimic the hand movements or voice commands.

where,

Z = Signal before filtering
X = It is a signal having low frequency noise
Y = It is a signal having high frequency noise
$Z'(s)$ = Signal after filtering
$G(s)$ = Low Pass Filter
$1 - G(s)$ = Complement of Low Pass Filter i.e. (High Pass Filter)

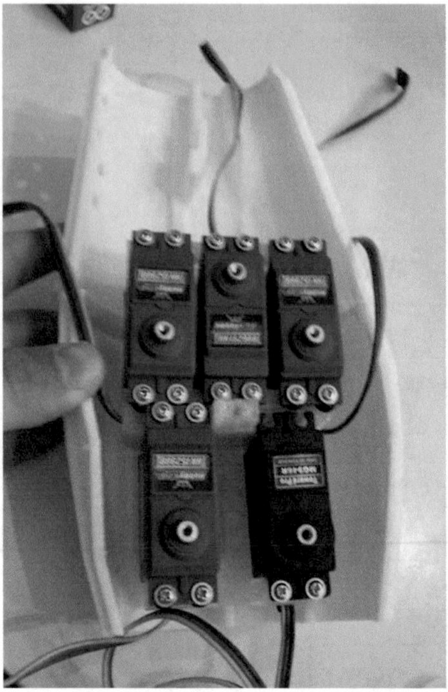

Fig. 9 Five servos assembly

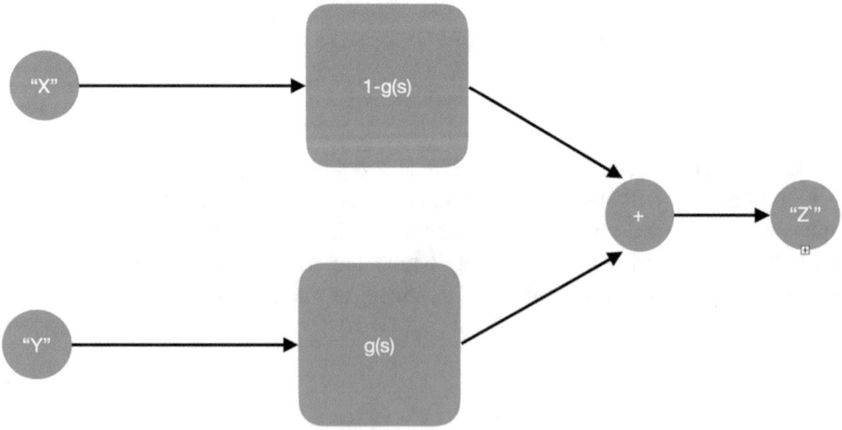

Fig. 10 Block diagram of complementary filter

$$\begin{aligned} \text{Filtered value} &= \text{X1*alpha} + \text{X2*}(1 - \text{alpha}) \\ &\text{X*}(1 - \text{g(s)}) + \text{Y*}(\text{g(s)}) \\ &= \text{X(s)} + \text{Y(s)} \\ &= \text{Z(s)} \end{aligned}$$

X(s) and Y(s) are complements of each other and $|Z'(s)| = 1$.

5 Result and Discussion

The process was implemented by using a VR headset worn by the user. The robotic arm is tested with both hand gesture control and voice commands and the results were satisfactory.

(A) **Tested by Internet Connectivity**:

Implemented WiFi communication using two Raspberry Pi B3 microprocessors and get appropriate results in controlling the robotic hand using WiFi communication.

(B) **Tested by XBEE Connectivity**:

Implemented ZigBee communication using two Xbee Pro and get appropriate results in controlling the robotic hand using Xbee communication.

(C) **Tested Voice Commands**:

The three cases indicating different actions of the Robot arm are achieved as shown in Figs. 11, 12 and 13.

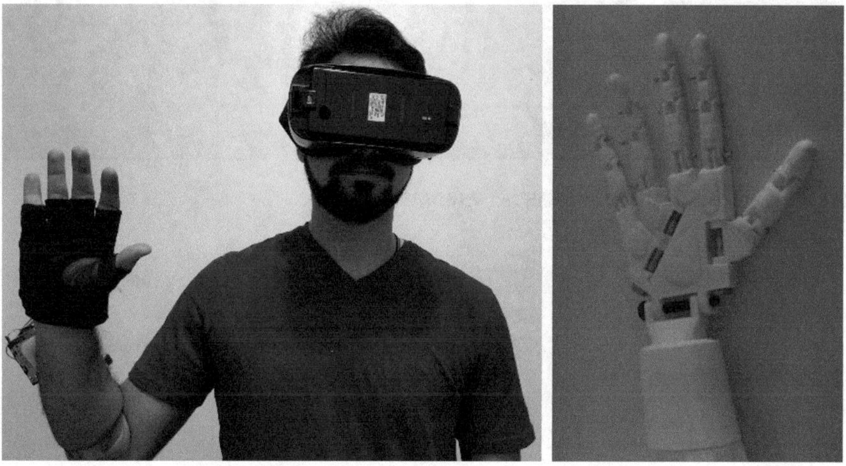

Fig. 11 User is executing the 'Release' movement

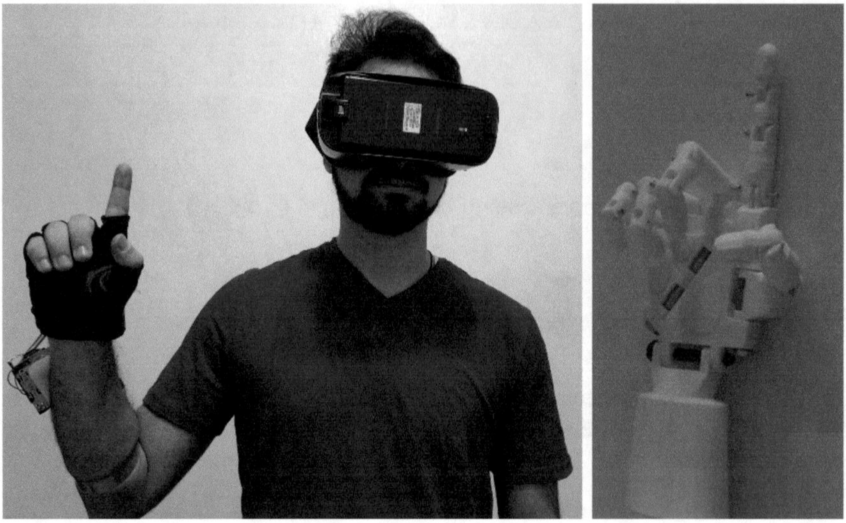

Fig. 12 User is executing the 'Point' movement

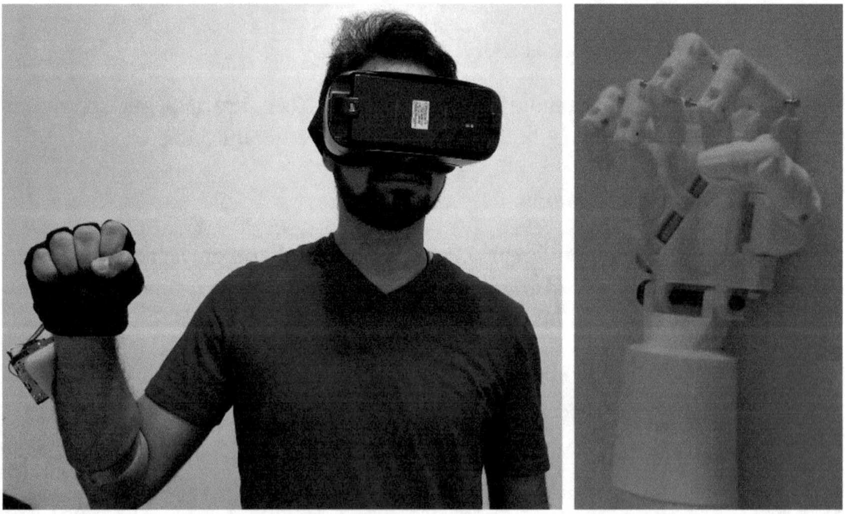

Fig. 13 User is executing the 'Grab' movement

CASE (1) Based on the gesture Hi or voice command 'Release', the arm mimics/replicates it.

CASE (2) Based on the gesture Point or voice command 'Point', the arm mimics/replicates it.

CASE (3) Based on the gesture Grab or voice command 'Grab', the arm mimics/replicates it.

Case 1

When the user does 'Release' movement, the system replicates the movement. Also, if the user says 'Release', then the system will recognize the voice command and it will be implemented onto the robotic arm, as shown in the Fig. 11.

Case 2

When the user does the 'Point' movement the system replicates the movement. Also, if the user says 'Point', then the system will recognize the voice command and it will be implemented onto the robotic arm, as shown in the Fig. 12.

Case 3

When the user does the 'Grab' movement, the system replicates the movement. Also, if the user says 'Grab', then the system will recognize the voice command and it will be implemented onto the robotic arm, as shown in the Fig. 13.

5.1 Application

The robotic system finds its application in various domains of our lives. First, it could be used like a cooking robot as in, the user could easily sit in his/her room and could control the robotic arms fixed in the kitchen. This helps the user easily cook food without being present in the kitchen physically, but he/she will still feel like standing in the kitchen. Second, the robotic system finds its application in dealing with hazardous materials (like nuclear waste and radioactive materials) which is usually done using humans by wearing radiation resistant suits, but they are never 100% secure. But by this manner, many lives could be saved which are being sacrificed every year. Further, the system could find its application in industries where precision tasks are required which could be done using robots only, and also while handling very heavy things. In addition, the user experience is close to reality due to implementation of Virtual Reality technology. Apart from this, implementation of voice recognition enables the use of this system for people with disability.

Our system is having two modes of use,—the first one being internet connectivity which gives it the capability to be controlled from anywhere in the world, and the other one is controlled using Xbee-2.4 GHz personal network communication. Thus, it allows the system to control and communicate even if the there is no internet. Moreover, our system uses Inmoov robotic hand which makes it really cost efficient in nature.

5.2 *Conclusion and Future Discussion*

Thus, we can conclude that by making use of current technologies like Virtual Reality and by making use of 360° camera we were able to control the robotic hand with great ease and satisfactory results were achieved. Also, as previously explained the robotic hand has great applications in handling hazardous and radioactive wastes. Therefore, the user does not have to be present in the high risk area, but still he/she is swiftly able to control the robotic arm with high precision. Voice recognition present in the robotic hand helps us develop a system which could be controlled via hands free mode via simplistic commands like "HI, Point, Grab", thus allowing it to be useful for disabled people. Also, implementation on ROS enables simpler amalgamation of nodes and its implementation onto the Raspberry Pi 3.

Acknowledgements The authors would like to thank Makerspace, New York University to provide support and resources to carry out our research and experiments.

References

1. Ding, X.: Humanoid Robot Technology. Science Press (2011)
2. Ogawara, K., Takamatsu, J., Kimura, H., et al.: Generation of a task model by integrating multiple observations of human demonstrations along time series. J. Inf. Process. Soc. Jpn. Comput. Vis. Image Media **43**,117–126 (2002)
3. Calinon, S.: Robot Programming by Demonstration. Springer, Berlin Heidelberg (2008)
4. Natonek, E., Flückiger, L., Zimmerman, T., Baur, C.: Virtual reality: an intuitive approach to robotics. In: SPIE Telemanipulator and Telepresence Technologies, vol. 2351, pp. 260–269, Boston, Oct–Nov 1994
5. Piguet, L., Fong, T., Hine, B., Hontalas, P., Nygren, E.: VEVI: a virtual reality tool for robotic planetary explorations. Virtual Reality World, vol. pp. 263–274, Stuttgart, Germany, February 1995
6. Hine, B., Hontalas, P., Fong, T., Piguet, L., Nygren, E., Kline, A.:VEVI: a virtual environment teleoperation interface for planetary explorations. In: SAE 25th International Conference on Environmental Systems, San Diego, CA, July 1995

FEA Evaluation of Ring-Shaped Soft-Actuators for a Stomach Robot

Ryman Hashem, Weiliang Xu, Martin Stommel and Leo K. Cheng

Abstract This paper evaluates four designs of circumferential pneumatic soft-actuator and shows its application to a soft stomach robot. The testing of the design is based on a finite element analysis of geometrical displacement and related pressurisation. In a biological human stomach, the antral contraction wave deformation is represented in a ring-shaped structure. The inspiration of such behavior of deformation leads to a proposal of a ring-shaped soft actuator. The proposed actuator includes a pneumatic system with multi-chambers and multi-layers to produce a deformation similar to that in the stomach organ. There are four proposed chamber designs: semicircular, cylindrical, ellipsoidal and semirectangular. The body of the actuator is made of soft material (silicone) with a high stress/strain relationship in order to exhibit large deformation behavior. In this article, the evaluation of four possible shapes of pneumatic chambers of the circumferential soft-actuator is examined and compared by Finite Element Analysis to simulate the displacement of each soft actuator. Two different methods are used in the experiments: (1) we applied the same pressure to all actuators and compare the displacements, (2) we applied different pressures to obtain the maximum pressure in each actuator before distortion and then examine the maximum displacement that can be achieved.

Keywords Soft robotics · Large deformation · Peristaltic actuation Circumferential actuator · Stomach robot

R. Hashem · W. Xu (✉)
Department of Mechanical Engineering, University of Auckland, Auckland, New Zealand
e-mail: p.xu@auckland.ac.nz

R. Hashem
e-mail: aabo845@aucklanduni.ac.nz

M. Stommel
Department of Electrical and Electronic Engineering, Auckland University of Technology, Auckland, New Zealand
e-mail: mstommel@aut.ac.nz

L. K. Cheng
Auckland Bioengineering Institute, University of Auckland, Auckland, New Zealand
e-mail: l.cheng@auckland.ac.nz

J.-H. Kim et al. (eds.), *Robot Intelligence Technology and Applications 5*, Advances in Intelligent Systems and Computing 751,
https://doi.org/10.1007/978-3-319-78452-6_38

1 Introduction

Soft robotics is an emerging field that aims to tune the current rigid robot example toward more delicate actuation. The aim for the soft robot field is to obtain a delicate interaction with the environment and the adaptability with different situations [1]. The invention of such technology provides new ideas that differ from the classic engineering strategies. With the soft robotic technology at hand, it is possible to imitate biological behaviour in a robot. As the soft robotic field is young, it lacks the fundamentals of the first principles, fabrication, and control techniques [1]. As a result, engineers are obtaining ideas and inspiration from nature and biology. The compliance of the biology system morphology is abstracted to the field of soft robotic. The abstract of these solutions from nature in the field of soft robotic can be seen in crawling robot [2], caterpillar robot [3], octopus arm [4] and elephant trunk [1]. These examples show the capability of a soft robot to imitate the muscular hydrostatic system in such a biology body [5, 6]. The capability to imitate such a motion from nature required a field that evolves multi-disciplinary from the material, chemical, mechanical biomedical and software engineering fields [7].

The direction of this field is mainly towards scientific research, while few applications are presented in the engineering context. Recently, several applications are intended to mimic human organs motion by biologically-inspired techniques such as swallowing robot [8]. This robot consists of a series of circumferential actuators stacked in 16 layers. Each layer included 4 embedded semirectangular chambers controlled pneumatically to present a large deformation [9]. Another approach is introduced to mimic the human stomach circumferential contracting muscles by pulling a wire on a soft surface that represents the structure of a stomach organ [10]. The mainstream of soft robotics actuators aims at bending actuator for crawling and grabbing objects, leaving a big gap in the literature regarding the circumferential soft actuator.

A physical gastric robot is a device that imitates the human stomach. Biology researchers and food technology industry are both interested in such an apparatus. The former requires stomach robot to investigate the motility of the human stomach, while the latter requires such a device as a test environment for a distinct food product. The stomach robot provides alternative solutions to an in-vivo operation. Currently, stomach robots are made of rigid rods and metal cylinders, which are unrealistic replica to the human stomach [10]. Therefore, a soft-bodied stomach robot is essential for more investigation and experiments on such a complex organ.

Soft ring pneumatic actuators (SRPA) are designed to handle large deformations and large mechanical strains. The result of such a high strain on the actuator allows unpredictable behavior, such as the soft material (silicone) employed in many projects which show a complex hyperelastic, viscoelastic nonlinear behavior [11]. Moreover, the process of fabrication SRPA is a time-consuming procedure. Hence, it is essential to comprehend the performance of the SRPA before fabrication. Thus, numerical models using finite element analysis (FEA) are promoted to predict actuator behavior at large deformation states, which leads to design efficient

SRPAs and soft robots. FEA enables us to represent the behavior of the actuator and examines the outcomes of varying parameters such as material stiffness or chamber dimensions, which reduce the need for a re-fabrication process.

This paper considers a circumferential pneumatic actuator consisting of multi-chambers to achieve a deformation on the inner ring of the actuator. A computer-aided design (CAD) and FEA will be used for designing and evaluating the chamber shape, and the deformation produced by the pneumatic force acting in the chambers.

2 FEA Design and Experimental Methods

2.1 Basic Structure and Working Principle

The motility of the human stomach is similar to that of the small intestine, both of which use a peristaltic movement to propel food particles and chyme from one position to another. However, this motion is not uniform. Instead, it relies on a natural

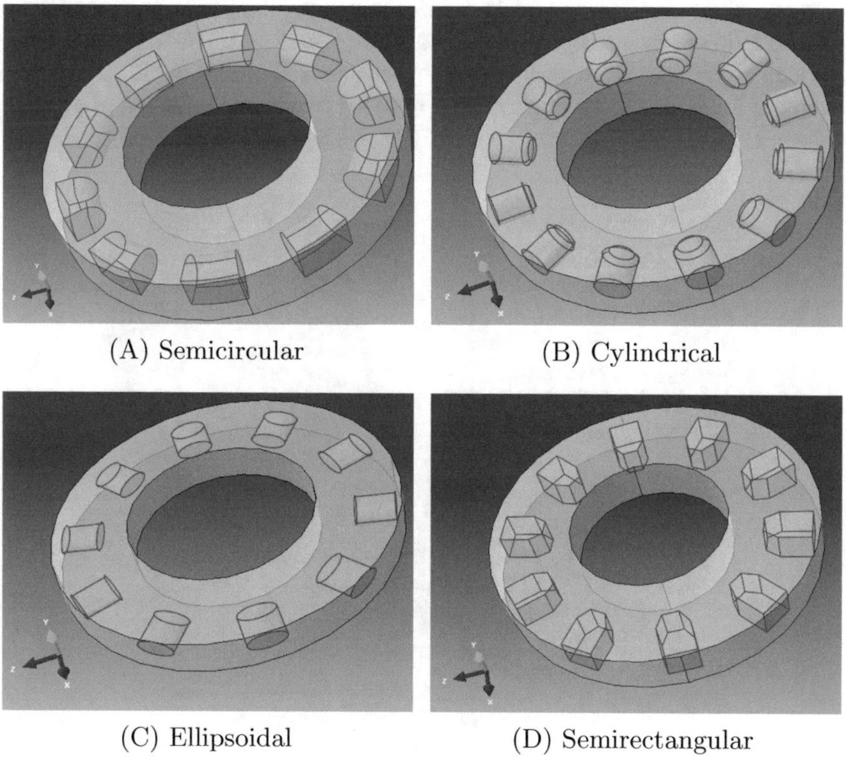

Fig. 1 A single layer of the four proposed chambers design

cycle rhythm [12]. In the human gastric system, the antral contraction wave (ACW) of electroactivity signal is three cycles per minute (3 cpm) [13]. Each cycle starts in the middle to the upper corpus area and spreads to the pylorus [14]. The rapid motion is a circumferential constriction that propagates toward the pylorus. The peristaltic movement ends at the pylorus which works as an electroactivity isolator, separating the ACW between stomach and duodenum [15]. The gastric pump, where the peristaltic movement occurs, shows the most challenging part of the stomach. The mechanical deformation of the gastric pump can be described as a decrease of circumferential diameter. These circles are along both stomach curvatures from the pacemaker until the pylorus. There is no typical geometry of the actual human stomach, and the geometry varies from a person to another influenced by the condition of the surrounding organs, the position of the body and the amount of meal ingested [16]. However, the average capacity of the human stomach is 1 L, and the average size is 10 cm⌀ at the widest point at the upper corpus, its great curvature about 34 cm long and has a pyloric gate about 1.5 cm⌀ [17]. The characteristics of the geometry of ACWs with respect to their distance from pylorus had been investigated [18]. The average amplitude of ACWs is 8 mm and the average width of ACWs is 10 mm [18]. In this work, the evaluation of different ring soft-actuators must exhibit a minimum displacement of 8 mm towards the inner surface of the actuators conduit (Fig. 1).

In a previous work, we designed a drawing that resembles a human stomach shape which is followed the geometry mentioned above [19], Fig. 2. The stomach drawing provided a symmetrical shape by extruding segments of circles in 3D. The antrum section of the stomach drawing has been used for developing a section of a gastric

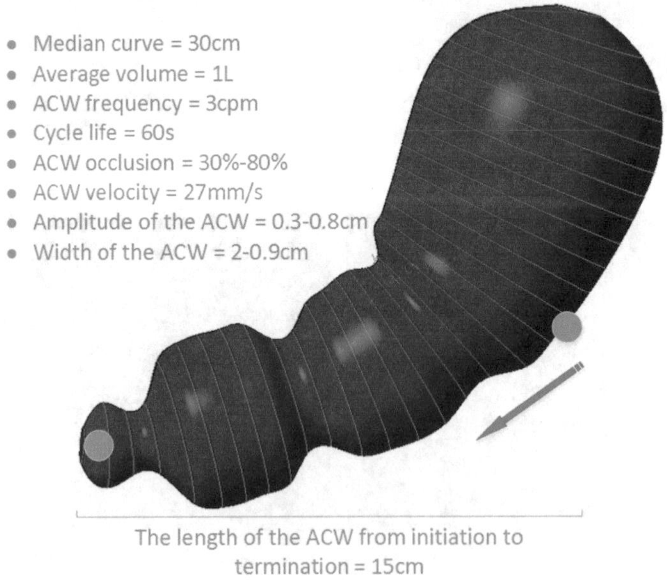

Fig. 2 Soft stomach robot and the ACW specifications of the conceptual conduit design

soft-robot stomach [19]. In this article, we used the same stomach drawing to evaluate the contraction of a single ring that is located in the top of the antrum. The geometry of the antrum from the drawing starts with a diameter of 50 mm⌀ and sweep down to the pylorus of 15 mm⌀ [19]. In this work, we selected the 50 mm⌀ ring and design a ring soft-robot actuator to evaluate the deformation of different chambers design.

The pneumatic technique is used for geometrical development of the gastric soft robot to achieve a satisfying deformation actuation similar to that of the biological behaviour. The aim is to develop a pneumatic system with inflatable chambers that represents a circumferential contraction similar to that in the human stomach. A pneumatic actuation provides high forces and generates large deformations inside the chambers, which makes it the preferred actuation system for the stated device [20]. The proposed ring soft-robot actuator consists of pneumatic chambers in a series which act perpendicularly on the inner surface that creates deformation by the pneumatic forces, Fig. 3.

A circumferential ring soft-actuator with multi-chambers, multi-layers and two degrees of freedom (DOF) is created in this article, while the method and the process of FEA testing are applicable to other design of soft robot actuators in general. Figure 1 shows a single layer form of these actuators for simplicity. The soft-actuators that will be tested have 3 layers toroidal shape, in which the pressure is controlled through pneumatic system. The actuator is covered by a shell to counter the unwanted DOFs and enforce the deformation to one a-axis direction.

Table 1 shows the geometry of a single chamber of each design. The measurement is in millimeter for height and width, and mm^3 for volume. The nodes number is representing the mesh of three layers ring actuator, the number of divisions for the edges (seeding) is 3.

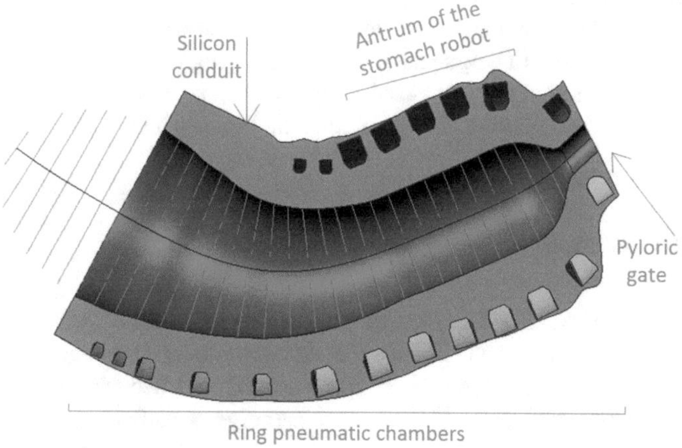

Fig. 3 The soft robot stomach conduit with a series of circumferential pneumatic chambers acting upon the inner surface to imitate the ACW behavior that observed in a human stomach

Table 1 The geometry of the four chambers represented in Fig. 1

	Height (mm)	Width (mm)	Volume (mm^3)	Nodes
(A) Semicircular	10	11	1000	137,366
(B) Cylindrical	10	10	833.3	208,579
(C) Ellipsoidal	10	15	1000	188,218
(D) Semirectangular	10	10	1000	162,533

In the stage of the actuation, the pneumatic system enforces the pressure in the chamber causing an axial deformation on the inner surface of the toroidal shape and a radial deformation between the layers. The deformation ratio is proportional to the pressure scale. A shell that is enveloped the actuator from outside is introduced to cover the actuator, which help to eliminate the backward axial direction leaving two DOF: one side of axial deformation (toward the centre of the toroidal chamber), and the deformation in the radial direction between neighbor chambers. The displacement of the different actuator design and the related pressure are investigated by nonlinear FEA system to emphasise the required actuator.

In previous work [19, 21], we studied the deformation of a ring actuator with a single chamber acting in a circumferential shape. It was found in a physical prototype that the deformation is not perfectly symmetrical due to the complex behavior of the soft silicone. Therefore, the multi-chamber design is introduced in this article to provide a better deformation profile on the inner surface. The embedded chambers are split along the conduit with a diameter of 70 mm⌀, which leaves 10 mm thickness between the inner surface (50 mm⌀) and the chamber's room. This thickness had been chosen based on previous experiments. Two of the chamber designs are inspired by previous work: semicircular chambers [19] and semirectangular chambers [8]. The other two chambers (cylindrical and ellipsoidal) are designed to provide a reasonable relation with neighboring chambers during the actuation.

The soft material that has been used in this work is silicone, Ecoflex 0030 which is a platinum-catalysed curing silicone with a hardness of below Shore A.

2.2 *FEA Method*

Abaqus is a commercial FEA software package and it is used in this article. The density of Ecoflex 0030 is 1064 kg/m, assumed isotropic. To simulate such a material and nonlinear elasticity in FEA, a hyperelastic mechanical behavior has been chosen in Abaqus with the third order of Ogden strain energy potential equation:

$$U \stackrel{\text{def}}{=} \sum_{i=1}^{N} \frac{2\mu_i}{\alpha_i^2}\left(\bar{\lambda}_1^{\alpha_i} + \bar{\lambda}_2^{\alpha_i} + \bar{\lambda}_3^{\alpha_i} - 3\right) + \sum_{i=1}^{N} \frac{1}{D_i}\left(J_{el} - 1\right)^{2i} \tag{1}$$

where:

$$\bar{\lambda}_i = J^{-\frac{1}{3}}\lambda_i \rightarrow \bar{\lambda}_1\bar{\lambda}_2\bar{\lambda}_3 = 1$$

The input coefficients are as follows:

$$\mu 1 = 0.001887; \alpha 1 = -3.848; \mu 2 = 0.02225; \alpha 2 = 0.663; \mu 3 = 0.003574; \alpha 3 = 4.225;$$

$$D1 = 2.93; D2 = 0; D3 = 0$$

These coefficients had been identified through the literature [22]. The code C3D10H is used to specify the model in Abaqus, which refers to hybrid elements that essentially design to simulate incompressible materials such as the behavior of a rubber. Tetrahedral elements are used to mesh the parts in this paper with node numbers as specified in Table 1.

The load is specified as a pneumatic pressure that acting in all internal chambers of the actuator cavity, and boundary restriction upon all the outer-walls leaving the inner conduit to deform freely. As the chambers shared the same conduit and they are made from the same material, the interaction between these chambers is ignored in this paper. The iteration is proceeded using the direct solver with 4 processors and full-Newton solver equation. The period of the experiment is 2 s with 20 iterations in total to apply a ramp pressure method over the time. To mesh the parts, an approximate global number of divisions (seeding) of size 3 is used in this work. It was noticed during the experiments that if the seeding size is smaller than three, the model become stiff to high distortions. The models will be assumed as a symmetrical problem in Abaqus.

In this work, standard quadratic elements are used with the hybrid formulation and reduced integration due to the hyperelastic behavior and nonlinearity of the Ecoflex 0003 silicone. Thus, will prevent issues linked with shear or volume locking, and allowing large deformations as is expected in this case.

3 Results

3.1 Experiment with a Fixed Pressure for the Four Models

In this experiment, we applied a fixed pressure to the four models. The pressure is a ramp from 0 to 70 kPa for the middle layer, and 40 kPa for the neighbor layers, both pressurised simultaneously within 20 steps (2 s), Fig. 4. The results are compared by the displacement of a selected node that moves in a straight line which is aligned to the x-axis. In Abaqus, the resultant model is brought to a cross section to recognize the mid point (node) that moves with the x-axis, Fig. 4. From that, data is collected for the displacement of that node for the 20 steps.

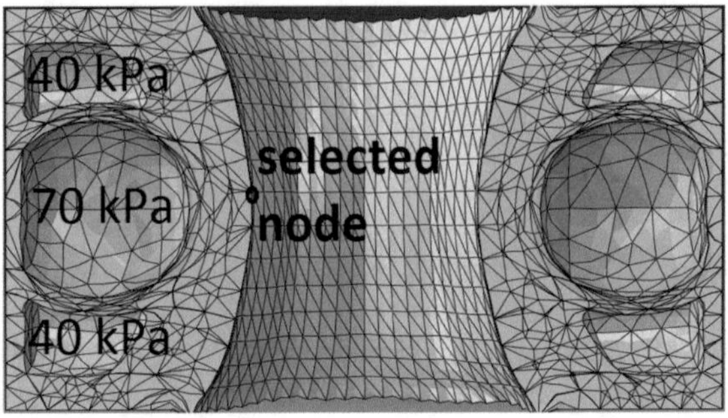

Fig. 4 The pressurised method for the three layers actuator and the selected node for measuring the displacement (cross section)

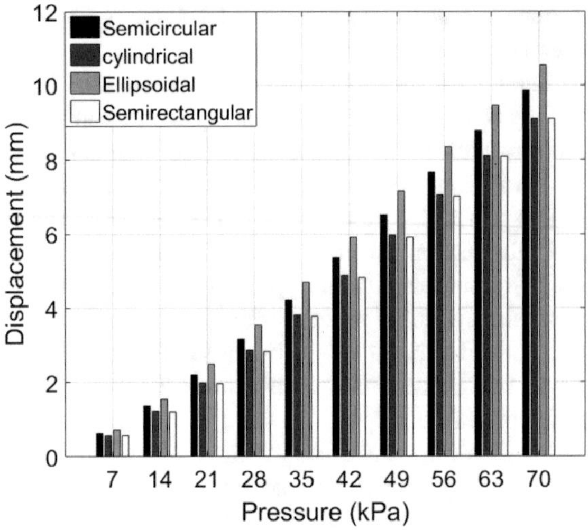

Fig. 5 The displacement of the proposed different chambers upon the same acting pressure

Figure 5 shows a comparison of the four models with displacement versus pressure, with a similar pressure applied to all actuators. It is noted that the models have a minor difference in term of 1–2 mm displacement and the models are responded to the pressure in a linear form. The highest deformation is recognised to the ellipsoidal chamber design of 10.4 mm followed by the semicircular actuator of 9.8 mm, then the semirectangular of 9.1 mm and lastly the cylindrical chamber of 9 mm.

Figure 6 shows the deformation of the chambers in each model. As mentioned before, the middle layer of chambers is the target deformable layer, while the upper and lower layer is a neighbouring chamber that is filled with a pressure of 40 kPa to resists the deformation along the y-axis and prolong the deformation on the x-axis. Moreover, the dark (red) spots on the middle horizontal line represent the magnitude of these actuators, which illustrates that the semicircular and the ellipsoidal models represent the highest magnitude and deformation.

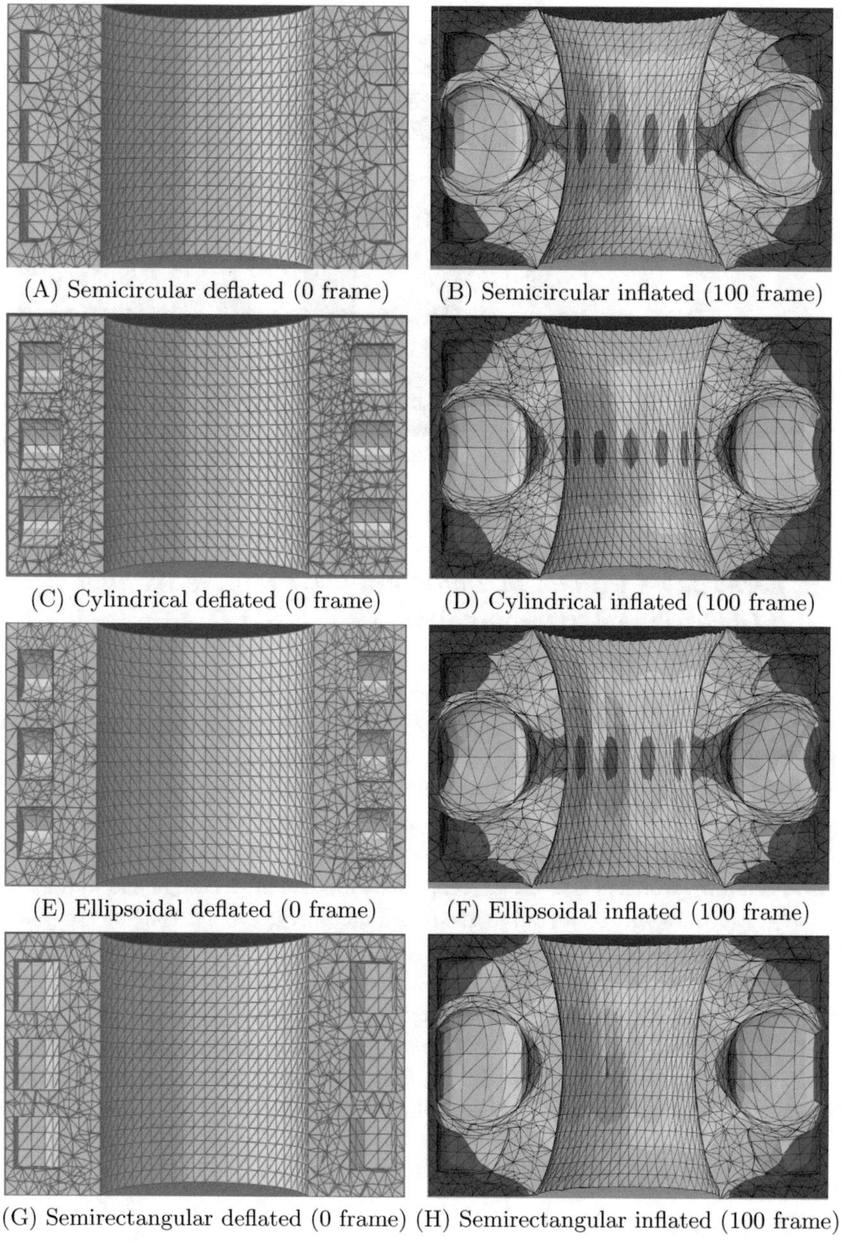

(A) Semicircular deflated (0 frame) (B) Semicircular inflated (100 frame)

(C) Cylindrical deflated (0 frame) (D) Cylindrical inflated (100 frame)

(E) Ellipsoidal deflated (0 frame) (F) Ellipsoidal inflated (100 frame)

(G) Semirectangular deflated (0 frame) (H) Semirectangular inflated (100 frame)

Fig. 6 Abaqus results of four models with three layers cross-section actuators by applying the same constant pressure with a maximum of 70 kPa. The left column represents the deflated state of 0 s and the right column represents the inflated state of 2 s. The rainbow colors represent the magnitude of the deformation (the red (dark) color is the maximum magnitude of the nodes)

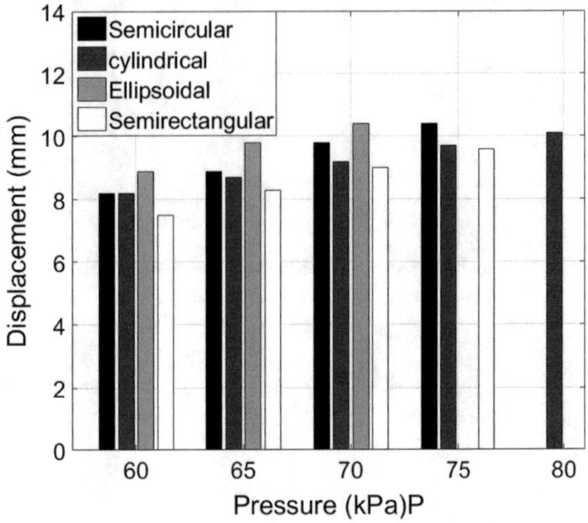

Fig. 7 The maximum pressure that can applied with the relative displacements of the actuators

3.2 *Experiment with a Maximum Applied Pressure Without a Distortion on the Actuators*

In this experiment, we applied a maximum pressure to the four models. The procedure followed is the same of the previous experiment (fixed applied pressure experiments). However, in this method, we increase the pressure in the middle layer only and then investigate the iteration of Abaqus system. The resultant model is included a smooth iteration of one step per iteration. When there is excessive distortion on the model, Abaqus system will try to solve the problem by dividing each step with multi-iterations, which leads to an incomparable model with a complex iteration system. Figure 7 shows the result of the maximum pressure applied to the four chambers design. The comparison of the pressure applied is from 60 to 80 kPa. The ellipsoidal shape has a maximum pressure of 70 kPa with a deformation of 10.4 mm. The semicircular design can handle a 74 kPa and represents a displacement of 10.4 mm. The cylindrical chamber has the highest pressure of 79 kPa but does not show the highest displacement, which is 10.1 mm. Finally, the semirectangular design is maximally inflated with 74 kPa and represents a displacement of 9.6 mm.

4 Discussion

The proposed 3 layers circumferential actuators had shown a large deformation of an average of 10 mm, and they are succeeded to cover the required average amplitude of the stomach contraction (ACWs) of 8 mm. The four models with different chamber configurations have a relatively small difference in displacements. All models exhibit

Fig. 8 A physical semicircular multi-chamber ring actuator design of 5 layers

a displacement that satisfies the requirements. However, the design that shows more flexibility with adding an extra pressure and exhibits an increase in displacement is the semicircular design. This flexibility is important in designing the actuator which provides a range of displacement that could be achieved compared to a fixed displacement that fits only the exact requirements. From that, the semicircular soft-robot ring actuator is selected for more investigation on the deformation.

The symmetrical assumption in Abaqus solver that had been made in the experiments may not represents the exact deformation behavior in the physical actuator. It was shown in a previous work that the behavior of a deformable silicone exhibits an unsymmetrical deformation towards the center of the circumferential actuator. Therefore, a physical prototype will be developed in order to validate the results of the FEA system. Figure 8 shows the physical prototype of 5 layers of semicircular chambers, and a shell of a silicone with the hard shoreA 50 that serves as a rigid shell. The inflation rhythmic and the pressure applied will follow the FEA method. The results of the displacement and further investigation will be discussed in future work. The second related issue that should be addressed is the width of the deformation. As the amplitude of the contraction is satisfied with an average of 10 mm displacement, the width of the contraction is not controlled in this work. It is a challenging task to control the width of the deformation because of the behavior of the pneumatic forces, as they spread in all direction leading to a balloon shape. As the neighbour chambers enforce the deformation in one direction, it was not enough to control the radial deformation (the width of the deformation) with a controllable rhythm. More investigation will be conducted towards the behavior of the radial deformation which will provide a method to control the displacement of the contraction width.

5 Conclusion

In this paper, we presented FEA evaluation for four designs of multi-chambers mulilayers soft robot ring actuators. The result deformation of the ring actuators in the four models has an average of 10 mm displacement. Two experiments had been investigated in the ring actuators: we applied the same pressure to all actuators for displacements comparison, and we applied a maximum pressure to each actuator without distorting the material to achieve the maximum displacement. The first experiment with a fix pressure applied resulted two models that exhibit a displacement about 10 mm: semicircular and ellipsoidal designs, and the other two models exhibit a displacement about 9 mm. On the other hand, the second experiment with a maximum pressure applied resulted three models that exhibit a displacement of 10 mm: semicircular, cylindrical and ellipsoidal models. From the experiments, one model shows a tolerance to an excessive pressure and a large displacement: semicircular design, this model will help to design a tolerant model that provides a minimum displacement of 8 mm with no excessive distortion on the material of the actuator. From that, a semicircular model has been developed for more investigation about deformation on a physical prototype.

Acknowledgements The work presented in this paper was funded by New Zealand Medical Technologies Centre of Research Excellence (MedTech CoRE).

References

1. Rus, D., Tolley, M.T.: Design, fabrication and control of soft robots. Nature **521**(7553), 467–475 (2015)
2. Shepherd, R.F., Ilievski, F., Choi, W., Morin, S.A., Stokes, A.A., Mazzeo, A.D., Chen, X., Wang, M., Whitesides, G.M.: Multigait soft robot. Proc. Natl. Acad. Sci. USA **108**(51), 20400–20403 (2011). Cited By (since 1996):51
3. Lin, H.T., Leisk, G.G., Trimmer, B.: Goqbot: a caterpillar-inspired soft-bodied rolling robot. Bioinspir. Biomim. **6**(2) (2011). Cited By (since 1996):24
4. Laschi, C., Cianchetti, M., Mazzolai, B., Margheri, L., Follador, M., Dario, P.: Soft robot arm inspired by the octopus. Adv. Robot. **26**(7), 709–727 (2012)
5. Trimmer, B.A.: New challenges in biorobotics: incorporating soft tissue into control systems. Appl. Bionics Biomech. **5**(3), 119–126 (2008). Cited By (since 1996):6
6. Kim, S., Laschi, C., Trimmer, B.: Soft robotics: a bioinspired evolution in robotics. Trends Biotechnol. **31**(5), 287–294 (2013). Cited By (since 1996):4
7. Cianchetti, M., Calisti, M., Margheri, L., Kuba, M., Laschi, C.: Bioinspired locomotion and grasping in water: the soft eight-arm octopus robot. Bioinspir. Biomim. **10**(3), 035003 (2015)
8. Dirven, S., Stommel, M., Hashem, R., Xu, W.: Medically-inspired approaches for the analysis of soft-robotic motion control. In: 2016 IEEE 14th International Workshop on Advanced Motion Control (AMC), pp. 370–375
9. Dirven, S., Chen, F., Xu, W., Bronlund, J.E., Allen, J., Cheng, L.K.: Design and characterization of a peristaltic actuator inspired by esophageal swallowing. IEEE/ASME Trans. Mech. **19**(4), 1234–1242 (2014)

10. Condino, S., Harada, K., Pak, N., Piccigallo, M., Menciassi, A., Dario, P.: Stomach simulator for analysis and validation of surgical endoluminal robots. Appl. Bionics Biomech. **8**(2), 267–277 (2011)
11. Trimmer, B.A., Lin, H.T., Baryshyan, A., Leisk, G.G., Kaplan, D.L.: Towards a biomorphic soft robot: design constraints and solutions. In: Proceedings of the IEEE RAS and EMBS International Conference on Biomedical Robotics and Biomechatronics, pp. 599–605 (2012). Cited By (since 1996):4
12. Cheng, L.K., Komuro, R., Austin, T.M., Buist, M.L., Pullan, A.J.: Anatomically realistic multiscale models of normal and abnormal gastrointestinal electrical activity. World J. Gastroenterol. **13**(9), 1378 (2007)
13. Benshitrit, R.C., Levi, C.S., Tal, S.L., Shimoni, E., Lesmes, U.: Development of oral food-grade delivery systems: current knowledge and future challenges. Food Funct. **3**(1), 10–21 (2012)
14. Cheng, L.K., O'Grady, G., Du, P., Egbuji, J.U., Windsor, J.A., Pullan, A.J.: Gastrointestinal system. Wiley Interdiscip. Rev.: Syst. Biol. Med. **2**(1), 65–79 (2010)
15. Ehrlein, H.J., Schemann, M.: Gastrointestinal Motility
16. Schulze, K.: Imaging and modelling of digestion in the stomach and the duodenum. Neurogastroenterol. Motil. **18**(3), 172–183 (2006)
17. Keet, A.D.: Infantile hypertrophic pyloric stenosis. In: The Pyloric Sphincteric Cylinder in Health and Disease, p. 107. Springer, Berlin, Heidelberg (2009)
18. Kwiatek, M.A., Steingoetter, A., Pal, A., Menne, D., Brasseur, J.G., Hebbard, G.S., Boesiger, P., Thumshirn, M., Fried, M., Schwizer, W.: Quantification of distal antral contractile motility in healthy human stomach with magnetic resonance imaging. J. Magnet. Reson. Imaging **24**(5), 1101–1109 (2006)
19. Hashem, R., Xu, W., Stommel, M., Cheng, L.: Conceptualisation and specification of a biologically-inspired, soft-bodied gastric robot. In: 2016 23rd International Conference on Mechatronics and Machine Vision in Practice (M2VIP), pp. 1–6, Nov 2016
20. Dirven, S., Chen, F., Xu, W., Bronlund, J.E., Allen, J., Cheng, L.K.: Design and characterization of a peristaltic actuator inspired by esophageal swallowing (2013)
21. Dang, Y., Cheng, L.K., Stommel, M., Xu, W.: Technical requirements and conceptualization of a soft pneumatic actuator inspired by human gastric motility. In: 2016 23rd International Conference on Mechatronics and Machine Vision in Practice (M2VIP), pp. 1–6, Nov 2016
22. Moseley, P., Florez, J.M., Sonar, H.A., Agarwal, G., Curtin, W., Paik, J.: Modeling, design, and development of soft pneumatic actuators with finite element method. Adv. Eng. Mater. (2015)

Part V
Machine Vision

Detection and Classification of Vehicle Types from Moving Backgrounds

Xuesong Le, Jun Jo, Sakong Youngbo and Dejan Stantic

Abstract Using unmanned aerial vehicles (UAV) as devices for traffic data collection exhibits many advantages in collecting traffic information. This paper introduces a new vehicle dataset based on image data collected by UAV first. Then a novel learning framework for robust on-road vehicle recognition is presented. This framework starts with conventional supervised learning to create initial training data set. Then a tracking-based online learning approach is applied on consecutive frames to improve the accuracy of vehicle recogniser. Experimental results show that the proposed algorithm exhibits high accuracy in vehicle recognition at different UAV altitudes with different view scopes, which can be used in future traffic monitoring and control in metropolitan areas.

1 Introduction

Although vehicle recognition has been an area of great recent interest in the machine-learning community [1], most recent works consider more lateral and frontal views, either from wide area monitoring imaging or from on-board-like camera views [2, 3]. No prior research study has used drone imagery to build an on-road vehicle recognition and we believe that this is the first work that vehicle recognition has been evaluated in the context of drone imagery.

X. Le · J. Jo (✉) · D. Stantic
School of Information and Communication Technology,
Griffith University, Southport, QLD 4222, Australia
e-mail: j.jo@griffith.edu.au

X. Le
e-mail: xuesongle@gmail.com

D. Stantic
e-mail: d.stantic@griffith.edu.au

S. Youngbo
Soletop, 409, Expo-Ro, Yuseong-Gu, Daejeon 34051, South Korea
e-mail: ybsakong@soletop.com

J.-H. Kim et al. (eds.), *Robot Intelligence Technology and Applications 5*,
Advances in Intelligent Systems and Computing 751,
https://doi.org/10.1007/978-3-319-78452-6_39

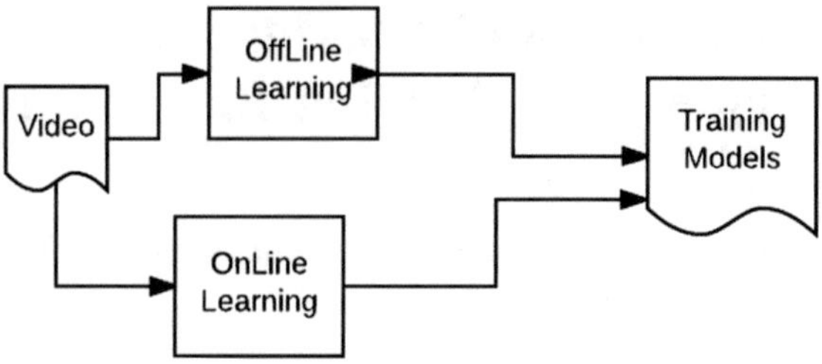

Fig. 1 The learning framework for vehicle recognition system consists of an offline learning and an online learning

One reason to explain the slow progress in this field is the limited availability of vehicles recognition datasets from drone imagery. Several datasets, are available for the evaluation of object detection tasks. UIUC image dataset [3] contains images of side views of cars for use in evaluating object detection algorithms. PASCAL VOC 2012 [4] benchmark provides one of the key datasets for vehicle classification of side views. BIT-Vehicle dataset [5] consists of vehicles belonging to 6 categories—Bus, Microbus, Minivan, Sedan, SUV and Truck with 150 images under each class. However, the dataset contains only frontal images of vehicles. None of those datasets is suitable for developing algorithms for vehicle recognition in drone imagery.

Within this context, the motivation for this work is twofold. First, the work aims to introduce VRDI (Vehicle Recognition in Drone Imagery), a new database designed to address the task of small vehicle detection and recognition in drone images in unconstrained environment. The vehicles to be recognised have different orientations, at different scales, occluded or masked. Second, a rapid automatic vehicle detection and recognition learning method is proposed. An overview of the complete learning framework can be seen in Fig. 1.

The remainder of this paper is structured as follows: it starts describing our new, VRDI dataset in Sect. 2. Section 3 details the off-learning portion of learning framework. Following that, a tracking-based online learning approach is presented in Sect. 4. In Sect. 5, we successfully apply the proposed learning algorithm to various real video sequences and analyse its performance. At last, we draw a conclusion in Sect. 6.

2 VRDI Dataset

The proposed dataset is a collection of drone images cropped from videos of urban traffic scenes in Korea. Videos are captured at 30 frames per second at a resolution of 1920 × 1080. Each video is sub-sampled to 960 × 540 and 480 × 270. As shown in the Fig. 2, images in this dataset are captured at different places and different time. These images contain changes in illumination condition, scale, the surface colour of vehicles and viewpoint. All vehicles in the dataset are divided into four categories: Bus, Sedan, SUV, and Truck. Each category contains 150 vehicles,

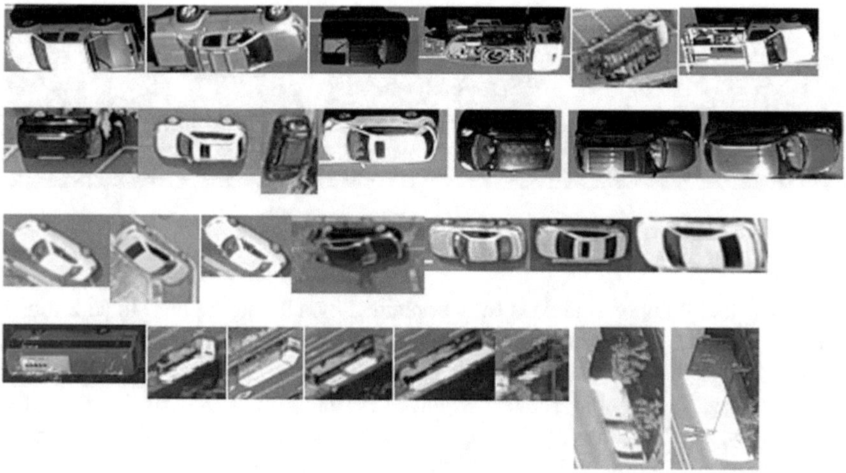

Fig. 2 Categorised sample images of the VRDI dataset

to provide a total of 600 vehicles. Annotations give information such as colour, probability of classification, tracking identification, location of each vehicle.

3 Offline Learning

As shown in Fig. 3, offline learning consists of three steps: selection of labelled "positive" examples and random "negative" examples, feature extraction and the conventional supervised multi-class learning. Each step is discussed in detail next.

3.1 Training Sample Selections

As the average aspect ratio of length to width for most vehicles is 2.375–1, each sample training image is resized to 152 by 64 pixel. ROIs in the images are selected

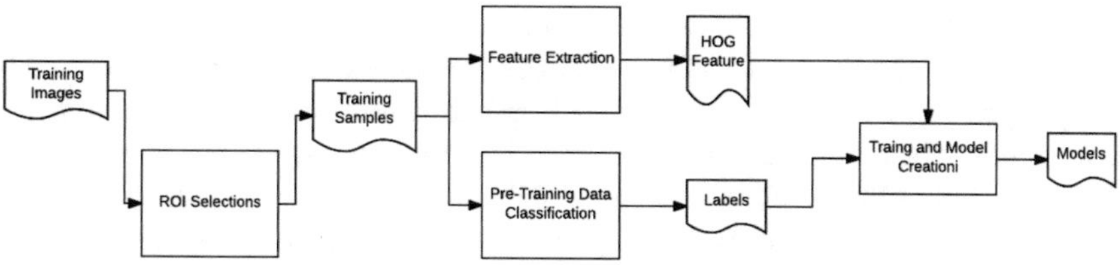

Fig. 3 The block diagram for offline learning portion

Fig. 4 8 rotations for a sample image at 1920 × 1080 resolution

in the way that body of vehicle is fully contained by a bounding box. In addition, as vehicles can move in arbitrary directions in the image. Thus, for each cropped vehicle, the ROI selection is performed at multiple directions. In this paper, each training sample image is rotated 8 times at interval of 20° at 3 resolutions respectively, resulting into 24 training images per cropped vehicle. Figure 4 shows 8 rotations for a sample vehicle at 1920 × 1080 resolution.

3.2 HOG Feature Extraction and Classification

The key component in vehicle classification is to select a good feature descriptor that can distinguish one vehicle from the other. Several popular shape-based algorithms feature descriptor have emerged in the literature, including Haar wavelets [6] and Gabor filter outputs [7], Haar-like feature [8], edge templates [9], histogram of oriented gradients (HOG) [10]. Within the class of the shape-based algorithms, HOG feature descriptor shows better performance in characterising object shape and appearance and it is considered one of the most accurate feature descriptor for visual classification problems. Our work uses only the HOG feature set to allow for classification on moving vehicles.

The HOG descriptor is initially predicted for pedestrian detection. The HOG are calculated by taking the orientations of the edge and histograms intensity in a local region. In this paper, the calculation of HOG is done as follows:

First, the gradient vector of each pixel in each local cell is calculated and grouped into a 9-bin orientation histogram. The orientation histogram ranges from 0 to 180 degrees, so there are 20° per bin. The contribution of each gradient vector to the histogram is given by its magnitude. Therefore, stronger gradients contribute more weight to their bins, and effects of small random orientations due to noise is reduced. This histogram pictures the dominant orientation of that cell.

Next, to reduce the light variations or shadowing problems, every 2 × 2 cells are grouped into a bigger unit called block and normalised based on all histograms in the block. The normalisation is performed by concatenating the histograms of the four cells within the block into a vector with 36 components (4 histograms × 9 bins per histogram). Then each component is divided by the magnitude of the vector. Same as in [10], blocks in a detection window are overlapped and a “50%

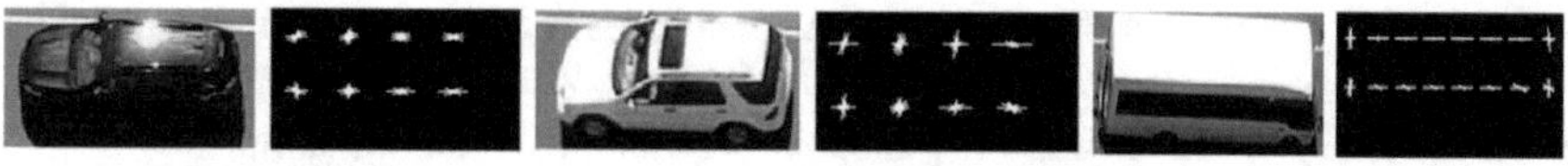

Fig. 5 HOG feature examples for describing different vehicles' appearance

overlap" rate is used here. Given above configuration, a 152 × 64 pixel detection window will be divided into 18 blocks across and 7 blocks vertically, for a total of 126 blocks. Each block contains 4 cells with a 9-bin histogram for each cell, for a total of 36 values per block. This brings the final vector size to 126 blocks per detection window × 4 cells per block × 9-bins per histogram that equals a total of 4536 values. Figure 5 gives a few examples of HOG features for the description of a vehicle's appearance.

In a multi-category context, HOG feature vectors are then classified by means of a multi-class Sequential Minimal Optimisation (SMO) [11]. Compared with the popular support vector machines (SVM) classifiers, SMO solves the quadratic programming (QP) problem that arises during the SVM training by breaking the large QP problem into a series of the smallest possible QP problems first and then solving each small QP problem analytically. In such way, it avoids using a time-consuming numerical QP optimisation as an inner loop.

4 Online Training

Due to the time-consuming work in labelling data and requirement of human interaction, it is unlikely to provide a good coverage over the space of possible appearance variations of the data during offline training. In this section, a tracking based online training approach is proposed to collect any false negative samples automatically. As shown in Fig. 6. It consists of four main steps: object extraction, tracking, recognition and model updating. In the object extraction stage, initial ROIs are identified using proposed background subtraction scheme and object localisation. Identified ROI are then classified either manually or selected from the

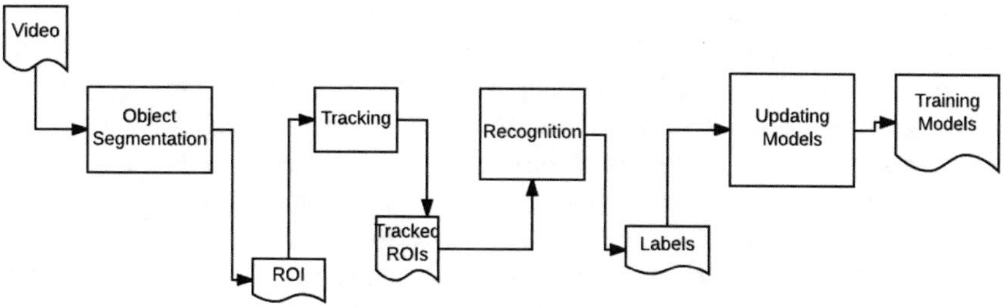

Fig. 6 The block diagram for online learning portion

offline training samples. Next, a tracking mechanism is used to assign multi-class probability map to the tracked ROIs. Finally, probabilities of each class are output from the SVM classifier and only those newly labelled training samples with high error probabilities are then used to retrain the classifier.

4.1 Initial ROIs Detection

The first step in online learning stage includes two sub-steps, object extraction using background subtraction and object localisation. The first sub-step improves the accuracy of the successive step of car recognition: it reduces the occurrence of false positives which may arise when the detection is performed on the whole image. Localisation of the object mask in the second step reduces the required processing time.

4.1.1 Object Extraction Using Background Subtraction

Extraction of independent moving object from a sequence taken by a non-stationary camera is the first step within the online training. Once a moving object is detected, the system can recognise and track it individually. In some situation such as a scene of a street, it is difficult to take images from a scene in which no moving objects exists in advance. Therefore, we have to estimate the background from the sequence of images in which some moving objects are included.

In this work, an adaptive backgrounding modelling method in RGB space is proposed. The background value of each pixel is estimated by an adaptive robust position estimator. Once the background is estimated, the detection of moving ROIs becomes straightforward. Given the fact that the point of view changes frame by frame due to the movement of the drone, homographic transformation is applied on every input image to compensate camera motion. Transformation (T) is found sufficiently quickly and exactly by the camera motion estimation method proposed in [12]. After camera motion compensation, a background approximation which is similar to the current static scene except where motion occurs is created based on a robust statistical model.

Since the duration of image sequences taken from the drone is usually short, we can assume that change of a pixel value is mainly caused by the moving object such as cars and by other factors like lighting changes where effects of moving elements of the scene are minimum. Therefore, the distribution of background modelling from the stationary signal can be described as a single Gaussian distribution. For this purpose, the first step is to apply change detection on each pixel in the last N frames to determine if it is a moving or stationary one. We can use a technique from robust statistic by considering the non-stationary signal as outliers. Such as, a new pixel x_i, is detected as background pixels if as a pixel value within 2.5 standard deviations of a distribution. Therefore, the parameter of Gaussian distribution can be estimated and used to verify background pixels.

Next, the background model is created using weighted moving average method. Given a more recent history of background pixel, denoted by $x = \{x_0, x_1 \ldots, x_i, \ldots x_{k-1}\}$ where $k < N$, the weight attached to each position in the history should be treated differently since higher weight is assigned to the more recent pixel value, and better adapt to the faster changes in the background. The weighted background pixel can be represented as

$$w(x) = \sum_{i=0}^{k-1} w_i x_i, \tag{1}$$

where both x and q are k-dimensional vector, w_i is the value of the weight attached to a background pixel in the *i*th measure. w_i can be derived in the following:

$$w_i = 1 - \left(\frac{1 - r^k}{1 - r}\right) r^i, \tag{2}$$

where $0 < w_i < 1$, $\sum_{i=0}^{k-1} w_i = 1$ $w_i = w_{i-1} r$, r is the diminishing rate between weight w_i and w_{i-1}.

4.1.2 Background Updating

Each time the parameters of the Gaussian are updated as a new frame arrives, the Gaussian for each pixel is evaluated using a simple heuristic to hypothesise which are most likely to be part of the "background process". If the current pixel value is classified as the background pixel, the weighted average background must be updated given the highest weight assigned to the current pixel value.

Using these adaptive robust estimators, the background image in each channel can be obtained by performing the adaptive estimation for all pixels. Then moving objects can be detected by taking the difference between the current image and the estimated background image. Example results for the scenes of a street are shown in Figs. 7 where N = 50, K = 10.

4.1.3 Object Localisation

The vehicle recogniser generally works well if perfect object masks can be generated. In practice, the precise object delineations are very difficult to be acquired due to the lack of perfect background subtraction, the existence of non-corresponding objects and environment effect such as light changes and registration errors. Selective search [13] is applied on the extracted ROIs to localise the

(a) Frame 43,44,45,46,47,48,49 Image sequence taken from a highway.

(b) Background modelling in R,G,B channel respectively.

(c) Frame 49 in R,G,B channel respectively

(d) Difference frame between frame 49 and background model in R,G,B channel respectively

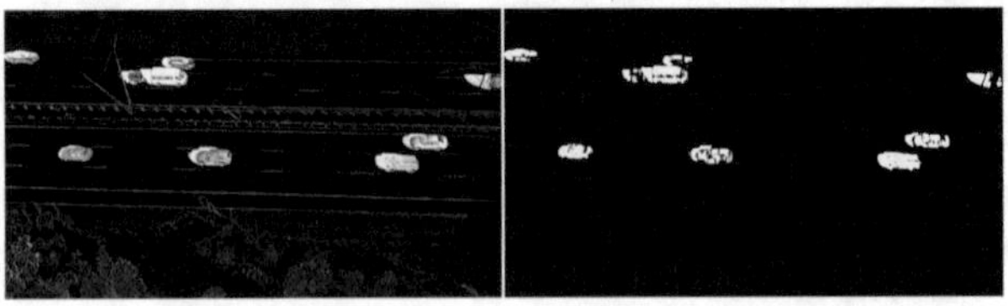

(e) Sum of difference frames in R,G,B channel

(f) Binary thresholding result from (e)

Fig. 7 Object extraction using background subtraction

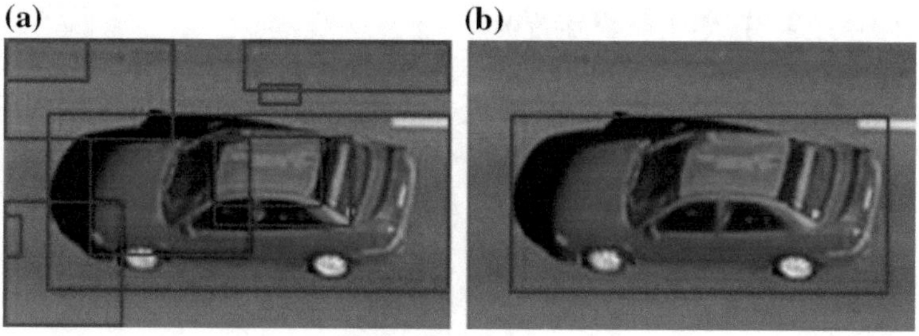

Fig. 8 **a** Top 10 region proposals. **b** Final selection

object accurately. The goal here is to find a bounding box where the desired parts of object masks are sufficiently close to each of the sides of the bounding box.

The selective search method assumes that all objects of interest share common visual properties that distinguish them from the background. Then a trained model can be used to detect a set of proposal regions that are likely to contain objects from an image. If high object recall can be reached with considerably fewer windows than used by sliding window detectors, significant speed-ups can be achieved. Figure 8a shows the top 10 region proposals arranged in decreasing order of objectness using selective search algorithm. As the final object mask is most likely to have the largest size among the top 10 region proposals. Then the final selection of the bounding box $\boldsymbol{B}$ can be defined as:

$$B = \max(sz(b_i)), i = 1\ldots, 10 \tag{3}$$

where sz is the size of each region proposal among the top 10 region proposals. Figure 8b shows the final selection of the bounding box $\boldsymbol{B}$ from Fig. 8a.

4.2 *ROI Tracking and Updating*

Once a final object mask is identified and classified, a tracking mechanism [14] is used to test segmented and tracked vehicle against an assigned label. Any false negative alarms and false positive samples are added to the original positive and negative samples respectively and re-trained. In such way, the time-consuming work of labelling an enormous amount of training samples is automated and the recognition accuracy is improved.

5 Experiment and Discussion

In this section, experimental results of the proposed method on urban traffic videos are presented to show the performance of our approach.

5.1 Performance Metrics

To quantify the performance of classification, the following metrics: precision, recall, and F measure are used.

Given the definition of TP, FP, FN, and TN in Table 1, we can then define the precision, recall and F measure for each class as follows:

$$\begin{aligned} \text{Precision} &= \text{TP}/(\text{TP}+\text{FP}), \\ \text{Recall} &= \text{TP}/(\text{TP}+\text{FN}), \end{aligned} \tag{4}$$

$$F_1 = 2(\text{Precision} * \text{Recall})/(\text{Precision}+\text{Recall}), \tag{5}$$

5.2 Experimental Results

To evaluate the performance of our classifier, three-folds Cross validation is applied on existing data set. The existing dataset is split into three-parts. The classifier is trained on two folds and tested on the held back fold. This process is repeated three times so that each fold of the dataset is given a chance to be the held back test set. Table 2 shows the results measured in precision, recall and F-Measure per class. The confusion matrix is shown in Table 3. It is of note that the result is very promising given the size of the dataset. Nearly 90% of testing instances are correctly classified.

Some vehicle classification results from the real-time videos are shown in Fig. 9, probability of classification for each class and tracking identification are displayed for each tracked vehicle. The labelled class is marked in red. As we can see, most of the vehicles have been correctly classified.

Table 1 Performance table for instances labelled with a class label A

	True label A	True not A
Predicted label A	True positive (TP)	False positive (FP)
Predicted not A	False positive (FN)	True negative (TN)

Table 2 Classification performance

TP rate	FP rate (%)	Precision (%)	Recall (%)	F-measure (%)	Class
82%	6	85	82	84	SED
85%	7	78	85	81	SUV
82%	1	93	82	88	TRU
100%	0	100	100	100	BUS
100%	0	100	100	100	BAC
Weighted avg. 0.897	3	90	90	90	

Table 3 Confusion matrix of precision and recall per class

a	b	C	d	e	Classified as
82%	17%	1%	0	0	I a = SED
14%	85%	1%	0	0	I b = SUV
12%	6%	82%	0	0	I c = TRU
0	0	0	100%	0	I d = BUS
0	0	0	0	100%	I e = BAC

Fig. 9 Annotations of frame 1, frame 4, frame 7 from testing video sequence include: probability of classification for each class and tracking identification

6 Conclusion

This work proposes VRDI, a new database for evaluating the recognition of vehicles in drone images. The images are split in five different categories. In addition to the dataset, a general learning framework for robust on-road vehicle recognition has been presented and a thorough quantitative evaluation has been proposed. The system has been evaluated on both real-world videos. The effectiveness of the proposed algorithm to robust vehicle recognition is demonstrated for a variety of real environments given current dataset.

References

1. Sivaraman, S., Trivedi, M.M.: A general active-learning framework for on-road vehicle recognition and tracking. IEEE Trans. Intell. Transp. Syst. **11**(2), 267–276 (2010)
2. Ballesteros, G., Salgado, L.: Optimized HOG for on-road video based vehicle verification. In: IEEE 22nd European Signal Processing Conference (EUSIPCO) (2014)
3. Agarwal, S., Awan, A., Roth, D.: Learning to detect objects in images via a sparse, part-based representation. IEEE Trans. Pattern Anal. Mach. Intell. **26**(11), 1475–1490 (2004)
4. Everingham, M., Van Gool, L., Williams, C.K.: The PASCAL visual object classes challenge 2012, VOC2012 results (Online). http://www.pascal-network.org/challenges/VOC/voc2012/workshop/index.html (2012)
5. Dong, Z., Pei, M., He,Y., Liu, T., Jia, Y.: Vehicle type classification using unsupervised convolutional neural network. In IEEE 22nd International Conference on Pattern Recognition (2014)
6. Papageorgiou, C., Poggio, T.: A trainable system for object detection. Int. J. Comput. Vision **38**(1), 15–33 (2000)
7. Jain, A.K., Ratha, N.K., Lakshmanan, S.: Object detection using Gabor filters. Pattern Recogn. **30**(2), 295–309 (1997)
8. Lienhart, R., Maydt, J.: An extended set of haar-like features for rapid object detection. In: IEEE International Conference on Image Processing (2002)
9. Zhao, G., Matti, P.: Dynamic texture recognition using local binary patterns with an application to facial expressions. IEEE Trans. Pattern Anal. Mach. Intell. **29**(6), 915–928 (2007)
10. Dalal, N., Triggs, B.: Histograms of oriented gradients for human detection. In: IEEE Computer Society Conference on Computer Vision and Pattern Recognition, CVPR 2005 (2005)
11. Platt, J.: Sequential minimal optimization: a fast algorithm for training support vector machines. Microsoft Res. (1998)
12. Le, X., Gonzalez, R.: A robust region-based global camera estimation method for video sequences. In: 2013 7th International Conference on Signal Processing and Communication Systems (ICSPCS) (2013)
13. Uijlings, J.R., Van, D.S., Gevers, K.E., Smeulders, A.W.: Selective search for object recognition. Int. J. Comput. Vision **104**(2), 154–171 (2013)
14. Henriques, J.F., Caseiro, R., Martins, P., Bati, J.: High-speed tracking with kernelized correlation filters. IEEE Trans. Pattern Anal. Mach. Intell. **37**(3), 583–596 (2015)

Multi-object Tracking with Pre-classified Detection

Siqi Ren, Yue Zhou and Liming He

Abstract Tracking-by-detection is a popular tracking framework nowadays. The paradigm determines that detections will bring huge impact on the final tracking result. Based on the idea, to improve the tracking precision, we propose a novel algorithm which will first divide the detections into false ones, high uncertainty and low uncertainty ones. After that, we first penalize the false detection boundingboxes and then we construct low and high uncertainty tree for different types of detections. For low uncertainty detections, we construct their tracking trees once, for high uncertainty detections, we adopt an improved MHT to delay the data association decision till the end of the sliding windows and make the decision with all the information in the sliding window. Experiments demonstrate that our algorithm can achieved competitive results with state-of-the-art trackers on some challenging datasets such as MOTChallenge2016 [11].

Keywords Multi-object tracking · MHT · Detection classification

1 Introduction

Multi-object tracking (MOT) is one of the most important tasks in computer vision. Although several decades of research has been done, it is still challenging due to occlusion, similar appearance, different viewpoints, camera motion and so on. Recently, tracking-by-detection approaches have shown a promising performance due to the development of object detectors [1, 16, 19]. After detecting all the object

S. Ren · Y. Zhou (✉) · L. He
Institute of Image Processing and Pattern Recognition, Shanghai Jiao Tong University, Shanghai, China
e-mail: zhouyue@sjtu.edu.cn

S. Ren
e-mail: rensiqi_stju@sjtu.edu.cn

L. He
e-mail: heliming@sjtu.edu.cn

J.-H. Kim et al. (eds.), *Robot Intelligence Technology and Applications 5*, Advances in Intelligent Systems and Computing 751,
https://doi.org/10.1007/978-3-319-78452-6_40

in the frame, the major issue is the data association problem. Object appearances models and object motion models are often used as important cues for data association.

Up to now, the strategies can be categorized into offline and online methods. The offline methods [4, 12, 15, 18] generate tracklets with past and future information. To make sure the detections are efficiently bound to the tracklets, these methods usually discretize the space of target locations to simplify the underlying optimization problem which may require huge computation due to the iterative associations for generating globally optimized tracklets. These offline approaches often get better results because they exploit the future information.

Online approaches [2, 9, 20] build trajectories based on frame-by-frame association using online information with high computational efficiency. However, they are more likely to be influenced by mis-detection, occlusion, abrupt change of object motion and produce many mistake as ID switches, track drift and fragmentation. Yoon et al. [21] formulates relative motion model to handle the problem like camera motion. Since most algorithms only set hsv histogram appearance as model, these methods may be prone to cause the ID switch and track drift.

In this paper, we propose a novel accurate, robust and efficient offline multi-object tracking framework based on the strategy for effective classification of the detections. We argue that some detectors may bring some unsuitable detections to trackers which may lead to poor tracking performance, and detections with low overlap ratio can obtain good results even with simple frameworks because they are less affected by the problem of occlusion. However, bounding boxes with high overlap ratio may lead to poor tracking results. In order to improve the tracking performance, we utilize the strength of MHT that we build a tree of potential track hypotheses for each candidate and prune the branches at the end of every sliding window. We perform a comprehensive experimental evaluation on [11]. The proposed method achieves impressive performance, demonstrating the effectiveness of our algorithm.

2 Proposed Method

As is shown in Fig. 1, we adopt a tracking-by-detection framework that the observations are obtained from DPM [1] which is provided by [11], then data association is left to be done. Before the data association, we first run a neural network which was motivated by [3] to identify the unsuitable detections and then penalize them. Then the rest of detections can be classified by overlap ratio. Low uncertainty detections are associated to the track trees by the strategy of the structural constraint event aggregation which was mentioned in [8]. High-uncertainty detections may suffer heavy occlusion, are hard to be associated to the right track trees. In accordance with the MHT [15] we build a tree of potential track hypotheses for each candidate target. Motivated by [10], the regularized least squares appearance models are trained online for each track hypothesis with rgb and hsv histograms of all candidates from the entire history of the track. We set a silding window which is τ frames long. When

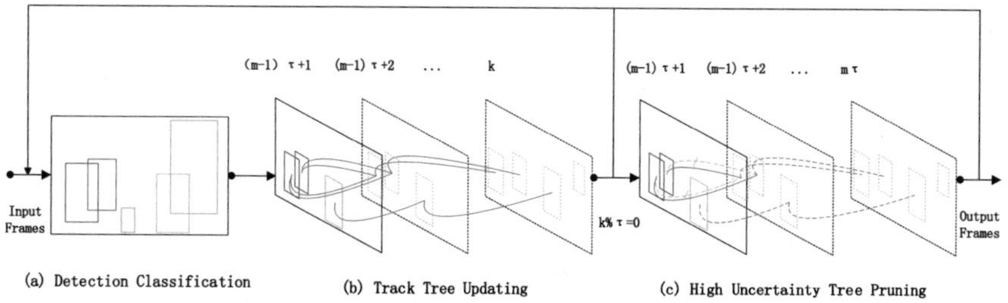

Fig. 1 Schematic illustration of the proposed method. **a** Given a set of detections D at frame k, we first classify the detections into false detections (yellow bounding boxes), low uncertainty detections (green bounding boxes) and high uncertainty detections (red bounding boxes), then penalize the false detections. **b** Track tree updating which will append both of the high uncertainty detections with each new detection spawning a separate branch. **c** Track tree pruning at the end of every sliding window

the current frame k can be divided by τ, we score all the hypotheses branches and only retain the branch with the highest score. We obtain our final tracklets up to the current window. In the rest of this section, we will describe the approach in more detail (Fig. 2).

2.1 Detection Classification

Some of the detectors may bring false detections to the trackers, so when we obtain all the bounding boxes in the frame, it's necessary for us to filter the detections for effectiveness. Figure 3 shows the result of our tracker with or without the false detection category. The result shows that it's helpful for increasing the tracking accuracy with the processing of false detections. Motivated by [3], we adopt an neural network as a filter and we thought the height of the detections in some of the positions of the screen are not only decided by the coordinates of their foots but also decided by the gray value of the ground under their foots. For example, when some of the bounding boxes are marked on the walls, others are marked on some part of the pedestrians, their height fit the scene, but they are false detections. If we take gray value into consideration, most of these boxes will be penalized. The framework of the network can been seen in Fig. 2. Let x and y be the coordinate of the middle of the bottom edge of the box, and the two coordinates inputs x' and y' of the network will be obtained by:

$$x' = \frac{x}{w} + 1 \tag{1}$$

$$y' = \frac{y}{w} + 1 \tag{2}$$

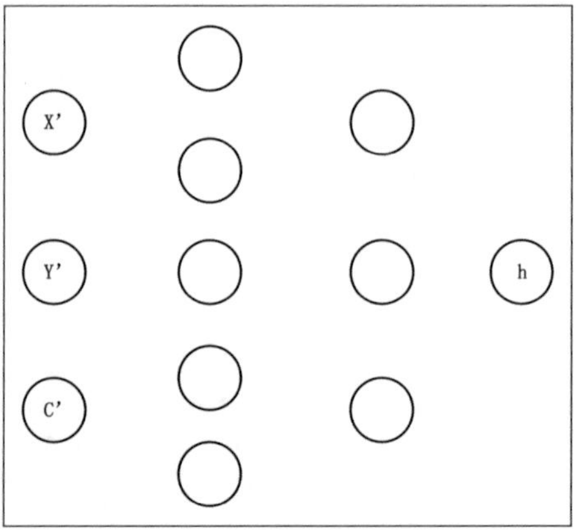

Fig. 2 The network we use to classify the detections

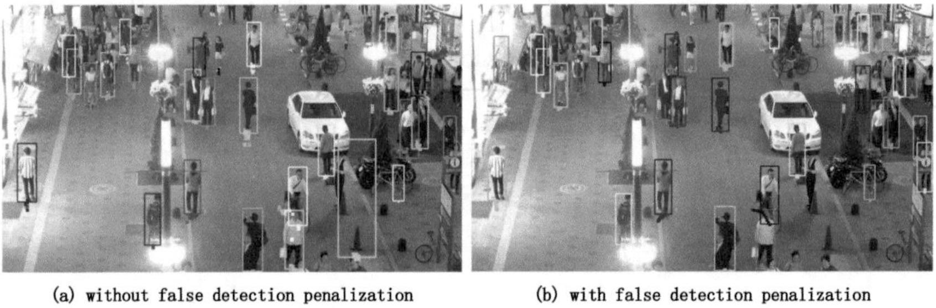

Fig. 3 The 15th frame's tracking result of the MOT16-03 sequence. **a** Our tracker without false detection category. **b** Our tracker with false detection category

and the third input c' is the mean gray value of the bottom edge of the boxes. The network will predict the height of the detections using these information. Bounding boxes beyond the confine of the height will be deleted.

For the rest of the detections, multiple comparisons of their overlap ratio will be carried out. The overlap ratio is formulated by:

$$o_{i,j} = \frac{area\left(B\left(d_i\right) \bigcap B\left(d_j\right)\right)}{area\left(B\left(d_i\right) \bigcup B\left(d_j\right)\right)} \tag{3}$$

whether a detection d_j is a high uncertainty or a low uncertainty detection is defined by:

$$O_j = \sum_{i=1, i \neq j}^{M} \left(c(i,j) = \begin{cases} 1 & \text{if } o_{i,j} > \mu \\ 0 & \text{if } o_{i,j} < \mu \end{cases} \right) \tag{4}$$

$O_j = 0$ represents for the low uncertainty condition. The others represents for the high uncertainty condition. For low uncertainty detections, we construct low uncertainty trees for them which don't need to be pruned. For high uncertainty detections, We adopt the strategy of MHT, and high uncertainty trees will be constructed. And we will keep multiple hypotheses active until current frame k is divided by τ.

2.2 Track Tree Updating

For low certainty detection, we adopt the structural constraint event aggregation [8] to evade the error caused by the global camera motion. In this section, we simply introduce the structural constraint event aggregation which we use for low-uncertainty tracklets updating. Readers can get more information in [8]. To update the track tree, we have to minimize the structural constraint cost function:

$$\widehat{A} = \underset{A}{argmin} \sum_{i \in N} \sum_{j \in L} \left(a^{i,j} \Omega_{i,j} + \sum_{\substack{p \in N \\ p \neq i}} \sum_{\substack{q \in L \cup \{0\} \\ q \neq j}} a^{p,q} \Theta^{p,q}_{i,j} \right) \tag{5}$$

where anchor assignment $a^{i,j}$ is a binary value. $a^{i,j}$ represents that the jth detection was associated with the ith track tree or not, $a^{i,0}$ stands for the case of mis-detected objects. N is the number of tracklets and L is the number of low uncertainty detections. The cost of the assignment is described by:

$$\Omega_{i,j} = F^l_s\left(T_i, d_j\right) + F^l_a\left(T_i, d_j\right) \tag{6}$$

Let T_i denote for the tracklets, $F^l_s\left(T_i, d_j\right)$ is the size cost and $F^l_a\left(T_i, d_j\right)$ is the appearance cost. We only use the rgb and hsv histograms for the appearance cost because of its convenient calculation and forthright. The structural constraint is formulated by:

$$\Theta^{p,q}_{i,j} = \begin{cases} F^l_s\left(T_p, d_q\right) + F^l_a\left(T_p, d_q\right) + F^l_c\left(d_q, b_p\right) & \text{if } q \neq 0 \\ \eta & \text{if } q = 0 \end{cases} \tag{7}$$

And the constraint cost $F^l_c\left(d_q, b_p\right)$ is the overlap ratio between the predict bounding box and the detected bounding box. The predicted bounding box is represented by b_p. We conduct all possible assignment events between the tracklets and the low uncertainty trees will be updated according to the scheme with the lowest structural constraint cost score.

For high uncertainty detections, we adopt the strategy of MHT by constructing high uncertainty trees. Based on linear motion model, if the tracklet's most suitable detection d_i's overlap ratio with detection d_j is greater than certainty threshold μ, the track tree will append both of the detections with each new detection spawning a separate branch. We then prune the track hypothesis branches at the end of each sliding window.

2.3 High Uncertainty Tree Pruning

Low uncertainty trees are the final data association results which don't need to be pruned. However, high uncertainty trees remain multiple branches. So, the strategy of pruning the tree plays an important role in data association. We mark and prune the high uncertainty branches every time the current frame k is divided by the sliding window length τ. After every pruning, only one branch with the highest score will be remained. And that's the final data association result at frame k. A branch score is denoted by:

$$F^u(i) = F_a^u(i) + F_m^u(i) \tag{8}$$

which is consist of appearance cost $F_a^u(i)$ and motion cost $F_m^u(i)$. As for the appearance cost, we adopt the modified multi-output regularized least squares classification which is mentioned in [10]. Blocking bounding box's rgb and hsv histograms are defined as appearance feature where c is the feature dimension. Let X_k be a $m \times c$ input matrix at frame k in sliding window w where m is the number of high uncertainty detections. Let R_k be a $c \times n$ regressor matrix where n is the number of high uncertainty tree branches. Let V_k denote the $m \times n$ response matrix. The response matrix V_{k+1} can be obtained by $X_{k+1} \times R_k$. After the marking, X_{k+1} will be used as training set. We set the whole track tree detections in the current sliding window as positive examples and other detections as negative examples. The regressor matrix can be trained as:

$$\min_{R_k} \sum_{t=(w-1)\tau+1}^{w\tau} \|X_k R_k - V_k\| + \lambda \|R_k\|_F^2 \tag{9}$$

Let $G(h_j)$ denotes the appearance score which was obtained from the classifier mentioned above for the detection j. The final score of the detection can be obtained by:

$$S(h_j|d_j \subseteq T_i) = \frac{e^{G(h_j)}}{e^{G(h_j)} + e^{-G(h_j)}} \tag{10}$$

Then the appearance score of the branches is designed followed log likelyhood ratio as:

$$F_a^u(i) = ln\frac{\prod_{t=1}^{k} S(h_i|d_i \subseteq T_i)}{\prod_{t=1}^{k} S(h_i|d_i \subseteq \phi)} \tag{11}$$

where $S\left(h_j|d_j \subseteq \phi\right)$ is a constant probability for the false-detection. We define the motion score as:

$$F_m^u(i) = ln\frac{\prod_{t=k-\tau}^{k} M\left(l_j|d_j \subseteq T_i\right)}{\prod_{t=k-\tau}^{k} M\left(l_j|d_j \subseteq \phi\right)} \tag{12}$$

where every objects motion score is written as:

$$M\left(l_j|d_j \subseteq T_i\right) = \frac{1}{2}\left(\frac{|h-h_d|}{2\left(h+h_d\right)} - \frac{|w-w_d|}{2\left(w+w_d\right)}\right) + \frac{1}{2}\cdot\frac{area\left(B\left(d_{i,d}\right)\bigcap B\left(d_j\right)\right)}{area\left(B\left(d_{i,d}\right)\bigcup B\left(d_j\right)\right)} \tag{13}$$

where h_d, w_d, and $B\left(d_{i,d}\right)$ are the predicted boundingbox's height, width and location. And $M\left(l_j|d_j \subseteq \phi\right)$ is a constant probability for the false-detection.

3 Experimental Evaluation

We implemented our approach in Matlab and tested it on a PC with 3 GHZ CPU and 8 GB memory. In order to evaluate the proposed algorithm, we use the MOT challenge dataset [11]. We use 7 MOT Challenge training sequences for training the network and tuning the parameters and 7 MOT Challenge testing sequences for experiments. We use public detector DPM [1] for multi-object detection. These sequences cover different difficulty levels of the multi-object-tracking problem. We follow the current evaluation protocols which include the multi object tracking accuracy (MOTA), multiple object tracking precision (MOTP), mostly tracked targets (MT), most lost targets (ML), track fragmentation (FM) and ID switches (ID Sw). The bold number indicates that it's the best value of the metric (Table 1).

3.1 Comparison of Our Tracker with Two or Three Categories

Table 2 demonstrates the performance of the algorithm with or without the false detection category on each testing sequence. There will be two detection categories if the tracker don't penalize the false detection boundingboxes. With the penalization, our tracking results have significant improvement on most of the main evaluating metrics for almost every sequences, especially on MOT16-03, MOTA dramatically grows from 36.4 to 46.7% and FAF decreases by 9.4. MOT16-03 is a crowded scenes recorded by fixed cameras and it has similar shooting angle and illumination condition with the training sequences MOT16-04. It proves that our tracker fits the condition well and can track multiple objects accurately. The rest of the sequences may have different illumination conditions, different shooting angles or recorded by

Table 1 The result of our method on the MOT challenge 2016 dataset

Tracker	MOTA	MOTP	FAF	MT (%)	ML (%)	FP	FN	ID Sw.	Frag	Hz
JPDA_m [7]	26.2	76.3	0.6	4.1	67.5	**3689**	130549	**365**	**638**	22.2
SMOT	29.7	75.2	2.9	5.3	47.7	17426	107552	3108	4483	0.2
GMPHD_HDA [17]	30.5	75.4	0.9	4.6	59.7	5169	120970	539	731	13.6
DP_NMS [14]	32.2	76.4	**0.2**	5.4	62.1	1123	121579	972	944	**212.6**
CEM [13]	33.2	75.8	1.2	7.8	54.4	6837	114322	642	731	0.3
TBD [6]	33.7	**76.5**	1.0	7.2	54.2	5804	112587	2418	2252	1.3
LINF1 [5]	**41.0**	74.8	1.3	11.6	51.3	7896	99224	430	963	4.2
Ours	40.7	75.2	1.6	**12.0**	**44.1**	9319	**97992**	773	1106	6.6

Table 2 Comparison of our method with two or three detection categories on the MOT challenge 2016 dataset

Sequence	our tracker	MOTA	MOTP	FAF	MT (%)	ML 9 (%)	FP	FN	ID Sw.	Frag
MOT16-01	two categories	30.1	71.9	0.9	17.4	43.5	421	**4013**	**33**	**51**
	three categories	**34.2**	**72.0**	**0.2**	17.4	43.5	**68**	4105	37	54
MOT16-03	two categories	36.4	74.3	13.2	**19.6**	**19.6**	19770	**46234**	4105	**457**
	three categories	**46.7**	**75.2**	**3.8**	18.2	20.9	**5699**	49638	**407**	569
MOT16-06	two categories	45.0	72.4	0.3	**14.9**	**49.8**	356	**5924**	**62**	114
	three categories	**45.1**	**72.6**	**0.2**	13.1	50.2	**253**	6010	66	**110**
MOT16-07	two categories	38.1	73.9	2.2	**11.1**	31.5	1114	**8916**	78	115
	three categories	**39.0**	**74.1**	**1.4**	9.3	31.5	**694**	9180	**75**	**113**
MOT16-08	two categories	27.7	78.8	1.9	9.5	**34.9**	1206	**10805**	**90**	**113**
	three categories	**28.0**	**79.1**	**1.6**	9.5	42.9	**989**	10957	94	121
MOT16-12	two categories	**37.1**	77.3	1.0	**17.4**	**44.2**	902	**4289**	**25**	42
	three categories	36.2	**77.5**	**0.8**	16.3	46.5	**763**	4501	26	**38**
MOT16-14	two categories	20.1	73.9	1.9	3.7	**58.5**	1435	**13281**	**59**	**95**
	three categories	**21.4**	**74.2**	**1.1**	3.7	60.4	**853**	13601	68	101

moving camera, and the MOTA and FAF may not increase by the largest margin, but the improvement is still obviously. And all MOTP of the test sequences is increase, all FAF and FP of the sequences are decrease. The results prove that our tracker is helpful for the trackers that have too many false trajectories especially when training and test sequences are with the similar shooting angle and similar illumination condition.

3.2 *MOT16 Result Comparison*

Table 1 summarizes the evaluation metrics of our methods and the other published multiple object trackers. Our tracker is better than other listed trackers on MT, ML and FN. Our MOTA is the second best of the listed trackers. Our final MOTA increases by 6.2% and reaches 40.7%, only 0.3% lower than the best tracker and our final MOTP is 0.4% higher than it. Our method deals with 6.6 frames on average, faster than [5]. The experiment demonstrates that our method can track multiple target accurately, robustly and efficiently. For those trackers whose detectors may bring too many false detections, our tracker may represents greater advantages. For more detailed results, please visit https://motchallenge.net/tracker/dMOT.

4 Conclusion

In this paper, we proposed a novel accurate, robust and efficient multi-object tracker based on detection classification. We divide detections into three subsets: false detections, low uncertainty detections and high uncertainty detections. False detections will be penalized after classification. For low uncertainty detections we adopt a simple association framework to get efficient and stable result. As for high uncertainty detection, we set a sliding window and adopt the improved MHT strategy with an online trained classifier to improve the discrimination of appearance model. The approach was demonstrated on the standard dataset [11] where it achieved impressive results in the field of multi-object tracking.

Acknowledgements The work is supported by National High-Tech R&D Program (863 Program) under Grant 2015AA016402.

References

1. Andrews, S., Tsochantaridis, I., Hofmann, T.: Support vector machines for multiple-instance learning. In: Advances in neural information processing systems, pp. 577–584 (2003)

2. Bae, S.H., Yoon, K.J.: Robust online multi-object tracking based on tracklet confidence and online discriminative appearance learning. In: Proceedings of the IEEE Conference on Computer Vision and Pattern Recognition, pp. 1218–1225 (2014)
3. Chen, J., Sheng, H., Zhang, Y., Xiong, Z.: Enhancing detection model for multiple hypothesis tracking. In: Proceedings of the IEEE International Conference on Computer Vision (2017)
4. Choi, W.: Near-online multi-target tracking with aggregated local flow descriptor. In: Proceedings of the IEEE International Conference on Computer Vision, pp. 3029–3037 (2015)
5. Fagot-Bouquet, L., Audigier, R., Dhome, Y., Lerasle, F.: Improving multi-frame data association with sparse representations for robust near-online multi-object tracking. In: European Conference on Computer Vision, pp. 774–790. Springer (2016)
6. Geiger, A., Lauer, M., Wojek, C., Stiller, C., Urtasun, R.: 3d traffic scene understanding from movable platforms. IEEE Trans. Pattern. Anal. Mach. Intell. **36**(5), 1012–1025 (2014)
7. Hamid Rezatofighi, S., Milan, A., Zhang, Z., Shi, Q., Dick, A., Reid, I.: Joint probabilistic data association revisited. In: Proceedings of the IEEE International Conference on Computer Vision, pp. 3047–3055 (2015)
8. Hong Yoon, J., Lee, C.R., Yang, M.H., Yoon, K.J.: Online multi-object tracking via structural constraint event aggregation. In: Proceedings of the IEEE Conference on Computer Vision and Pattern Recognition, pp. 1392–1400 (2016)
9. Kieritz, H., Becker, S., Hübner, W., Arens, M.: Online multi-person tracking using integral channel features. In: 2016 13th IEEE International Conference on Advanced Video and Signal Based Surveillance (AVSS), pp. 122–130. IEEE (2016)
10. Kim, C., Li, F., Ciptadi, A., Rehg, J.M.: Multiple hypothesis tracking revisited. In: Proceedings of the IEEE International Conference on Computer Vision, pp. 4696–4704 (2015)
11. Milan, A., Leal-Taixé, L., Reid, I., Roth, S., Schindler, K.: Mot16: A benchmark for multi-object tracking. arXiv:1603.00831 (2016)
12. Milan, A., Leal-Taixé, L., Schindler, K., Reid, I.: Joint tracking and segmentation of multiple targets. In: Proceedings of the IEEE Conference on Computer Vision and Pattern Recognition, pp. 5397–5406 (2015)
13. Milan, A., Roth, S., Schindler, K.: Continuous energy minimization for multitarget tracking. IEEE Trans. Pattern Anal Mach. Intell. **36**(1), 58–72 (2014)
14. Pirsiavash, H., Ramanan, D., Fowlkes, C.C.: Globally-optimal greedy algorithms for tracking a variable number of objects. In: 2011 IEEE Conference on Computer Vision and Pattern Recognition (CVPR), pp. 1201–1208. IEEE (2011)
15. Reid, D.: An algorithm for tracking multiple targets. IEEE Trans. Autom. Control **24**(6), 843–854 (1979)
16. Ren, S., He, K., Girshick, R., Sun, J.: Faster R-CNN: towards real-time object detection with region proposal networks. In: Advances in neural information processing systems, pp. 91–99 (2015)
17. Song, Y.m., Jeon, M.: Online multiple object tracking with the hierarchically adopted gm-phd filter using motion and appearance. In: IEEE International Conference on Consumer Electronics-Asia (ICCE-Asia), pp. 1–4. IEEE (2016)
18. Wang, B., Wang, L., Shuai, B., Zuo, Z., Liu, T., Luk Chan, K., Wang, G.: Joint learning of convolutional neural networks and temporally constrained metrics for tracklet association. In: Proceedings of the IEEE Conference on Computer Vision and Pattern Recognition Workshops, pp. 1–8 (2016)
19. Wang, X., Yang, M., Zhu, S., Lin, Y.: Regionlets for generic object detection. In: Proceedings of the IEEE International Conference on Computer Vision, pp. 17–24 (2013)
20. Xiang, Y., Alahi, A., Savarese, S.: Learning to track: online multi-object tracking by decision making. In: Proceedings of the IEEE International Conference on Computer Vision, pp. 4705–4713 (2015)
21. Yoon, J.H., Yang, M.H., Lim, J., Yoon, K.J.: Bayesian multi-object tracking using motion context from multiple objects. In: 2015 IEEE Winter Conference on Applications of Computer Vision (WACV), pp. 33–40. IEEE (2015)

Autonomous 3D Model Generation of Unknown Objects for Dual-Manipulator Humanoid Robots

Adrian Llopart, Ole Ravn, Nils A. Andersen and Jong-Hwan Kim

Abstract This paper proposes a novel approach for the autonomous 3D model generation of unknown objects. A humanoid robot (or any setup with two manipulators) holds the object to model in one hand, views it from different perspectives and registers the depth information using a RGB-D sensor. The occlusions due to limited movement of the manipulator and the gripper itself covering the object are avoided by switching the object from one hand to the other. This allows for additional viewpoints leading to the registration of more depth information of the object. The contributions of this paper are as follows: 1. A humanoid robot that manipulates objects and obtains depth information 2. Tracing the hand movements with the robots head to be able to see the object at every moment 3. Filtering the point clouds to remove parts of the robot from them 4. Utilizing the Normal Iterative Closest Point algorithm (depth points, surface normals and curvature information) to register point clouds over time. This method will be applied to those pointclouds that include the robots gripper for optimal convergence; the resultant transform is then applied to those point clouds that describe only the segmented object 5. Changing the object from one hand to another 6. Merging the resulting object's partial point clouds from both the left and right hands 7. Generating a mesh of the object based on the triangulation of final points of the object's surface. No prior knowledge of the objects is necessary. No human intervention nor external help (i.e visual markers, turntables …) is required either.

A. Llopart (✉) · O. Ravn · N. A. Andersen
AUT Group, Department of Electrical Engineering, DTU, Anker Engelunds Vej 1, 2800 Kgs. Lyngby, Denmark
e-mail: adllo@elektro.dtu.dk

O. Ravn
e-mail: or@elektro.dtu.dk

N. A. Andersen
e-mail: naa@elektro.dtu.dk

A. Llopart · J.-H. Kim
RIT Lab, School of Electrical Engineering, KAIST, 291 Daehak-ro,, Yuseong-gu Daejeon, Republic of Korea
e-mail: johkim@rit.kaist.ac.kr

J.-H. Kim et al. (eds.), *Robot Intelligence Technology and Applications 5*, Advances in Intelligent Systems and Computing 751,
https://doi.org/10.1007/978-3-319-78452-6_41

Keywords Humanoid robot · 3D model creation · Point cloud processing

1 Introduction

With the increase of social and service robotics, the demand for Human-Robot-Object collaboration has risen considerably. Therefore, robots necessarily have to understand their surroundings to be able to interact with them. Over the past years, special emphasis has been put on robots capable of adjusting to dynamically changing environments, especially when dealing with object recognition and manipulation. Novel research has been proposed for a more rapid and precise detection of known objects. Despite this, robots must also be able to cope with unknown objects: being able to model them becomes a key feature for faster detection, recognition and manipulation in the future.

1.1 Related Work

Many approaches to object recognition deal with identifying the 6 DoF pose estimation of the object based on the correspondence grouping of a set of points with a previously generated model [1, 2]. Other approaches use the synthetic data of 3D models to train Convolutional Neural Networks (CNN) for object detection [3–5]. Finally, some research has been done lately on the generation of grasping poses based solely on the object's point cloud or model [6]. Consequently, a previous 3D model of the model must be known.

Generating 3D models from unknown objects can be accomplished in many ways, each of which has their own advantages and inconveniences. The objects placement, when generating the model, can be divided into three main groups: a. static objects with the camera moving around it, b. static camera with object on top of a turntable which is rotated, c. non-static object and camera.

Concerning the first type, the major issue arises when transforming the camera poses as it revolves around the object. The usage of markers is a widespread solution but requires human intervention, not only for positioning the markers but also to move the camera. The generation of the model can then be achieved by ray-casting the objects silhouette from every view onto a 3D regular grid (volumetric image) as proposed by Denkowski [7]. A more common approach when using markers is to apply an Iterative Closest Point (ICP) algorithm to the point clouds extracted from the depth images of every viewpoint [8].

When dealing with the second type of object placement, namely doing so on a turntable with a fixed camera, the problem with a moving frame disappears.

However, the necessity for a human or any other external agent to spin the turntable limits the autonomy of the model generation. Once again, to keep track of the objects viewpoint (in other words, how much the table has rotated), some approaches continue to use markers [9], whilst others rely on SIFT features [10] or the ICP algorithm with loop closure [11].

The last objects placement, whilst being the most difficult to track, allows for the maximum autonomy. Particularly, assuming a robot that wants to autonomously model unknown objects it has recently grasped, to be later used for a faster object recognition and detection. In this case, the objects frame will be inconstant (because the manipulator and gripper holding the object will be moved to try and view it from different perspectives), and so will be the camera frame (presuming the camera is mounted on the head of the robot, it needs to move to be able to track the manipulator's end effector and the object). Krainin et al. [12] propose the usage of a Kalman filter for the camera to keep track of the gripper plus utilizing a modified ICP algorithm (that takes into account sparse feature matching, dense color matching and prior state information) and loop closure to generate smooth object models. Even though this methodology could clearly be considered *state-of-the-art* in terms of autonomously generating 3D models, it still requires the help of external agents: the problem of grasping an object with a manipulator is that parts of the object will always be occluded. For this reason, the paper suggests leaving the object on a table or any other surface and regrasp it from another position to finish registering the point clouds.

1.2 Basic Methodology

The approach proposed in this paper will try to make the registration process of unknown objects as autonomous as possible. The only assumption required is that the robot has an unknown object grasped in its gripper. Plenty of research has been conducted in this field; specifically, a method to achieve this has been previously proposed by Llopart et al. [13].

The pipeline starts with a dual-manipulator humanoid robot holding an object in one hand. The arm and gripper will be moved in a way that a. the end effector will always be inside the field of view of the robot whilst the head is also tracking the object (Sect. 2), b. the difference between end poses is very small to allow better NICP convergence (Sect. 3.2), c. the total amount of viewpoints will try to cover the entire 360° around the object.

The RGB-D sensor located in the robot's head will provide depth images for each step, which are then converted to point clouds. The point clouds are then filtered with respect to the distance to the RGBD sensor and to the robot's surface (Sect. 3.3). A radius sparse outlier process will also take place. The resulting point clouds will then be transformed into the grippers frame to become invariant to movement (Sect. 3.4). The point cloud's normals and curvatures will be estimated and used during the NICP registration (Sect. 3.5). One of the major problems previous approaches had was that convergence failed when registering symmetrical or low-detailed objects. To solve this, the NICP is done to the point clouds that include those parts of the robot that are invariant to the object, i.e. the grippers. This allows for additional information and evident better results. The obtained transformations are then applied to the point clouds that represent only the object. The occlusions and lack of viewpoints that occur with one manipulator can be solved by changing the object from one hand to another, and carrying out the same procedure again (Sect. 3.6). This results in two partial models of the object. These two point clouds must be merged to obtain one complete model (Sect. 3.7). Finally, a Moving Least Squares (MLS) algorithm is carried out to smoothen the surfaces of the model and remove artifacts. A mesh of it is then built based on a greedy surface triangulation algorithm (Sect. 3.8).

The experimental setup, results and discussion is found in Sect. 4. Conclusions and future work are dealt with in Sects. 5 and 6, respectively.

2 Tracking the Gripper's Movements

The key concept proposed in this paper is registering depth information of the object from different perspectives. To achieve this, the robot rotates and translates the object continuously. Thus, it becomes essential that the robot can follow the movements of the end effector instantly with its head, where the depth sensor is located. Krainin et al. [12] propose the usage of a Kalman filter, taking as input the previous time step mean and covariance, the clouds representing the manipulator and object, and joint angles reported by the encoders of the manipulator.

A different and much simpler approach to achieve gripper tracking is to constantly monitor the position (x, y, z) of both end effectors (left and right grippers). This information must be then transformed into two angles: pan and tilt, corresponding to the 2 DoF the robot's head has. Two simple equations solve this problem:

$$pan\_angle = atan2(y, x) \tag{1}$$

$$tilt\_angle = atan2(z, \sqrt{x^2 + y^2}) \tag{2}$$

This allows the robot to rapidly switch from following the left end-effector to the right one by simply changing the input coordinates to those of the selected gripper.

3 Model Creation

This section describes the core concepts of the proposed methodology: the disabling of collision checking between manipulator and object, the multiple filtering steps of the depth data from the RGB-D sensor, the transformation to the grippers frame, the registration of the multi viewpoint data to generate the objects model, changing the object from one hand to another to create two semi-full models of the object, their merging procedure and, finally, the creation of a mesh for the model.

3.1 Disabling Collision Checking

Before the proposed pipeline can be executed, the collision checking between the objects point cloud and the robots links must be removed. The *MoveIt* libraries for ROS will be used to control the manipulators and to set their poses. An octomap of the environment will be continuously generated to evaluate whether an action can be executed or if the movement will end up colliding with the surroundings. The only issue with this approach is that the object's point cloud (which will be held by the robot) will also be included in the octomap, thus denying any possible robots movement because the gripper will already be in a collision state with the object. Hence, by removing the collision check between the robots links and the part of the octomap that represents the object, the manipulators will be free to move.

3.2 Selecting Poses

Due to the robot's configuration and limitations, setting up multiple poses for all viewpoints becomes a tedious task with poor results. For that reason, instead, one manipulator pose is set and two specific revolute joints in the wrist are rotated, as seen in Fig. 1. By rotating in a step-wise manner both wrist joints, the robot is able to take 360° depth images of the objects. This allows for better control of the difference between poses, smoother transitions and, most importantly, leaves the object in a quasi-static state, where the center of the object barely moves, thus removing the requirement of a pre-alignment step prior to the NICP procedure. For every rotational step, the filtering and registration procedures take place. Some end effector poses are shown in Fig. 2.

3.3 Filtering of Sensor Data and Robot Links

Depth images are obtained from a RGB-D sensor and transformed into 3D point clouds. These represent the environment that surrounds the object. Yet, some filtering and segmentation must be done to be able to extract only the important information (the objects shape) from the large quantity of points the sensor outputs.

The first step is to limit the range in which useful information is found: considering that the maximum reach of the manipulators (when they are fully stretched out) is of around 1 m, it makes no sense to keep point cloud data which is further away

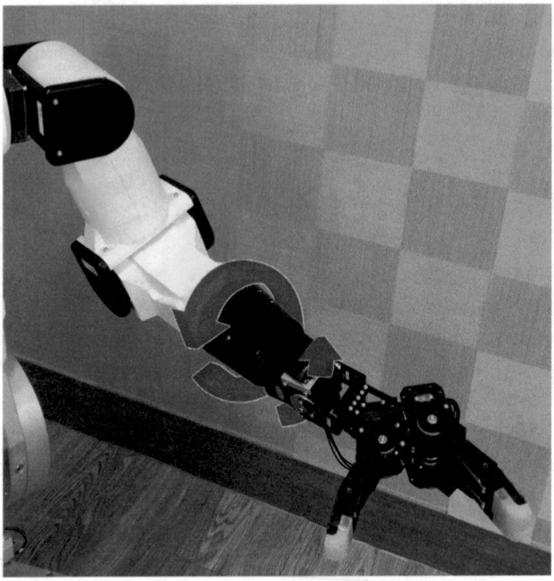

Fig. 1 The red arrows represent the rotational wrist joints

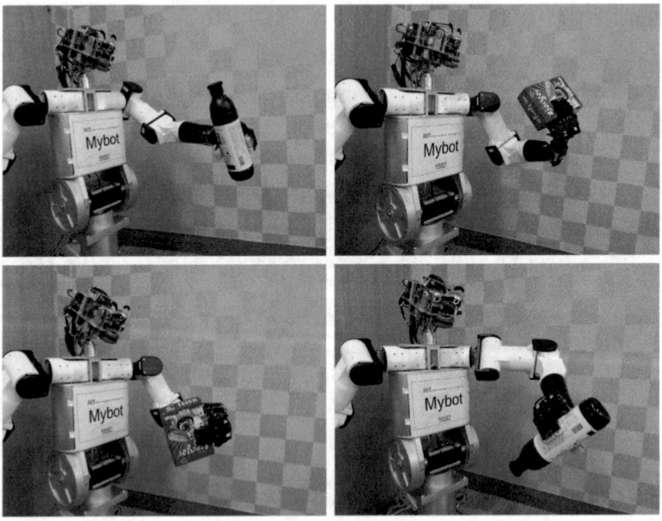

Fig. 2 Different manipulator poses to see the objects from all possible viewpoints

because the object is sure to not be outside that region. The depth sensor does not output points closer to 5 cm, thus, if a point is in that range, it is surely an error and must be discarded too. Removing so many points reduces drastically the processing time and improves performance in future steps.

Then, those points that represent some parts of the robot will be filtered out too. This means that two point clouds will result from this process. The former will have all points belonging to the robots links removed, thus, only the segmented object data will be visible. The latter, will have most of the robot's link's points also removed, except those that represent the gripper itself. As mentioned in Sect. 1, modeling symmetrical objects usually ends in bad results during the registration process due to poor convergence when applying the NICP algorithm. For this reason, the entire pipeline proposed in this paper will use point clouds that include the gripper since they add information that removes the symmetry problems when registering clouds.

All small noise and artifacts that are still present in the point clouds are minimized using a radius sparse outlier filter [14]. Those points that do not have a certain minimum number of neighbors inside a radial threshold will be eliminated from the cloud.

3.4 Transforming Points to the Gripper Frame

To be able to register point clouds, it is necessary that they have the same frame and have been slightly pre-aligned before the NICP algorithm is applied. Doing this increases greatly the chances of convergence.

The problem with multi viewpoint registration is that even if the base frame of the data is the same (the camera frame), due to the movement of the end-effector, the point clouds will never be the pre-aligned. To achieve this, an initial alignment based on local feature descriptors (e.g. FPFH) could be applied, but this process might produce bad results in itself, specially when dealing, once again, with symmetrical objects, where key feature density is low (Fig. 3).

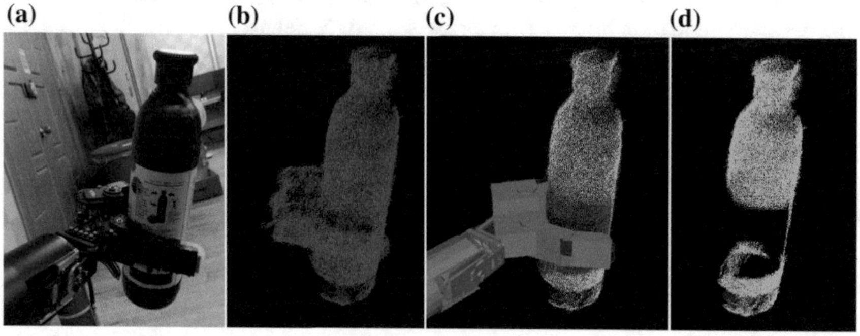

Fig. 3 From left to right. **a** RGB image of robot grasping bottle **b** Full model after registering point clouds from different viewpoints using only one hand. Both object and robot links are included as red points **c** Simulated gripper on top of point cloud. The green points correspond only to the object's geometry **d** Final partial model of the object

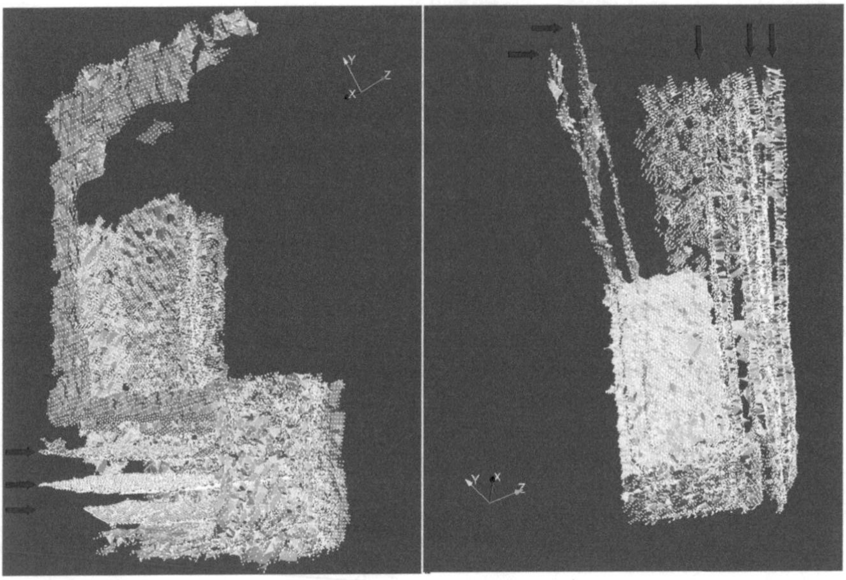

Fig. 4 Results of building a model without ICP registration: the accumulated drift renders the model unusable. The red arrows show those planes that have translational drift errors. A correct model achieved using registration based on the same data is shown in Figs. 5 and 6

A simpler solution is to transform all received point clouds to the grippers frame. The advantages of this approach are that the objects points will be invariant to the hands movement (because once the object is grasped, it does not move in relation to the hand grasping it) and so no pre-alignment will be necessary. In fact, in the optimal scenario, this solution will allow to build the 3D model without the need for registration. Nonetheless, the reality is that small drift errors will accumulate over time between point clouds rendering the generated model useless (Fig. 4). Hence, the registration process becomes a high necessity to be able to correct for those small translational and rotational errors.

3.5 *Normal Iterative Closest Point (NICP) Registration*

The backbone of the proposed pipeline is the constant registration of point clouds over time from different viewpoints. These point clouds are the result of the objects segmentation, have the same reference frame (the gripper) and are roughly pre-aligned. To fuse these point clouds in the correct manner, the Normal Iterative Closest Point (NICP) algorithm will be used. NICP minimizes an augmented error metric (based on point coordinates and surface normals and curvature) during the least squares formulation of the alignment problem to find the best data association (trans-

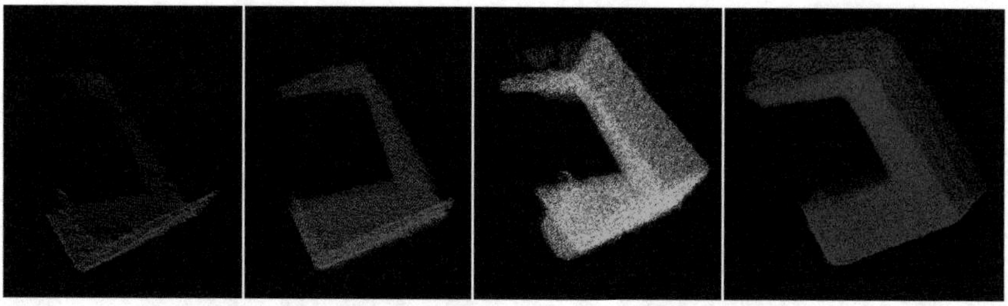

Fig. 5 NICP registration procedure of a box over several end effector poses

form) between two sets of point clouds. After that, a smoothing process (MLS) is performed to obtain better results. Generally, the second point cloud will include most of the information from the first, but with additional details due to modifying slightly the perspective. It is important that the changes from one point cloud to the next are very small so that the NICP algorithm can converge. If the differences are too great (for instance, by having rotated the hand a full 180° so that the opposite side of the object is being viewed), the NICP will fail to converge or give poor transforms which will accumulate over time, leading to suboptimal results. This is the main reason for using small rotational steps in the joints when changing the end effector's poses.

For even better results, a surface smoothing algorithm is applied. By estimating the normals of the point cloud's surface, and using the Moving Least Squares (MLS) algorithm as proposed in [11, 12, 14], the resulting normals will be aligned producing a more precise model of the object, with less noise, occlusions and "double walls" artifacts. The general leaf size during the smoothing process is set to 2 cm, however depending on the geometry of the object, this value must be changed. For objects with sharper sides (e.g. boxes) this values should be lowered (1 cm) so as not to round off the edges too much. For initially already curved surfaces, like bottles or balls, this value can be slightly increased.

3.6 Changing Hands

Occlusions are a big problem when modeling objects. These are due to a limited movement of the end effector (which does not allow the object to be seen all around) or links of the robot always blocking the view of part of the object (specifically the fingers and palm of the gripper) (Fig. 6).

A way to solve this is by grabbing the object from a different position and starting the registration process all over again. The results from the last registration will be merged with those of the previous registration to achieve a complete model. The fact that the system must not depend on external help, makes the grasp change difficult. If, for instance, a table could be used, the robot would have to simply place the object on

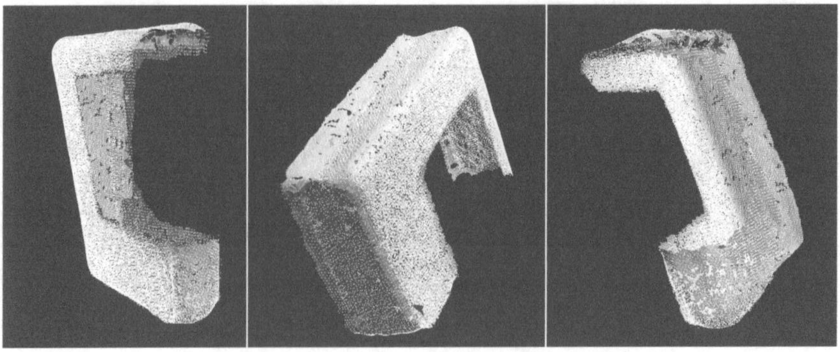

Fig. 6 Three views of the partial model generated by constructing a mesh from the resulting point cloud in Fig. 5

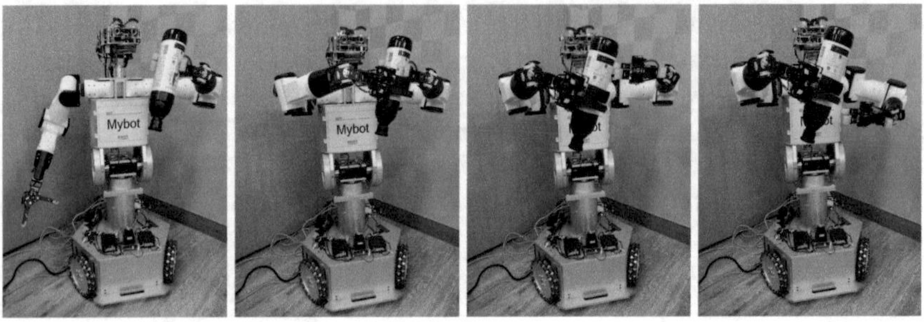

Fig. 7 *Mybot* humanoid robot changing object (bottle) from left to right hand

it and re-grab it from a different position [12]. In spite of this, a second manipulator can be used: it is important that the second grasp allows for additional viewpoints and for the registration of the object's opposite side. For that, the second grasp must be rotated 180° with respect to the initial one. To do so, and due to the movement limitations of both manipulators, during the exchange one hand will be facing the opposite direction as the other, as seen in Fig. 7b. In the current approach, the poses of the end effectors have been predefined. For an additional degree of autonomy, new grasping poses could be found based on the geometry of the partial model already created using the *agile_grasp* package [6], as seen in Llopart et al. [13].

Finally, for the two sides of the object to be merged together, it is necessary that they are related to the same frame. For this reason, when changing hands, the registered point cloud resultant from the movements of the first gripper must be transformed to the frame of the second gripper.

3.6.1 *Mybot* Characteristics

The proposed pipeline has been tested out on the MyBot humanoid robot, developed in the Robot Intelligence Technology (RIT) Laboratory at KAIST (Figs. 2 and 7). It includes an Odroid XU board, Ubuntu 16.04 and ROS Kinetic. A Xtion ASUS RGB-D sensor is mounted on the 2 DoF head of the robot. It also has two 7 DoF arms with 3 finger grippers each. Finally, the torso of the robot includes another 2 DoF for panning and tilting.

3.7 Merging Both Partial Models

Having two, almost complete, models of the object (one for each hand used) is not enough. For the model to be useful in future real life scenarios, it is necessary that it represents the object correctly from every angle. For this reason, both point clouds must be merged.

As aforementioned, both point clouds are related to the same frame. Generally, these point clouds will have an offset due to precision errors during the changing hands process. Hence, it is necessary to pre-align them. The Fast Point Feature Histogram (FPFH) descriptors can be calculated for both clouds. In this case, the FPFH based on *OpenMP* is used since its multi-threaded implementation produces results at a faster rate (6–8 times) than the normal FPFH descriptors estimation. When the features from both clouds are matched, a transformation between both models is found that allows them to be roughly aligned.

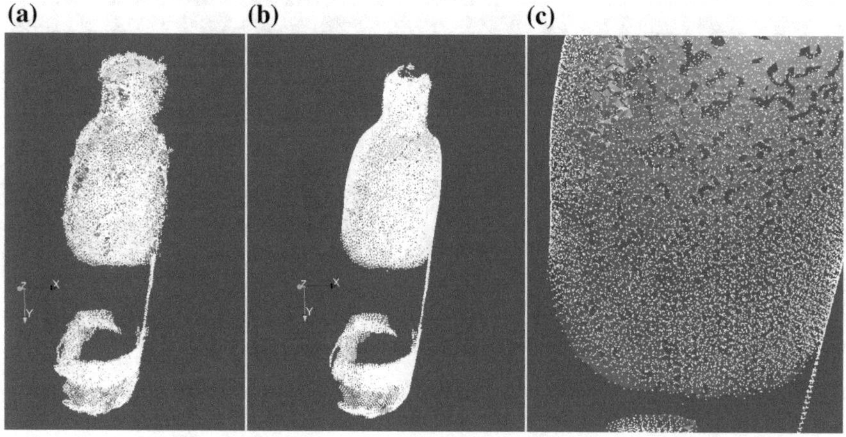

Fig. 8 Mesh result from Fig. 3, from left to right. **a** Without prior smoothing **b** With prior smoothing **c** Detail of mesh

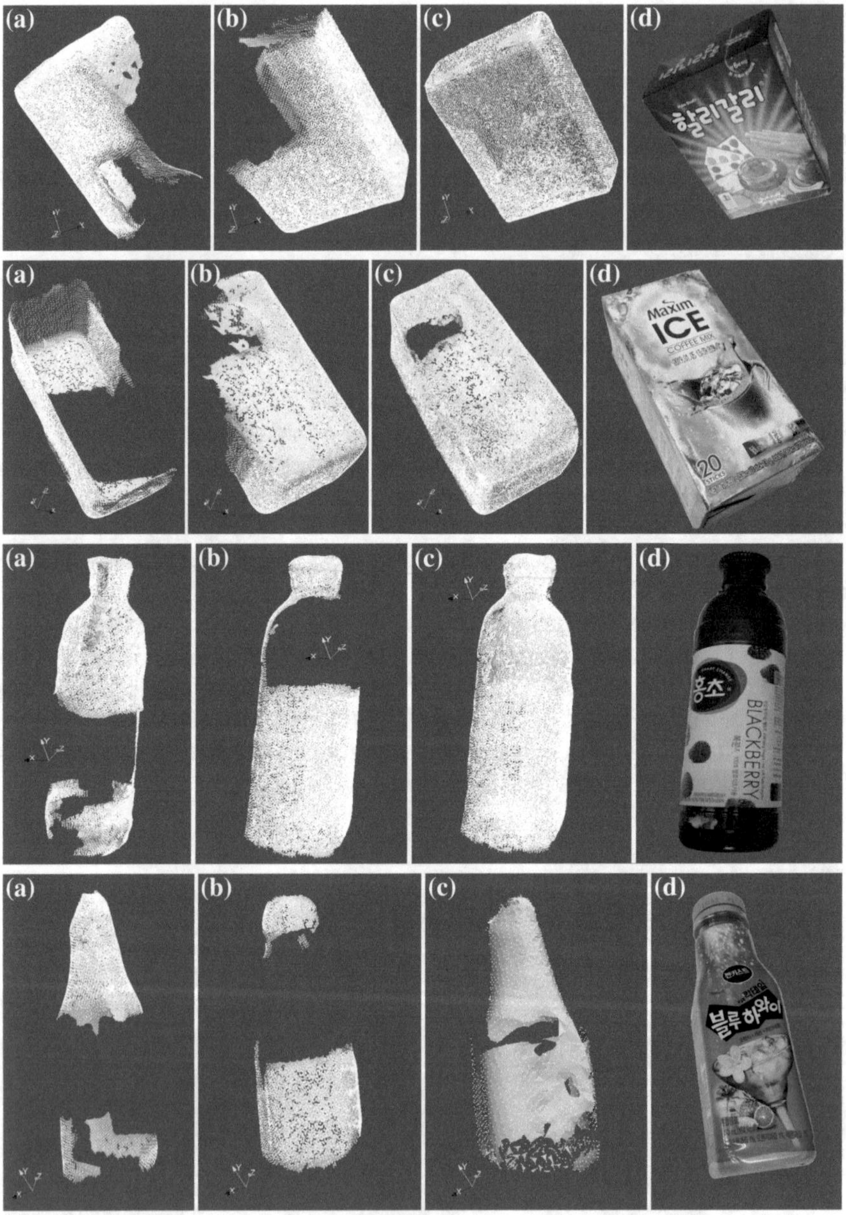

Fig. 9 From left to right. **a** Partial model created using left gripper **b** Partial model created using right gripper **c** Full model after merge **d** Real RGB image of model (for viewing purposes only)

Then, by using the NICP algorithm, and setting the parameters so that the final convergence is achieved by small rotations and translations, a full and complete model of the object is achieved (Fig. 9c). Once again, the Moving Least Squares algorithm is applied to reduce noise and errors during the merging process.

3.8 *Building the Mesh of Model*

For better visualization purposes, a greedy surface triangulation algorithm is run on the resulting point cloud with normals which results in a triangle mesh of the object. The maximum search radius and nearest neighbors values are set to 0.025 and 500, respectively, but may be altered depending on the desired number of resulting triangles. The file will be stored with a *.vtk* format [15]. Some results are seen in Figs. 6, 8 and 9.

4 Experimental Results

The experiments were carried out on the *MyBot* humanoid robot. Four different unknown objects (two boxes and two bottles) were placed in one of the grippers and the process was started. The rest of the pipeline was completely autonomous. Some partial models of objects are seen in Figs. 6, 8, 9a, b. Full models are seen in Fig. 9c.

As seen in these figures, the proposed pipeline correctly outputs 3D models of unknown objects. Their overall shapes and sizes are very similar to the real ones, which is a great result considering the geometrical symmetries in all test cases. A way of assessing the overall success of the results is difficult to find. In spite of this, the measurement discrepancies between real object and generated model will be evaluated (Table 1).

It is seen that the size difference between real objects and models generally stays below the 6 mm mark: likewise, the percentage error never surpasses 9%. Concerning the overall shape of the models, the meshed result clearly matches the real surface

Table 1 Geometric measurement comparison between real object and model

Object	Red box			Blue box			Black bottle		Blue bottle	
Dimension	w	h	d	w	h	d	ø	h	ø	h
Object (cm)	12.8	17.5	5.6	8.1	16.9	6.9	7.0	28.7	5.7	18.4
Model (cm)	12.7	17.9	6.1	8.5	17.3	6.6	7.6	28.8	6.1	18.2
Error (cm)	0.1	0.4	0.5	0.4	0.4	0.3	0.6	0.1	0.4	0.2
Error (%)	0.78	2.29	8.93	4.94	2.37	4.35	8.57	0.35	7.02	1.09

of the objects, except for the last model of Fig. 9 where the chosen value for the smoothing process was set a bit to high, thus flattening parts of the geometry. Despite obtaining positive results, in some cases occlusions do still occur, hence the lack of details.

A way to achieve better results is by correctly differentiating between points that represent the object or the robot. When filtering the robots surface from the initial point cloud, an overall minimum distance threshold was selected. By fine-tuning the distance between robot and points, more precise results would be achieved. For instance, by reducing the threshold for the finger links, less object points would be filtered out, and, consequently, more details would appear in the final model (the red points in Fig. 3c would be green).

The major issues when building full 3D models occur when merging the partial results. As aforementioned (Sect. 3.7), the pre-alignment is done using FPFH descriptors followed by an NICP algorithm. However, if the partial models do not represent almost the entirety of the object, the alignment process will fail, rendering the final model unusable. Additionally, even if the merging process finishes correctly, usually, some parts of the model will still have not been modeled. These normally describe those surfaces of the object that were in contact with the grippers during the registration process, thus, being occluded. Another cause is the mechanical limitations of the manipulators to see the object from certain viewpoints. A good example of occlusion errors (holes in the model) in the final result are the second and fourth models in Fig. 9c.

5 Conclusions

Full 3D models of unknown objects are generated through the proposed pipeline. Registering the point clouds of different views whilst holding the object with one manipulator, enables the robot to create a partial (due to occlusions) model of it. By switching the object from one hand to another and repeating the process, a new partial model is obtained which will not have the occlusion errors of the first one (and vice versa). Merging both partial models accomplishes the removal of errors and the generation of a full 3D model of the object that can be later used for detection, recognition or manipulation purposes. The results show that correct models are being generated for diverse objects in spite of the geometrical symmetries. Figure 10 shows possible grasping locations based off the geometry of the 3D model, which can be stored and applied whenever the object has to be manipulated.

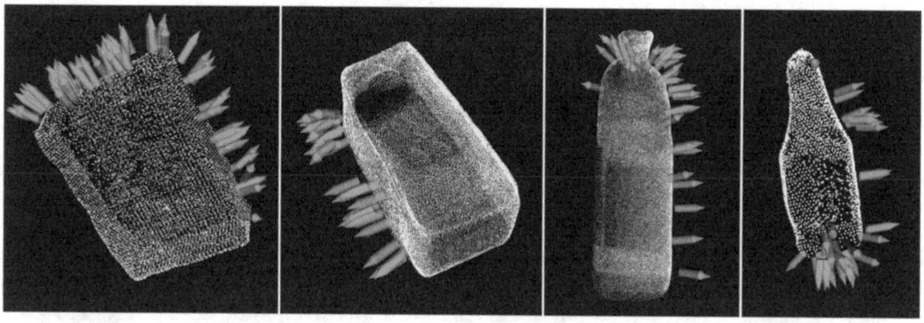

Fig. 10 Grasping poses for the generated 3D models based off solely the geometry of the point cloud and the grippers dimensions, as proposed by Pas et al. [6]

6 Future Work

For better results, a loop closure detection system, as proposed in [11, 12, 16], could be implemented. Additionally, the poses from which the object is seen have been pre-selected by the user. These do not take into account occlusions nor geometry of the object, sometimes leading to redundant point clouds that add no information or, on the contrary, lack of viewpoints to generate a full model. Consequently, it would be interesting to add the *Next best view* concept presented in [12] and [17]. The ultimate goal of the proposed approach is to be combined with the pipeline introduced by Llopart et al. [13] to autonomously detect, segment and manipulate unknown objects using a humanoid robot.

References

1. Aldoma, A., Tombari, F., Stefano, L.D., Vincze, M.: A global hypotheses verification method for 3D object recognition. In: 2012, European Conference on Computer Vision (ECCV). Lecture Notes in Computer Science, vol. 7574. Springer
2. Zhu, M., Derpanis, K., Yang, Y., Brahmbhatt, S., Zhang, M., Phillips, C., Lecce, M., Daniilidis, K.: Single image 3D object detection and pose estimation for grasping. In: Proceedings International Conference on Robotics and Automation (ICRA), Hong Kong, China, 31 May–5 June 2014
3. Sarkar, K., Varanasi1, K., Stricker, D.: Trained 3D models for CNN based object recognition. In: Proceedings International Conference on Computer Vision (ICCV), Santiago, Chile, 13–16 Dec 2015
4. Peng, X., Sun, B., Ali, K., Saenko, K.: Learning deep object detectors from 3D models. In: Proceedings International Conference on Computer Vision (ICCV), Santiago, Chile, 13–16 Dec 2015
5. Gupta, S., Arbelaez, P., Girshick, R., Malik, J.: Aligning 3D models to RGB-D images of cluttered scenes. In: Proceedings International Computer Vision and Pattern Recognition (CVPR), Boston, MA, USA, 7–12 June 2015
6. Pas, A., Platt, R.: Using geometry to detect grasp poses in 3D point clouds. In: Proceedings International Symposium on Robotics Research (ISRR), Genova, Italy, Sept 2015

7. Denkowski, M.: GPU accelerated 3D object reconstruction. Procedia Comput. Sci. **18**, 290–298 (2013)
8. Mihalyi, R.-G., Pathak, K., Vaskevicius, N., Fromm, T., Birk, A.: Robust 3d object modeling with a low-cost RGBD-sensor and AR-markers for applications with untrained end-users. Robot. Auto. Syst. **66**, 1–17 (2015)
9. Xie, J., Hsu, Y.-F., Feris, R., Sun, M.-T.: Fine registration of 3d point clouds fusing structural and photometric information using an RGB-D camera. In: J. Visual Commun. Image Represent. **32**, 194–204 (2015)
10. Foissotte, T., Stasse, O., Escande, A., Wieber, P.-B., Kheddar, A.: A two-steps next-best-view algorithm for autonomous 3D object modeling by a humanoid robot. In: Proceedings International Conference on Robotics and Automation (ICRA), Kobe, Japan, 12–17 May 2009
11. Jaiswal, M., Xie, J., Sun, M.-T.: 3D object modeling with a Kinect camera. In: Proceedings Signal and Information Processing Association Annual Summit and Conference (APSIPA), Chiang Mai, Thailand, 6–9 June 2014
12. Krainin, M., Henry, P., Ren, X.: Manipulator and object tracking for in-hand 3d object modeling. Int. J. Robot. Res.
13. Llopart, A., Ravn, O., Andersen, N., Kim, J.-H.: Generalized framework for the parallel semantic segmentation of multiple objects and posterior manipulation. In: Proceedings International Conference on Robotics and Biomimetics (ROBIO), Macau, China, 5–8 Dec 2017
14. Rusu, R., Marton, Z.C., Blodow, N., Dolha, M., Beetz, M.: Towards 3d point cloud based object maps for household environments. Robot. Auto. Syst. **56**(11), 927–941 (2008)
15. Marton, C., Radu, R., Beetz, M.: On fast surface reconstruction methods for large and noisy datasets. In: Proceedings International Conference on Robotics and Automation (ICRA), Kobe, Japan, 12–17 May 2009
16. Weise, T., Wismer, T., Leibe, B., Gool, L.V.: In-hand scanning with online loop closure. In: Proceedings International Conference on Computer Vision Workshops (ICCV Workshops), Kyoto, Japan, 27 Sept–4 Oct 2009
17. Krainin, M., Curless, B., Fox, D.: Autonomous generation of complete 3D object models using next best view manipulation planning. In: Proceedings International Conference on Robotics and Automation (ICRA), Shanghai, China, 9–13 May 2011

Adaptive Planar Vision Marker Composed of LED Arrays for Sensing Under Low Visibility

Kyukwang Kim, Jieum Hyun and Hyun Myung

Abstract In image processing and robotic applications, two-dimensional (2D) black and white patterned planar markers are widely used. However, these markers are not detectable in low visibility environment and they are not changeable. This research proposes an active and adaptive marker node, which displays 2D marker patterns using light emitting diode (LED) arrays for easier recognition in the foggy or turbid underwater environments. Because each node is made to blink at a different frequency, active LED marker nodes were distinguishable from each other from a long distance without increasing the size of the marker. We expect that the proposed system can be used in various harsh conditions where the conventional marker systems are not applicable because of low visibility issues. The proposed system is still compatible with the conventional marker as the displayed patterns are identical.

1 Introduction

Planar vision marker systems that are generally made of printed black and white two-dimensional (2D) patterns have been widely used for various robotic applications including information encoding, image processing, augmented reality [1], and localization [2]. Several marker patterns such as quick response (QR) codes, AprilTags, and ArUco [3–5] were proposed. These marker systems are widely used

K. Kim · J. Hyun · H. Myung (✉)
Urban Robotics Laboratory, Korea Advanced Institute of Science and Technology,
291 Daehak-ro, Daejeon 34141, Republic of Korea
e-mail: hmyung@kaist.ac.kr
URL: http://urobot.kaist.ac.kr

K. Kim
e-mail: kkim0214@kaist.ac.kr
URL: http://urobot.kaist.ac.kr

J. Hyun
e-mail: jimi.hyun@kaist.ac.kr

J.-H. Kim et al. (eds.), *Robot Intelligence Technology and Applications 5*,
Advances in Intelligent Systems and Computing 751,
https://doi.org/10.1007/978-3-319-78452-6_42

as they can be easily generated, printed, and installed at a low cost, and they require low maintenance. They can provide encoded information and the camera pose can be estimated by utilizing appropriate computer vision techniques. Passive markers also have their own benefits—they can be deployed without electricity which is essential in active beacon systems such as Radio Frequency (RF) beacons [6]; however, it is difficult to use the passive marker systems in low visibility environments in which finding the markers in the image is hard or nearly impossible. Robots operating with indoor marker systems can face failure of lightings. Outdoor robots such as drone landing systems [7] or unmanned underwater vehicles (UUV) that rely on the markers [8] cannot operate at night or at sea with high turbidity.

Active markers have been used to overcome the limitations of the passive markers. Applications that use light emitting diodes (LEDs) as a point indicator or a light communication medium for docking the underwater robots were proposed [9, 10]; however, these methods require additional specialized hardware modules for LED-based communication, and they lose the merits of the 2D planar markers such as the encoded information or camera position calculation.

In this paper, we propose a planar 2D marker composed of LEDs to overcome the aforementioned problems. The planar vision markers are well-established systems that contain useful information but are not observable in low visibility environments. The LEDs are observable in low visibility environments as they are light sources, but they are not compatible with the conventional marker-based systems. By combining the two systems, we propose a system that exhibits the merits of both systems. A matrix of LEDs is used to display a dot image of $n \times n$ markers ($n = 8$ in this paper). The image is processed to recognize the glowing red lights and converted into a binarized image. Finally, the detected regions are converted into an image with size similar to the printed planar markers. Experiments are designed and performed to show that the LED markers are feasible and that they are compatible with the printed planar marker systems but they can still be operated in low visibility environments where the printed markers cannot be used.

2 Materials and Methods

The proposed active marker node comprises an LED matrix for the planar marker display, a distance sensor, and a communication module to detect the proximity of the robot. The functions of the distance sensor and the communication module are described in Sects. 2.3 and 2.4, respectively. The 3D-printed casings were designed to hold the display and sensors at appropriate positions. The circuit diagram and an image of the developed node are shown in Fig. 1.

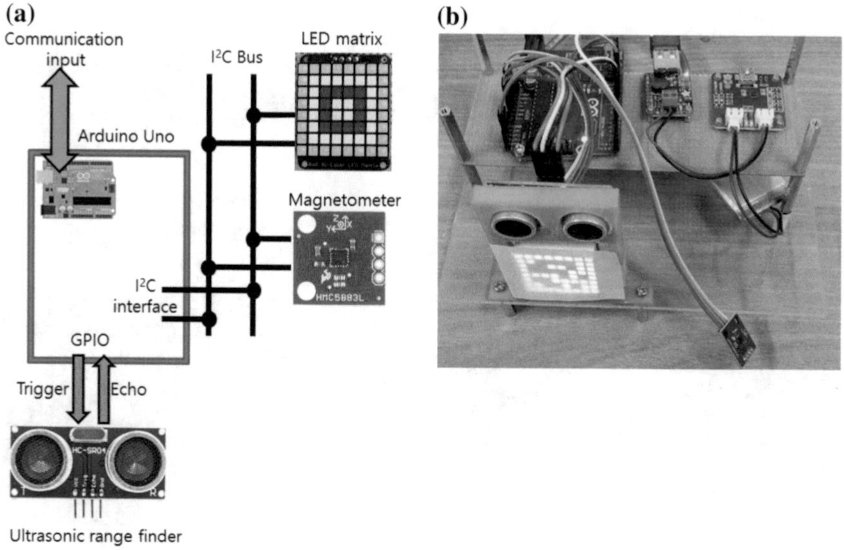

Fig. 1 **a** Circuit diagram of an active marker node. GPIO stands for general purpose input and output. **b** Image of the developed marker node

2.1 *Image Processing Algorithms for the LED Marker Detection*

The LED marker is a bright red-colored square LED matrix, which is easily detectable. The acquired raw image was split into RGB channels, and the R channel grayscale image was converted into a binary image by thresholding. Image erosion and dilation were performed to remove the noise from the image. The contours in the image were detected and passed to a polynomial approximation function, and the contour with four corners was selected as a marker. A perspective transformation was used to convert the detected rectangle with an arbitrary shape and size into a square of 80×80 pixels. The square was divided into an 8×8 grid (10×10 pixels per grid), and each grid was color coded as white or black based on the ratio of the black pixels to the white pixels in the grid, reconstructing the final marker image matrix. The developed image processing algorithm is shown in Fig. 2.

2.2 *Experiments Under the Low Visibility Conditions*

The key feature of the proposed marker is its detectability in low visibility environments. There are many low visibility cases such as high smoke concentration in the atmosphere, high turbidity water, or environments where no lightings are

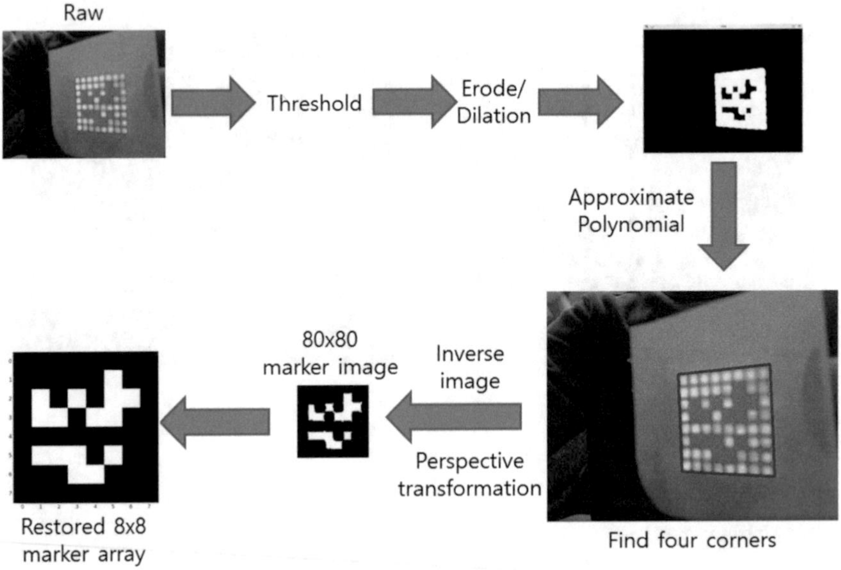

Fig. 2 Overall image processing algorithm for marker detection

available. The active marker node was placed in a completely dark room, and we checked whether the displayed markers could be recognized appropriately. For further examination, a marker detection experiment was conducted under fog generated by a fog machine.

The major application of the printed marker tag is the estimation of the camera pose by calculating the extrinsic/intrinsic parameters of the camera with the real world/image marker coordinates. The proposed marker system was tested for the same task. The marker was assumed to be fixed at the origin (0, 0, 0 in the real world coordinate) with a ruler and protractor for the ground truth measurement. The distance and angle of the camera were calculated by using the solvePnP function of the OpenCV library [11]. The camera was calibrated by a robot operating system (ROS) camera calibration package. The results of the previous experiment showed that a printed marker of size 25 × 25 cm is visible at a distance of 1.0–1.5 m from the camera. A marker of size 3.2 × 3.2 cm was used in this research, so the experiments were carried out at a condition of 1/8 scale (approximately 10 cm) of the conventional setting. The marker and the camera were fixed at the given points (marker at the origin and camera 10 cm away from the marker), and the camera was placed to face the marker directly. Translation and rotation vectors were calculated, and the distance along the x-direction and the yaw angle from the marker were compared with the ground truth (10 cm and 0° meaning front facing). The conventional printed 2D marker with the same pattern was also tested for comparison. The experimental setup is shown in Fig. 3.

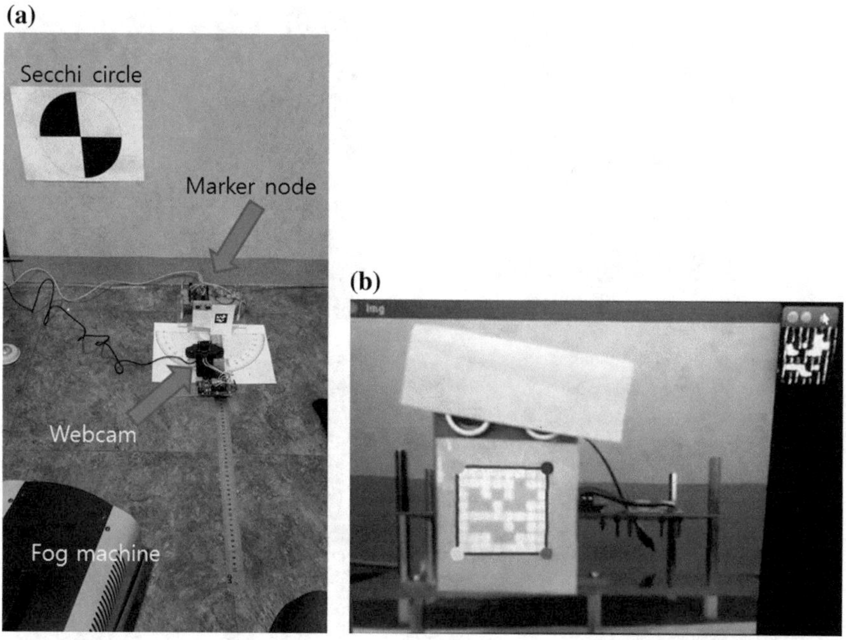

Fig. 3 **a** Experimental setup. **b** Image acquisition and marker recognition under clear bright situation

The properties of the LEDs were also used for a better identification of the markers. It is rather difficult to generate large-sized markers with the LED matrix compared to the printed markers. Instead, we used the fact that the LED lights can be detected from a long distance even if the size of the display is small. Each LED matrix blinks at its own frequency, so the robots can recognize the small marker from a long distance where the details of the markers are not recognizable. When the robot is close enough to the marker, the details of the marker are recognizable. The distance sensor attached to each active marker system recognizes the nearby object, and the marker stops blinking for easier image processing.

3 Result and Discussion

3.1 Reconstruction of LED Matrix Displayed Markers

The results of the marker restoration from the LED image is shown in Fig. 4. The marker detected from the image of the LED matrix (Fig. 4d was identical to the pattern of the original marker (Fig. 4a). The result depicted in Fig. 4c shows that the binarized image of the LED marker is similar to those of the paper markers. Although the outer form of the shapes of the markers is different, the intermediate

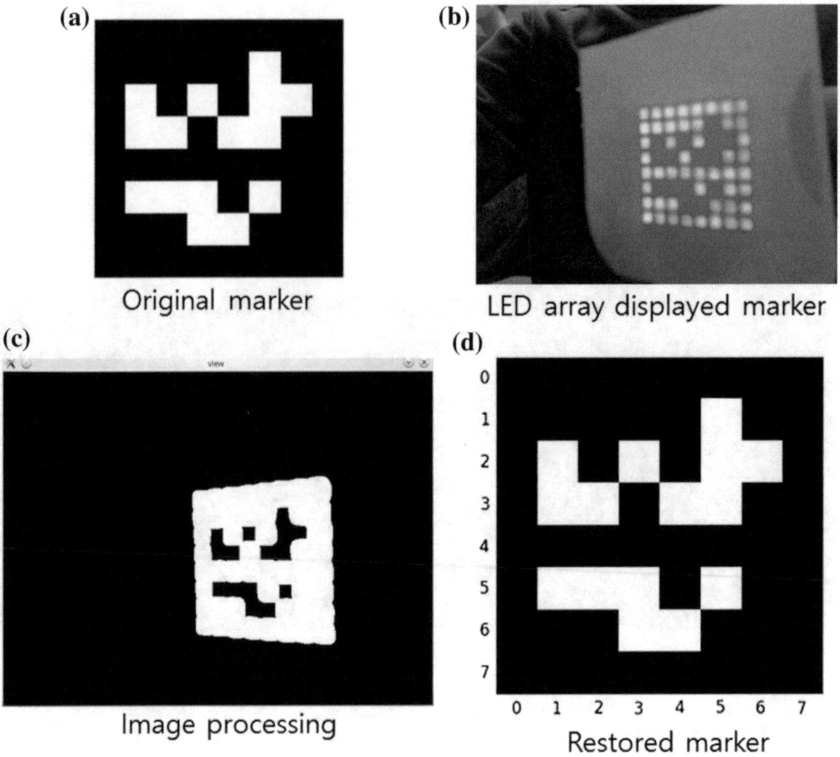

Fig. 4 Restoration of the marker from LED marker. **a** Original AprilTag marker image. **b** LED displayed marker. **c** Contour image of the marker detected during image processing. **d** Restored marker from an inverted contour image, which is identical to the original marker

results from the image processing operation of the LED markers and that of the printed planar markers are identical. This shows the compatibility of the LED markers with the conventional printed planar markers. After simple color filtering and thresholding, previously developed methods and algorithms such as the camera pose estimation, augmented reality, and localization for the printed markers can be directly applied. As mentioned earlier, the markers composed of a few LEDs cannot be used in the mentioned applications because they do not contain many information and adequate corners and areas for the geometric calculation, which is required for the pose estimation.

3.2 Detection Under the Low Visibility Environments

The LED marker was placed in a dark room and foggy environment to test the detectability in a low visibility environment. The results are shown in Fig. 5.

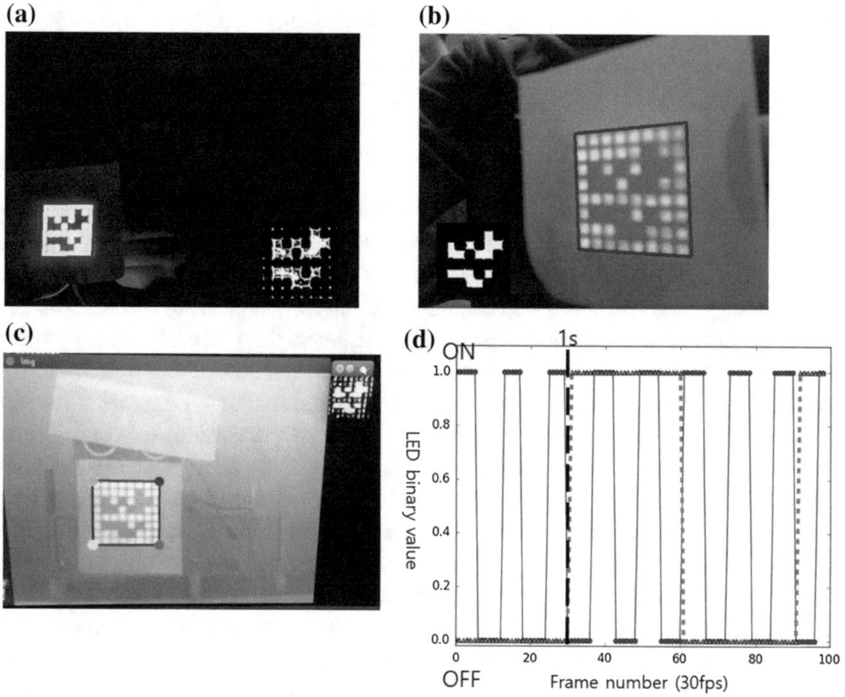

Fig. 5 Marker ID recognition under low visibility environment. **a** Marker recognition in dark environment. Perspective transformation result is indicated at the right corner. **b** Marker recognition in the normal environment. Perspective transformation result is indicated at the left corner. **c** Marker recognition in foggy environment. **d** Blinking detection plot of 200 and 1,000 ms settings. Blue circled line indicates 200 ms blinking while green triangle dashed line shows 1,000 ms blinking. The value of 1 was given when the contour were found at the frame number (30 fps speed). Black long dashed line indicates 1 s time interval

Reflection caused by the LED lights can be a source of error. Smooth reflective surfaces such as clean metals that are near the marker can be integrated with the contours. The translucent sheet and the blue 3D printed casing covering the edges of the LED matrix, shown in Fig. 1b, were designed to prevent the formation of an adjacent reflecting point by avoiding the direct attachment of the LED matrix to foreign materials.

The blinking frequency was detected by placing the marker 2 m away from the camera in the dark room. Only one marker was assumed to be at sight. The frame per second (FPS) of the camera stream and the number of the frames in which the majority of the pixels in the selected contour coordinates are white (LEDs turned on) were counted. The ratio of the white frame counts to the FPS indicates the period of the blinking LED and the inverse of the period is the frequency. The blinking delays (time that the LEDs are turned on/off) of 1,000 and 200 ms (LED turned on for a given time and turned off for the same duration) were tested

Table 1 Distance and angle measurements using LED marker and printed marker

	Ground truth	LED	Printed
Distance (cm)	10	13.06 ± 0.07	12.4 ± 0.04
Angle	0°	4.89° ± 1.06	4.84° ± 0.7

and the frequency calculator successfully calculated the frequency of the blinking LEDs (average and standard deviation of 229 ± 19 ms in the case of 200 ms blinking and 989 ± 133 ms in the case of 1,000 ms, respectively).

A printed planar marker of approximately 25 × 25 cm needs to be observed at a similar distance. For the detection of a longer distance, the size of the planar marker has to be gradually increased. However, only an LED matrix of size 4 × 4 cm can generate similar observability compared to the printed markers. Decreasing the size of the active marker is a great advantage because it requires waterproofing processes, unlike the passive printed markers, and the ease of sealing increases with the decrease in the size. Decreasing the size can be an advantage in the case of cleaning. Underwater structures suffer from the sea adherences such as seaweeds and barnacles [12]. A smaller size of the marker decreases the area to be cleaned and monitored and the area covered by coatings, preventing these adherences. In addition, a smaller size can help minimize the modification of the original underwater structure on which the marker has to be attached. The pose and the position of the camera were calculated for testing the compatibility of the proposed LED markers with the conventional printed 2D marker systems. The difference in the distance and the angle from the ground truth data was measured and the results are shown in Table 1.

The root mean square errors (RMSEs) of the distance and angle measured by using the LED marker were 3.0 cm and 4.89°, respectively, whereas the RSMEs of the distance and angle measured by using the printed marker were 2.4 cm and 4.84°, respectively. The scaled down experiment showed that the proposed marker could be used like the conventional printed markers. Although there were some errors in the distance measurement, a measurement similar to that of the conventional printed markers was obtainable. The measurement made using the LED marker had a slightly higher error than that using the printed markers. The boundary of the recognized LED markers becomes slightly larger than the original marker because the light scattered by the surface of the marker cover is also detected during the thresholding processes. We prospect that this is the cause of the increased error. In addition, the size of the marker used in this research is relatively smaller than that used in the real world applications. Even small errors at the boundary can affect the overall results significantly when compared to the large-sized markers. Because the purpose of this research is to show that the LED markers can function like the conventional 2D markers by using the same algorithms, we did not elaborate further on the accuracy.

The similar experiment was conducted in a foggy environment to compare the printed marker and the LED marker in the same environment. Because the turbidity measurement sensor works only in water, the exact scale of the thickness of fog was not measurable. Instead, the smoke generated by the fog machine (HN-1200 W, Hanasound. Inc., Seoul, Republic of Korea) [13] was filled until the Secchi circle with a diameter of 20 cm becomes invisible at a distance of approximately 2 m (about 4 NTU according to [14]). According to [15], normal seawaters generally showed 1–10 which is similar to the value used in this research. A brown background wall shown in Fig. 3b becomes invisible (Fig. 5c), indicating the increase in the turbidity. Controlling and maintaining the turbidity of a small gap (10 cm) between the marker and a camera with the fog generated by the fog machine is difficult because the fog tends to rise to the ceiling. Thus, the markers were moved to a distance of 30 cm for the homogenous turbidity between the overall environment and the region between marker and camera, and we checked whether their contour edge is detectable. The proposed marker and the conventional marker exhibited similar functionalities in normal environments. However, the detection ability of the proposed marker and that of the conventional marker differed in a foggy environment. The number of frames at which contours are detected in a foggy environment within the given time was also compared. Among the 327 frames passed, the LED marker detected contours in all the 327 frames. Meanwhile, the contour of the conventional printed marker was detected in only 215 frames (65.7% of success rate). The printed marker was not observable for a few seconds when the fog was belched from the fog machine. However, the light from the LED marker was still observable. In a high turbidity environment, the printed markers are undetectable or they can be wrongly recognized if the grids of the markers are faded. Although the proposed system does not recognize the accurate shape of the marker, it still has a higher visibility and is equipped with the blinking system, which allows long distance detection or high turbidity detection.

4 Conclusions and Future Work

We have proposed a new active marker system displaying the conventional planar patterned marker with the LED matrix for easier recognition of markers in low visibility environments. The image processing algorithm for the marker restoration and long-range detection by using the blinking frequency were also proposed. Overall experiments showed that the proposed marker is compatible with the conventional printed markers but has higher visibility in various environments.

In the future work, we will consider elaborating the overall system for better detectability and removing errors caused by light reflection and scattering. Also applications of the developed marker at various conditions such as a structural health monitoring [16], marker based construction assistance [17], or marker-guided landing of the Unmanned Aerial Vehicles [18] under low visibility environment will be developed.

Acknowledgements This work was partially supported by Korea Minister of Ministry of Land, Infrastructure and Transport (MOLIT) through a U-City Master and Doctor Course Grant Program. This work was also supported by grant No. 10043928 from the Industrial Source Technology Development Programs of the MOTIE (Ministry Of Trade, Industry and Energy), Korea.

References

1. Toyoura, M., Aruga, H., Turk, M.: Detecting markers in blurred and defocused images. In: Proceedings of the 2013 International Conference on Cyberworlds (CW 2013), pp. 183–190, Yokohama, Japan (2013)
2. Jung, J., Li, J., Choi, H., Myung, H.: Localization of AUVs using visual information of underwater structures and artificial landmarks. Intel. Serv. Robotics. **10**, 67–76 (2017)
3. Kim, K., Choi, D., Lim, H., Kim, H., Jeon, J.S.: Vision marker-based in situ examination of bacterial growth in liquid culture media. Sensors **16**, 2179 (2016)
4. AprilTag. https://april.eecs.umich.edu/software/apriltag.html. Accessed 12 Mar 2017
5. ArUco: A minimal library for Augmented Reality applications based on OpenCV. https://www.uco.es/investiga/grupos/ava/node/26. Accessed 12 Mar 2017
6. Jung, J., Lee, S., Myung, H.: Indoor mobile robot localization and mapping based on ambient magnetic fields and aiding radio sources. IEEE Trans. Instrum. Measur. **64**, 1922–1934 (2015)
7. Jung, S., Koo, J., Jung, K., Kim, H., Myung, H.: Vision-based autonomous landing system of an unmanned aerial vehicle on a moving vehicle. J. KROS **11**, 262–269 (2016)
8. Han, K., Lee, Y., Choi, H.: Vision based framework for autonomous docking of underwater robots using multi-purpose self-similar landmark. J. CICS **4**, 71–72 (2012)
9. Lee, P., Jeon, B., Kim S.: Visual servoing for underwater docking of an autonomous underwater vehicle with one camera. In: Proceedings of the OCEANS, pp. 677–682, San Diego, USA (2003)
10. Sohn, K.: A study on the short-range underwater communication using visible LEDs. J. Korea Soc. Mar. Eng. **37**(4), 425–430 (2013)
11. OpenCV library. http://opencv.org/. Accessed 26 Mar 2017
12. Hull cleaning. http://hullcleaning.com/. Accessed 12 Mar 2017
13. Fog machine catalog. http://hanainsound.co.kr/product/list.html?cate_no=508. Accessed 26 Mar 2017
14. Manuals from Dakota City Soil and Water Conservation District. http://www.dakotaswcd.org/pdfs/mm_watclar_fs.pdf. Accessed 26 Mar 2017
15. Jang, I., Won, D., Baek, W., Shin, C., Lee, S.: Turbidity Characteristics of Korean Port Area. JKAIS **16**(12), 8889–8895 (2015)
16. Shin, J., Jeon, H., Choi, S., Kim. Y., Myung H.: Laser pose calibration of ViSP for precise 6-DOF structural displacement monitoring. SSS **18**(4), 801–818 (2016)
17. Choi, S., Myeong, W., Jeong, Y., Myung, H.: A vision-based 6-DOF displacement measurement method for assemble of PC bridge structures using a planar marker. In: Proceedings of the International Conference on Robot Intelligence Technology (RiTA), pp. 501–509, Bucheon, Korea (2015)
18. Jung, S., Koo, J., Jung, K., Kim, H., Myung, H.: Vision-based autonomous landing system of an unmanned aerial vehicle on a moving vehicle. JKROS **11**(4), 262–269 (2016)
19. Choi, S., Myeong, W., Myung, H.: A vision-based 6-DOF displacement measurement system using multiple planar markers for precise assemble of PC members. In: Proceedings of the International Conference on Control, Automation and Systems (ICCAS), pp. 1036–1039, Gyeongju, Korea (2016)
20. Li, J., Liu, H.: Design optimization of amazon robotics. AC I S **4**, 48–52 (2016)

Development of Robust Recogintion Algorithm of Retro-reflective Marker Based on Visual Odometry for Underwater Environment

Pillip Youn, Kwangyik Jung and Hyun Myung

Abstract In this paper, we propose the robust algorithm of retro-reflective marker recognition algorithm based on visual odometry. Retro-reflective is used in order to distinguish markers under the low visibility underwater environment. The existing marker recognition algorithm estimates 6-DOF pose only if camera captures the whole marker image. To overcome this weakness we proposed the robust recognition algorithm based on visual odometry in this paper. The recognition algorithm is tested in real sea experiment.

Keywords Autonomous underwater vehicles · Robot vision · Marker recognition · Underwater vision · 6-DOF pose estimation · Visual odometry Retro-reflective material · Unmanned underwater vehicle · Robot vision system

1 Introduction

Unmanned underwater vehicles (UUV) is widely used for investigation and exploration of ecosystem, unknown ocean floors. The need of UUV is increasing because of the construction and monitoring of offshore construction industry growth. The existing work method depends on divers' ability. So it is very dangerous and low efficient. And it has limitation about the depth of water.

P. Youn · K. Jung (✉)
Department of Civil and Environmental Engineering, KAIST (Korea Advanced Institute of Science and Technology), Daejeon 305-701, Republic of Korea
e-mail: ankh88324@kaist.ac.kr

P. Youn
e-mail: pillibi@kaist.ac.kr

H. Myung
Department of Civil and Environmental Engineering and Robotics Program, KAIST (Korea Advanced Institute of Science and Technology), Daejeon 305-701, Republic of Korea
e-mail: hmyung@kaist.ac.kr

J.-H. Kim et al. (eds.), *Robot Intelligence Technology and Applications 5*,
Advances in Intelligent Systems and Computing 751,
https://doi.org/10.1007/978-3-319-78452-6_43

The most important technology is to help the robot to carry out its works in underwater environment is localization. The selection of appropriate sensors in localization under the underwater environment is important. The common sensors in localization system such as GPS can not be used because of disruptions of the underwater environment. Because it is difficult to navigate an underwater robot in 3D dynamic complex underwater environment with common sensors, the research in the robot localization using various underwater sensor systems is needed [1, 2].

In the underwater environment sonar and camera are well used sensors. The sonar sensor is more robust because it is affected by light and turbidity less than camera. But it is expensive and can not provide high quality images. The underwater images captured camera suffer from poor visibility. The reasons are absorption and scattering. The absorption reduces the light energy. The scattering causes changes in the light direction. Despite the weaknesses the camera can capture abundant and high quality information over a short distance [3, 4].

The proposed algorithm in this paper recognize retro-reflective markers using visual information captured camera. And the visual odometry makes the algorithm robust.

2 Methodology

2.1 Retro-reflective Marker

In this research, the markers are made of the retro-reflective material. The retro-reflective material is a kind of microprism in the form of tiny glass beads. So the retro-reflective material reflects incident light to the same direction of the incident light (See Fig. 1a). The absorption causes the light energy shortage in the underwater. So UVV flashes light using halogen lamp. The camera installed UVV captures reflected lights by retro-reflective markers. Figure 1 shows the marker performance between normal marker and retro-reflective marker. In this research the 28 markers are produced and used (See Fig. 2).

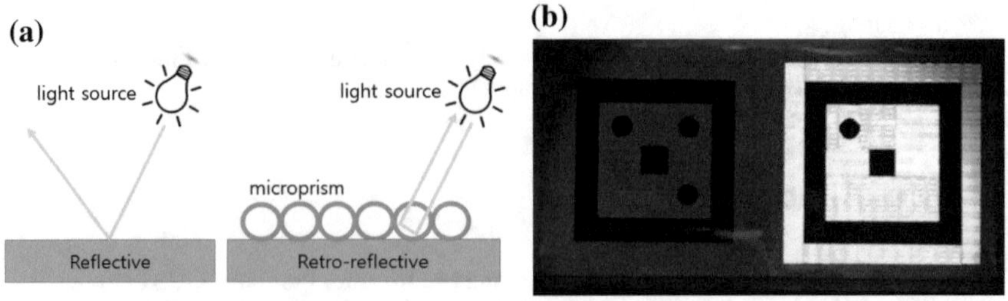

Fig. 1 **a** Retro-reflective material's feature. **b** Comparison of light reflections between a normal marker and a retro-reflective marker

Fig. 2 28 markers produced using retro-reflective materials

2.2 Marker Recognition Algorithm

System Architecture The proposed algorithm system consists of four main parts (See Fig. 3). The first part is 'pre processing'. This part get the input image from camera installed on UVV, then processes in order to increase the quality of input images. The second part is 'marker recognition' that detects and distinguish marker ID. The third part is '6-DOF Pose estimation' that calculate the global 6-DOF pose using transformation matrix. The last part is 'visual odometry' to help tracking marker. The calculated global 6-DOF pose data is transmitted to the UVV.

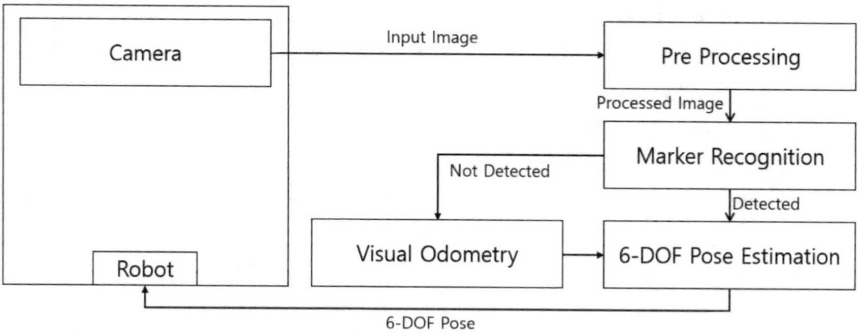

Fig. 3 Overall system architecture

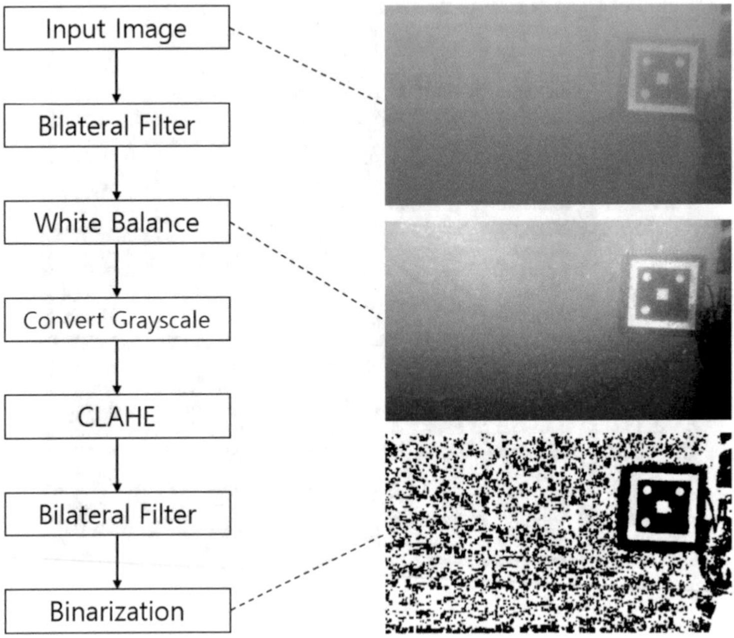

Fig. 4 Flow chart and results of pre-processing

Pre-processing It is difficult to get information from underwater images because of poor visibility different from to use common images. The light attenuation, the absorption and floating matters make to process the underwater images difficult [3]. Therefore pre image processing is needed to overcome underwater physical properties.

The pre-processing is shown in Fig. 4. The bilateral filter removes noise while preserving edges of input image. So this filter enhances the sharpness. To improve image appearance by discarding unwanted color casts, a white balance is used. The white balance inhibits ambient light for revising to actual color. CLAHE is 'Contrast Limited Adaptive Histogram Equalization'. It enhances contrast keeping sharpness of object. After all process pre-processing part converts image to binary image.

Marker Recognition Algorithm After pre-processing, labeling is carried out using the blob algorithm [5]. The labeling algorithm finds the marker location on image by recognizing the adjacent areas having the same binary value. Then marker point extraction module and sub pixel extraction module extract 12 feature points of recognized marker [1] (Fig. 5).

6-DOF Pose Estimation With the pixel information extracted marker recognition algorithm, 6-DOF pose estimation module estimates the position of camera install on UVV. Then the global coordinate is calculated with transformation matrices. Figure 6 shows the result.

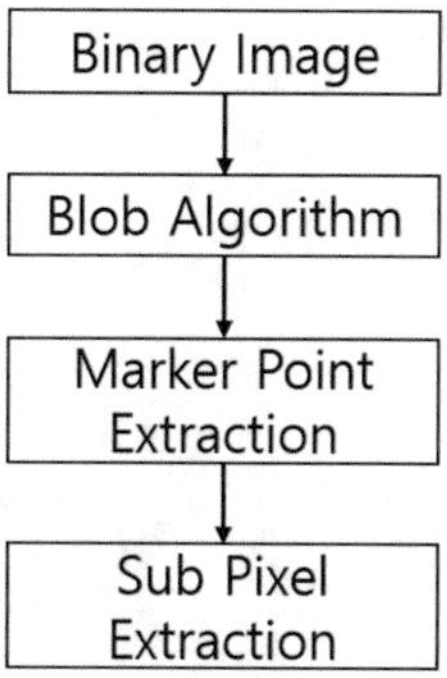

Fig. 5 Flow chart of marker recognition algorithm

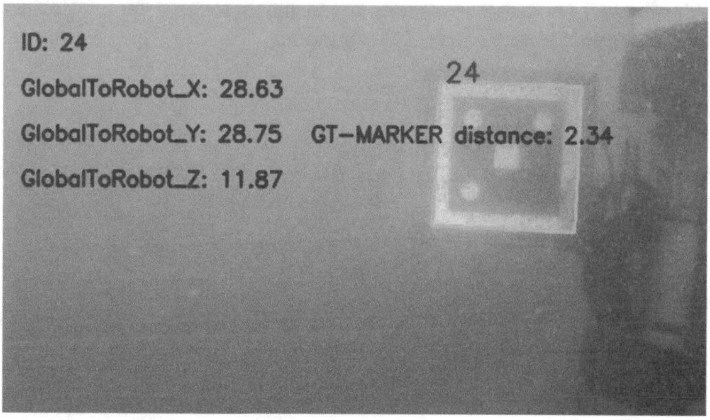

Fig. 6 Results of the 6-DOF pose estimation

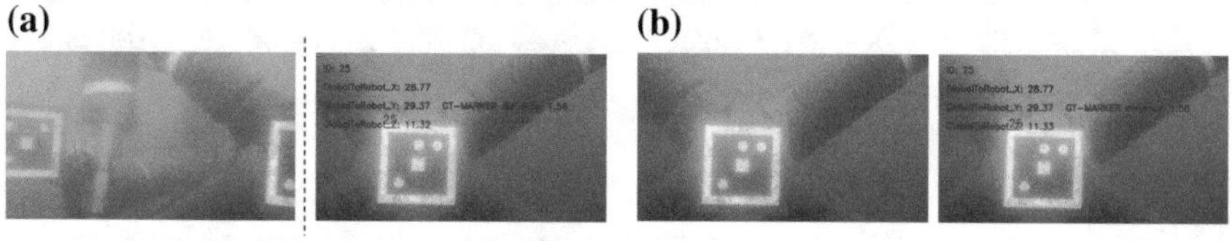

Fig. 7 **a** The portion of the marker image case. **b** Many floating matters case

Visual Odometry The marker recognition algorithm of this paper does not detect the marker, when the camera captures the portion of the marker image or there are many floating matters (See Fig. 7).

To overcome these weak points, this paper proposes visual odometry. When the marker detected, the feature points are detected by FAST corner detection algorithm. Then KLT tracker tracks feature points calculating point-correspondences. With point correspondences information, essential matrix is estimated by 5 point

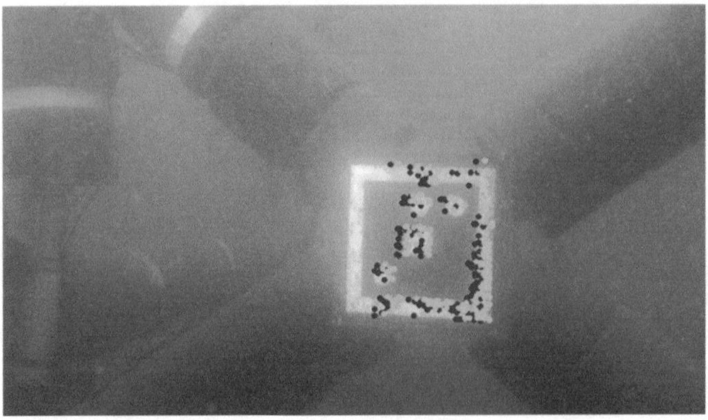

Fig. 8 Result of visual odometry estimation. Green points indicate feature points of previous frame. Blue points indicate feature points of current frame

algorithm [6]. After estimating essential matrix, the translation vector and rotation vector are calculated. Figure 8 shows the feature points tracked.

3 Result

3.1 Experiment

KIRO's P-SURO AUV has 6 halogen lamps [7] (See Fig. 9a) and SIDUS SS465 waterproof camera is used for real sea experiment. The markers were installed on inside and outside of offshore construction (See Fig. 9b). The location of the experiment is Yeongil Bay, Pohang, Korea.

(a) **(b)**

Fig. 9 **a** KIRO's P-SURO AUV. **b** Offshore construction

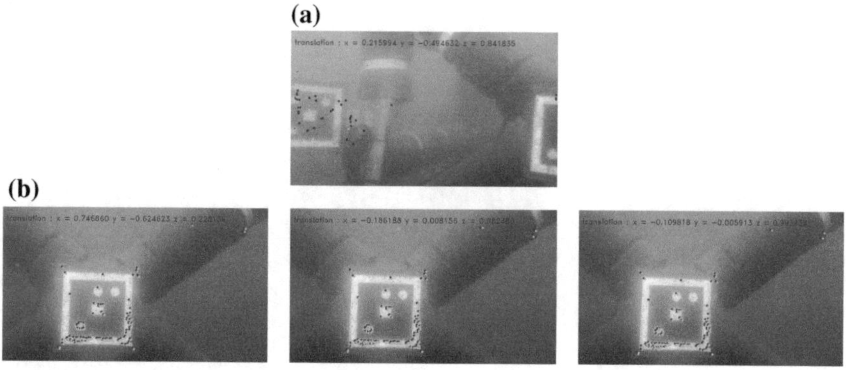

Fig. 10 **a** To overcome Fig. 7a. **b** To overcome Fig. 7b

3.2 *Conclusion*

Figure 10 shows results of the proposed algorithm. The weak points mentioned of existing algorithms not applied visual odometry are that it can not detect the marker when the camera captures the portion of the maker and many floating matters. Figure 10 shows the improved algorithm calculate the translation between frames. The improved algorithm applied visual odometry overcomes the limitation of existing algorithm.

Acknowledgements This research was supported by grant No. 10043928 from the Industrial Source Technology Development Programs of the MOTIE (Ministry Of Trade, Industry and Energy), Korea. The students are supported by the Korea Ministry of Land, Transport and Maritime Affairs (MLTM) as U-City Master and Doctor Course Grant Program.

References

1. Jung, K., Youn, P., Choi, S., Lee, J., Kang, H.J., Myung, H.: Development of retro-reflective marker and recognition algorithm for underwater environment. In: 2017 14th International Conference on Ubiquitous Robots and Ambient Intelligence (URAI), Jeju, pp. 666–670 (2017)
2. Kim, D., Lee, D., Myung, H., et al.: Intell. Serv. Robot. **7**, 175 (2014). https://doi.org/10.1007/s11370-014-0153-y
3. Ancuti, C., Ancuti, C.O., Haber, T., Bekaert, P.: Enhancing underwater images and videos by fusion. In: 2012 IEEE Conference on Computer Vision and Pattern Recognition, Providence, RI, pp. 81–88 (2012)
4. Jung, J., Li, J.H., Choi, H.T., et al.: Intell. Serv. Robot. **10**, 67 (2017). https://doi.org/10.1007/s11370-016-0210-9
5. Parker, J.R.: Algorithms for Image Processing and Computer Vision. Wiley (2010)
6. Nister, D.: An efficient solution to the five-point relative pose problem. IEEE Trans. Pattern Anal. Mach. Intell. **26**(6), 756–770 (2004)
7. Li, J.H., et al.: Development of P-SURO II hybrid AUV and its experimental study. In: OCEANS-Bergen, 2013 MTS/IEEE, pp. 1–6. IEEE (2013)

A Robust Estimation of 2D Human Upper-Body Poses Using Fully Convolutional Network

Seunghee Lee, Jungmo Koo, Hyungjin Kim, Kwangyik Jung and Hyun Myung

Abstract We present an approach to efficiently detect the 2D human upper-body pose in RGB images. Among the system for estimating the joints position, the method using only RGB camera sensor is very cost-effective compared to the system with high-priced sensors such as a motion capture system. In this work, we use semantic segmentation using a fully convolutional network to estimate the upper-body poses of each skeleton and choose the location coordinate using joint heatmaps. The architecture is designed to learn joint locations and their association via the sequential prediction process. We demonstrate the performance of the proposed method using various datasets.

1 Introduction

Human body pose estimation has been actively researched with the aim of fast and reliable skeleton extraction results for decades. It has been applied to gaming, animation, surveillance, medical diagnose, security, and human-computer

---

S. Lee · J. Koo · H. Kim · K. Jung · H. Myung (✉)
Department of Civil and Environmental Engineering, KAIST, Daejeon 34141, Korea
e-mail: hmyung@kaist.ac.kr
URL: http://urobot.kaist.ac.kr

S. Lee
e-mail: seunghee.lee@kaist.ac.kr
URL: http://urobot.kaist.ac.kr

J. Koo
e-mail: jungmokoo@kaist.ac.kr

H. Kim
e-mail: hjkim86@kaist.ac.kr

K. Jung
e-mail: ankh88324@kaist.ac.kr

H. Myung
Robotics Program, KAIST, Daejeon 34141, Korea

J.-H. Kim et al. (eds.), *Robot Intelligence Technology and Applications 5*,
Advances in Intelligent Systems and Computing 751,
https://doi.org/10.1007/978-3-319-78452-6_44

interaction (HCI) [1, 2]. Pose estimation is mainly used for the gesture recognition. Perceiving human gesture has an important role for worldwide usage without using various languages but with body language and also for deaf people signing communication in a good way. Also, it can be used for the surveillance system for CCTV for the crime or the alarming system for the emergency of elder people living alone. With the accurate skeleton extraction, correcting the posture in a medical way could be possible. Such requires keep the research of the robust and fast skeleton extraction.

Two different approaches for automatic human body pose estimation methods are categorized as follows: graph-based and machine learning-based methods. The first category is the graph-based method where geodesic distance is mostly used. A geodesic distance is a straight line from the curved place [3]. 3D joints are represented easily in the graph-based approach. Plagemann et al. [4] find points which are under the maximum geodesic distance, and these points are identified as the body points with body descriptors. Schwarz et al. [5] attempts to find pose estimation using anatomical landmarks in a depth geodesic graph and inverse kinematics. The skeletal graph is also used in the graph-based approach. Straka et al. [6] used a skeletal tree-like graph as the human body.

The other category is the machine learning-based method where there is no pre-required information about human. Shotton et al. [7] presented a per-pixel classification of the joints and estimation of joint position both using random forests. Hernández-Vela et al. [8] used graph cut optimization for the image segmentation over the work per-pixel classification.

However, pose estimation is still considered as a difficult problem since 230 human joints movements are not fully evident. Moreover, clothing and body shape differences cause a variety of situations. Partial occlusions due to self-articulation, for example, hand covering the face or the frontal portion of the body, or occlusions due to external objects may cause ambiguities in the obtained results [9]. Only using graph-based pose estimation requires model calibration before starting the predictions of the joints which mean it is not adequate for the real-time application [10]. However, learning-based pose estimation doesn't require prior calibration and can be trained with vast images for robustness.

Robots should be able to perceive human in order to understand the environment and interact. Hence most interacting robots would look at a person in a similar view of the human, which means that robots would see the person in a close look, so not a full body in a distance, but a part of the person, for example, upper body, or lower body. In this study, we focused on the estimation of the upper-body pose which gives some idea about partly shown body estimation.

In the next section, we review related works on skeleton extraction. The following sections explain pose estimation, introducing our architecture. Finally, we demonstrate pose estimation experiment result and discuss the conclusion.

2 Related Research Trends

2.1 Skeleton Extraction Using Depth Information

Under the development of Kinect system from Microsoft, depth-based automatic human pose estimation system without marker has become a great study [11]. 640 * 480 image at 30 frames per second is obtained from the Kinect camera. Depth information is invariant to color and texture. Haritaoglu et al. [12] divided the blob using geometrical characteristics and decided different joints (head, hands, and feet). Similarly, Fujiyoshi et al. [13] also predict the blob of a head, hands, and feet with the characteristics without template model. Guo et al. [14] determine the location of full body points by using distance function for fitting the model. Neural networks [15] and genetic algorithms [16] have also been used to obtain the complete position of all of the joints of the person.

The most well-known depth using method is from Shotton et al. [7] inferring 20 joints for every pixel classification in every single depth image from Kinect sensor. A randomized decision forest is performed for the pixel classification. A random decision forest is operated by building decision trees when training and giving the output for the body part classes. A weighted Gaussian kernel with mean shift is used to generate 3D joint position proposals. The classifier has been trained by using a vast amount of database for the variety of motions and different body shape-people.

OpenNI library is also widely used for skeleton estimation. 15 skeleton joints are estimated by depth-edge counting local descriptors [17]. Canny edge detector extracts depth edges from the depth map, and statics about edge pixels are computed in each patch. Then the locations of skeletal joints are searched by the approximate nearest neighbors (ANN) algorithm for matching the patch descriptors. By the time all the patch descriptors are compared, the body joint locations are determined one-by-one.

2.2 Skeleton Extraction Using RGB Image Information

Using image data for inferring human joints in the image has been suspicious for variation of color difference and human body shape. However, emerging of the advanced hardware has made it possible to train a vast number of datasets using the neural network. Many skeleton extraction methods use graphic processing units for the increase of robustness and frame rate [8, 18–20].

Recently, Deep Convolutional Neural Networks (DCNNs) using RGB images have achieved excellent performance on human pose estimation [21]. DeepPose [22] used regression learning for the body joint locations with the convolutional network. Tompson et al. [23, 24] suggested multi-resolution DCNN for the better accuracy by reducing the pooling effect. Chen et al. [25] proposed convolutional network dependent on pairwise spatial joints relationships. Chu et al. [26] used the relationship among body joints at the feature level.

Evaluation of the graphical methods is mostly held with renowned public pose estimation data sets. There are various benchmarks for human pose estimation. The Frames Labeled in Cinema (FLIC) dataset is upper body images obtained from Hollywood movies, and it consists of 4000 training images and 1000 test images [27]. The Leeds Sports Poses (LSP) dataset consists of sports images with 14 body joints and contains 11,000 training images and 1000 testing images [28]. The Image Parse (PARSE) dataset [29], the MPII Human Pose Dataset [30], the Buffy dataset [31], and so on are also popularly used.

3 Pose Estimation Algorithm

In this study, segmentation based learning classifier is used for the estimation of the upper-body human pose. The total number of joints is seven, and the joints are a head, left shoulder, left elbow, left hand, right shoulder, right elbow, and right hand.

3.1 Fully Convolutional Network

For human upper-body segmentation, we deploy a network architecture that is based on and initialized by segmentation network of Long et al. [32]. They cast the problem of deep classification. They adapt contemporary classification networks such as Alexnet [33] into fully convolutional networks and fine-tune [34] to the segmentation task so they can pass their learned representations. FCN is trained end-to-end and predicts pixel-wise segmentation result. Dense feedforward and backpropagation computation enable both learning and inference on the whole image at one time.

Our segmentation network is focused on the human-body oriented and trained on our upper-body pose dataset. Details on the network architecture and its training procedure are provided in Sect. 4.2 (Fig. 1).

3.2 Pose Estimation Using Joint Confidence Map

For human pose estimation, we formulate localization of 2D keypoints as estimation of 2D confidence maps, where each map contains information about the likelihood of a certain body joint is present at a location.

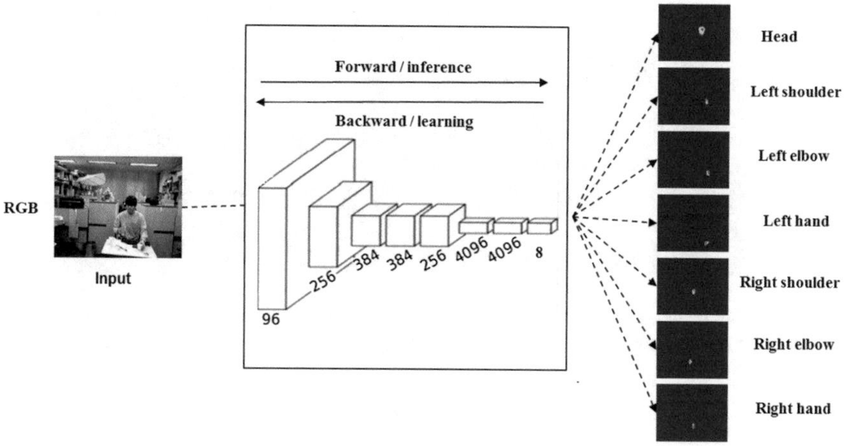

Fig. 1 The architecture of the proposed method. We show a convolutional architecture based on segmentation training and learning. The network results in heatmaps of each joint which are the probability maps of each joint existence

Given the image feature representation produced by the encoder, an initial score map is predicted and calculated the central moment of the region. A complete overview over the network architecture is located in Sect. 4.

4 Evaluation

4.1 Datasets for Experiments

For training dataset, there is no available dataset that provides segmentation image of human upper-body. Therefore, we complement them with a new dataset by ourselves.

Our dataset is built upon 3 different persons performing 12 different scenarios including cleaning the table, making a cereal in a bowl, reading a book, etc.

For each new scenario, we randomly sample a new camera location, which is roughly located at the probable place where a camera of the humanoid robot Mybot, developed in the Robot Intelligence Technology Laboratory at KAIST, would take place.

In total, our dataset provides 433 images for training and 58 images for evaluation with a resolution of 640 $\times$ 480 pixels. All samples come with full annotation of 7 joints: head, left shoulder, left elbow, left hand, right shoulder, right elbow, and right hand. Figure 2 shows a sample of the dataset.

Fig. 2 Our new dataset provides segmentation image with 7 classes: head, left shoulder, left elbow, left hand, right shoulder, right elbow, and right hand

4.2 Training and Testing Method

We used an ASUS Xtion pro to estimate the joints poses. The camera is fixed at the height of Mybot, and the movement of the camera downward or above is about ± 15. For the computation, Intel i5 3.0 GHz quad-core CPU is used, and two GTX1080 were used for training and testing. We define and implement our model using the Caffe [35] libraries and DIGITS for deep learning.

It was trained for upper-body joints segmentation with a batch size of 1 and using Adam solver [36]. The network architecture is initialized using weights of FCN-AlexNet and fine-tuned with a pre-trained model by PASCAL-VOC dataset [37]. The learning rate was 1×10^{-4}, and sigmoid decay was used for the learning rate policy.

4.3 Experimental Results

In this section, we present our numerical results with our dataset. We trained the network for 30 epochs, which takes an average of 0.5 h on the Nvidia DIGITS.

We show in Fig. 3 our results on the upper-body image. Each heatmap gives the probability map of each joint's existence. The red is the highest confidence, and the blue is the lowest confidence. With the segmentation result according to heatmaps of each joint, final pose estimation result is obtained as shown in Fig. 4. Different View and environment dataset were tested as the test.

For the accuracy evaluation, we have set the ground truth manually for the random test images as we don't have ground truth dataset. As the Table 1 describes, the average of error length of all the joints with the proposed algorithm is about 45.25 mm, and all errors are below 72 mm. The error length of the left elbow was the highest. One reason might be because there were many datasets which are occluded the left elbow by working on the desk.

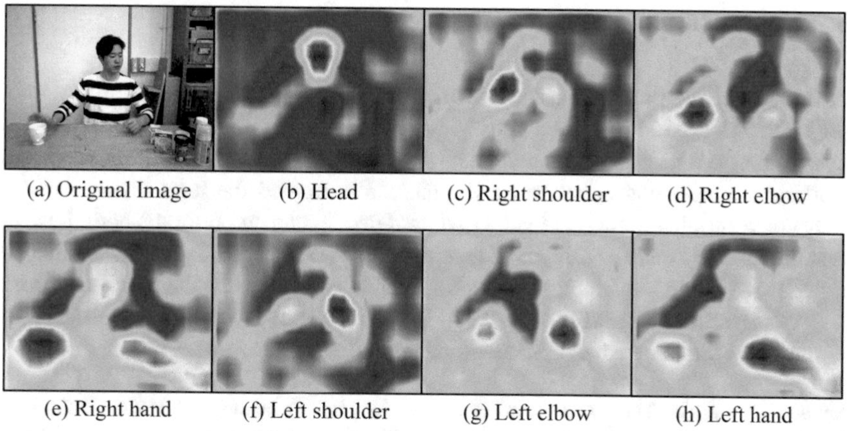

Fig. 3 Heatmap results of our method on each joint. We see that the method is able to provide the confidence map of each joint. Red color in the map represents the higher confidence, and blue color represents the lower confidence [38]

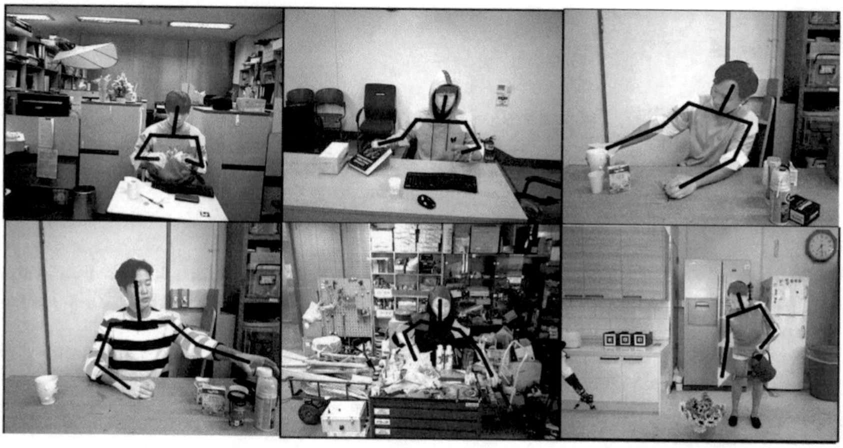

Fig. 4 Upper-body pose estimation results on different environment and person [38]

Table 1 Experimental results of the distance errors for each joint for the test dataset

Head	Left shoulder	Left elbow	Left hand	Right shoulder	Right elbow	Right hand
6.35	43.15	56.43	50.05	33.11	72.09	55.57

5 Conclusion

In this paper, we proposed a method of recognizing the pose attitude through the segmentation-based articulated body joints estimation for upper-body RGB images. In addition, we made upper-body image dataset on our own consisting of mostly working and cleaning situation on the table. We trained the joint features with a fully convolutional network to get a part confidence map and infer the body pose by calculating the central moment of each joint's confidence map. This method is expected to be applicable to a real robot including Mybot. In the future, a further research is needed to enable joint position estimation even when harsh articulation of the body joints and radical change of the human orientation appear.

Acknowledgements This work was supported by the Technology Innovation Program, 10045252, Development of robot task intelligence technology, supported by the Ministry of Trade, Industry, and Energy (MOTIE, Korea). The students are supported by Korea Minister of Ministry of Land, Infrastructure and Transport (MOLIT) as U-City Master and Doctor Course Grant Program.

References

1. Aggarwal, J., Cai, Q.: Human motion analysis: a review. Comput. Vis. Image Underst. **73**, 428–440 (1999)
2. Moeslund, T., Hilton, A., Krüger, V.: A survey of advances in vision-based human motion capture and analysis. Comput. Vis. Image Underst. **104**, 90–126 (2006)
3. Oxford Dictionaries. http://oxforddictionaries.com/definition/english/VAR. Accessed 16 Oct 2017
4. Plagemann, C., Ganapathi, V., Koller, D., Thrun, S.: Real-time identification and localization of body parts from depth images. In 2010 IEEE International Conference on Robotics and Automation (ICRA), pp. 3108–3113 (2010)
5. Schwarz, L.A., Mkhitaryan, A., Mateus, D., Navab, N.: Human skeleton tracking from depth data using geodesic distances and optical flow. Image Vis. Comput. **30**, 217–226 (2012)
6. Straka, M., Hauswiesner, S., Rüther, M., Bischof, H.: Skeletal graph based human pose estimation in real-time. In: BMVC, pp. 1–12 (2011)
7. Shotton, J., Sharp, T., Kipman, A., et al.: Real-time human pose recognition in parts from single depth images. Commun. ACM **56**, 116 (2013)
8. Hernández-Vela, A., Zlateva, N., Marinov, A., Reyes, M., Radeva, P., Dimov, D., Escalera, S.: Graph cuts optimization for multi-limb human segmentation in depth maps. In: Computer Vision and Pattern Recognition (CVPR), pp. 726–732 (2012)
9. Droeschel, D., Behnke, S.: 3D body pose estimation using an adaptive person model for articulated ICP. In: Intelligent Robotics and Applications, pp. 157–167 (2011)
10. Kim, H., Lee, S., Lee, D., Choi, S., Ju, J., Myung, H.: Real-time human pose estimation and gesture recognition from depth images using superpixels and SVM classifier. Sensors **15**(6), 12410–12427 (2015)
11. Jain, H., Subramanian, A., Das, S., Mittal, A.: Real-time upper-body human pose estimation using a depth camera. In: Computer Vision/Computer Graphics Collaboration Techniques, pp. 227–238 (2011)
12. Haritaogalu, I.: W4S: A real-time system for detecting and tracking people in 2 1/2-D. In: European Conference on Computer Vision (1998)

13. Fujiyoshi, H., Lipton, A.J., Kanade, T.: Real-time human motion analysis by image skeletonization. IEICE Trans. Inf. Syst. **87**(1), 113–120 (2004)
14. Guo, Y., Xu, G., Tsuji, S.: Tracking human body motion based on a stick figure model. J. Vis. Commun. Image Represent. **5**(1), 1–9 (1994)
15. Ohya, J., Kishino, F.: Human posture estimation from multiple images using genetic algorithm. In: Pattern Recognition, 1994. Vol. 1-Conference A: Computer Vision and Image Processing. Proceedings of the 12th IAPR International Conference, vol. 1, pp. 750–753 (1994)
16. Takahashi, K., Uemura, T., Ohya, J.: Neural-network-based real-time human body posture estimation. In: Neural Networks for Signal Processing X, 2000. Proceedings of the 2000 IEEE Signal Processing Society Workshop, vol. 2, pp. 477–486 (2000)
17. Presti, L.L., La Cascia, M.: 3D skeleton-based human action classification: a survey. Pattern Recogn. **53**, 130–147 (2016)
18. Zhang, Z., Seah, H.S., Quah, C.K., Sun, J.: GPU-accelerated real-time tracking of full-body motion with multi-layer search. IEEE Trans. Multimedia **15**(1), 106–119 (2013)
19. Shotton, J., Girshick, R., Fitzgibbon, A., Sharp, T., Cook, M., Finocchio, M., … Blake, A.: Efficient human pose estimation from single depth images. IEEE Trans. Pattern Anal. Mach. Intell. **35**(12), 2821–2840 (2013)
20. Ganapathi, V., Plagemann, C., Koller, D., Thrun, S.: Real time motion capture using a single time-of-flight camera. In: Computer Vision and Pattern Recognition (CVPR), pp. 755–762 (2010)
21. Yang, W., Ouyang, W., Li, H., Wang, X.: End-to-end learning of deformable mixture of parts and deep convolutional neural networks for human pose estimation. In: IEEE Conference on Computer Vision and Pattern Recognition, pp. 3073–3082 (2016)
22. Toshev, A., Szegedy, C.: Deeppose: Human pose estimation via deep neural networks. In: IEEE Conference on Computer Vision and Pattern Recognition, pp. 1653–1660 (2014)
23. Tompson, J., Goroshin, R., Jain, A., LeCun, Y., Bregler, C.: Efficient object localization using convolutional networks. In: IEEE Conference on Computer Vision and Pattern Recognition, pp. 648–656 (2015)
24. Tompson, J.J., Jain, A., LeCun, Y., Bregler, C.: Joint training of a convolutional network and a graphical model for human pose estimation. In: Advances in Neural Information Processing Systems, pp. 1799–1807 (2014)
25. Chen, X., Yuille, A.L.: Articulated pose estimation by a graphical model with image dependent pairwise relations. In: Advances in Neural Information Processing Systems, pp. 1736–1744 (2014)
26. Chu, X., Ouyang, W., Li, H., Wang, X.: Structured feature learning for pose estimation. In: IEEE Conference on Computer Vision and Pattern Recognition, pp. 4715–4723 (2016)
27. Sapp, B., Taskar, B.: MODEC: multimodal decomposable models for human pose estimation. In: IEEE Conference on Computer Vision and Pattern Recognition, pp. 3674–3681 (2013)
28. Johnson, S., Everingham, M.: Clustered pose and nonlinear appearance models for human pose estimation (2010)
29. Ramanan, D.: Learning to parse images of articulated bodies. In: Advances in Neural Information Processing Systems, pp. 1129–1136 (2007)
30. Andriluka, M., Pishchulin, L., Gehler, P., Schiele, B.: 2D human pose estimation: new benchmark and state of the art analysis. In: IEEE Conference on computer Vision and Pattern Recognition, pp. 3686–3693 (2014)
31. Ferrari, V., Marin-Jimenez, M., Zisserman, A.: Progressive search space reduction for human pose estimation. In: Computer Vision and Pattern, pp. 1–8 (2008)
32. Simonyan, K., Zisserman, A.: Very deep convolutional networks for large-scale image recognition (2014). arXiv:1409.1556
33. Krizhevsky, A., Sutskever, I., Hinton, G.E.: Imagenet classification with deep convolutional neural networks. In: Advances in Neural Information Processing Systems, pp. 1097–1105 (2012)

34. Donahue, J., Jia, Y., Vinyals, O., Hoffman, J., Zhang, N., Tzeng, E., Darrell, T.: DeCAF: a deep convolutional activation feature for generic visual recognition. In: International Conference on Machine Learning, pp. 647–655 (2014)
35. Jia, Y., Shelhamer, E., Donahue, J., Karayev, S., Long, J., Girshick, R., … Darrell, T.: Caffe: Convolutional architecture for fast feature embedding. In: The 22nd ACM International Conference on Multimedia, pp. 675–678 (2014)
36. Kingma, D., Ba, J.: Adam: A method for stochastic optimization (2014). arXiv:1412.6980
37. Everingham, M., Eslami, S.A., Van Gool, L., Williams, C.K., Winn, J., Zisserman, A.: The pascal visual object classes challenge: a retrospective. Int. J. Comput. Vision **111**(1), 98–136 (2015)
38. Koo, J., Lee, S., Kim, H., Jung, K., Oh, T., Myung, H.: Human upper-body pose estimation using fully convolutional network and joint heatmap. In: IEEE/RSJ International Conference on Intelligent Robots and Systems (2017)

Surface Crack Detection in Concrete Structures Using Image Processing

Mi-Hyeon Cheon, Dong-Gyun Hong and Donghwa Lee

Abstract This paper proposes a surface crack detection method in concrete structures using image processing. Although many alternative methods using image processing have been suggested, non-crack parts are frequently detected as cracks, which make it difficult to obtain sufficiently accurate results. The proposed crack detection algorithm uses additional noise filtering steps after using the noise filtering proposed by the existing study. Finally, experimental results of the algorithm proposed by this study show cracks more clearly and indicate more effective noise removal.

1 Introduction

The recent increase in the number of natural disasters, including earthquakes, necessitates accurate safety inspections of building structures. However, visual inspection is still the dominant method. Inspectors check cracks and corrosions with the naked eye and mark risky elements on drawings. In the case of a large structure, such a method is time consuming, and the accuracy of the results tends to deteriorate. In addition, personnel expenses to hire many inspecting workers are also burdensome [1, 2].

Although many alternative methods using image processing have been suggested to solve these problems, shadows of images, stains, and other non-cracks are frequently detected as cracks, which makes it difficult to obtain sufficiently accurate results [3]. This paper intends to advance the existing studies of crack detection by

M.-H. Cheon · D.-G. Hong · D. Lee (✉)
School of Computer & Communication Engineering, Daegu University,
201 Daegudaero, Jillyang, Gyeongsan, Gyeonbuk, Republic of Korea
e-mail: leedonghwa@daegu.ac.kr

M.-H. Cheon
e-mail: cjsal95@daegu.ac.kr

D.-G. Hong
e-mail: qwas9789@daegu.ac.kr

J.-H. Kim et al. (eds.), *Robot Intelligence Technology and Applications 5*,
Advances in Intelligent Systems and Computing 751,
https://doi.org/10.1007/978-3-319-78452-6_45

image processing and to propose an image-processing-based system that shows more effective performance.

2 Proposed Crack Detection Algorithm

Various types of cement are used for real-world concrete structures, and foreign substances are often mixed into the cement. In addition, crack images can become blurry depending on the resolution. Thus, in order to consider such variables, many cement crack images taken in as diverse environments as possible were used to detect cracks. Figure 1 shows the proposed crack detection algorithm.

2.1 *Image Preprocessing*

In this study, all pixel values were scanned, and the darkest pixel value was selected as a preliminary step before applying noise filtering. In general, a crack has the darkest value among all of the pixel values. However, as sample images have different degrees of brightness, the cracks do not have a uniform pixel value. Accordingly, the darkest value is selected from the entire area, and this value is adopted as a reference value. In this study, the darkest value found in the images was replaced by the minimum value, and the brightest value was replaced by the maximum value. This process made crack parts stand out and the remaining parts less noticeable (Fig. 2a).

As a next step, median-filtering was used to remove salt-and-pepper noise [4], of which the result is shown in Fig. 2b. The image was black/white reversed to apply noise filtering.

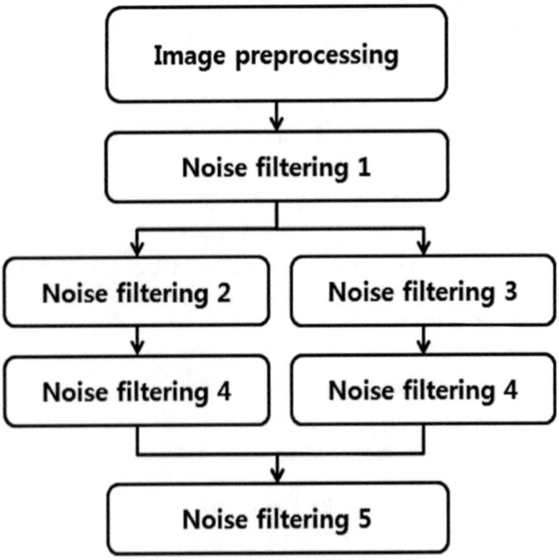

Fig. 1 Proposed crack detection algorithm

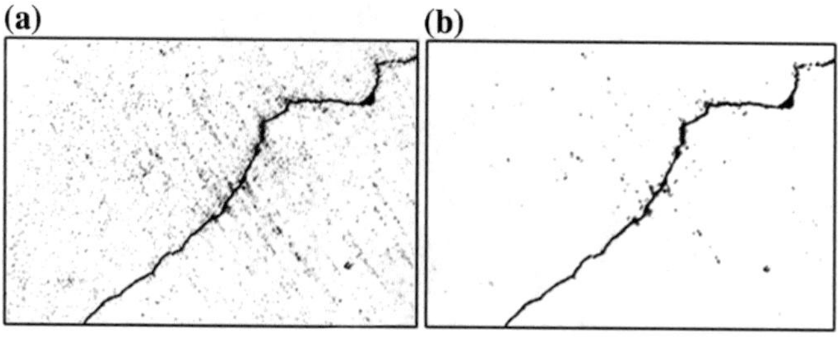

Fig. 2 Preprocessing before applying noise filtering

2.2 *Noise Filtering*

This study considered five types of noise filtering to remove noises. Two images were created in the process of noise removal. After noise filtering was conducted for each image, the two images were combined, and then noise was filtered again from the combined image.

Noise filtering 1 considered the fact that cracks must have more than one darkest value (Fig. 4a). As the final image was black/white reversed in Sect. 2.1, cracks need to have more than one darkest value in the reversed image (Fig. 3). Accordingly, pixel values were successively scanned, and any scanned pixels that did not have a pixel value of 255 were considered to be noise and were removed.

In noise filtering 2, a single matrix was created, and the entire area was scanned. Then, in case the number of pixels detected as edges in the matrix did not exceed a certain number, they were to be considered as noise and removed (Fig. 4b). In this study, a 4 × 4 matrix was applied, and if the number of pixels detected as edges was eight or less, they were removed as noise.

Noise filtering 3 was applied to the image to which Noise filtering 1 had been applied. This step considered edges that were detected intermittently in the image as

Fig. 3 Noise and crack

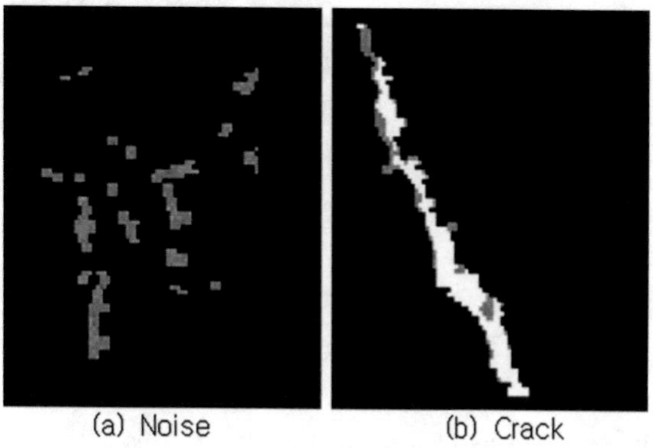

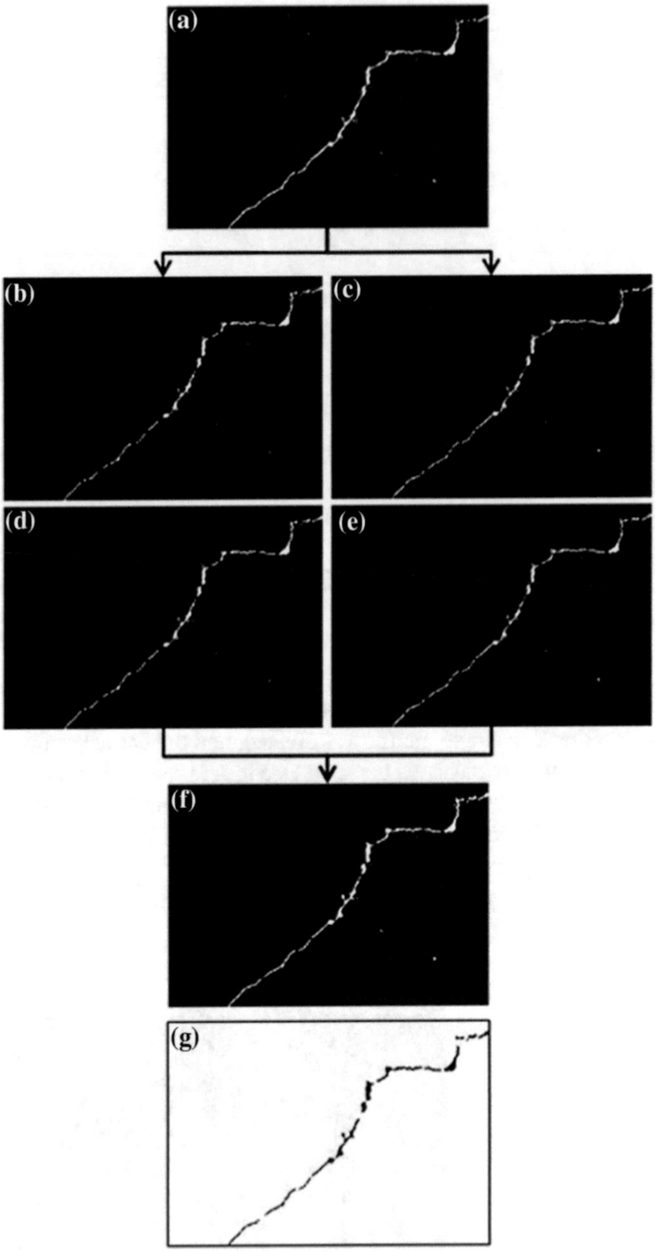

Fig. 4 Noise filtering process

noise, and removed them (Fig. 4c). In this study, when the number of successive edges was less than three, they were considered to be noise and were removed.

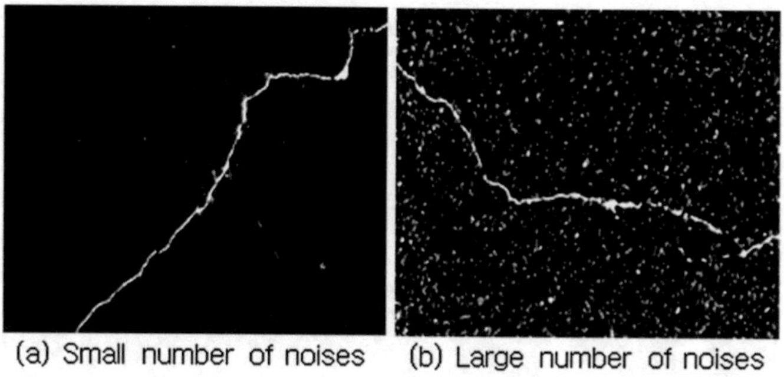

Fig. 5 Difference in number of noises according to image

Noise filtering 4 was applied to the images processed by noise filtering 2 (Fig. 4b) and 3 (Fig. 4b), respectively. This noise filtering needs to be applied when the gray values, which are middle values except for 0 and 255 in the image, occupy more than half the entire area. As shown in Fig. 5, the left image concentrates pixels with gray values around the crack, while the right image shows a wide distribution of noises with gray values. Normally, in an original image, areas with gray values indicate foreign substances. In order to remove them, all pixel values of the image are scanned, and if the values except for 0 and 255 occupy more than half the entire area, they are considered to be noise and are removed.

When the above noise removal process was completed, two images (Fig. 4d, e) were created and combined.

As a final step, noise filtering 5 detected an outline. In case the detected outline was not a line, it was considered to be noise and was removed (Fig. 4f, g).

3 Discussion and Conclusions

Figure 6 shows the final images, which were black/white reversed to display the detection results clearly. Figure 6a, b represent the results of the algorithm proposed in reference [3], and Fig. 6c, d are the final images produced by the algorithm proposed in this paper.

Reference [3] could detect cracks as small as 0.32 mm, but it also detected surface stains or small holes as cracks and thus left them as noise [3]. To solve this problem, this study applied an additional new noise filtering after using the noise filtering proposed by the existing study. As shown in Fig. 6, in comparison with the images of detection results produced by the existing algorithm (Fig. 6a, b), the final images of the algorithm proposed by this study (Fig. 6c, d) show cracks more clearly and indicate more effective noise removal.

However, some crack parts in the final images of this study are broken as a result of noise removal. This needs to be improved in further work.

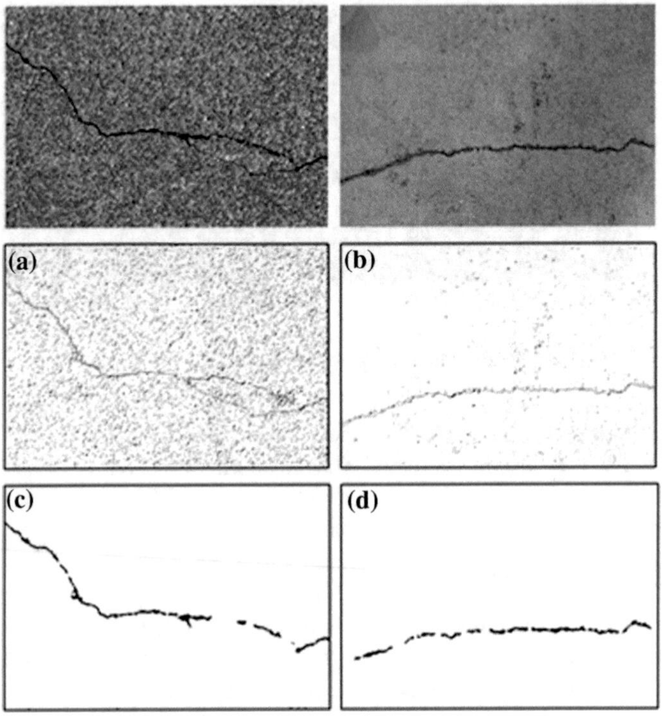

Fig. 6 Comparison of crack detection results

References

1. Kang, J., Oh, Y., Um, D.: The crack information acquisition of concrete object by digital image processing. J. Korean Soc. Civil Eng. D **22**(5D), 1001–1010 (2002)
2. Kim, H., Lee, J., Moon, Y.: Crack detection on concrete bridge by image processing technique. In: Proceedings of the Institute of Electronics Engineers of Korea, pp. 30, no. 1, pp. 381–382 (2007)
3. Kim, Y.: Development of crack recognition system for concrete structure using image processing method. J. Korean Inst. Inf. Technol. **14**(10), 163–168 (2016)
4. Youm, M.-K., Yun, H.-S., Kim, H.-B., Lee, C.-K.: Crack detection of structure by OpenCV. In: Conference of the Korean Society for Geospatial Information Science, pp. 227–228 (2016)

Evaluation of High-Speed Image Processing for Low Latency Control of a Mechatronic System

Joshua Lues, Gourab Sen Gupta and Donald Bailey

Abstract The use of artificial vision in control loops as a feedback element or a reference signal for the control of automation systems requires low latency. Several artificial vision system realizations are compared in terms of latency of novel event detection. The target application is a semi-automated foosball table. Novel event detection latency ranged from 26.4 to 260.7 ms. The system with the lowest latency has a camera connected directly to an FPGA SoC (System on Chip), where streamed vision preprocessing is performed in hardware. Finally, an SoC based control platform with low overall latency is briefly discussed.

1 Introduction

Accurate, low latency, low noise data acquisition is very important in controlling a fast, dynamic system [1]. This is the case regardless of what system is being controlled—if it operates at a high speed, obtaining reliable measurements with minimal latency is imperative. On these grounds, a method to obtain, process and utilize visual data (image data) with minimal overall system latency is required when visual feedback is used in the context of a robotic control system.

This paper will discuss control system latency and its importance both generally, and for the specific example of high-speed image processing, for control of an automated foosball table. The proposed vision system can be used as either the reference signal (the position of the ball which the foosball table rods are trying to

J. Lues (✉) · G. S. Gupta · D. Bailey
School of Engineering and Advanced Technology, Massey University,
Palmerston North 4442, New Zealand
e-mail: j.lues@massey.ac.nz; josh.guitar123@gmail.com

G. S. Gupta
e-mail: g.sengupta@massey.ac.nz

D. Bailey
e-mail: d.g.bailey@massey.ac.nz

J.-H. Kim et al. (eds.), *Robot Intelligence Technology and Applications 5*,
Advances in Intelligent Systems and Computing 751,
https://doi.org/10.1007/978-3-319-78452-6_46

match for interception), or both the reference and the feedback sensor element in the control loop.

The performance of several different implementations of artificial vision systems will be compared. These systems have been designed for high speed control. As such, latencies have been minimized where possible. The aim of the experiments was to determine the novel event recognition latency involved with each of the candidate test systems.

2 Background

2.1 Semi-automated Foosball Table Control System

In this research, a semi-automated foosball table shown in Fig. 1, in which a human opponent plays against the table, will be the test platform for the vision and control schemes. Therefore, most of the hardware and software discussed will be somewhat focused on this application.

Foosball, also known as table soccer, is a miniature soccer game played on a small table with a field area of approximately 1.1 m × 0.7 m. The game uses 1 ball, and involves two to four players. It is played one-on-one or two-on-two, with players from the two teams standing and operating on both sides of the table. The players manually operate small rigid puppets (foosmen), attached to spinning and sliding metal rods to kick the ball into the opponent's goal, and to defend their own

Fig. 1 Semi-automated foosball table

goal from opponent's kicks. There are four rods per team (eight rods in total) and between two and five foosmen per rod. It is a fast paced, dynamic game with ball speeds in excess of 10 m/s, and therefore requires very fast reflexes from the human players.

The table used in our experimentation is semi-automated. On one side, the human players have been replaced with a mechatronic control system comprising actuators to slide and rotate the metal rods. The rods are controlled by 8 stepper motors: 4 for the linear (sliding) movement of the rods, and 4 for the rotation (spinning). This is shown in Fig. 2.

The camera faces upwards from beneath the table, and tracks the ball through the base of the table which is made of glass. The human player plays against the automated control system; both try to defend their own goal and shoot into the opponent's goal. The control system, therefore, is required to track the position of the ball with sufficient speed, low latency, and sufficient resolution to accurately calculate where the ball is on the playing field.

2.2 Latency in Control Systems

Control systems are designed to accept some bounded input and provide some bounded output, both within a specified range [2]. Many dynamic control systems are run in closed-loop mode. This means that the system output (controlled

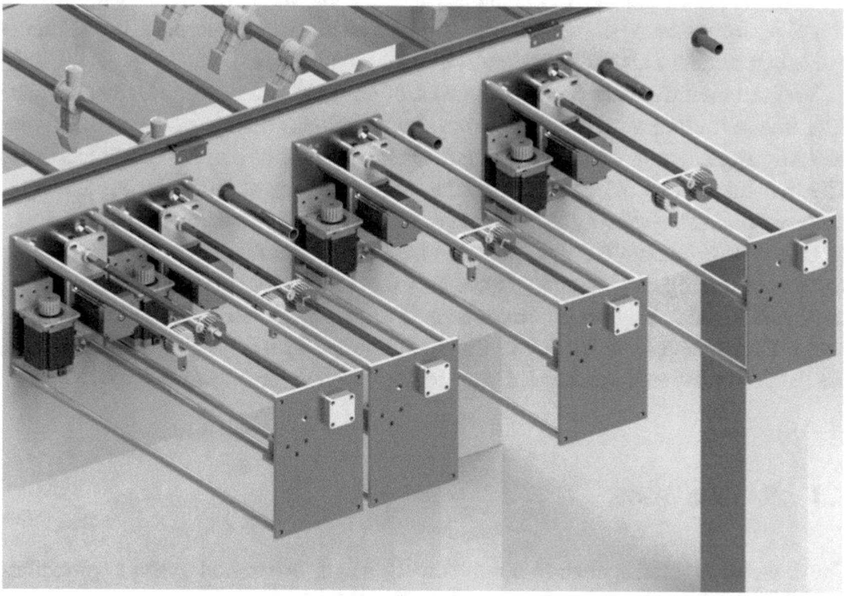

Fig. 2 CAD model of mechatronic actuator modules

variable) feeds back, via some sensor, into the controller of the system, and the system uses that feedback to make some changes to its control input [1].

Within a control system, sensor latency can be viewed as the time between an event occurring, and the data associated with the event being captured by the measurement system and passed to the controller [2]. In the experiments documented here, these events are the movement or appearance of an object. The image is captured with a camera and image processing is used to provide the position of the object.

In the control of an automated foosball table, the position of the ball is measured, and fed into the control algorithms to control the positions of the foosmen to intercept the ball. This is the feedback element of the positional control. In practice, the control is a little more complex than simple positional control. The ball's current and past positions are used to predict the motion of the ball which determines the position at which the rods controlling the foosmen will need to arrive when, or just before, the ball reaches the interception point. The system therefore represents a type of tracking control system with the predicted ball intercept position as the reference signal and the foosmen's position as the controlled variable [1].

In control systems, two important parameters which determine stability and robustness of a system are phase and gain margins [2]. The phase and gain margins indicate how much external disturbance (subtracted phase and added gain) the system can withstand without becoming unstable (oscillating) or failing entirely [2]. Latency plays a strong part in this problem. If the sensor feedback is delayed by some amount, then this delay corresponds to a linear phase delay with frequency, with the delay proportional to the sensor and measurement system latency. As this appears within the loop gain, the latency reduces the phase margin, and reduces the actual performance of the system making it more difficult to control. If the latency exceeds a critical value, the system will become unstable.

Vision based systems typically consist of some image capture hardware (camera), image transmission interface, image processing system (hardware or software compute engine), communication interface and actuator or output control hardware. The transmission of the data, processing, and communication delays all add up to cause substantial system latency. Some of the latency is caused by the sheer volume of data being transmitted and processed. Various methods including pipelining, stream-processing, and simply increasing the processing power of the compute engine, have been used to improve the performance of frame-based vision systems over time. However, the latency is ultimately limited by the system frame rate, since the data is transferred from the sensor to the compute engine serially.

2.3 Related Work

The literature includes several arrangements which use vision systems to capture and utilize motion data of various objects. The limiting factors of latency are mostly system-dependent; however, the general principle is that excessive latency can lead to instability. This is the primary motivation for efforts to minimize latency.

The most similar system [3], was also an automated foosball table, which operated at 200 fps with a measurement system latency of 17.5 ms. The camera transmitted 8-bit monochrome images with a resolution of 657 × 446 to a PC. This was done via Gigabit Ethernet. The system's inability to defend against certain shots was determined to be due to latency in controlling the rods.

An FPGA SoC was used by Cizek et al. [4] for hardware acceleration of complex image processing tasks for vision-based navigation systems. The authors used a frame rate of 60 fps at 640 × 480 resolution. For their application, they claimed an image processing latency of approximately 17 ms.

Two different realizations of high speed control of UAVs using stereo vision were discussed by Barry et al. [5]. They achieved 2 ms image processing latency with the FPGA implementation and a worst case 16.6 ms image processing latency with the ARM processor implementation. Given the lower data rate of 320 × 240 at 120 fps (approximately equivalent to 640 × 480 at 30 fps), and the lower computational power required for the matching algorithm, the latency achieved in this work appears reasonable. Hardware based processing gave approximately an eightfold improvement over the equivalent ARM based processing. These reported latencies are only for the image processing, and not the complete system latency for object detection (including image capture). It does give a representation of the effectiveness of hardware image processing platforms such as FPGAs compared to traditional systems, such as the ARM processor.

Anderson [6] demonstrated a 60 fps real time stereo vision system achieved through custom hardware with around 32 ms latency. Given that the system was achieved through custom hardware and software 32 ms latency was reasonably effective for the real time autonomous control of their robotic manipulator. However, it is likely that the latency and throughput requirements for their application are less demanding than an automated foosball table due to the relatively slow speeds at which their robot manipulator was actuated.

De la Malla and Lopez-Miliner [7] demonstrated that humans use both predictive and online visual cues to catch flying objects. It makes sense that this framework could be applied or at least researched in the context of robotics, especially interactive/humanoid robotics.

Cigliano et al. [8] discusses a distributed computing and control platform for a robotic arm and gripper assembly to catch a ball thrown at it. This represents a tracking control system. Latency was simply estimated and used to synchronize events between the various subsystems, rather than latency being minimized.

Many others [9–12] also discuss the use of dynamic vision sensors (DVS) for event based image processing systems with extremely low latencies, ranging from 12 to 15 µs. However due to the lower resolution of these sensors, between 128 × 128 and 240 × 180 pixels, and the lower level of configurability, they do not fully match the target application of automating a foosball table. The foosball table requires higher resolution due to the size of the ball, and the precision required for interception by the foosmen. They do, however, reflect the importance of high-speed, low latency vision-type sensors in the control of robotics.

Based on the findings of this review, it was determined that for our application, some experimentation was required to determine the true system latencies, as opposed to optimistic estimates based purely on image processing speeds.

3 Experiment Design

The experiments were conducted using the following experimental plans:

3.1 Aim

The aim of the experiment was to determine the latency of several different realizations of artificial vision (object tracking) systems. Each system was tailored to reduce latency as much as possible, and to make the comparisons as fair as possible. These methods included:

1. Where possible, using the same resolution for all the systems—640 × 480;
2. Using the same image format and color depth for all the image streams;
3. Not displaying the images when performing the actual tests, as this can add some processing overhead;
4. Minimizing the amount of data sent in the communication systems—as would be done on similar implementations;
5. Using the highest baud rate available if serial communication was required;
6. Using the maximum available framerate.

3.2 Methodology

The following steps are common for all implementations of the vision system.

1. Capture images at given resolution and best frame rate
2. Colorspace conversion (YUV for colored object segmentation)
3. Filtering—color, morphological
4. Blob analysis and centroid calculation
5. Produce an "Object recognized" output signal.

The experiment was carried out for each of the systems with the following steps:

1. An LCD display unit on an STM F429ZI is set to display a target object after a random amount of time between 5 and 10 s. The display latency is fixed, and would be the same for all experiments.
2. When the STM unit displays the object, it starts a high precision timer on a separate microcontroller (an Arduino Uno).
3. The vision system being tested watches for the target object to appear on the display—in this case just a uniform blue screen.
4. Once the object is correctly recognized, the "object recognized" signal is output. Where possible this output signal is in the form a logic high written to a GPIO pin. On systems where this was not practical, a serial command was output via a USB port to the microcontroller.
5. Once the microcontroller receives the object recognized signal, it immediately stops the timer and outputs the timer value. This is the total latency from the time the object was displayed until the time the object was recognized by the vision system and the signal sent to the microcontroller, closing the loop.

The overall latency is the combination of several components:

$$L_{total} = L_{imagedisplay} + L_{imagecapture} + L_{transfer} + L_{processing} + L_{comms} \quad (1)$$

3.3 Test Platforms

Figure 3 shows the key hardware for each of the test systems:

1. Altera DE1-SoC FPGA development board with a Terasic D5M camera module. The camera is capable of capturing image at resolutions of up to 5 MP. At 640 × 480 resolution, the system can capture images at approximately 127 fps.
2. CMUcam5 (PixyCam) embedded vision platform. The system is capable of color object tracking at 50 fps with a resolution of 320 × 240. It communicates with a microcontroller via either SPI, I2C or UART. I2C was used in this experiment.
3. PS3 EYE. This camera can capture images at around 75 fps at 640 × 480 resolution and greater than 100 fps at 320 × 240 resolution. It transmits the image data via USB 2.0. Note, the PS3 EYE was used for 2 experiments.
4. Logitech C920 Pro Webcam. This webcam is internally limited to 30 fps. However, it is capable of many different resolutions, including 640 × 480. It also transmits the image data over USB 2.0.

Additionally, a quad-core i7 Windows laptop (Asus N550-JV) was used for the MATLAB based image processing for the PS3 EYE and the Logitech C920

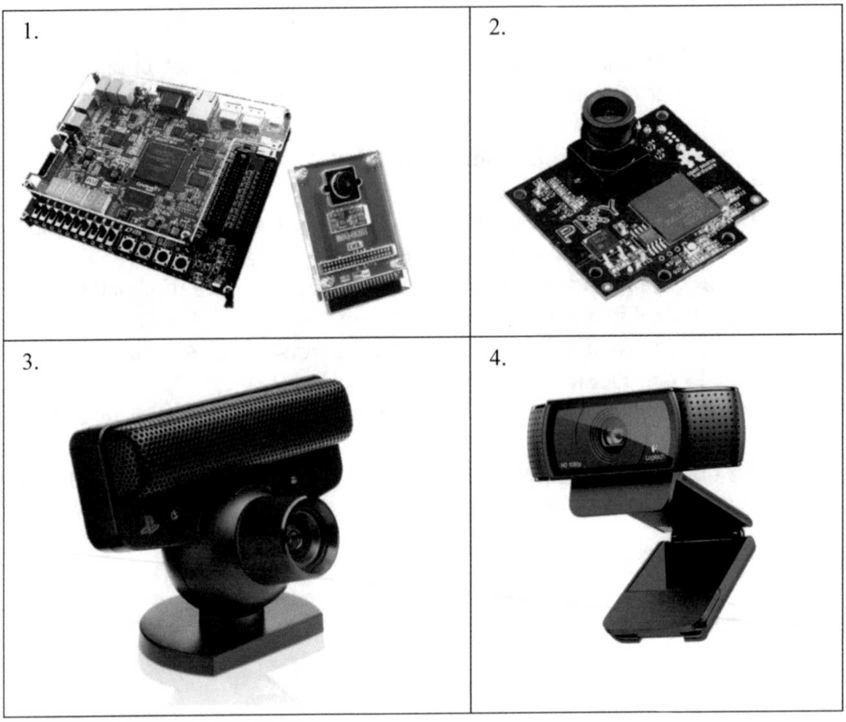

Fig. 3 Main hardware used for testing

webcam, an Arduino Uno was used for the high precision timer, and an STM-F429ZI discovery board was used as the display unit.

3.4 Data Analysis

The experiments consisted of 20 samples for each test platform, resulting in a total of 100 samples. Minitab was used to analyze the data and produce a visual representation of the results.

4 Results

The boxplot in Fig. 4 shows the latency for the 5 experiments. The latency data is presented in Table 1.

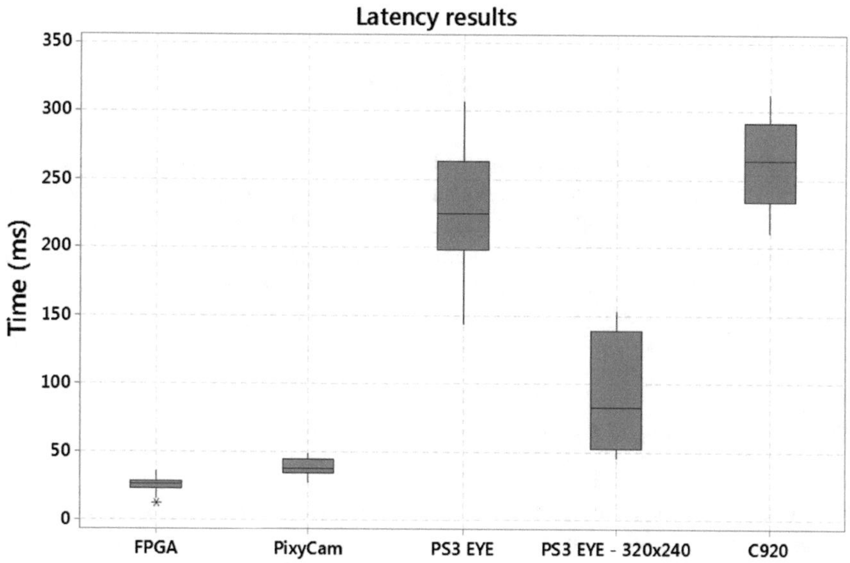

Fig. 4 Visualization of latency for test platforms

Table 1 Latency data for the test platforms

Hardware	Mean latency (ms)	Std dev	Min	Max	Range	Resolution	Nominal frame rate
FPGA	26.4	5.4	12.9	36.8	23.9	640 × 480	60
PixyCam	39.7	6.5	28.3	49.6	21.3	320 × 240	50
PS3 EYE 640 × 480	229.2	48.4	144.4	307.4	163	640 × 480	75
PS3 EYE 320 × 240	94.6	43.5	46.6	154.2	107.6	320 × 240	>100
Logitech C920	260.7	31.3	211.5	313.2	101.7	640 × 480	30

The reason for the PS3 EYE test being performed at 320 × 240 and 640 × 480 was to test it against the specified resolution, as well as to compare it with the PixyCam, which operates at a resolution of 320 × 240. It can be seen in Fig. 4 that as the average latency of novel event recognition increased, the variability increased too. This would be very unhelpful if implemented in a robot control system that requires prediction because the latency, upon which predictions are partially based, would vary significantly, making it difficult to accurately estimate the ball motion parameters such as velocity, heading angle, and curvature. This is also true with changing ball motion parameters, such as in acceleration or deceleration situations, which a constant velocity prediction model may not be able to account for.

5 Discussion

5.1 Immediate Usefulness of Results

It is impossible to have zero system latency, therefore experimentally determining the novel event recognition latency is an important step towards characterizing the system for accurate modelling and future development purposes. It also serves as a benchmark to measure future improvements. Additionally, as mentioned before, it is useful to have an accurately measured novel event recognition latency as it can assist greatly with predictive compensations for unavoidable latency.

5.2 Comparison of Performance

Firstly, it should be noted that these experiments were performed with the FPGA system running at less than half of the maximum image capture and processing speed of which it is capable. This was done for display synchronization and development reasons, however future work will see the frame rate increased to above 120 fps and the latency (and variability of latency) should, thus, decrease proportionately, due to the streamed nature of the processing.

It can be seen from the results that, even with the currently capped frame rate, the FPGA system still yielded the best performance. 26 ms latency is roughly equivalent to the latency experienced by the human eye when detecting new stimulus. Given a reasonably efficient algorithm, however, the overall response time of a control system implemented in an FPGA SoC, as is proposed in future work, should be no higher than 30 ms. This is drastically lower than the average human response time of approximately 180 ms.

The PixyCam, which yielded the second lowest latency was only running at a resolution of 320 × 240, so cannot truly be considered a close contender, given that the specifications for this application are a minimum resolution of 640 × 480. However, it was included for reference as it is representative of a simple system for embedded image processing where high resolution is not required. It appears that the lower resolution and the application specific hardware are the primary reasons for the PixyCam's low latency.

The PS3 EYE, running at 640 × 480 at 75 fps, yielded a higher latency than the PixyCam, despite the higher frame rate. The experimental data demonstrates the importance of efficient image processing algorithms. Because the image processing algorithms were implemented in MATLAB, there were processing overheads (within the operating system and within MATLAB) which likely slowed down the processing.

Finally, the test system using the Logitech webcam, which was internally limited to 30 fps, yielded the poorest performance of the tested systems. The latency was the highest (an order of magnitude higher than the FPGA system), and the most variable. It is likely that the combination of inefficient processing, combined with relative timing of the event and the image being captured was the cause of the latency. This, in a highly dynamic control system, would likely result in poor system stability.

5.3 Automated Foosball

In the context of automated foosball, and of robotic control in general, it is obvious from the data that the best choice from the systems tested would be the FPGA based image processing platform. The low latency is desirable from a control perspective [13]. Additionally, based on human systems, "predictive control is essential for the rapid movements commonly observed in dexterous behavior" [14]. In foosball, there are very high speeds involved, relative to the size of the playing field. This means that high-speed image processing is imperative if any level of response is to be possible. Not only does the sensing need to be excellent, so do the actuation techniques. Using reconfigurable FPGA hardware to create stepper motor controllers, or general-purpose actuator control, could be an extremely effective method to solve the problems involved with the automated foosball table.

By combining low latency image processing (streamed image processing on Altera DE1 SoC FPGA), and low latency communication (high speed AXI bridge from FPGA to the embedded ARM cores), artificial vision can be used as both a reference signal and a feedback element in a control loop. The system becomes even more powerful when actuator control and other general system functions (for example safety interlocks and user inputs) are also implemented within the FPGA. This results in comparatively low overall system latency for all elements of the control loop. Additionally, the use of an FPGA SoC enables scalability at no cost of additional latency or scheduling issues present in conventional microcontrollers.

6 Conclusions

The latencies ranged from 26.4 to 260.7 ms, with the FPGA system yielding the lowest latency.

The experimentally determined latency values will be useful in future work for the prediction capabilities of the system, as well as for modelling purposes.

For the FPGA system to be a completely vision-controlled closed loop system, the camera could be used for tracking both the position of the foosmen, and for tracking the ball.

Future work on this platform will include the control of all user inputs, interlocks, and all 8 stepper motors using FPGA hardware. The FPGA will communicate with software running on the embedded ARM core which will perform all strategy and trajectory calculations. Finally, all additional functionalities will still need to meet the overall system latency requirements.

References

1. Franklin, G.F., Powell, J.D., Emami-Naeini, A.: Feedback Control of Dynamic Systems, 7th edn. Pearson, New Jersey (2015)
2. Engelberg, S.: A Mathematical Introduction to Control Theory, 2nd edn., vol. IV. Imperial College Press, London (2015)
3. Janssen, R., Verrijt, M., de Best, J., van de Molengraft, R.: Ball localization and tracking in a highly dynamic table soccer environment. Mechatronics **22**, 503–514 (2012)
4. Cizek, P., Faigl, J., Masri, D.: Low-latency image processing for vision-based navigation systems. In: International Conference on Robotics and Automation, Stockholm, pp. 781–786 (2016)
5. Barry, A.J., Oleynikova, H., Honegger, D., Pollefeys, M., Tedrake, R.: Fast onboard stereo vision for UAVs. In: Vision-Based Control and Navigation of Small Lightweight UAVs, IROS Workshop (2015)
6. Anderson, R.L.: A low-latency 60 Hz stereo vision system for real-time visual control. In: International Symposium on Intelligent Control, Philadelphia, pp. 165–170 (1990)
7. de la Malla, C., Lopez-Moliner, J.: Predictive plus online visual information optimizes temporal precision in interception. J. Exp. Psychol.: Hum. Percep. Perf. **41**(5), 1271–1280 (2015)
8. Cigliano, P., Lippiello, V., Ruggiero, F., Siciliano, B.: Robotic ball catching with an eye-in-hand single camera system. IEEE Trans. Control Syst. Technol. **23**(5), 1657–1671 (2015)
9. Berner, R., Brandli, C., Yang, M., Liu, S.-C., Delbruck, T.: A 240 × 180 10 mW 12 μs latency sparse-output vision sensor for mobile applications. In: Symposium on VLSI Circuits, Kyoto, pp. C186–C187 (2013)
10. Conradt, J., Berner, R., Cook, M., Delbruck, T.: An embedded AER dynamic vision sensor for low-latency pole balancing. In: International Conference on Computer Vision Workshops, Kyoto, pp. 780–785 (2009)
11. Mueller, E., Censi, A., Frazzoli, E.: Low-latency heading feedback control with neuromorphic vision sensors using efficient approximated incremental inference. In: Annual Conference on Decision and Control, Osaka, pp. 992–999 (2015)
12. Censi, A., Strubel, J., Brandli, C., Delbruck, T., Scaramuzza, D.: Low-latency localization by active LED markers tracking using a dynamic vision sensor. In: International Conference on Intelligent Robots and Systems, Tokyo, pp. 891–898 (2013)
13. Honegger, D., Oleynikova, H., Pollefeys, M.: Real-time and low latency embedded computer vision hardware based on a combination of FPGA and mobile CPU. In: International Conference on Intelligent Robots and Systems, Chicago, pp. 4930–4935 (2014)
14. Wolpert, D.W., Flanagan, J.R.: Motor prediction. Curr. Biol. **11**(18), pp. R729–R732 (2001)

Author Index

J.-H. Kim et al. (eds.), *Robot Intelligence Technology and Applications 5*,
Advances in Intelligent Systems and Computing 751,
https://doi.org/10.1007/978-3-319-78452-6

Printed in the United States
By Bookmasters